Quantum Computing, Cyber Security and Cryptography

S. B. Goyal · Vidyapati Kumar ·
Sardar M. N. Islam · Deepika Ghai
Editors

Quantum Computing, Cyber Security and Cryptography

Issues, Technologies, Algorithms, Programming and Strategies

Springer

Editors
S. B. Goyal
Chitkara University Institute of Engineering and Technology
Chitkara University
Rajpura, Punjab, India

Sardar M. N. Islam
ISILC
Victoria University
Melbourne, VIC, Australia

Vidyapati Kumar
Department of Mechanical Engineering
Indian Institute of Technology Kharagpur
Kharagpur, West Bengal, India

Deepika Ghai
School of Electronics and Electrical Engineering
Lovely Professional University
Phagwara, Punjab, India

ISBN 978-981-96-4947-1 ISBN 978-981-96-4948-8 (eBook)
https://doi.org/10.1007/978-981-96-4948-8

Mathematics Subject Classification: 68Q12, 68M25, 81P68, 81P94, 94A60, 94A40

This Springer imprint is published by the registered company Springer Nature Singapore Pte Ltd.
The registered company address is: 152 Beach Road, #21-01/04 Gateway East, Singapore 189721, Singapore

Foreword

Cybersecurity is perhaps one of the fields under immense pressure due to the development of quantum computing. Organizations anticipate using quantum capabilities to maintain the privacy and security of valuable information and technologies as these latter evolve at the technological frontier. But with these, come great risks, especially with the traditional cryptographic underpinnings of cybersecurity now running the risk of being vulnerable to quantum computers.

This edited volume, Quantum Computing, Cyber Security and Cryptography: Information and Computing Resources, Science, Technology, Security and Intelligence, IT and Computing, is a timely input on Quantum Computing and Cybersecurity. The volume collects views from scholars who hail from academic institutions and technology firms, with their insights illuminating diverse aspects of quantum technologies and the future of cryptographic approaches. All together the set of chapters of the current book offers the vast outline of how quantum computing can affect information security and what threats and benefits can be expected.

Across this book, we explore several major issues. Analyses of quantum algorithms that can transform the field of cryptanalysis, and post-quantum cryptography solutions which have been developed for the purpose of protecting digital assets are outlined in each chapter. In addition to the general background of the topics discussed in the book, it identifies the prospects for practical applications for different industries including financial, medical, and security ones.

The book has a total of 18 chapters and each chapter is devoted to the discussion of one of the topics within the general topic of quantum computing and cybersecurity. Initial chapters comprise simple topics such as quantum programming, cryptographic technologies, and relative cybersecurity prospects of quantum computing, while the subsequent chapters discuss sophisticate subjects as the prospects of quantum computing in the field. Quantum computers and quantum communications technologies are explained in each chapter along with new guard solutions for defending against them.

Being editors, we have assisted the authors to write this volume in a way that future readers will find the most contemporary developments in the area of quantum computing and overcome challenges in cyber security. It is our wish that this book triggers more works, generates new thoughts and ideas and expands the broader discussion on the future of secure digital communication in a quantum era.

Our sincere thanks to all the authors for their vital inputs. It has been their professionalism and commitment that have led to the creation of this book, and the results presented herein should serve as a solid basis for further investigation and practices in the given discipline.

Last, we would like to thank all institutions who supported this work, and City University in particular to where we strive to contribute to the development of information technology education and research. We also expressly accept our gratitude to the team at Springer who have remained with us and ensured the publication of this book.

Consequently, this book aims to be a handy reference for researchers, practitioners, and policymakers operating in the field of quantum computing and cybersecurity. We hope it will be the starting point for additional evolution toward further securing our digital future against quantum technologies.

Sandeep Kautish
Director—AIT CSE
Chandigarh University
Sahibzada Ajit Singh Nagar, Punjab,
India
Sandeepkautish.e17021@cumail.in

Preface

The rapid advancement of quantum computing represents both tremendous opportunities and serious risks for our increasingly digital world. As quantum computers continue scaling up in power, they will be capable of breaking the cryptography underlying most of today's secure communications and data storage. This poses an existential threat to cybersecurity as we know it. At the same time, quantum technologies open up new capabilities for generating true random numbers, distributing cryptographic keys, and developing secure encryption algorithms against attacks from classical and quantum computers. The field of post-quantum cryptography is advancing novel cryptosystems based on mathematics problems outside what quantum algorithms can easily crack. Quantum key distribution leverages properties of quantum physics to exchange keys with inherent privacy.

This edited volume brings together leading experts across academia, industry and government to provide a holistic reference on the security implications of quantum computing. It examines the latest research and strategies for transitioning to a "quantum-safe" future across applications like classified communications, financial systems, blockchain, and healthcare data.

The book begins with foundational overviews of quantum computing concepts, fundamental algorithms, programming tools, and quantum cryptoanalysis basics. These chapters equip readers with the technical context to understand quantum advantages in codebreaking and hacks. The book then dives into post-quantum cryptography primitives, including lattice-based, hash-based, multivariate and code-based systems, with comparisons of their strengths and limitations. Practical topics such as quantum key distribution protocols, random number generation and post-quantum algorithm integration give insights into building real-world quantum-resistant solutions. The edited volume concludes by connecting quantum technologies to concrete cybersecurity challenges like national security, financial networks, privacy, and the Internet of Things. Experts provide strategic guidance, best practices, and recommendations for organizations to assess and manage risks from quantum computing. The policies, standards, and innovations needed to foster broad collaboration across stakeholders in quantum-safe security are also discussed.

This book provides an essential reference for students, researchers, and professionals aiming to prepare our data, devices, and critical systems for the coming quantum computing revolution. It balances accessible explanations of technical topics with practical roadmaps and frameworks for cybersecurity in a quantum world. We hope the insights compiled in this volume contribute to the responsible advancement of quantum technologies for the benefit of society.

Rajpura, Punjab, India — S. B. Goyal
Kharagpur, India — Vidyapati Kumar
Melbourne, Australia — Sardar M. N. Islam
Phagwara, India — Deepika Ghai

Contents

Quantum Computing: Transforming Cybersecurity in the Quantum Era

Nisheeth Saxena and Akriti Nigam

1 Introduction

The Quantum Computation (QC) concept was first proposed by Richard Feynman [1] in 1982, following his observation of the exponential complexity required to simulate a quantum system's behavior with classical computers. Quantum computing is a rapidly evolving transformational technology with the potential to significantly affect businesses, society, and the humanities in the years to come, underscoring its profound significance. The ongoing development of quantum computers, a journey spanning several decades, is a testament to human ingenuity and holds immense potential, making it one of the most significant advances in the field of quantum computing, science and technology. In the current era of NISQ technology, we are working with a limited number of qubits, typically between 50 and 100, for our calculations. This underscores the current constraints and the need for further advancements. Fault-tolerant quantum computers remain a distant goal since decoherence and noise in quantum gates restrict the depth of quantum circuits. Yet, even with these constraints, they have the potential to outperform today's conventional machines. The goals of quantum computing include designing algorithms based on quantum mechanical phenomena that are remarkably quicker than classical computers while cracking the same problem and augmenting classical computers to enhance their capabilities. Quantum advantage is anticipated to manifest itself in situations when real-world problems cannot be handled effectively by even the most powerful classical supercomputers. Drug research, healthcare, fraud detection in financial transactions, secure communication, and machine learning are just a few

N. Saxena (✉) · A. Nigam
Department of Computer Science and Engineering, Birla Institute of Technology, Mesra Ranchi, Jharkhand 835215, India
e-mail: nisheeth.somnath@gmail.com

A. Nigam
e-mail: akriti@bitmesra.ac.in

S. B. Goyal et al. (eds.), *Quantum Computing, Cyber Security and Cryptography*,
https://doi.org/10.1007/978-981-96-4948-8_1

of the key areas where quantum computers have been used. Quantum computing is the future of computing. Quantum computers will acquire a quantum advantage, or the capacity to solve problems efficiently that are thought to be beyond the grasp of today's most influential classical computers [2]. The most significant computer revolution will shortly be brought about by quantum computing in conjunction with AI, ML, and Deep Learning (DL). Quantum technologies include quantum systems, quantum hardware, quantum communication, quantum sensing, quantum machine learning, quantum cryptography and cybersecurity, and other related technologies. Because quantum computing is interdisciplinary, it can include mathematics, computer science, physics, chemistry, electrical, mechanical, and material engineering. The industry is ecstatic about the intriguing application advances in a variety of disciplines, which include finance and logistics optimization. In other words, quantum computing is multidisciplinary and can impact several fields. Examples of some critical killer use cases are modeling lithium–sulfur material to enhance lithium ion batteries' performance in automobiles, modeling the expected values in financial instruments such as non-performing assets and loan risk profiling to name a few.

While IBM offers its quantum computer hardware through its cloud services, D-Wave quantum computers have been around for over a decade. Also, IBM, Google, Microsoft, Intel, Rigetti, etc., have invested heavily in developing quantum computers and their applications, making them accessible to physicists, computer scientists, and information system researchers. Quantum volume is a statistic used to assess a quantum computer's capacity and mistake rate. The IBM Candor processor family has comprised nearly 1024 qubits of actual quantum computers on the cloud since 2023. Figure 1 shows the quantum computing timeline since the 1980s, showcasing all significant developments till now and beyond. Figure 2 depicts the dedicated timeline highlighting specific breakthroughs that had implications for cryptography and cybersecurity from the 1935s to 2019 and after.

In today's cryptography, algorithms such as RSA and Elliptic Curve Cryptography (ECC), whose robustness lies in solving some difficult mathematical problems such as prime factorization problems and discrete logarithm problems, respectively, are the backbones of digital transactions, digital certificates, data encryption, and physical authentication devices like Radio Frequency Identification (RFID) cards. As

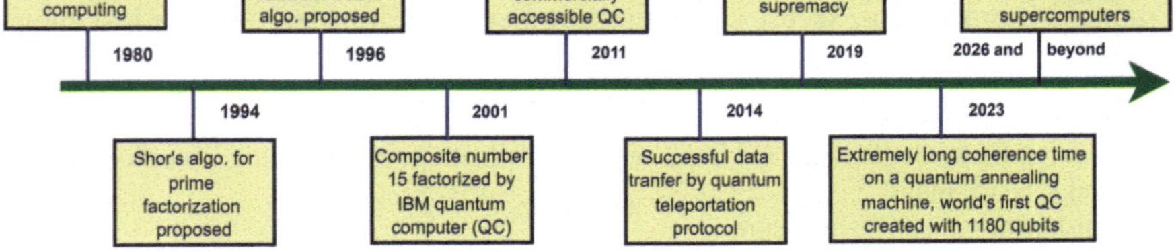

Fig. 1 Timeline highlighting specific breakthrough in quantum computing since the 1980s (Figure by author)

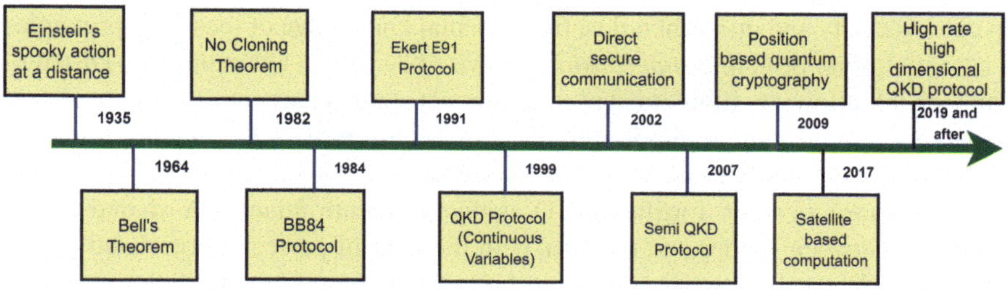

Fig. 2 Timeline highlighting specific breakthrough that had implications for quantum cryptography and cybersecurity since the 1935s (Figure by author)

the quantum computing wave rises, these mathematical problems will be solved easily, and these cryptographic protocols will soon be rendered completely obsolete, putting digital security in deep danger. In this way, quantum computing supremacy will fundamentally change the cybersecurity and secure communication landscape and make the world vulnerable to cyberattacks. With this technology sabotage by quantum computing, governments, and private companies must take preventative measures to avoid a future disaster in the form of information breaches, hacked confidential transactions, and loss of trust in the digital world [3].

This chapter offers a brief outline of the exciting and novel field of quantum computing from the standpoint of a novice, inviting her to delve into this fascinating world. Section 2 unveils an overview of critical quantum computing ideas, such as superposition, entanglement, and the two most essential quantum gates. Section 3 outlines the foundational quantum computing algorithms, which are crucial for understanding and advancing the field of quantum computing, and briefly discusses two necessary quantum communication protocols: teleportation and superdense coding. Section 4 explores the practical applications of quantum computing in various fields, emphasizing cybersecurity and cryptography. This section aims to demonstrate the tangible and significant real-world relevance of quantum computing. Finally, Sect. 5 provides the conclusion.

2 An Overview of Essential Quantum Computing Principles

The most basic component of quantum computing is a `qubit`. It may be expressed in Dirac's ket notation as $|0\rangle$ and $|1\rangle$ for qubit states 0 and 1, respectively. Any subatomic particle, such as a nucleus, electron, or photon, can realize a qubit. A phosphorous atom's outermost electron can be used as a qubit. An n number of qubits may contain the same amount of information as a 2^n number of classical bits (cbits), according to [4].

The three critical principles in quantum computing are *Superposition*, *Entanglement*, and *Interference*. To understand precisely the impacts of quantum computing on cybersecurity and different fields, fundamental knowledge of these key principles is essential. First principle *superposition* plays a key role in Quantum Key Distribution (QKD), allowing qubits to stay in numerous states concurrently, and precisely measuring a qubit state is impossible without disturbing its state, making it impossible for eavesdroppers to intercept transmitted keys without detection. The second principle, *entanglement*, fortifies QKD against quantum attacks. A shared key is generated between sender and receiver with the help of perfectly correlated entangled pairs. When there is an attempt to breach the shared key, the interception is detectable, thus raising an alert of a security threat. The third principle, *interference*, is attributed to the *wave-particle* dual nature of matter. It finds several applications in quantum computing, e.g., detecting attempts of interception in QKD and thus assuring secure, shared key transfer, Quantum Error Correction (QEC), enhancing the security of quantum cryptography, fast searching a targeted key in an unordered list, etc. All three principles are described in brief in Sects. 2.1, 2.2, and 2.3 as follows.

2.1 Superposition: Hazy Scrambling of Qubit States

Quantum superposition is a basic notion in quantum computing. Quantum computers, as contrast to classical computers, utilize qubits, which may be in a superposition state, achieving a condition of 0 and 1 at the same time [5]. The square of the absolute values of the qubit's amplitudes determines its probability of falling apart into one of the two states, $|0\rangle$ and $|1\rangle$, in `Dirac` notation, following an observation or measurement. A qubit $|\psi\rangle$ is characterized as

$$|\psi\rangle = \alpha|0\rangle + \beta|1\rangle \tag{1}$$

Here α, β are the complex amplitudes of qubit states $|0\rangle$, $|1\rangle$, respectively. Probabilities of measuring $|0\rangle$ is $|\alpha|^2$ and that of $|1\rangle$ is $|\beta|^2$, while $|\alpha|^2 + |\beta|^2 = 1$.

2.2 Entanglement: Perfect Correlation Among Qubits, Even Apart at a Distance of Light years

Quantum Entanglement, a bizarre yet fundamental phenomenon in quantum mechanics, has baffled and intrigued physicists since its discovery. Coined as "**Spooky action at a distance**" by Albert Einstein, the essence of this quantum property defies our conventional understanding of the world, hinging on the peculiar idea that two particles can become so deeply linked that the state of one directly and instantaneously, influences the state of the other, regardless of the distance separating them. This puzzling concept found its roots in the early 20th century amid the rise of quantum theory. It is a hallmark of quantum mechanics that arises when a group of particles exhibits correlation or association, even when they are later separated over vast distances. When qubits are entangled, their spin or polarization precisely correlates. Qubits are composed of spinning electrons or polarized photons. Quantum computers use microwaves or lasers to orient the qubits into a state or alignment where they can be entangled or correlated. All the quantum theory predictions must be verified experimentally in the laboratory. In the laboratory, it took nearly 70 years to prove the quantum mechanics paradigm: entanglement, recognized by Einstein, Podolsky, Rosen (**EPR** paradox), and Schrodinger, as a new resource as accurate as energy [6]. In entanglement, it appears as though there is a concealed cable connecting the two parts of the entangled pair. The non-local nature of quantum entanglement is evident in the way interactions with the first half of an entangled pair directly impact the other half, regardless of the distance between them. When creating quantum algorithms for quantum computers, the remarkable non-local character of quantum physics in entanglement is crucial. Quantum cryptography promises virtually unhackable communication by utilizing the principle of entanglement. If an eavesdropper tries to intercept the communication, the entangled particle's state will change, revealing the intruder's presence.

2.3 Interference: Controlling and Manipulating Quantum States

Quantum interference permits quantum systems to be biased toward their targeted state. The goal is to produce an interference pattern in which the routes leading to incorrect answers interfere destructively and cancel one another, while the ones leading to correct answers interfere constructively and reinforce each other. Interference plays a central role in the form of amplitude amplification in Grover's algorithm to seek out the key targeted element. In Quantum Fourier Transform (QFT), it helps identify the potential elements that lead to the speedup of quantum algorithms over their classical analogues [7].

2.4 Quantum Gates: Building Blocks of Quantum Circuits and Quantum Algorithms

Quantum circuits can be realized with the help of quantum gates, which are reversible, unlike classical gates which are irreversible in nature. The two most basic gates in quantum computing are the `Hadamard(H)` gate and the controlled `not(CNOT)` gate. The H gate is used for superposition of qubits:

$$H|0\rangle = \frac{1}{\sqrt{2}}(|0\rangle + |1\rangle) = |+\rangle \quad ; \quad H|1\rangle = \frac{1}{\sqrt{2}}(|0\rangle - |1\rangle) = |-\rangle \tag{2}$$

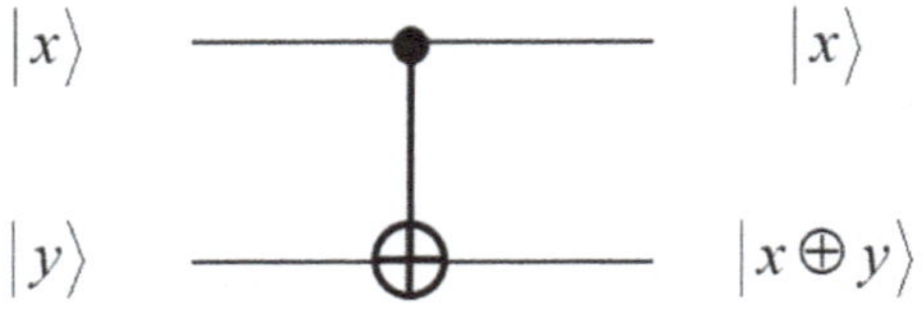

Fig. 3 Action of CNOT gate on two qubits $|x\rangle$ and $|y\rangle$, where $|x\rangle$ is control and $|y\rangle$ is target qubit (Figure by author)

One of the most essential gate, the *CNOT* gate operates on two qubits $|x\rangle$ and $|y\rangle$ as follows (Fig. 3): If $|x\rangle$ is in equal superposition of $|0\rangle$ and $|1\rangle$, i.e., $|x\rangle = \left(\frac{|0\rangle+|1\rangle}{\sqrt{2}}\right)$ and $|y\rangle = |0\rangle|y\rangle = |0\rangle$, then this transformation produces 2 qubits EPR form or Bell state or perfectly entangled state:

$$CNOT\left[\frac{1}{\sqrt{2}}(|0\rangle + |1\rangle)|0\rangle\right] = CNOT\left[\frac{1}{\sqrt{2}}(|00\rangle + |10\rangle)\right] = \frac{1}{\sqrt{2}}(|00\rangle + |11\rangle) \quad (3)$$

Such correlation between the two components of an entangled pair provides the basis for quantum *teleportation* [8]. Both H and CNOT gate are used in the design of *teleportation* and *superdense coding* quantum circuits.

A Quantum Permutation Pad (QPP) is a cryptographic protocol applicable in both traditional and quantum computing domains. It applies quantum physics principles to improve cybersecurity and data safety. CNOT gate, one of a prominent two-qubit quantum gate, designated as a universal quantum gate when combined with H gate, generates random permutations and creates random keys that set a pipeline to implement QPP and classical One-Time Pad (OTP) encryption schemes. Furthermore, QKD is viable only because of the application of quantum mechanical principles like superposition and entanglement, which can only be realized over hardware by quantum gates.

3 Quantum Computing Algorithms and Protocols: Foundations and Applications in Cybersecurity

This section briefly discusses some fundamental quantum computing algorithms that prove quantum advantage over classical computing algorithms. To emphasize their relevance to cybersecurity, Shor's algorithm and Grover's algorithm are placed at the beginning of the section so that it would immediately focus the reader on the quantum threats to current cryptographic systems.

3.1 Shor's Algorithm: Shaking the Foundations of Public-Key Cryptography

For classical computers, the prime factorization problem, i.e., breaking down a large composite number into its prime factors, is a computationally hard problem requiring exponential time [9]. This problem is the backbone of security of most of the security algorithms such as RSA. Shor's factorization algorithm exploits the inherent parallelism of quantum computers to solve this problem in polynomial time with a high degree of success probability. Shor's algorithm is strong evidence for showing a quantum advantage over classical computers. Shor's algorithm applies Fast Fourier Transform (FFT) to detect the period of a function using a quantum computer [10]. Suppose $N = p.q$ is a composite number which is a multiplication of two large prime integers p and q. The most basic Euclid's algorithm requires $O(\sqrt{N})$ number of operations, which is quite inefficient for factoring large integers. The best classical algorithm for factorization takes time of the order $O\left(e^{(\log N)^{1/3}(\log\log N)^{2/3}}\right)$, which is exponential in nature. Shor's approach has polynomial run time complexity of order $O(\log(N)$ for factoring a big composite integer N on a quantum computer. The time complexity of the following order is achieved for quantum gates:

$\left(O\left((\log N)^2(\log\log(N))(\log\log\log(N))\right)\right)$ while applying fast multiplication. Table 1 shows the time complexity comparison of some quantum algorithms with their classical counterparts.

3.2 Grover's Algorithm: Speeding up Brute-Force Attacks on Private (Symmetric)-Key Cryptography

For sorted or structured data most prominent binary search algorithm takes time $O(1)$ in the best case and $O(\log(n))$ time in the worst case. We resort to a linear search algorithm whose time complexity is $O(n)$ for unstructured or unsorted data. For extensive data set, this time is inappropriate. Grover's algorithm can improve the unstructured database search quadratically, i.e., in time $O(\sqrt{n})$ in comparison to $O(n)$ on classical computers. It uses a quantum oracle to mark the searched state, and then it uses a diffuser that amplifies the amplitude of the marked state to enhance its measurement probability. The main idea behind it is amplitude amplification based on *interference* principle. Amplitude amplification increases the phase of the searched element and decreases the phase of all other elements hence increasing the measurement probability of the target searched element [11].

Table 1 The table shows the theoretically proven quantum advantages of some quantum algorithms over their classical counterparts (Table by author)

S. no	Quantum algorithm	Classical complexity	Quantum complexity	Implications on cybersecurity
1	Deutsch–Jozsa (DJ)	$O\left(2^{n-1}+1\right)$	$O(1)$	For data points relevancy in quantum communications [14], square QKD based on a special DJ algorithm using GHZ states [15]
2	Bernstein–Vazirani (BV)	$O(n)$	$O(1)$	Based on BV a novel entanglement-based QKD protocol is proposed [16]; different kinds of attacks are proposed on block ciphers [17]
3	Simon's	$\Omega\left(2^{n/2}\right)$	$O(n)$	A precursor to Shor's algorithm; speeds up classical attacks based on finding key collisions; gives a lethal threat to symmetric-key cryptography
4	Grover's	$O(n)$	$O(\sqrt{n})$	Speedy brute-force attacks due to fast searching and a dangerous threat to Hash functions and symmetric-key cryptography
5	Shor's	$O\left(e^{(\log N)^{1/3}(\log\log N)^{2/3}}\right)$	$O(\log n)$	It solves prime factorization problem in reasonable time and breaks down RSA and ECC encryption; inspires quantum-resistant cryptography

3.3 Deutsch–Jozsa (DJ): First Breakthrough Proving Quantum Advantage in Principle

A single query to the quantum oracle is needed for the Deutsch algorithm to ascertain if a given function is balanced or constant. At the same time, a classical computer requires two queries on a classical oracle to perform the same task [12].

Input: one qubit $|0\rangle$ or $|1\rangle$.

Output: one qubit $|0\rangle$ or $|1\rangle$.

The function can take only two forms—either constant or balanced, i.e., exactly half of them gives **0,** while the other half gives **1** [13]. Input is a uniform linear combination of the n qubit computational basis states: $\{x_{n-1}, x_{n-2}, x_{n-3} \ldots, x_1, x_0\}$, with $x_i = 0, 1$.

In case of classical system, number of queries: $2^{n-1} + 1$.

In case of quantum system, number of queries: 1.

3.4 Bernstein–Vazirani (BV): Pioneering the Way for Quantum Complexity Theory

BV, a limited variant of DJ, is another quantum algorithm which demonstrates Quantum advantage and it gives a proof-of-principle showcasing the power of quantum computing in solving complex computational problems. This algorithm allows us to reveal a secret number stored inside a quantum oracle (Conceptual Black Box). A quantum oracle takes n bits secret string as input and reveals the results as 0 or 1. For an n bit string, classical algorithm will require 2^n computations to reveal the secret number. Even if the classical computers use a clever method of finding the secret string by ANDing the secret string with n bit string with one 1 and all 0's starting from LSB and ending at MSB, it will require n computations [18]. This is a special case of Deutsch–Josza with $f(x) = s.x$, where f is a hidden Boolean function which takes n qubit input string x. But with the help of the BV algorithm, it requires only one step to find the secret number and is independent of the size of the number [18]. Bernstein–Vazirani proposed an algorithm in 1993 which depicted clear-cut classical computing vs. quantum computing parting even when minor errors are permissible. Deutsch–Jozsa proved a deterministic quantum advantage. If minor errors are permitted in the computation, both quantum and classical versions of algorithms run in time $O(1)$ demonstrating no distinction. But BV algorithm showed clear difference even when minor errors are allowed in computation, indicating quantum advantage. The BV algorithm took $O(n)$ time complexity on classical computers, while $O(1)$ time complexity on a quantum computer. They further proposed Quantum Fourier Transform (QFT), with the help of which Peter Shor developed his phenomenal quantum factorization algorithm in 1994 [10]. Earlier successful implementations of Grover's algorithm on 2-qubit systems was done by Jones et al. at Oxford University [19, 20]. Paul Kwiat et al. from Los Alamos National Lab demonstrated successful execution of the Grover algorithm utilizing conventional optical interferometer [21].

BV algorithm finds some remarkable relevances in cryptography. It can breach symmetric-key cryptosystems, pluck out error traces in Quantum Error Correction (QEC) codes, and can solve some particular optimization problems in a single shot or query [22].

3.5 *Simon's Algorithm: A Stepping Stone and Inspiration Behind Many Quantum Algorithms*

A problem that a quantum computer can solve significantly faster than a conventional computer was reported by Simon in 1994. On that problem classical computer has best—case complexity of the order of: $\Omega(2^{n/2})$, while quantum computer has worst—case complexity of the order of: $O(n)$. Since the upper bound (worst-case complexity) on the quantum computer [23] is of lower order than the lower bound (best-case complexity) on the conventional computer, it is an obvious illustration of the quantum advantage. One input cannot be transformed into two separate outputs; however, a function can transfer two different inputs to the same output. It means that 2:1 is acceptable but 1:2 is not. The central objective of Simon's problem is to find the periodicity of the function given that the function is of the type 2:1. Shor successfully implemented Simon's result while developing his quantum factorization algorithm. His insight was in realizing that determining the periodicity of a function and finding the prime factors of a large composite integer are indeed isomorphic problems. Researchers in 2004 successfully tapped the potential of Simon's result on a quantum system to determine the periodicity of a function [23].

Simon's algorithm is a pillar on which stands Quantum Fourier Transform (QFT) and Shor's algorithm, which are applied further to break down many cryptographic primitives, launching destructive key recovery attacks and solving hidden subgroup problems [24].

3.6 *Teleportation Protocol: A Common Playground for Classical Communication and Quantum Entanglement; Revolutionizing Secure Communications*

It is a technique to transmit qubits from one location to another that may be light years apart in the universe. According to the `no-cloning` theorem in quantum computation, creating an independent and identical copy of a qubit is impossible. Suppose Alice plays the role of a sender of the qubits, and Bob that of the receiver. Telemon is the third person responsible for creating an entangled qubit (**ebit**), while qubit pair formed by qubit A and qubit B is shared with Alice and Bob making it a `3-qubits` system. Alice has a qubit $|\psi\rangle$, which she needs to transmit, and another one A from entangled pair. Bob has only one qubit B from entangled pair [25, 26]. Alice measures the first two qubits (now classical bits after measurement) she possesses and sends them to Bob over the classical channel. Since Alice uses the classical channel for passing qubits, which can be done at the maximum speed of light, Einstein's theory of relativity still holds good. Bob can obtain the original qubits by Alice's bits after applying appropriate transformations on his qubit B, which is the part of entangled pair [27].

Quantum teleportation has the ability to transform the cybersecurity environment by delegating extremely secure data transmission and enabling the implementation of quantum-safe internet. Its immanent relationship with quantum entanglement allows it to securely station QKD and generate unbreakable encryption. Scaling the teleportation protocol to higher dimensions to satisfy real-world demands remains a pipedream due to a number of technological issues that must be addressed [28]. Secure information transfer can be realized via a secure quantum communication network, which lays its basis on the principles of quantum teleportation protocol (QTLPP). Information sensing and eavesdropping problems inherent in classical networks can be neutralized by installing ultra-robust quantum communication networks. QTLPP raises hope for realizing a universal quantum internet, which is ultra-reliable and highly secure compared to its classical counterpart [29]. Safeguarding confidential information will require incorporating quantum teleportation into cybersecurity procedures, as quantum computing technology advances. An easy to understand schematic diagram of quantum network is shown in Fig. 4.

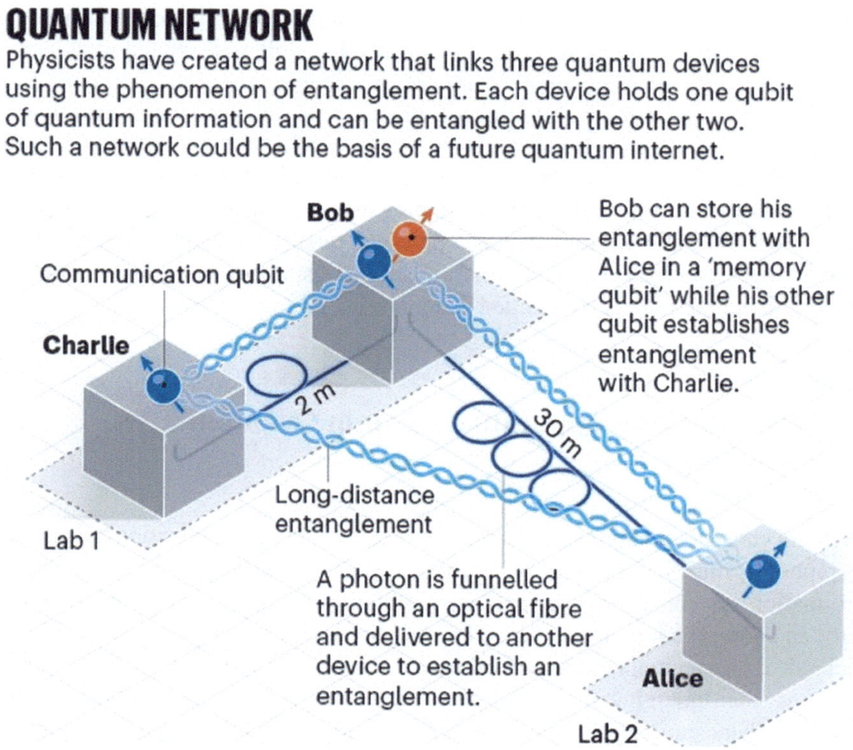

Fig. 4 A schematic diagram of quantum network (Figure by Davide Castelvecchi [30]; in public domain)

3.7 *Superdense Coding (SDC) Protocol: Mobilizing the Strength of Quantum Entanglement for Communication*

Superdense coding is a quantum communication protocol where many classical bits (`cbits`) can be transmitted by only a few numbers of qubits. Speaking specifically, SDC is a robust protocol that transmits *two* `cbits` just with the help of *one* `qubit`, meaning saving 50% of the classical information. It enhances channel volume and information security by enabling the transmission of two cbits with one qubit, since information transmission reduces to half. With practical uses in a variety of quantum communication contexts, its unique ability to resist network hacking makes it a fundamental component of secure quantum secret coding [31]. Comparing this capability to classical communication channels, the channel capacity is virtually doubled. SDC's ability to optimize information transport might completely change how information is sent across networks and allow for more economical use of resources [32]. In conclusion, SDC offers prospects for increasing the effectiveness and capacity of quantum communication channels and networks while enhancing information transmission security. Furthermore, ongoing efforts are needed to resolve the related weaknesses and open the door for its potential future uses in cybersecurity. Here Alice and Bob share entangled bits (`ebits`) pair or a Bell pair. SDC involves the preparation of ebits pair, encoding, and decoding the message. SDC is the underlying principle of secure quantum communication [33]. Since classical information is encoded in quantum form for transmission, eavesdropping and intercepting messages are eliminated [34]. SDC may be regarded as a reversed version of teleportation protocol. In SDC *two* `cbits` of information is transmitted using *one* `qubit`, whereas, in teleportation, *one* qubit is transmitted using *two* `cbits` of information.

4 Some Emerging Application Areas in Quantum Computing

Quantum encryption plays a crucial role in safeguarding communications. Quantum encryption is impenetrable, can detect malicious parties instantly, and five pillars of information system security, namely confidentiality, integrity, availability, authenticity, and non-repudiation, can be well protected. It solves the fundamental problem of classical encryption, i.e., the key exchange problem with robustness. This section discusses some emerging application areas of quantum computing, emphasizing our focus on cybersecurity and cryptographic applications. The following Fig. 5 summarizes some key emerging application areas in quantum computing, while Fig. 6, which is a subsection of Fig. 5 highlights quantum encryption and its key roles in securing communication:

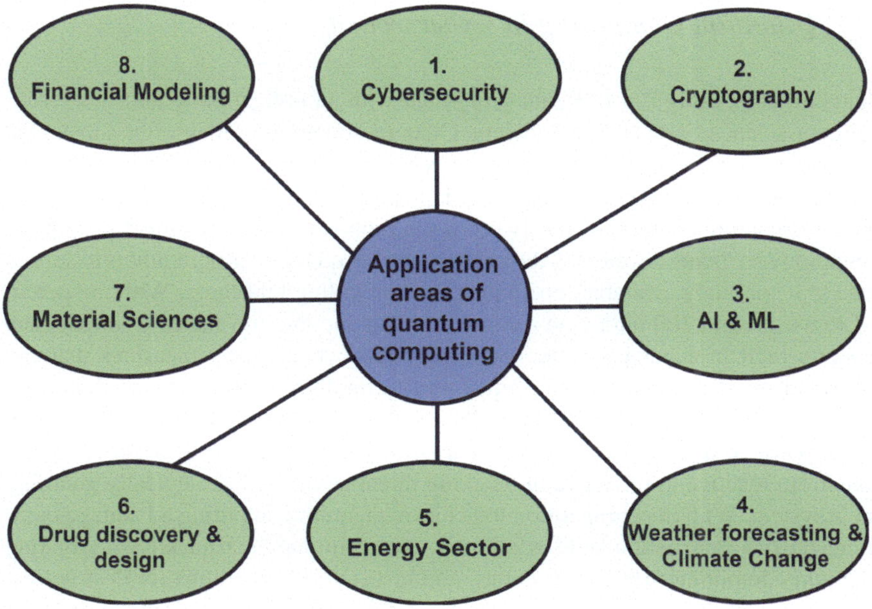

Fig. 5 Key Emerging applications areas of quantum computing (Figure by author)

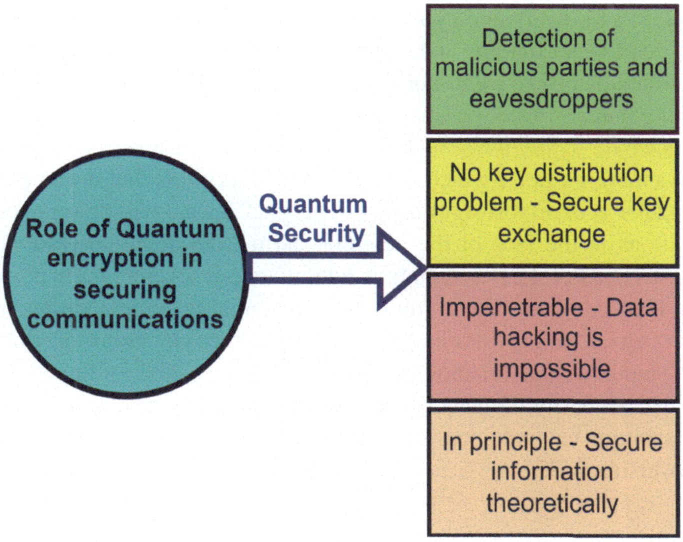

Fig. 6 A subsection of Fig. 5 highlighting quantum encryption and its key roles in securing communications (Figure by author)

4.1 Quantum Computing in Cybersecurity

Quantum computing is a fast-enhancing field with an obligation to alter numerous facets of science, industry, and society. Quantum computing may offer substantial advantages over traditional approaches in cybersecurity, making it a vital and imperative application [35]. The development of quantum microchips opens the possibility of using quantum computing to solve classical problems. As algorithms for quantum technology are being developed, the industry invites quantum programmers to submit code to its publicly available, open quantum computing interfaces. With the potential to use around 100 qubits on a computer chip and the prevalence of intellectual property theft in cyberspace, the need for robust cybersecurity measures that can withstand the disruptive effects of quantum technology is more urgent than ever. Cybersecurity is vulnerable to fraud, illicit activity, and vandalism after the advent of the wave of quantum computing. Threat intelligence can shift security operations from reactive to proactive in blocking threat actors and strategically guarding one's system. With the competition to achieve quantum computing advantage escalating, the risk from nation-state individuals, cybercriminals, and hacktivists targeting quantum computing firms, universities, and research organizations for destruction, espionage, and financial gains is on the rise. This underscores the critical need for all of us to play our part in implementing and maintaining enhanced cybersecurity measures to protect our networks, hardware, software, and data from these increasingly prevalent and diversified cyber assaults. Since there are few products on the market and insufficient funding and resources for quantum computing development, we should immediately go right. For this reason alone, we should consider security while creating and constructing quantum hardware and software components [36]. The importance of "*Quantum security*" for cybersecurity cannot be overstated. The goal of this emerging field, which is to develop algorithms that are impervious to quantum attacks, is of utmost significance. This goal is what will preserve our digital communications in the face of the potential harm from quantum computers. The *urgency* of this need is clear, as quantum computers can severely harm conventional encryption. Protecting our digital future from potential hacking attempts requires a proactive strategy. Leading this charge are influential organizations like the National Institute of Standards and Technology (NIST). They advocate for the establishment of a robust cybersecurity framework that can withstand the increased computational power of quantum technologies. The importance of these preventative measures cannot be overstated [37].

4.1.1 Quantum Computing Threats to Cybersecurity

The Aadhaar case in the year 2018 in India, the Alibaba breach a year later, and the Yahoo data breach in 2013, which resulted in an estimated three billion accounts being compromised, all exemplify the devastation that may arise in the digital age when people let their guard down [38]. They also highlight the irony of firms professing their commitment to privacy with one hand while keeping the back door ajar with the other, a fact that should make us all more critical and vigilant about our online privacy. These are just a few of the many clear instances of how businesses have been breached, leading cybersecurity professionals to deeply ponder the potential disruptive effects that quantum computers may have on the cryptographic systems in use today.

Conventional cybersecurity depends mainly on cryptographic techniques that cannot keep up with the processing efficiency of quantum computers. Algorithms such as *RSA*, *ECC*, and *Diffie–Hellman*, which are backbones of safe and robust internet communications, are vulnerable to quantum assaults because they rely on the complexity of problems such as *prime integers factorization* and *discrete logarithms*, which quantum computers can solve with high efficiency in polynomial time. Quantum computers leverage phenomena such as *superposition*, *entanglement* and *interference* to perform many calculations simultaneously, which can break these cryptographic schemes. These vulnerabilities have ushered in an urgent call for *post-quantum cryptography*—a new suit of algorithms that can resist quantum attacks.

4.1.2 Enhancing Cybersecurity with Quantum Technologies

The rise of quantum computers presents a grave threat to existing cybersecurity protocols. Their unmatched processing power gives them the potential to dismantle widely used encryption techniques, as they can handle complex models and solve complicated mathematical problems. This threat looms large, even though the technology needed to break current encryption standards is not yet available. It would require devices of a significantly larger scale than those currently in use. The US-based certification authority NIST has expressed concern over this and called for creating "*quantum-safe*" encryption techniques [39]. This project, with its potential to implement these innovative algorithms in the upcoming decades, could significantly impact the future of cybersecurity. This action was partially prompted by an unexpected 2015 National Security Agency (NSA) announcement. The NSA suggested preparing for the switch to quantum-resistant algorithms rather than significantly investing in switching to Grade B elliptic curve algorithms. This is supported by the fact that establishing a new encryption standard may take 5–10 years. The strategy is not just preventive, but also crucial, encouraging early preparation for quantum-resistant techniques [40]. The worry is that the existence of a sufficiently potent quantum computer may remain a secret until it becomes accessible. Data security is a severe danger because of this obscurity, which might make it simple for malicious parties to decrypt data using outdated techniques.

4.1.3 The Future of Quantum Computing in Cybersecurity

The research on quantum computing at major scientific organizations, government agencies, and research educational institutions is moving forward quickly. This development raises worries about its potential to undermine conventional cryptography and render data encryption techniques outdated. The imminent threat of quantum computing to existing encryption cannot be exaggerated. Researchers are on the brink of needing to develop entirely new encryption methods to counter these powerful machines. Quantum computers are poised to effortlessly handle classical encryption algorithms like *RSA* and *elliptic curves*, drastically reducing the time it takes to crack security keys from years to mere hours [41]. While it's true that experts predict it could be more than a decade before quantum computers can crack existing encryption, the time for cybersecurity preparations is now. The potential threat of quantum computing to existing encryption demands immediate action. Organizations are strongly advised to implement defense-in-depth strategies, prioritize data protection during both transmission and storage, and most importantly, remain adaptable in the face of new threats. The ever-evolving nature of cybersecurity requires a malleable approach to prepare for the future. Proactive cybersecurity strategies, such as network segmentation, use of 5*G* private networks, Zero Trust architectures, and re-encryption of outdated information using modern technology, are designed to address current cybersecurity issues and provide a robust defense against a range of threats. Importantly, these strategies also prepare us for the advent of quantum computing [42]. It is critical to keep up with the most recent advancements in quantum computing and their implications for cybersecurity as the field develops. Quantum-resistant algorithms are a crucial advancement in cybersecurity, requiring continuous research and debate to ensure their effectiveness in various applications. Organizations can upgrade their security protocols and prepare themselves more for attacks by knowing about the developments in quantum computing. Resilience against present and future difficulties in cybersecurity maneuvers may be ensured by emphasizing agility and a proactive approach. Stay updated on cybersecurity and quantum technologies developments to protect your digital assets and uphold confidence in an ever-connected world.

4.1.4 Expansion of Cybersecurity with Quantum Computing

Quantum computing offers both advantages and disadvantages for cybersecurity. It poses a threat as well as unique techniques for improving it. QKD employs quantum mechanics' fundamental principles to provide secure communication by allowing two parties to produce a unique random secret key that can be utilized for data encryption and decryption. The distinctive element of QKD is that it is proved theoretically safe; any attempt of espionage may be detected since it modifies the fragile and delicate quantum states of communicating qubits. Quantum computing can generate unique random numbers, which is essential in constructing safe cryptographic keys. Quantum computers can enhance the randomness of quantum data since they can generate *truly* random numbers, unlike classical computers, which can produce only *pseudo*-random numbers, thus strengthening the foundations of cryptographic protocols since they use randomness inherently.

4.1.5 Preventive Actions for Corporations

Organizations like NIST have initiated efforts to regulate Post-quantum Cryptography (PQC) algorithms to admit the potential quantum menaces. These algorithms are designed to be secure against both quantum and classical computers and are based on intractable mathematical problems that do not presently have efficient solutions on quantum machines. Lattice-based cryptography, hash-based signatures, and multivariate quadratic equations are some of the front runners in this new era of cryptography. The transition to a post-quantum era is filled with challenges. One of the major apprehensions is the "*harvest-now, decrypt-later*" risk, where adversaries collect scrambled information with the intent of decoding it once quantum computing becomes widely available in the future. The utter immensity of upgrading the worldwide internet architecture also requires international bonding and remarkable expenditure and funding. Moreover, post-quantum algorithms tend to require massive key sizes and heavy computational resources, which could negatively alter the efficiency of systems that implement them.

Organizations should implement a defense-in-depth cybersecurity policy that includes preparation for quantum assaults. A vital aspect of this policy is the adoption of quantum-resistant algorithms, which involves assessing the security infrastructures and establishing a culture of awareness about security through regular teaching, training, and guidance. These precautionary steps will be essential to reducing risks and ensuring the integrity and security of sensitive data as quantum technologies advance [43]. The advancements and preventive steps that Infosys is making in quantum computing is shown in Fig. 7.

4.2 Quantum Computing in Cryptography

Quantum computing, with its remarkable computational capabilities, is set to revolutionize the field of cryptography. Quantum computing, utilizing qubits and quantum principles such as *superposition*, *entanglement*, and *interference*, revolutionizes computing by reducing years spent on complex calculations to a few hours, significantly improving the efficiency of computers. This efficiency could potentially expose vulnerabilities in cryptographic techniques that rely on the complexity of these calculations, leaving us in awe of the power of quantum computing. The technique that uses quantum mechanical processes to carry out cryptographic operations

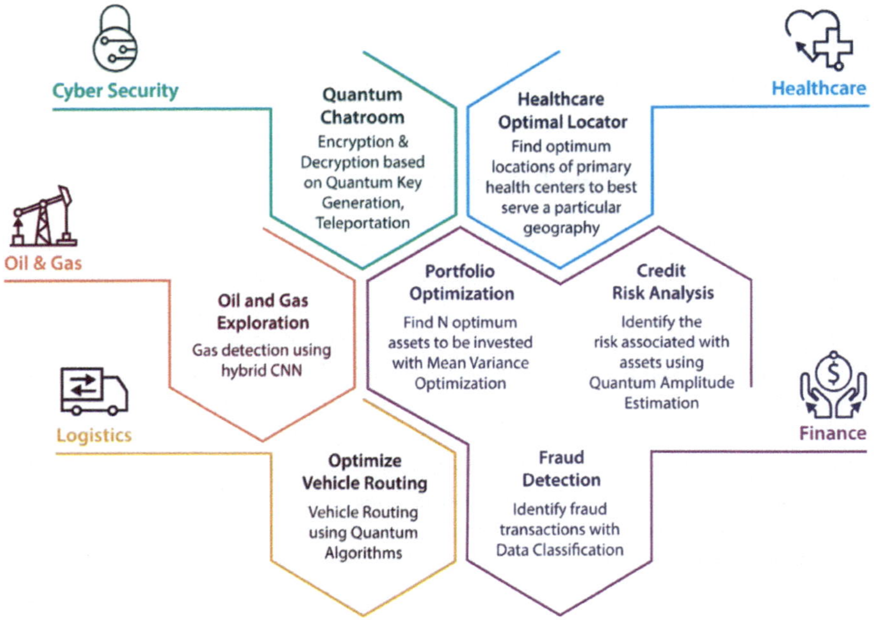

Fig. 7 The advancements and preventive steps taken by Infosys in quantum computing [44] (In public domain)

is known as *quantum cryptography*. It's possible to think about quantum cryptography as a potential development over classical cryptography. Quantum cryptography has transformed communication by ensuring secure data transmissions, eliminating the drawbacks of classical cryptography. By utilizing the *no-cloning* theorem of quantum states, quantum cryptography creates dependable and secure cryptographic systems. According to the *no-cloning* theorem, an unknown quantum state cannot be replicated precisely. Quantum encryption is based on this unique property of quantum states [45]. The field of quantum cryptography is relatively young. A quantum variant of the factorization technique based on the QFT was described by Peter Shor in 1994 [10]. Unlike its traditional predecessor, which took exponentially longer (intractable) time to solve the factorization problem, this technique completed the task in polynomial time. It presented a risk of breaking the foundation of our most widely used cryptographically secure system, the RSA cryptosystem, in the future. In the age of developing quantum technology, the threat to the security of conventional cryptosystems cannot be overlooked. Our task is to create and construct cryptosystems using quantum technologies that are impervious to even the might of quantum computers. Even while there aren't currently fully working quantum computers that can rupture current encryption, breakthroughs in the field reveal that substantial progress is being made. Research suggests that to breach present encryption techniques, future quantum computers successfully could need an order of magnitude more computing power than existing ones. However, given how quickly science progresses, there are questions about when these capabilities may become accessible.

4.2.1 Threat to Current Encryption and Decryption

Quantum computing could revolutionize encryption methods like RSA and elliptic curves, potentially rendering them obsolete and insufficient for data protection due to their faster computational capabilities. These sophisticated computers function differently from classical systems because they are based on the concepts of quantum physics. When completely developed, they could decrypt popular encryption techniques, making digital communications just as insecure as if they hadn't been protected. Quantum computers, though not fully advanced, are yet to fully undermine traditional encryption techniques, indicating a need for significant technological advancements in the field. It is reasonable to be concerned that quantum computing may ultimately outwit current encryption defenses. Analyzing digital cryptography, its current uses, and its vulnerabilities is essential, given the possible future danger to the security of online communication and business transactions. Therefore, the transition to PQC will be essential for protecting sensitive data in the upcoming years. One of the major risks posed by quantum computing is to asymmetric cryptography, particularly techniques like *RSA*, which rely on the difficulty of factoring huge composite integers. Shor's technique explains how a sufficiently powerful quantum computer might factor these composite integers in a fraction of the time classical computers require, possibly cracking encryption in days rather than years. This underscores the urgent need for new encryption algorithms to counter quantum threats, demanding immediate action from the cryptography community. While symmetric cryptographic algorithms (e.g., *AES*) are less vulnerable than their asymmetric counterparts, they are not immune to quantum threats. Grover's algorithm allows quantum computers to search through encrypted data more efficiently, reducing the key length's security [11]. This means that symmetric keys will need to be longer to retain their security against quantum-enabled attacks. The necessity of immediate adjustments in today's cryptographic strategies is clear, as businesses must act swiftly to adapt to this new reality.

4.2.2 Key Exchange Problem and Quantum Key Distribution (QKD) Protocols

QKD, a groundbreaking development, harnesses quantum physics principles to fortify cryptographic security, offering a more secure key exchange method than classical techniques. This promising innovation, which shields information with quantum attributes, is a vital step towards future secure communication systems [41]. A theoretically secure solution to the traditional key exchange problem, which is vulnerable to *man-in-the-middle* attacks, is provided by the QKD protocol [46].

According to Stephen Wiesner [47], polarized light might be used to store information about the value of unforgeable monetary notes. Every banknote has 20 light traps, a conventional number, and a database with polarization states to help identify it. Eventually, this notion blossomed into the idea of quantum cryptography. Quantum cryptography holds the potential to enhance the security and confidentiality of our data in the future, given the rapid advancement of quantum technology. Bennett and Brassard's first Basic QKD protocol, BB84 [48–50], was released in 1984. Two basis pairs, which are sets of mutually orthogonal states, were employed in this protocol: one diagonal (45° and 135°) and one horizontal/vertical (45°). Using conventional communication, Alice and Bob cannot firmly identify Eve as evil. A measurement procedure modifies the system's initial state in a quantum system. Using a one-time pad, a method of encryption where a key is used only once, Alice (the sender) and Bob (the recipient) may safely and covertly communicate once they are certain that Eve the eavesdropper is not there. A schematic diagram of BB84 QKD protocol is shown in Fig. 8.

Artur Ekert developed the QKD Protocol, often known as the E91 protocol, in 1991 [45], expanding on this concept. In this protocol, Charlie creates an entangled pair (EPR Bell pair) and transmits Bob and Alice a single qubit. Eve, Alice, and Bob test Bell's inequalities [53, 54] to confirm the existence of an eavesdropper. A protocol termed B92, after Bennet, was introduced in 1992. It applies to two *non-orthogonal* states: D for 1 and H for 0. Compared to BB84, the B92 protocol is more straightforward to implement. While it may be used with continuous variables, BB84 [51, 55] offers more security. QKD makes secure quantum communication possible,

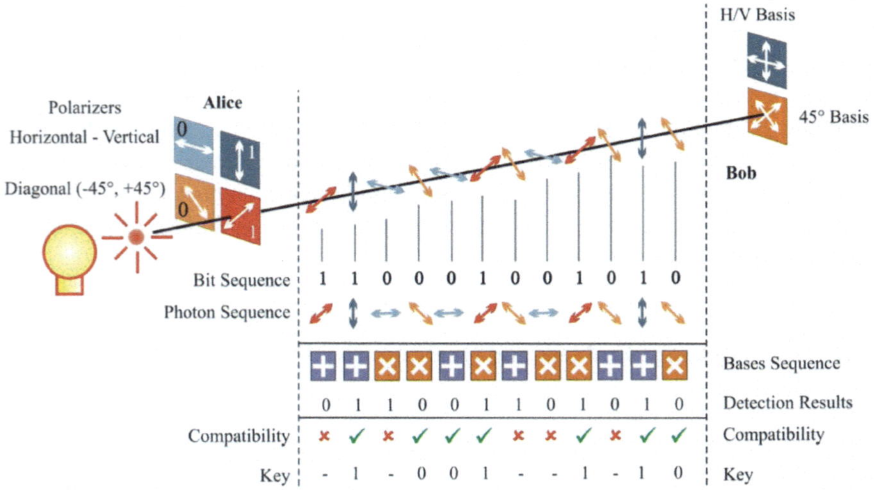

Fig. 8 Polarization of photons is used to accomplish key exchange in the BB84 protocol (In public domain-reprinted from Vasileios Mavroeidis et al. [51] and derived from Vadim Makarov PhD thesis [52])

which is essential for commercializing upcoming quantum technology. A summary of important quantum cryptographic protocols is as follows:

Summary of important QKD protocols [56] is as follows:

a. **BB84** (1984) is the oldest QKD protocol that securely communicates a private key for one-time pad encryption, ensuring information gain only occurs when the states are not orthogonal.
b. The **E91** protocol is a quantum encoding approach that employs photon-entangled pairs to produce secure communication keys while detecting eaves-dropping attempts and ensuring perfect correlation between observations.
c. Polarized photon pairs and decoy states are used in the **BBM92** protocol (1992) to transmit non-orthogonal quantum signals in a safe manner.
d. Through Z basis measurements, the **B92** protocol (1992) provides unconditional security for non-orthogonal quantum state transmission over noisy and lossy channels.
e. By removing photon polarization or entangled photons, the **MSZ96** protocol (1996) encrypts a cryptographic key bit using four non-orthogonal quantum states in a faint optical environment.
f. In comparison to the **BB84** protocol, the six-state protocol (1998) is a quantum cryptography technique that provides improved error detection and noise resistance.
g. A straightforward, effective technique for distributing quantum keys, the **DPS** protocol (2002) use successive pulses to create keys quickly and is resistant to assaults via photon-number splitting.
h. A quantum encryption technique called the **Decoy State** Protocol (2003) combines a variety of bit error rates and intensity levels to identify and thwart photon-number-splitting attacks.
i. The quantum key distribution technique **SARG04** (2004) was created to fend off attenuated laser pulses and photon-number-splitting assaults, especially in cases when data is received by an imperfect detector from a weak Poissonian source.
j. All that is needed for secure communication utilizing weak coherent light pulses is a client-side random number encoder and high-speed key transfer, thanks to the **COW** protocol (2005).
k. **Random polarization rotations** are used in the 2006 three-stage quantum cryptography protocol to encrypt data, allowing for key exchange and single photons as well as multi-photon quantum cryptography and defense against *man-in-the-middle* attacks.
l. By incorporating a quantum bit error rate and minimal index transmission error rate, the **KMB09** protocol (2009) improves the transmission distances between Alice and Bob, particularly for higher dimensional photon states.
m. With the use of multicore fibers or free-space lines, **HDQKD** technology provides secure communication by encoding quantum information in large dimensions and transferring it across great distances.

n. The **T12** protocol makes QKD more feasible, it eliminates several idealizations and incorporates features that possibly elevate the system's key rate.

4.2.3 Post-quantum Cryptography (PQC): Preparing for the Future

The pressing need to develop cryptographic systems that can withstand the ominous threats from quantum computers has given rise to post-quantum cryptography, also known as Quantum-Resistant Cryptography (QRC). This crucial field of study is dedicated to shielding conventional computer systems from the formidable computational powers of quantum computing, a potential threat that cannot be ignored. In the 1990s, mathematician Shor's groundbreaking discovery sent shock waves through the field of cryptography, revealing the immense potential of quantum computers to swiftly compromise Public-Key Encryption (PKE) schemes. This discovery gave rise to a worldwide search for the conceptualization, design, realization, and implementation of post-quantum cryptography technologies by cryptographers. Post-quantum cryptography standards are still in the development stages [46]. Encryption systems are reinvented by post-quantum cryptography, which guarantees system security against quantum and conventional computers while improving the reliability and security of encryption. Companies like Utimaco are spending on post-quantum cryptography to offer quantum-resistant solutions that shield systems from possible adversarial attacks based on quantum computers [57]. The following Table 2 summarizes how PQC will affect contemporary cryptography.

Below, we discuss some *unconventional* cryptography techniques that resist quantum computing power attacks:

I. Lattice-Based Cryptography (LBC):

Quantum computers are revolutionizing cryptographic methods, with LBC emerging as a critical area of focus due to their potential to break existing cryptographic methods. LBC, an infinite grid, involves identifying specific points on a lattice. This task becomes more complex in higher dimensions, while a set of points represents private and public keys. In LBC, an exhaustive examination of all potential solutions would be necessary to decipher the private and public keys. Even with the quantum computer's ability to calculate faster, this is still a challenging task unlikely to be completed in a reasonable amount of time [58]. The NIST process utilizes LBC for

Table 2 Table depicting the impact of Quantum Computing on classical cryptographic algorithms (Table by author)

Cryptographic algorithms	Post-quantum computing
RSA	No more trustworthy
Diffie–Hellman	No more trustworthy
ECC	Strong but highly vulnerable
ECDSA	No more trustworthy
DSA	No more trustworthy
AES-256	Strong but prone to vulnerability
SHA-256	Strong but prone to vulnerability

key encapsulation, a key-based cryptographic primitive, to enhance the security of NIST-approved cryptographic schemes.

II. Hash-Based Cryptography (HBC):

Data in plaintext, which refers to the original, unencrypted form of the data, transforms into a distinct, fixed-length code known as a *digest* by way of a *hash* function. Cryptographic hash functions frequently employed are *SHA2*, *SHA3*, and *Blake2*. While hash-based signature schemes are indeed powerful tools, it's crucial to remember that they expand on one-time signature techniques. This expansion, if not managed carefully, can potentially jeopardize security if utilized repeatedly. By generating *Lamport* keypairs and allocating each public key to a leaf of a *Merkle* hash tree—the root of which serves as the master public key—Merkle's approach makes it possible to sign numerous messages. It makes secure verification of immense data structures possible [59]. The Merkle root is a single value that encapsulates all transaction data, while the Merkle tree, at its most significant level, represents a higher level of transaction hashing. *Leighton–Micali*, a signature scheme, is a critical component in the LMS, alongside SPHINCS, a stateless protocol designed for blockchain systems. NIST standardizes CRYSTALS-Dilithium, FALCON, and SPHINCS+ digital signature algorithms for quantum-safe signatures, ensuring the security and reliability of signatures.

III. Code-Based Cryptography (CBC):

Code-based cryptography is a discipline that concentrates on creating cryptographic systems with error-correcting codes, a notion pioneered by McEliece and Niederreiter in the latter part of the 1970s and the early 1980s. These systems are recognized for their resilience to assaults by quantum computers, making them a key topic in the wider discipline of post-quantum cryptography [59].

IV. Quantum-Safe Network Design:

A quantum-safe network illustrates the growing need to implement a quantum-safe architecture that is resistant to attack, as quantum computing poses a mounting danger to current cryptography standards. The emphasis would be on the need for businesses and organizations to enhance their cybersecurity infrastructure to guard against possible quantum computing intrusions [60].

4.2.4 Development of Quantum-Resistant Algorithms: Problems and Solutions

Researchers are urgently pouring their efforts into creating quantum-resistant algorithms, a necessary response to the potential vulnerabilities that quantum computing may introduce. This project, led by organizations such as the NIST, is focused on developing new cryptographic protocols that can withstand quantum computing attacks. A variety of techniques, including *hash-based* signatures and *lattice-based* encryption, are under scrutiny to ensure secure data transfer in the post-quantum era [61]. One growing concern is the potential threat of quantum computing power becoming widely accessible. This could enable malicious actors to gather encrypted material for future decryption. The "*Harvest Now, Decrypt Later*" method underscores the urgency for businesses to adopt quantum-safe encryption [62]. This is crucial to protect sensitive information that may remain classified for an extended period. It's vital to act now to prevent potential security breaches that could exploit quantum computing advancements. To address these issues, researchers are analyzing *Post-quantum Cryptography* (PQC). It is being assessed whether algorithms based on *code-based* cryptography, *hash-based* cryptography, and *lattice-based* cryptography can repel quantum attacks. These solutions aim to build reliable structures that ensure data integrity and confidentiality in a post-quantum era.

4.3 Quantum Computing in Artificial Intelligence (AI) and Machine Learning (ML)

AI uses quantum computing, a quantum mechanics-based technology that allows numerous computations to be performed simultaneously, making it more trustworthy for complicated and real-life AI applications. Similarities exist between AI and quantum computing: the former may give error correction methods on quantum computers, while the latter offers quicker training times and processing capacity. Support vector machines and clustering in supervised and unsupervised learning, respectively, are two examples of successful AI implementations [63]. NASA is interested in using quantum computing to tackle complicated issues in robotics, air traffic control, navigation, and communication, which are ultimately connected with AI. Conversely, Google believes that quantum computing can resolve complex scientific calculations related to AI and ML. According to Hartmut Neven, Google Engineering Director of Research, "*ML and AI is about building better models of the world to make more accurate predictions*" in his blog [64]. Some applications of AI and quantum computing in a pipeline format are summarized in Fig. 9.

Although quantum computing in artificial intelligence is promising, there is still much to learn about making this promise a reality [65].

Quantum Machine Learning (QML) is an emerging subject that blends quantum computing and ML, placing it at the crossroads of the two. QML is a field that

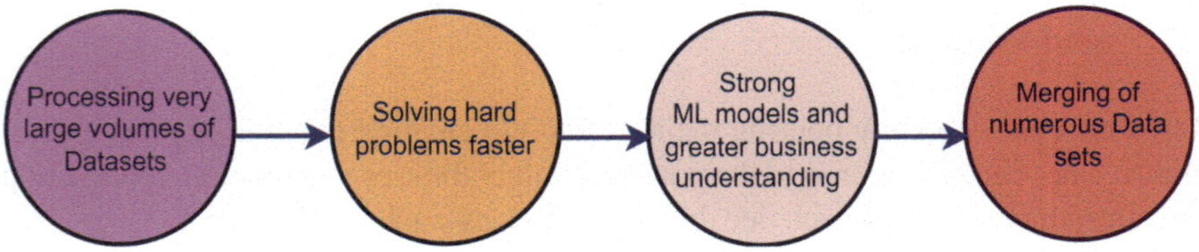

Fig. 9 AI and quantum computing applications (Figure by author)

Table 3 Several domains of quantum and classical computing

	Type of data (second letter)	
Algorithm type (first letter)	CC	QC
	CQ	QQ

investigates ideas at the intersection of ML and quantum computing [66–69]. QML aims to improve the performance of Classical Machine Learning (CML) algorithms by utilizing natural quantum mechanical phenomena such as *superposition*, *entanglement*, and *quantum parallelism* [70]. In ML, a vast number of data vectors are examined and categorized in multidimensional spaces according to their labeling or other attributes. Scalability in qubits and noise reduction in quantum circuits are two major design issues for quantum computers; in other words, building large-qubit, resilient quantum computers is a challenging and demanding endeavor. However, it is anticipated that the NISQ computers will be available soon. These NISQ era quantum computers will be at-par in power with the theoretical versions of quantum computers [71]. The training phase of ML models involves numerous samples of data, and models may consume a significant amount of time to be trained in order to yield correct answers [72]. ML is a powerful and exciting application area for quantum technologies [73]. Present classical methods demand substantial computational resources, and training charges a significant amount of time in several cases. ML algorithms learn from experience, using data instances and samples, without a user's interventions or programmer's unambiguous instructions—the machine figures out its peculiar logic. *Supervised*, *Unsupervised*, and *Reinforcement* learning are the three primary classes of the traditional machine learning paradigm, while *Semi-supervised* learning is occasionally included as a fourth class.

The quantum and classical computing have several domains depending upon types of data and algorithms. The input and the processing part can be classical, or quantum giving four possible combinations [74]. Here C represents the classical version of data or algorithm, while Q represents the Quantum version of data or algorithm as mentioned in the Table 3.

Nowadays, main focus of research is in block QC, i.e., processing classical data using quantum algorithms. Even though a fully functional, noise-tolerant quantum computer should be created in a few years, it shouldn't stop us from enjoying the thrill of coming upon brand-new research opportunities in QML [75].

4.4 Quantum Computing in Weather Forecasting and Climate Change

Scientists have long desired to predict the weather accurately, but doing so requires calculating many variables, making typical models extremely complicated. Simulating complex molecules on classical computers involves various obstacles. More robust climate models might be built utilizing quantum computers with tremendous processing power. The `Haber-Bosch` process absorbs 3–5% of global natural gas output to produce ammonia fertilizer [76]. Understanding how bacteria organically make ammonia with considerably less energy and replicating the `FeMoCo` molecule—the major cofactor of natural nitrogenase—would be tremendous steps forward. This replication may result in the construction of more effective catalysts for green hydrogen generation and carbon removal, significantly reducing energy consumption and greenhouse gas emissions, and thereby considerably advancing both sectors. Quantum simulation holds the potential to significantly reduce the cost of creating green hydrogen by accelerating the discovery of novel catalysts. These catalysts can effectively split water molecules, replacing the costly platinum, and improving the efficiency of current conversion methods. Enhanced catalysts play a crucial role in reducing carbon emissions [76]. They improve binding efficiency and cut down on costs when it comes to removing carbon from the environment at its source. Quantum simulation is not just a buzzword, it is a powerful tool that can solve complex issues such as simulating `Haber-Bosch` catalytic processes, a significant challenge in the field of `renewable` energy and catalysis.

4.5 Quantum Computing in Energy Sector

Quantum computers have proven useful in fields like high energy physics, nuclear science, quantum chemistry, economics, communications, and optimization, enabling applications in nuclear energy, waste management, fossil energy, and carbon management. Fast and efficient simulations of complex energy system processes to discover the optimal retrofit option for CO_2 absorption from exhaust gases at a pilot size power plant in South Korea, for instance, superstructure-based methodologies have been used [77]. Quantum computing and quantum sensing are used at the Argonne National Laboratory to address challenging issues in nuclear theory and experimentation. Nuclear vitrification in nuclear waste management calls for computational fluid dynamics modeling, which can be assisted by quantum algorithms. Sustainable quantum computing is a bright and optimistic area of research involving quantum computing technologies to improve environmental health while minimizing the lousy environmental impact, reducing footprints, and prioritizing energy efficiency [78]. Some quantum computing applications in energy sector are shown in Fig. 10.

4.6 Quantum Computing in Drug Discovery and Design

Quantum computing has the potential to transform medicinal research, lowering prices and saving time for pharmaceutical corporations. Researchers hope it will help them better understand medication interactions while saving time, resulting in

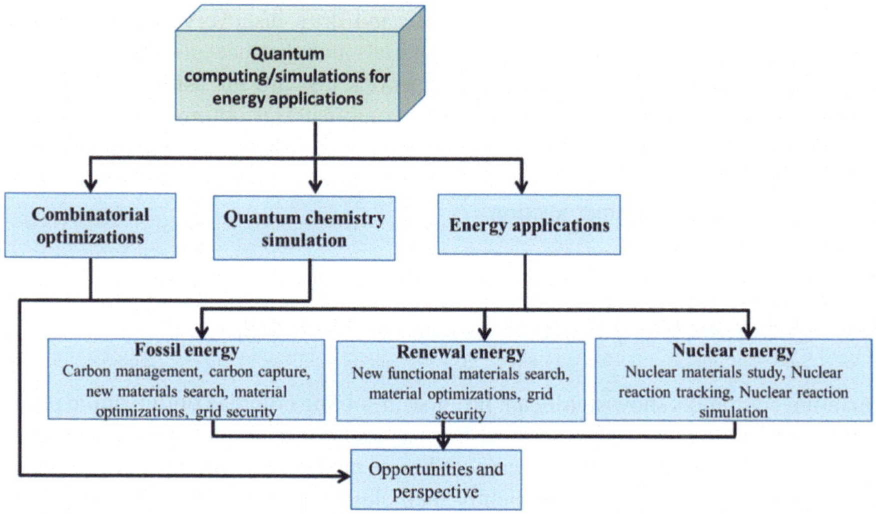

Fig. 10 Energy sector applications of quantum computing (Figure by author)

more medicinal product discoveries and increased pharmaceutical sector efficiency. Quantum computers are transforming drug research using quantum simulation and through algorithmic advances in machine learning. These developments allow for faster and more exact molecular system characterization, which might help biochemical research. Despite progress, exact and fast simulation remains a challenge [79]. Computational chemistry is a vital tool in the pharmaceutical industry. With the capacity to do previously unachievable calculations, quantum computing is a fast-evolving technology with the potential to modify computational capabilities in many disciplines of chemical study significantly. This view-point emphasizes the near-term significance of quantum computing of molecules to pharmaceutical challenges. Blunt et al. provide fresh estimates of the quantum computing cost of modeling successively bigger embedding sections of a pharmaceutically relevant covalent protein–drug combination using the antineoplastic medication Ibrutinib [80].

4.7 *Quantum Computing in Material Sciences*

Quantum computers, using quantum mechanical phenomena like superposition and entanglement, could outperform classical computers for certain problems. However, noise and imperfections in materials can obscure these effects. Advances in synthesis, characterization, and modeling drive future progress. As multi-qubit systems become more complex and research is accelerated by government and industrial investment, new challenges and opportunities for materials science emerge [81]. Quantum computing can simulate molecular interactions at an unprecedented level of detail, enabling breakthroughs in materials science and drug discovery. This capability could accelerate the development of new materials and pharmaceuticals by allowing researchers to model complex chemical reactions that are infeasible for classical computers [82]. Using quantum computers, combinatorial optimization problems and material search problems might be resolved with reduced computational costs and more precision owing to the special scaling features of quantum systems deriving from entanglement and superposition.

4.8 Quantum Computing in Financial Modeling

Quantum computers show significant promise in solving complex optimization problems, such as those found in logistics, finance, and artificial intelligence. Quantum annealing, in particular, is being explored as a method for finding optimal solutions to these problems more efficiently than classical approaches. The subject of quantum finance is expanding at an astonishing rate, in part because of conceptual advancements that promise enormous speedups for commonly used algorithms and experimental advances in quantum technology that are exceeding all expectations. The notion of using quantum mechanics to finance is not new; in fact, there are certain well-known financial issues that have direct quantum–mechanical equivalents. The Black–Scholes–Merton (BSM) formula, for instance, can be used to represent the arbitrage connections that lead to the construction of the Schrödinger equation and estimation of prices in stock market. Quantum processes may be used to represent the whole financial market, with key numbers for finance like the covariance matrix emerging naturally [83]. Quantum algorithms for Monte Carlo integration, gradient estimation, stochastic modeling, optimization, and machine learning are applicable to various financial problems and can provide quadratic speedup over their classical analogues [84].

5 Conclusions

This chapter provides an introduction to the amazing and emerging area of quantum computing; while reading this article, a novice in this field may get a glimpse of some essential aspects of quantum computing and its implications on emerging and critical areas, including AI and ML, weather forecasting and climate change, energy sector, and so on, while giving particular emphasis on quantum computing aspects of cybersecurity and cryptography in the most vital form. The world is on the transition edge of quantum disruption. Quantum computers are fundamentally different from ordinary computers in that they operate more like nature. So, ideally, we can utilize them to better understand how nature works. Contrary to the speculations of many, quantum computers will not replace classical computers, but both technologies are there to complement each other. They are not universally faster but they are superior for particular types of calculations. The fact is that we have all these quantum phenomena like superpositions, entanglements, and interference available to us simultaneously to exploit their inherent computational parallelism. In quantum computers, the calculation acceleration is not in the form of operations, but the number of operations to arrive at the solutions is exponentially minute compared to classical computers. Quantum computers can be used to leverage the power of classical computers.

Quantum computing, a technology that operates on the principles of quantum mechanics, is set to topple the terrain of cybersecurity. As we approach the quantum era, it is crucial to understand the transformative impact it will have on cybersecurity practices, threat management, and the overall uprightness of digital infrastructures. The intermingling of quantum computing and cybersecurity is both a risk and an opportunity. While quantum computing poses a clear and present threat to current cryptography standards, it is also propelling the invention of more robust quantum-resistant algorithms and security protocols. As we prepare for the quantum future beyond NISQ era, we must find a middle way between migrating to quantum-resistant systems and the urgency for practical, implementable solutions. The business, governments, and individuals must stay informed and proactive as the fault-tolerant quantum age arrives. The contest to retain confidentiality, integrity, and availability of data in the post-quantum world has already begun, and it is one we cannot afford to lose. As cybersecurity professionals in the current era, we must follow some proactive actions and strategies. These actions may include grasping quantum computing basics, comprehending quantum threats in-depth, moving towards post-quantum cryptography protocols, and remaining updated and knowledgeable about state-of-the-art quantum technology. So, let us all be ready for post-quantum cybersecurity administration and management for a safer world. Hopefully, quantum computers will solve many unsolvable problems, which will benefit all humankind in almost every dimension of life.

References

1. R.P. Feynman, Simulating physics with computers. Int. J. Theor. Phys. **21**(6), 467–488 (1982). https://doi.org/10.1007/BF02650179
2. Y. Kanamori, S.-M. Yoo, W. Pan, F. Sheldon, A short survey on quantum computers. Int. J. Comput. Appl. **28** (2006). https://doi.org/10.2316/Journal.202.2006.3.202-1700
3. A. Majot, R. Yampolskiy, Global catastrophic risk and security implications of quantum computers. Futures **72**, 17–26 (2015). https://doi.org/10.1016/j.futures.2015.02.006
4. T. Hey, *Quantum computing: An introduction* (CERN European Organization for Nuclear Research, 1998)
5. M.L. Li, H. Yang, X. Guo, Research on quantum computing technology and application, in *2019 International Conference on Modeling, Analysis, Simulation Technologies and Applications (MASTA 2019), Hangzhou, China 2019* (Atlantis Press, 2019), pp. 176–180. https://doi.org/10.2991/masta-19.2019.30
6. R. Horodecki, P. Horodecki, M. Horodecki, K. Horodecki, Quantum entanglement. Rev. Mod. Phys. **81**, 865–942 (2009). https://doi.org/10.1103/RevModPhys.81.865

7. J.V. Gael, *The Role of Interference and Entanglement in Quantum Computing* (2005). [Online]. Available: https://pages.cs.wisc.edu/jvangael/pubs/msthesis.pdf
8. S.E.A. Bennett, *Entanglement-Assisted Capacity of a Quantum Channel and the Reverse Shannon Theorem* (2001). https://doi.org/10.48550/arXiv.quant-ph/0106052
9. V.D.W. Bacon, Recent progress in quantum algorithms. Commun. ACM **53**, 84–93 (2010). https://doi.org/10.1145/1646353.1646375
10. P.W. Shor, Polynomial-time algorithms for prime factorization and discrete logarithms on a quantum computer. SIAM J. Comput. **26**(5), 1484–1509 (1997). https://doi.org/10.1137/S0097539795293172
11. L.K. Grover, A fast quantum mechanical algorithm for database search, in *Proceedings of the Twenty-Eighth Annual ACM Symposium on Theory of Computing* (1996). https://doi.org/10.1145/237814.237866
12. D. David, Quantum theory, the church-turing principle and the universal quantum computer. Proc. R. Soc. Lond. **A400**, 97–117 (1985). https://doi.org/10.1098/rspa.1985.0070
13. D. Deutsch, R. Jozsa, *Rapid solution of problems by quantum computation* (University of Bristol, 1992). https://doi.org/10.1098/rspa.1992.0167
14. D. Tosh, O. Galindo, V. Kreinovich, O. Kosheleva, Towards security of cyber-physical systems using quantum computing algorithms, in *2020 IEEE 15th International Conference of System of Systems Engineering (SoSE)* (2020), pp. 313–320. https://doi.org/10.1109/SoSE50414.2020.9130525
15. K. Nagata, T. Nakamura, A. Farouk, Quantum cryptography based on the Deutsch-Jozsa algorithm. Int. J. Theor. Phys. **56**, 09 (2017). https://doi.org/10.1007/s10773-017-3456-x
16. M. Ampatzis, T. Andronikos, Qkd based on symmetric entangled Bernstein-Vazirani. Entropy **23**(7), 870 (2021). [Online]. https://doi.org/10.3390/e23070870
17. H. Xie, L. Yang, Using Bernstein-Vazirani algorithm to attack block ciphers. Designs, Codes Crypt. **87**, 05 (2019). https://doi.org/10.1007/s10623-018-0510-5
18. E. Bernstein, U. Vazirani, Quantum complexity theory. SIAM J. Comput. **26**(5), 1411–1473 (1997). https://doi.org/10.1137/S0097539796300921
19. M. Mosca, J.A. Jones, R.H. Hansen, Implementation of a quantum search algorithm on a quantum computer. Nature **393**(6683), 344–346 (1998). https://doi.org/10.1038/30687
20. J. Jones, Fast searches with nuclear magnetic resonance computers. Science **280**(5361), 229–229 (1998). https://doi.org/10.1126/science.280.5361.229
21. M.J.E.A. Kwiat, Optical implementation of Grover's algorithm: it's all done with mirrors. Quant. Commun. Comput. Meas. **2** (2002). https://doi.org/10.1007/0-306-47097-7_41
22. Quantum Algorithm (6): Bernstein-Vazirani Algorithm. https://quantumpedia.uk/quantum-algorithm-6-bernstein-vazirani-algorithm-f371ea83256e. Accessed 18 Nov 2024
23. M.S. Tame, B.A. Bell, C. Di Franco, W.J. Wadsworth, J.G. Rarity, Experimental realization of a one-way quantum computer algorithm solving Simon's problem. Phys. Rev. Lett. **113**, 200501 (2014). https://doi.org/10.1103/PhysRevLett.113.200501
24. Lecture 8: Simon's algorithm and applications to cryptography. https://groups.uni-paderborn.de/fg-qi/courses/UPBINTROQUANTUM/S2020/notes/Lecture%208%20-%20Simon's%20algorithm%20and%20applications%20to%20cryptography.pdf. Accessed 18 Nov 2024
25. M.A. Nielsen, I. Chuang, *Quantum Computation and Quantum Information* (2010). https://doi.org/10.1017/CBO9780511976667
26. E. Rieffel, W. Polak, *Quantum Computing: A Gentle Introduction*, 1 edn (The MIT Press, Cambridge England, Massachusetts) (2011). https://mitpress.mit.edu/9780262526678/quantum-computing/
27. D. Bouwmeester, J.-W. Pan, K. Mattle, M. Eibl, H. Weinfurter, A. Zeilinger, Experimental quantum teleportation. Nature **390**, 575–579 (1997). https://doi.org/10.1038/37539
28. G.V. Studios, Teleportation of quantum particles and impact (2024). [Online]. Available: https://www.gsdvs.com/post/teleportation-of-quantum-particles-and-impact
29. R. Uppal, *Quantum Teleportation Breakthroughs Paying Way for Unhackable Global Quantum Internet—International Defense Security & Technology* (2024). [Online]. Available: https://idstch.com/technology/quantum/quantum-teleportation-breakthroughs-paying-way-for-unhackable-global-quantum-internet

30. D. Castelvecchi, *Quantum network is step towards ultra secure internet* (2021). Available: https://doi.org/10.1038/d41586-021-00420-5
31. D. Arias, I. García Rodríguez de Guzmań, M. Rodríguez, E.B. Terres, B. Sanz, J. Gaviria de la Puerta, I. Pastor, A. Zubillaga, P. García Bringas, Let's do it right the first time: Survey on security concerns in the way to quantum software engineering. Neurocomputing (2023). [Online]. Available: https://doi.org/10.1016/j.neucom.2023.03.060
32. F. Bellaiche, *On superdense coding—quantum bits* (2024). [Online]. Available: https://www.quantum-bits.org/?p=2694
33. S. Pirandola, J. Eisert, C. Weedbrook, A. Furusawa, S. Braunstein, Advances in quantum teleportation. Nat. Photonics **9**, 641–652 (2015). https://doi.org/10.1038/nphoton.2015.154
34. IBM, *Qiskit organization: superdense-coding* (2017). Available at https://qiskit.org/textbook/ch-algorithms/superdense-coding.html. Accessed: 01 May 2022
35. The role of quantum computing in cybersecurity. https://fastercapital.com/topics/the-role-of-quantum-computing-in-cybersecurity.html. Accessed 05 July 2024
36. N. Kilber, D. Kaestle, S. Wagner, *Cybersecurity for quantum computing* (2021). https://doi.org/10.48550/arXiv.2110.14701
37. K. Keplinger, Is quantum computing becoming relevant to cybersecurity? Netw. Secur. **2018**(9), 16–19 (2018). https://doi.org/10.1016/S1353-4858(18)30090-4
38. Yahoo! data breaches. https://en.wikipedia.org/wiki/Yahoo!databreaches. Accessed 05 Aug 2024
39. Nist releases first 3 finalized post-quantum encryption standards. https://www.nist.gov/news-events/news/2024/08/nist-releases-first-3-finalized-post-quantum-encryption-standards. Accessed 05Aug 2024
40. H. Holland, Quantum computing and US Cybersecurity: a case study of the breaking of RSA and plan for Cryptographic Algorithm transition (2024). https://digitalcommons.unl.edu/honorstheses/681/
41. W. Easttom, *Modern Cryptography: Applied Mathematics for Encryption and Information Security* (Springer International Publishing, 2020). https://doi.org/10.1007/978-3-031-12304-7
42. J.R. Lindsay, Surviving the quantum cryptocalypse. Strateg. Stud. Q. **14**(2), 49–73 (2020). https://www.airuniversity.af.edu/Portals/10/SSQ/documents/Volume-14_Issue-2/Lindsay.pdf
43. An approach to quantum resilience: navigating to a new frontier. https://www.techuk.org/resource/an-approach-to-quantum-resilience-navigating-to-a-new-frontier.html. Accessed 05 Aug 2024
44. Infosys is all set to bell Schrodinger's cat with quantum computing labs. https://www.infosys.com/services/incubating-emerging-technologies/offerings/quantum-computing.html. Accessed 05 Aug 2024
45. A. Ekert, J. Rarity, P. Tapster, G. Palma, Practical quantum cryptography based on two-photon interferometry. Phys. Rev. Lett. **69**, 1293–1295 (1992). https://doi.org/10.1103/PhysRevLett.69.1293
46. W. Stallings, *Cryptography and Network Security: Principles and Practice*, 6th edn (Prentice Hall Press, USA, 2013). https://archive.org/details/isbn_9789332518773
47. S. Wiesner, Conjugate coding. SIGACT News **15**(1), 78–88 (1983). https://doi.org/10.1145/1008908.1008920
48. C.H. Bennett, G. Brassard, An update on quantum cryptography. Adv. Cryptol. **196**, 475–480 (1985). https://doi.org/10.1007/3-540-39568-7_39
49. C. Bennett, G. Brassard, L. Salvail, J. Smolin, Experimental quantum cryptography. J. Cryptol. **5**, 3–28 (1992). https://doi.org/10.1007/BF00191318
50. C.H. Bennett, G. Brassard, Quantum cryptography: public key distribution and coin tossing. Theor. Comput. Sci. **560**, 7–11 (2014). https://doi.org/10.48550/arXiv.2003.06557
51. V. Mavroeidis, K. Vishi, M. Zych, A. Jøsang, The impact of quantum computing on present cryptography. Int. J. Adv. Comput. Sci. Appl. **9**, 03 (2018). https://doi.org/10.14569/IJACSA.2018.090354
52. V. Makarov, Quantum Cryptography and Quantum Cryptanalysis, Ph.D. dissertation, Norwegian University of Science and Technology Faculty of Information Technology, NTNU (2007). http://www.vad1.com/publications/phd-thesis-makarov-200703.pdf

53. G. Brassard, Brief history of quantum cryptography: a personal perspective, in *IEEE Information Theory Workshop on Theory and Practice in Information-Theoretic Security-Awaji Island, Japan* (2005), pp. 19–23. https://doi.org/10.1109/ITWTPI.2005.1543949
54. A. Nanda, D. Puthal, S.P. Mohanty, U. Choppali, A computing perspective of quantum cryptography [energy and security]. IEEE Consum. Electron. Mag. **7**(6), 57–59 (2018). https://doi.org/10.1109/MCE.2018.2851741
55. M. Niemiec, A.R. Pach, Management of security in quantum cryptography. IEEE Commun. Mag. **51**(8), 36–41 (2013). https://doi.org/10.1109/MCOM.2013.6576336
56. M. Sabani, I.K. Savvas, D. Poulakis, G. Makris, Quantum key distribution: Basic protocols and threats, in *Proceedings of the 26th Pan-Hellenic Conference on Informatics, PCI 2022, Athens, Greece*, ed. by. N.N. Karanikolas, C. Troussas, M. Vassilakopoulos, S.D. Kollias, C. Douligeris, I. Voyiatzis (ACM, 2022), pp. 383–388. https://doi.org/10.1145/3575879.3576022
57. The quantum revolution vs. cybersecurity: Ensuring resilience with post quantum cryptography. https://utimaco.com/news/blog-posts/ensuring-resilience-post-quantum-cryptography. Accessed 05 Aug 2024
58. D.R. Stinson, *Cryptography: Theory and Practice* (4th ed.) (Chapman & Hall/CRC, 2017). https://doi.org/10.1201/9781315282497
59. J.P. Aumasson, *Serious Cryptography: A Practical Introduction to Modern Encryption* (No Starch Press, USA, 2017) ISBN13: 9781718503847
60. B. Forouzan, D. Mukhopadhyay, *Cryptography and Network Security (SIE)* (McGraw Hill Education (India) Private Limited, 2011) ISBN 978-0-07-287022-0
61. J. Katz, Y. Lindell, *Introduction to Modern Cryptography*, 3rd ed (Chapman & Hall/CRC, 2020) ISBN 9781584885511
62. Make the world quantum safe. https://www.ibm.com/quantum/quantum-safe. Accessed 05 Aug 2024
63. N. Abdelgaber, C. Nikolopoulos, Overview on quantum computing and its applications in artificial intelligence, in *2020 IEEE Third International Conference on Artificial Intelligence and Knowledge Engineering (AIKE)* (2020), pp. 198–199. https://doi.org/10.1109/AIKE48582.2020.00038
64. Launching the quantum artificial intelligence lab. https://research.google/blog/launching-the-quantum-artificial-intelligence-lab/. Accessed 05 Aug 2024
65. V. Moret-Bonillo, Can artificial intelligence benefit from quantum computing? Prog. Artif. Intell. **09** (2014). https://doi.org/10.1007/s13748-014-0059-0
66. M. Schuld, I. Sinayskiya, F. Petruccionea, *An introduction to quantum machine learning* (Citeseer, 2014). https://doi.org/10.1080/00107514.2014.964942
67. M. Schuld, N. Killoran, Quantum machine learning in feature Hilbert spaces. Phys. Rev. Lett. **122** (2019). https://doi.org/10.1103/PhysRevLett.122.040504
68. M. Schuld, M. Fingerhuth, F. Petruccione, Quantum machine learning with small-scale devices: implementing a distance-based classifier with a quantum interference circuit. Quant. Phys. (2017). https://doi.org/10.48550/arXiv.1703.10793
69. J. Biamonte, P. Wittek, N. Pancotti, P. Rebentrost, N. Wiebe, S. Lloyd, Quantum machine learning. Nature **549** (2016). https://doi.org/10.1038/nature23474
70. J. Yin, Y.H. Li, S.K. Liao, M. Yang, Y. Cao, L. Zhang, J.G. Ren, W.Q. Cai, W.Y. Liu, S.L. Li, R. Shu, Entanglement-based secure quantum cryptography over 1,120 kilometres. Nature **582**(7813), 501–505 (2020). https://doi.org/10.1038/s41586-020-2401-y
71. J. Preskill, Quantum computing in the Nisq era and beyond. Bull. Am. Phys. Soc. **2**, 79 (2018). https://doi.org/10.48550/arXiv.1801.00862
72. F. Phillipson, *Quantum Machine Learning: Benefits and Practical Example*, vol. 2561 (2020). ceur-ws.org. http://ceur-ws.org/Vol-2561/paper5.pdf
73. N. Neumann, F. Phillipson, R. Versluis, Machine learning in the quantum era. Digitale Welt **3**, 24–29 (2019). https://doi.org/10.1007/s42354-019-0164-0
74. M. Schuld, F. Petruccione, *Supervised Learning with Quantum Computers* (Springer, 2019). https://doi.org/10.1007/978-3-319-96424-9

75. T.M. Khan, A. Robles-Kelly, Machine learning: quantum vs classical. IEEE Access **8**, 219275–219294 (2020). https://doi.org/10.1109/ACCESS.2020.3041719
76. A quantum advantage in fighting climate change. https://www.bcg.com/publications/2020/quantum-advantage-fighting-climate-change. Accessed 05 Aug 2024
77. H.P. Paudel, M. Syamlal, S.E. Crawford, Y.-L. Lee, R.A. Shugayev, P. Lu, P.R. Ohodnicki, D. Mollot, Y. Duan, Quantum computing and simulations for energy applications: review and perspective. ACS Eng. Au **2**(3), 151–196 (2022). https://doi.org/10.1021/acsengineeringau.1c00033
78. V. Sood, R.P. Chauhan, Quantum computing: impact on energy efficiency and sustainability. Exp. Syst. Appl. **255**, 124401 (2024). [Online]. Available: https://www.sciencedirect.com/science/article/pii/S0957417424012673
79. Y. Cao, J. Romero, A. Aspuru-Guzik, Potential of quantum computing for drug discovery. IBM J. Res. Dev. **62**(6), 6–1 (2018). https://doi.org/10.1147/JRD.2018.2888987
80. N.S. Blunt, J. Camps, O. Crawford, R. Izsák, S. Leontica, A. Mirani, A.E. Moylett, S.A. Scivier, C. Sünderhauf, P. Schopf, J.M. Taylor, N. Holzmann, Perspective on the current state-of-the-art of quantum computing for drug discovery applications. J. Chem. Theory Comput. **18**(12), 7001–7023 (2022). PMID: 36355616. https://doi.org/10.48550/arXiv.2206.00551
81. V. Lordi, J.M. Nichol, Advances and opportunities in materials science for scalable quantum computing. MRS Bull. **46**, 589–595 (2021). https://doi.org/10.1557/s43577-021-00133-0
82. M.W. Rose, Exploring the potential of quantum computing: From theory to practical applications. Int. J. Issue Sci. **1**(1), 21–27 (2024)
83. R. Orús, S. Mugel, E. Lizaso, Quantum computing for finance: overview and prospects. Rev. Phys. **4**, 100028 (2019). https://doi.org/10.1016/j.revip.2019.100028
84. D. Herman, C. Googin, X. Liu, Y. Sun, A. Galda, I. Safro, M. Pistoia, Y. Alexeev, Quantum computing for finance. Nat. Rev. Phys. **5**(8), 450–465 (2023). https://doi.org/10.1038/s42254-023-00603-1

Cutting-Edge Quantum Computing Technologies: A Focus on Cybersecurity Applications

Ajay Kumar Badhan, Abhishek Bhattacharjee, Ramandeep Sandhu, Rita Roy, and Deepika Ghai

1 Introduction

The information technology industry has started a revolutionary journey into quantum computing in an unwavering quest for computational efficiency and problem-solving capabilities. It explores the exciting world of computing technologies, where quantum bits replace standard bits as the primary unit of measurement and where the limits of computing are redefined by the laws of quantum mechanics.

The fundamental ideas of quantum mechanics, a field of physics that studies the behavior of substances and liveliness at the tiniest sizes, are used by quantum computing. The computers that are built for quantum utilize special qualities of qubits, as opposed to traditional computers, which represent information in the form of 0s or 1s. Due to the superposition and entanglement processes, these quantum bits allow a quantum computer to process exponentially more data at once, achieving previously unheard-of levels of computational power. Qubits can be in numerous conditions concurrently because of the principle of superposition, increasing the

A. K. Badhan · A. Bhattacharjee · R. Sandhu (✉)
School of Computer Science and Engineering, Lovely Professional University, Phagwara, India
e-mail: ramandeepsandhu887@gmail.com

A. K. Badhan
e-mail: ajay.27337@lpu.co.in

A. Bhattacharjee
e-mail: abhishek.27306@lpu.co.in

R. Roy
Department of Computer Science and Engineering, Gitam Institute of Technology (Deemed to be University), Pune, India
e-mail: ritaroy1311@gmail.com

D. Ghai
School of Electronics and Electrical Engineering, Lovely Professional University, Phagwara, India
e-mail: money.ghai25@gmail.com

S. B. Goyal et al. (eds.), *Quantum Computing, Cyber Security and Cryptography*,
https://doi.org/10.1007/978-981-96-4948-8_2

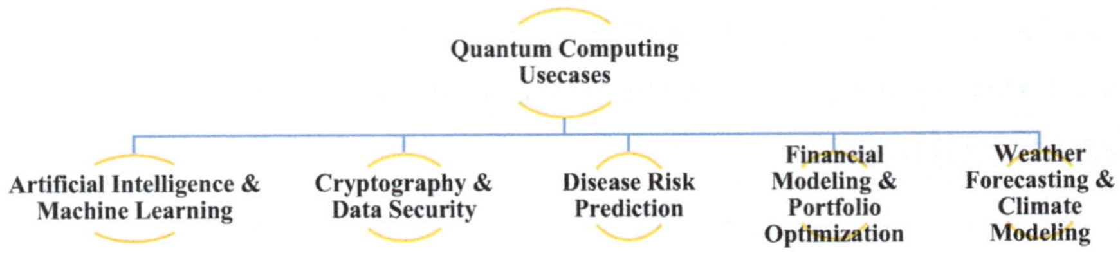

Fig. 1 Quantum computing use cases

possibilities for computation. An additional degree of complexity and connectedness is introduced by entanglement, a phenomenon in which qubits become coupled and the state of one quantum bit instantly affects the state of another. Quantum parallelism is made possible by the combination of superposition and entanglement, which allows computers that are designed for quantum to investigate several solutions to a problem simultaneously.

Quantum Computing has been utilized in several different areas. Figure 1 represents some of the most common areas where Quantum technology is implemented. The most compelling implementation is performed in the field of Artificial Intelligence. Fevrier et al. [1] provide their views on the mixture of Quantum Computing and Deep Learning in Computational Intelligence. The algorithms in quantum demonstrate notable speedups over traditional ones, suggesting their capabilities to solve present intractable problems. The purpose is to identify key works at the intersection of Quantum Computing and Deep learning, offering insights into the evolving field of computational intelligence. Similarly, Stefano et al. [2] outline how the advent of small-scale quantum computing devices has led to the convergence of neural networks in machine learning and artificial intelligence. The author examines new ideas that incorporate features of artificial neural networks into quantum structures. The combination of these technologies has the potential to be transformative, along with the role that near-term quantum hardware will play in enhancing quantum machine learning capabilities and highlighting the interesting implications.

The advent of quantum computing in the field of cryptography and data security has a lot of potential to revolutionize the field. In this, author [3] discusses the adaptability of quantum computing with a focus on how well it performs high-intensity jobs in fields such as data encryption, machine learning, data search, etc. The efficiency benefits over classical computing are highlighted, along with the network security applications where key distribution and management are the main focus areas. Similarly, [4] study analyzes the vulnerabilities added to cryptography to investigate how quantum computing affects the security of 5G mobile telecommunications. The author proposes a phased approach to recommended changes, safeguarding the integrity of 5G, 4G, and 3G systems against large-scale quantum computing. The research leverages the backwards-compatibility of 5G security to provide an easy shift to post-quantum secure systems, guaranteeing long-term security.

In the field of healthcare as well quantum plays a pivotal role. Like in the paper [5], the author addresses the critical need for precise disease subtype classification, mainly in non-small cell lung cancer (NSCLC), focusing on Glandular cancer and Squamous cell carcinoma. The author makes use of genetic factor expression data from 104 patients curated from famous datasets and proposes an approach that combines classical and quantum machine learning models. The technologies incorporated are univariate statistics and XGBoost, alongside a novel data representation method called QCrush. In a similar fashion, the author of the paper [6] addresses the global diabetes crisis, presenting a prediction tool grounded on the PIMA Indian Diabetes dataset. Making use of the techniques, two prediction models were developed and evaluated. The DL model outperformed with accuracy measures of 0.95, showcasing a 1.06% improvement over existing work, while the QML model demonstrated satisfactory performance aligned with current literature. The proposed models aim to assist medical practitioners in reducing diabetes-associated lethality.

In the monarchy of finance, the integration of quantum computing marks a paradigm shift, promising transformative capabilities to tackle complex computational problems. In [7], the author highlights the imminent transformative impact of quantum computing on the finance sector. Focusing on stochastic modeling, optimization, and machine learning, it delivers a comprehensive overview of significant applications. The paper delves into how adapted quantum solutions can enhance efficiency in financial tasks such as derivative pricing, risk modeling, and fraud detection. Similarly, the paper [8] delves into the impact of emerging paradigms like fractals, quantum physics, and blockchain. The author conducts a SWOT analysis and evaluates the potential threats and opportunities associated with these technological advancements. The investigation emphasizes the interconnectedness of these advancements, contributing to rapid shifts in economic paradigms.

Embarking on the frontier of weather forecasting and climate modeling, quantum computing stands as a game-changer, offering unparalleled capabilities to enhance precision and efficiency in predictions. In the paper [9], the author delves into the growing interest in quantum computers and algorithms across scientific disciplines. The main focus is on exploring the latent utility of quantum computing in numerical meteorological conditions and climate prediction. The integration of quantum algorithms for resolving non-linear gap equations is demonstrated using a modest non-linear model. Thus, escalating the complexity of atmospheric simulations, quantum computing holds the latent to redefine the landscape utilizing remarkable computational power and speed.

2 Related Work

Many significant algorithms have been developed in the modern era, and most of them agree that using quantum mechanics to your advantage yields significant (exponential, polynomial, super-polynomial) improvements over traditional algorithms. This implies that Quantum methods may be able to unravel some complex topics that are presently thought to be unsolvable using standard methods. On the other hand, Deep Learning (DL) algorithms provide a collection of machine learning methods aimed at teaching computers to process inputs via hierarchical structures to identify patterns and anticipate outcomes. These inputs can be in the form of sound, graphics, or plain text, among other formats. The information processing systems of human intelligence serve as an inspiration for deep learning [1]. The computational models known as neural networks have led to noteworthy progress in the arenas of artificial intelligence (AI) and machine learning (ML). Concurrently, a new paradigm in information processing has been established by the recent appearance of the first small-scale quantum computing systems. As in the conventional arena, there are many different architectures for Quantum Neural Networks (QNNs). Here is a summary of a few that have received a lot of interest: Quantum dissipative neural networks, quantum dissection Al neural networks, quantum convolutional neural networks, generative models, and quantum perceptron models. Despite the encouraging advancements, computational power and sufficient training regimens have mostly kept AI approaches as scholarly tools, making the implementation of learning algorithms on current devices all but impossible. This emphasizes a basic difficulty with all learning models, which is effectively teaching them. To overcome this difficulty, ideas like barren plateaus, avoiding plateaus, and optimization algorithms are essential [2].

Across a wide range of scientific fields, quantum computers and quantum algorithms have attracted a prodigious deal of interest and attention recently. This essay aims to provide an instructive, non-technical overview of the essential concepts of quantum computing. In particular, we investigate how quantum computers can prove to be useful instruments for mathematical weather prediction. We incorporate a simple non-linear model into our discussion by employing a newly created quantum technique for solving non-linear differential equations. In addition to outlining the benefits provided by quantum computers, we also explore the difficulties in applying them to practical issues, especially those requiring "big data" situations like weather forecasting [9].

Using the expressive power of quantum computing to solve challenging issues has been increasingly popular recently. There is great potential for quantum computers to surpass traditional ones and transform many different industries. This impetus is reinforced by ongoing research and related difficulties in fields such as global energy, materials, economics, chemistry, sentiment analysis, weather analysis, cloud computing, and cryptography. As the quantum finance review points out, the banking industry stands to benefit the most from quantum computing, especially in areas like portfolio optimization. It's critical to debunk the myth that quantum computing just speeds up classical algorithms; rather, it requires a radical redesign of computational techniques to make them compatible with quantum hardware. Reduced computational complexity is not a guarantee of quantum computing, particularly in use cases about finance. Before deciding whether or not a quantum speedup is feasible, a comprehensive assessment is essential [10].

Quantum computing is a relatively novel technology that can handle computationally demanding tasks that are necessary in many different domains, such as prime factorization, cybersecurity, data mining from large datasets, machine learning, artificial intelligence, and many more applications. It is interesting to note that a quantum computer can solve every computational problem that a classical computer can, usually more quickly. Among them, network security and quantum computing are closely related, with key distribution and management emerging as essential applications used in a variety of scenarios [3].

Advancements in Deep Learning (DL) and Machine Learning (ML) have advanced significantly in the past several years. These methods have resulted in the creation of models with applications in the military, aerospace, banking, healthcare, and agriculture, among other fields. Nevertheless, these models require millions of parameters to learn as the number of features rises, which poses difficulties to effective training by placing a significant computing load on the system. On the other hand, because of their capacity to calculate several states at once using current technologies, quantum computers have demonstrated promise in successfully tackling such problems. They take advantage related to the superposition and entanglement followed by interference, which are basic features of quantum physics. Qubits are endowed with intrinsic features that allow them to exist in numerous states concurrently (superposition), and show extreme correlation over countless distances by exhibiting bias in favor of the anticipated result, i.e., interference. Thus, the field of research could advance toward Artificial General Intelligence (AGI) with the use of quantum computing [6].

Controlling nuclear spin dynamics precisely offers a powerful means of implementing quantum methods in fluid and solid-state nuclear magnetic resonance (NMR) systems, with a focus on fluid-state NMR. The NMR platform has several benefits over other quantum information processing systems, such as extended coherence times, accurate manipulation, and advanced quantum control techniques. These characteristics allow precise control over quantum systems with up to 12 qubits. For the past 50 years, quantum info processing, which includes communication, quantum-related computing, and simulation, has made considerable use of liquid-state NMR spectroscopy, leading to significant breakthroughs and advances [11].

Today, a wide range of governmental and non-governmental domains, including banking, healthcare, defence, transportation, automotive, navigation, weather forecasting, and others, use different symmetric-based cryptography techniques or algorithms to make sure that software systems are resilient and secure. However, there is a serious risk associated with the growth of huge qubit-based systems, which could render many cryptography techniques. Under such circumstances, it becomes imperative to give quantum computing-based security solutions top priority. These days, human confidence, credibility, usefulness, and reliability are the many factors affecting software longevity [12].

The development of quantum-based software programs and algorithms has advanced significantly in recent years, bringing quantum computing one step closer to becoming a reality. An important first step toward the realization of quantum advantage in the Noisy Intermediate Scale Quantum (NISQ) is the demonstration of quantum supremacy. This refers to the fact that quantum systems are more effective than traditional systems at solving real-world issues. Realizing the advantage in the

NISQ era is hindered by issues like quantum decoherence and qubit interconnectedness, which predict the arise when quantum devices scale up in the upcoming years. There have been major ongoing improvements in all aspects of quantum computing research, which is still a science that is developing quickly and is very relevant [13].

Many industries deal with optimization issues daily, with multi-knapsack optimization being one such issue that many industries face frequently. With the arrival of quantum computing, there is now a new approach to solving computationally demanding problems that may lead to faster and better solutions for particular problem classes. The effectiveness of quantum methods is evaluated at different hyperparameters. Among those compared are quantum annealing and gate-based algorithms such as Quantum-Approximate Optimization Algorithm (QAOA) and Variational Quantum Eigen solver (VQE). Even after accounting for simulation time, it is found that VQE has longer runtimes than QAOA. These runtimes point to the necessity of techniques meant to cut down on the quantity of traditional iterations. Nevertheless, the benefits of VQE, such as noise-resistant quantum circuits tailored for hardware, might not materialize until true quantum hardware is used. Therefore, VQE deserves more investigation into the field of quantum optimization algorithms, even with its lower approximation rates and longer runtimes [14].

Researchers have been deeply involved in the noisy intermediate-scale quantum (NISQ), concentrating mainly in quantum computing systems with loads of qubits. Currently, one of the most critical challenges is to develop applications that can stably perform interesting nontrivial tasks on quantum devices subjected to large quantum noise. Many quantum computing approaches, such as variational quantum algorithms, error mitigation plans, and other compilation techniques, are put forth to address this problem. By characterizing and mitigating mistakes and implementing noise-resistant algorithms, these techniques hope to push the limits of potential applications for quantum devices to improve their capabilities. Furthermore, effective classical simulation is essential for the design of quantum algorithms and their verification, fault-tolerant validation, and other associated applications, and its progress is directly tied to the progress of quantum devices [15].

Quantum computing customizes the properties of quantum procedures to carry out designs. They use quantum bits (qubits) that can be in any linear principle of superposition of these conditions, to enable simultaneous computing along numerous paths. This is in contrast to traditional systems, which use bits limited to one or two conditions at a time. The systems exhibit potential for effectively resolving computational issues that are thought to be beyond the capabilities of classical computers. Unfortunately, a lot of the material on quantum computing is buried in theoretical physics, which makes it difficult for researchers in other fields to find. The use of quantum computers for the examination of classical data is denoted as "quantum-enhanced machine learning" or "quantum machine learning." However, there exist more points of convergence between machine learning and quantum computing. Physics challenges can be solved using data from quantum systems and classical machine learning. Further work is being done on simplifying traditional machine learning to deal with data, which consists of both quantum states for the input and output. A quantum-inspired classical recommendation system algorithm was recently devised, important for the creation of a new session of "quantum-inspired algorithms." These classical procedures, which are performed on traditional systems, use concepts from quantum computing to theoretically outperform the most advanced classical algorithms now in use [16].

Information is carried by individual atoms in quantum computing systems. Nuclear magnetic resonance (NMR) is included for historical and educational purposes. Quantum information is also stored and operated upon using quantum light, or photons. A quick summary of some upcoming technologies that might be important in the future rounds of this section. Superconducting qubits are the most widely used type of quantum computing device design and have attracted the greatest amount of commercial interest to date. On the other hand, electrical currents are the main means of information transportation in conventional computers, which are based on semiconductor technologies. More variety can be found in quantum computing technology, which uses a wider variety of physical systems to carry out information [17].

The noisy intermediate-scale quantum era includes the current circuit-based quantum computers. Even with these advances, the best-performing qubits available today remain noisy, which means that there is a discernible probability that quantum gates may produce false positives, usually on a per-mill basis. As such, circuits have to be confined to a certain number of gates to guarantee consistent results. While error-correcting methods exist, they frequently need more qubits than can be implemented in devices of intermediate size. Hybrid quantum–classical algorithms have arisen to address this difficulty. By working with classical computers, these algorithms take advantage of quantum computers' ability to measure circuit states efficiently. This allows them to modify circuit parameters and minimize the objective function iteratively. Small quantum circuits can be used thanks to this continuing interaction, which lessens the impact of possible mistakes. Two algorithms are the quantum approximation optimization algorithm (QAOA) and the variational quantum Eigen solver (VQE). The traditional optimizer in both scenarios was the Constrained Optimization BY Linear Approximation (COBYLA) method. When the gradient is unknown, COBYLA samples the cost function and uses linear approximations to determine the gradient [18].

3 Applications

Recent years have seen the advent of quantum computing as a front-line technology that is attracting interest from a wide range of scientific fields. Quantum computing offers revolutionary answers to challenging issues in a variety of fields, including weather forecasting, machine learning, encryption, and optimization. Moreover, it

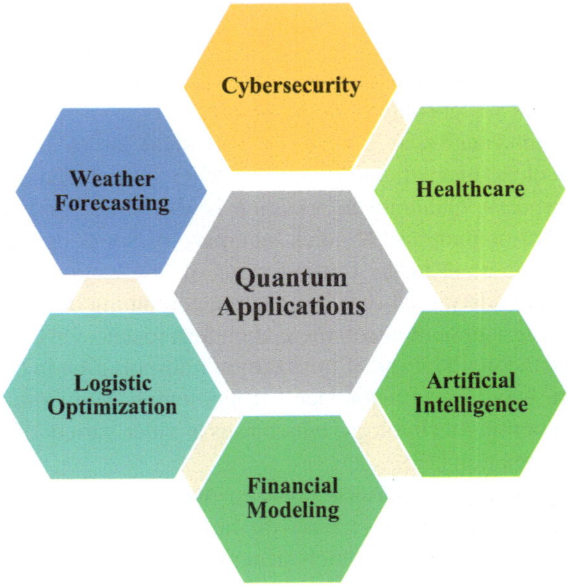

Fig. 2 Different applications where quantum computing is implemented

holds the key to changing Conventional computational methods and tackling important difficulties in domains ranging from finance to artificial intelligence. The systematic diagram for the various applications where quantum computing is implemented is presented as follows:

Figure 2 represents the key areas where quantum computing has widespread outages. In this section, the elaborated discussion is presented on various applications:

I. **Cybersecurity**:

The first and foremost is related to cybersecurity. In this field, quantum computing can transform the sector by breaking widely used encryption algorithms. Some of them are represented as:

1. **Shor's Algorithm**: The most common method implemented in this field is Shor's algorithm. It is mainly designed to factorize a large number. However, it has a significant implication for cryptography because several classical encryption schemes depend on the difficulty of factoring huge numbers [19]. The formulas used for the application of Shor's algorithm were presented in a stepwise manner as follows:

   Steps for Shor's Algorithm

   a. In the first step, the period of a function is determined by utilizing the Quantum Fourier Transform (QFT). The formula is as follows [19]:

$$\Phi_K = \frac{1}{\sqrt{P}} \sum_{m=0}^{P-1} e^{i\alpha_{mk}|\Phi_j}$$

where

- P ⇒ number to be factored
- $|\Phi_j \Rightarrow$ quantum state representing function values.
- $\alpha_{jk} = \frac{2\pi jk}{N} \Rightarrow$ phase angle.

b. After determining the period of a function, the modular exponentiation is evaluated using the formula:

$$g(y) = b^y \text{mod} N$$

where

- b represents the base for modular exponentiation
- y represents the exponent
- N represents the modulus.

c. In the third step, the algorithm finds the quantum ordering using the formula:

$$r = \text{Order}(b, N)$$

where

- b represents the base for modular exponentiation
- N represents Modulus.

d. In the final step, the algorithm factors "**N**" into primes once the order "**r**" is found. It is represented as:

$$N = p * q$$

where

- p and q represent the prime factors of N.

2. Quantum Key Distribution

It is a technique that makes use of quantum physics to protect communication channels and facilitate the sharing of secret keys between parties. The behavior of quantum states, in particular the no-cloning theorem [20], which asserts that an arbitrary unknown quantum state cannot be fully replicated, is the essential idea behind QKD. Since the algorithms of quantum pose a potential threat to traditional public key cryptography. The author [21] proposes leveraging quantum key distribution (QKD) for enhanced security, introducing protocols that establish secure shared keys. Some of the insights, along with the formulas used in QKD, are:

a. The first step is that QKD relies on the transmission of quantum states in superposition to encode the information. It is presented as:

$$|\varnothing\rangle = \alpha|0\rangle + \beta|1\rangle$$

where

- $|\varnothing\rangle \Rightarrow$ represents Quantum state encoding information
- $\alpha, \beta \Rightarrow$ represents complex probability amplitudes.

b. After performing the superposition, now Quantum entanglement is implemented using entangled qubits for secure key distribution. It is represented as:

$$|\varphi^{+}\rangle = \frac{1}{\sqrt{2}}(|00\rangle + |11\rangle)$$

where

- $|\varphi^{+}\rangle \Rightarrow$ represents the entangled state of two qubits.

c. In the third step, measurement outcomes are used that reveal the basis used by the sender, forming the shared secret. It is provided as:

$$\mathrm{P}(|0\rangle) = |\alpha|^2$$

$$\mathrm{P}(|1\rangle) = |\beta|^2$$

where

- $\mathrm{P}(|0\rangle), \mathrm{P}(|1\rangle) \Rightarrow$ represents the probabilities of measurement outcomes
- $\alpha, \beta \Rightarrow$ represents complex probability amplitudes.

d. In the fifth step, we randomly select a basis to encode each bit. The basis can be either rectilinear or diagonal.
e. After performing the above step, entanglement swapping is employed using the BBM92 protocol.
f. The final step comprises the determination of the Quantum bit fault rate. It quantifies errors in the keys due to eavesdropping or noise in the quantum channel. The formula is as follows:

$$\mathrm{QBER} = \frac{\text{Number of Errors}}{\text{Total number of Qubits}}$$

3. Lattice-Based Cryptography

It is an arrangement based on the difficulty of lattice glitches. There are several different types of problems, such as the Learning with mistakes problem, Ring Knowledge with errors problem, knowledge with errors in ideal lattices, module lattices, etc., that resulted in the development of this technique. In the paper [22], the author emphasizes the research for developing safety of lattice-based cryptosystems that depend on resolving the closest vector problem (CVP) and the shortest vector problem (SVP). The effective use of the Hash Sieve sieving technique is covered, along with the creation of the modular software library for improved performance. The experimental findings have shown that the developed Hash Sieve method outperforms a conventional example found in the literature in terms of running time. The shortest vector problem seeks the shortest non-zero lattice vector. It is presented as:

$$\mathrm{SVP}(n, \beta) \rightarrow \text{minimize}\, ||v||\, \text{subject to}\, ||v|| \leq \beta$$

where

- $n \rightarrow$ presents the measurement of the lattice
- $\beta \rightarrow$ presents the length bound

Similarly, the CVP aims to find the lattice vector closest to the given target. The general representation of it is as follows:

$$\mathrm{CVP}(n, t) \rightarrow \text{find}\, v \ni ||v - t||\, \text{is minimized}$$

where

- $n \rightarrow$ represents a dimension of the lattice
- $t \rightarrow$ represents the target vector.

II. **Healthcare**

The nexus between quantum computing and healthcare has generated a great deal of curiosity and creativity recently. The unmatched processing capacity of quantum computing presents enormous potential for transforming many facets of healthcare, especially in the area of molecular simulations. This work investigates how quantum computing can improve molecular simulations, opening up new possibilities for drug development, customized treatment, and comprehension of intricate biological processes.

1. **Quantum Computing: A Revolution in Processing Capabilities**

Through the application of quantum mechanics, computations can be carried out exponentially faster than with traditional computers. Quantum bits can occur in more than one state at once, in distinction to traditional bits, which can only exist in one form. This property allows massive quantities of data that must be processed in parallel. This computer may handle complicated problems tenfold quicker than its classical counterparts thanks to its intrinsic parallelism.

2. **Molecular Simulations**: **Unraveling the Complexities of Life**

   Understanding the behavior of biological molecules, including proteins, DNA, and chemical compounds, is largely dependent on molecular simulations. Researchers can learn more about the underlying causes of diseases, the activities of drugs, and biochemical processes by simulating molecular interactions and dynamics. However, because of their computational cost and the intrinsic quantum character of chemical interactions, classic computational approaches are not always able to adequately represent complicated molecular systems.

3. **Encouraging the Development of New Drugs**

   Accelerating the process of finding and developing new drugs is one of the most promising uses of quantum computing in the healthcare industry. The superiority of quantum computers in replicating the quantum behavior of atoms and molecules enables more accurate drug-target interaction simulation. Quantum simulations decrease the timeframe for drug research by identifying possible therapeutic candidates more quickly by precisely modeling molecule structures and characteristics.

4. **Personalized Medicine and Treatment Optimization**

   Quantum computing also makes personalized medicine possible by making it easier to simulate how different patients would react to different therapies. Healthcare professionals can customize therapies based on each patient's own molecular profile, maximizing efficacy and reducing side effects, by combining patient data with quantum simulations. This customized strategy has the potential to significantly improve patient outcomes for a wide range of illnesses, such as genetic disorders, neurological diseases, and cancer.

5. **Understanding Complex Biological Systems**

   At the molecular level, quantum computing also helps to simplify the intricate workings of biological systems. Through the simulation of complex biochemical pathways and interactions between proteins, scientists can clarify the mechanisms underlying diseases and discover new targets for treatment. Furthermore, the investigation of quantum phenomena in biological systems is made possible by quantum simulations, which shed light on basic mechanisms like photosynthesis and enzyme reactions.

6. **Difficulties and Prospects**

   The integration of quantum computing in healthcare is confronted with various hurdles, such as algorithm development, hardware restrictions, and data security issues, despite its immense promise. It will take multidisciplinary cooperation between computational biologists, quantum physicists, and medical professionals to overcome these obstacles. Further investigation and advancement are also needed to fully utilize quantum computing in the medical field [23].

Some of the basic protocols of Quantum Key Distribution algorithms that are utilized in healthcare are [24]:

a. **BB84 Protocol**: It's the protocol proposed in 1984 by Bennett and Brassard. It encodes qubits in polarized photons and guarantees that any eavesdropping will be recognized by the Heisenberg Uncertainty Principle. In this protocol, there will be a key in every bit that will be normalized to one of the basic sets, and each such qubit has been encoded. They are represented as:

    - Orthogonal Wave Address Digits (Initial H): $\in \{|0\rangle$ and $|1\rangle\}$
    - Diagonal basis: $|+\rangle$ and $|-\rangle$
    - If the source and destination choose matching bases, they will obtain the same bit. If they choose different bases, then the bit will be discarded

    The mathematical representation of the state of each qubit is represented as a linear combination of the basis states:

    $$|\varphi\rangle = \text{alpha}|0\rangle + \text{beta}|1\rangle$$

    where alpha and beta are multifaceted numbers such that $|\text{alpha}^2| + |\text{beta}^2| = 1$

b. **E91 Protocol**: It was proposed by Arthur Ekert in 1931. It was the first quantum cryptography protocol that used quantum entanglement for secure key exchange the source and destination share common entangled particles, and they check the correlation of their measurement outcomes to detect eavesdropping. The entanglement states are represented as [25]:

    $$|\varphi\rangle = \frac{1}{\sqrt{2}}(|01\rangle - |10\rangle)$$

    When the source and destination measure their qubits in the same basis, the results are perfectly correlated. Any deviation from this correlation indicates the presence of an eavesdropper.

From a healthcare point of view, BB84 can be used for securing the transmission of patient data between hospitals and research centers. The E91 can protect sensitive data that is exchanged between telemedicine platforms and remote patient monitoring devices. Unauthorized access to the entangled particles will alter the correlation and signal a breach.

III **Weather Forecasting**

With quantum computing's ability to do complex computational tasks faster than traditional computers, it offers an exciting path forward for improving weather forecasting skills. Some of the possible methods and insights on how quantum computing may be used in weather forecasting are presented below:

1. **Quantum Machine Learning**: It handles huge volumes of data and finds complex outlines more accurately. The result of this is that efficient weather forecasting models can be generated. In [26], the author draws attention to the expanding importance of quantum machine learning across several fields, such as climate change estimation and its forecasting. It explores current advances and trends in the field of machine learning in quantum, using practical applications to demonstrate its superiority compared to conventional methods. The techniques implemented in this are Quantum Neural Networks or Quantum Support-Vector Machines. The algorithm is used for cataloguing tasks. The basic formula implemented is [27]:

$$\text{QSVM loss} = \sum_{i=1}^{N} (1 - y_i.f(x_i))^2 + \lambda tr\left(X.X^T\right)$$

where

- $N \rightarrow$ represents the number of information points in the training datasets
- $y_i \rightarrow$ represents the true label of the ith data-point
- $f(x_i) \rightarrow$ represents the foreseen label for the ith information point using the QSVM
- $1 - y_i \cdot f(x_i) \rightarrow$ It calculates the variance between a true label and the forecast label. The difference is squared to penalize misclassification.
- $\lambda \rightarrow$ It represents the regulation constraint that manages the trade-off between minimizing the classification error, maximizing the boundary
- $X \rightarrow$ It represents the feature's matrix of the training dataset
- $X^T \rightarrow$ represents the transposition of the feature matrix

2. **Quantum Simulation**: Weather patterns can be predicted more accurately due to the ability of quantum simulators to represent complicated atmospheric events. In [28], the author delves into the possibilities of the Variational Quantum Eigen solver (VQE) in modeling complicated wave functions in polynomial time, with a special emphasis on its use in condensed matter physics and quantum chemistry. It tackles problems such as scalability and robustness to noise, emphasizing the necessity of parallelization among quantum computers and measurement system optimization. The basic formula implemented for the Variational Quantum Eigen solver (VQE) is as follows:

$$E(\theta) = (\varphi(\theta)|H|\varphi(\theta))$$

where

- $E(\theta)$ → represents the expectancy-value of the energy of the quantum system, parameterized by θ
- $(\varphi(\theta))|$ → denotes the trial tendency function organized by a quantum circuit with variational parameters θ
- H → It represents the Hamiltonian operator of the quantum structure, representing the overall energy observable.

3. **Quantum Data Analysis**: Weather forecasting involves massive, high-dimensional datasets, which may be handled using quantum data processing techniques to produce more accurate forecasts. In [29], making use of the Inherent exponential speedup of quantum computers the abstract suggests developing quantum principle component analysis this one may effectively reduce dimensionality unsupervised by taking high dimension data and converting it into lower dimensional space while keeping important characteristics it classifies current QPCA techniques and makes recommendations for improvement toward more effective quantum-based algorithms. The systematic formula for Principal Component Analysis or Quantum Clustering Algorithm is given as:

$$U = \text{argmax} Tr(U^T XX^T U) \text{subject to } UU^T = I$$

where

- U → It represents the unitary transformative matrix that we aim to optimize. Transformation is applied to the quantum state encoding the data
- X → It is the data matrix, represented in the form of a quantum state or a quantum circuit. The matrix encapsulates the representation of the classical data points
- U^T → It denotes the transpose of the unitary matrix in the content competing unitary matrix corresponds to the reversible transformation
- $Tr(A)$ → It calculates the trace of a matrix A, which is the totality of its diagonal rudiments. In the context of QPCA, the trace operation quantifies the information retained in the transformed data.

4. **Quantum Optimization**: The algorithms related to optimization can improve the efficiency of numerical weather prediction models by optimizing parameters and reducing computational costs. The basic technique implemented is Quantum Annealing or Quantum Variational Optimization. The formula is represented as:

$$M(t) = P(t) \sum_{a=1}^{N} \sigma_i^x + Z(t) H_{\text{problem}}$$

where

- $M(t) \rightarrow$ represents the total Hamiltonian at time t.
- $P(t)$ and $Z(t)$ $\rightarrow$ These are the parameters that regulate the relative importance of the problem Hamiltonian, the driver Hamiltonian, respectively.

5. **Quantum Sensing**: The technologies in Quantum sensing can enhance data collection methods, providing more precise and comprehensive inputs for weather forecast models.

IV. **Artificial Intelligence**

Artificial Intelligence is one of the most essential tools for enhancing cybersecurity strategies. Due to the evolving threat landscape of intrusions, data at risk is more extensive than previously anticipated. So, to overcome it, many different algorithms are used to detect threats, predict future attacks, and automate security operations. Some of the techniques of AI in cybersecurity are provided below with specifications:

1. **Intrusion Detection and Prevention Systems (IDPS)**

   It uses machine learning and pattern recognition to analyze network traffic for attack indicators. The main purpose is to identify and act upon malicious activities quickly. The anomaly detection in the network traffic is performed using a mathematical framework based on probability distribution and statistical models. Let us assume that a set of features is provided and represented as $X = \{x_1, x_2, \ldots\ldots\ldots, x_n\}$ representing network data, the likelihood of observing a feature vector "B" under a normal distribution $P(B)$ is computed as:

$$P(B) = \sum_{1}^{n} P(b_i))$$

   where $P(b_i)$ is the probability of feature "*i*" based on the historical data. When the observed probability is much less than the threshold δ, then the system assumes it to be an intrusion. But if the $P(B) < \delta$ then it means an anomaly is detected [30].

2. **Phishing detection using Natural Language Processing (NLP)**

   The technique used by the phishing attack is to use the deceptive emails and websites that appear legitimate to encourage some form of user activity, such as divulging confidential information. By using the natural language processing techniques, one can perform a content and structural inspection to detect these types of attacks. Since an email or a specific document consists of a sequence of words such as $W = \{w_{1,} w_2, \ldots, w_n\}$, the NLP models, such as transformer-based architecture, encode semantic meaning in them. Each word w_i is embedded into a high-dimensional vector space [31].

$$\text{Embedding}(w_i) = Ew_i \in Rd$$

where "d" is the embedding dimension. The email is then fed into a SoftMax classifier, labeling it as phishing or non-phishing based on its semantic content:

$$P(\text{phishing}|W) = \frac{\exp(wT * \text{Embedding}(W))}{j\exp(w_jT * \text{Embedding}(W))}$$

If $P(\text{phishing}|W) > 0.5$, then the email is determined to be a phishing attempt.

V. **Financial Modeling**

Cybersecurity financial modeling involves determining, quantifying, and predicting the financial impact of various cyber risks, prioritizing security investments, and conducting cost-benefit analysis for different cyber defence strategies. With increasing threats, organizations face risks like data breaches, service disruptions, and reputational damage. Thus, mathematical models are essential for predicting potential losses and evaluating security strategies. Some of the financial modeling techniques used in cybersecurity are as follows:

1. **Measuring the Impact of Cyber Risks**: **Loss Distribution Models**

   It is used for the evaluation of the financial impact of cyber incidents. The main aim is to expose the expected economic loss (L) caused by cyber incidents and support investment decisions. The total loss (L) can be framed as [32]:

   $$L = \lambda \cdot S$$

   where "λ" represents the product of the frequency of cyber incidents per year and "S" represents severity, i.e., average loss per incident.
2. **Calculation of Return on Security Investments (ROSI)**

   It is a core financial metric for evaluating the effectiveness of cybersecurity investments. It helps organizations decide the return on security investments by comparing costs with financial benefits gained from risk mitigation. The mathematical formulation is provided as follows [33]:

   $$\text{ROSI} = \frac{\text{Risk Exposure-Cost of security investment}}{\text{cost of security investment}}$$

   where

   - Risk exposure $\Rightarrow$ It is the exposure factor without the security controls.

- Cost of Security Investments $\Rightarrow$ It includes the expenses for installing security controls such as hardware, software, and personnel.

4 Conclusion

To sum up, Quantum Computing Technologies provides a thorough overview of the state of quantum computing now and its possibilities for the future. The chapter explores a diversity of topics related to quantum computing, such as its foundation, uses, and difficulties. The book chapter inspects the high-tech developments in the subject, ranging from quantum algorithms to quantum hardware design. Moreover, a brief discussion on how quantum computing can affect various industries, including weather forecasting, healthcare, banking, and cybersecurity, is presented. It highlights how quantum technologies have the potential to change the world and stresses the importance of current research and growth. Furthermore, insights on how quantum methods might be integrated into current computational frameworks and the consequences of quantum computing on conventional computing paradigms are explored. In summary, the chapter highlights the importance of quantum computing as a game-changing technology with far-reaching effects on computing and other fields. The ideas and information discussed will help to shape the route of quantum computing in the future.

5 Future Scope

The book chapter paves the way for a plethora of fascinating prospects and developments in the subject going forward. The ongoing development of quantum hardware represents an important area of future research. Researchers should anticipate more qubits, faster coherence time, and better error correction methods as quantum systems get more complex and scalable. Apart from this, the Quantum Key Distribution has the potential to revolutionize safe communication by building on the principles of quantum mechanics, such as the principle of superposition and the predicament of developing security that cannot be achieved with classical methods. The QKD technique is a milestone in this area, allowing for the secure dispersal of encoding keys, with plans for the development of QKD networks on a country-wide scale and even over satellites. Furthermore, based on entanglement-based protocols and quantum repeaters, quantum secure communication networks will also be available for long-range transmission of data. In addition, quantum cryptography that is independent of the devices would eliminate implausibility due to bad hardware. Naturally, there are other use cases like quantum-resistant blockchain architectures, quantum-powered identity authentication, or quantum-safe clouds been planned, which will fortify the security of the distributed systems and cloud infrastructures. These developments will ultimately transform the way digital security is protected and unlock new opportunities to secure critical data in a post-quantum world.

References

1. F. Valdez, P. Melin, A review on quantum computing and deep learning algorithms and their applications. Soft. Comput. **27**(18), 13217–13236 (2023). https://doi.org/10.1007/s00500-022-07037-4
2. D. Fisica, Quantum computing models for artificial neural networks (2021), pp. 1–9
3. N. Sharma, R. Ketti Ramachandran, The emerging trends of quantum computing towards data security and key management, in *Archives of Computational Methods in Engineering*, vol. 28, no. 7 (Springer Netherlands, 2021), pp. 5021–5034. https://doi.org/10.1007/s11831-021-09578-7
4. C.J. Mitchell, The impact of quantum computing on real-world security: a 5G case study. Comput. Secur. **93**, 1–27 (2020). https://doi.org/10.1016/j.cose.2020.101825
5. S. Jain, J. Ziauddin, P. Leonchyk, S. Yenkanchi, J. Geraci, Quantum and classical machine learning for the classification of non-small-cell lung cancer patients. SN Appl. Sci. **2**(6), 1–10 (2020). https://doi.org/10.1007/s42452-020-2847-4
6. H. Gupta, H. Varshney, T.K. Sharma, N. Pachauri, O.P. Verma, Comparative performance analysis of quantum machine learning with deep learning for diabetes prediction. Complex Intell. Syst. **8**(4), 3073–3087 (2022). https://doi.org/10.1007/s40747-021-00398-7
7. D. Herman, et al., A survey of quantum computing for finance (2022). http://arxiv.org/abs/2201.02773
8. N.R. Mosteanu, A. Faccia, Fintech frontiers in quantum computing, fractals, and blockchain distributed ledger: paradigm shifts and open innovation. J. Open Innov. Technol. Mark. Complex. **7**(1), 1–19 (2021). https://doi.org/10.3390/joitmc7010019
9. F. Tennie, T.N. Palmer, Quantum computers for weather and climate prediction the good, the bad, and the noisy. Bull. Am. Meteorol. Soc. **104**(2), E488–E500 (2023). https://doi.org/10.1175/BAMS-D-22-0031.1
10. J. Singh, K.S. Bhangu, Contemporary quantum computing use cases: taxonomy, review and challenges. Arch. Comput. Methods Eng. **30**(1), 615–638 (2023). https://doi.org/10.1007/s11831-022-09809-5
11. T. Xin, et al., Nuclear magnetic resonance for quantum computing: techniques and recent achievements. Chinese Phys. B **27**(2) (2018). https://doi.org/10.1088/1674-1056/27/2/020308
12. H. Alyami, et al., The evaluation of software security through quantum computing techniques: a durability perspective. Appl. Sci. **11**(24) (2021). https://doi.org/10.3390/app112411784
13. S.S. Gill et al., Quantum computing: a taxonomy, systematic review and future directions. Softw. - Pract. Exp. **52**(1), 66–114 (2022). https://doi.org/10.1002/spe.3039
14. A. Awasthi, et al., Quantum computing techniques for multi-knapsack problems, in *Lecture Notes in Networks and Systems* (2023), pp. 264–284. https://doi.org/10.1007/978-3-031-37963-5_19
15. H.L. Huang, et al., Near-term quantum computing techniques: variational quantum algorithms, error mitigation, circuit compilation, benchmarking and classical simulation. Sci. China Phys. Mech. Astron. **66**(5) (2023). https://doi.org/10.1007/s11433-022-2057-y
16. V. Kulkarni, M. Kulkarni, A. Pant, Quantum computing methods for supervised learning. Quantum Mach. Intell. **3**(2), 1–22 (2021). https://doi.org/10.1007/s42484-021-00050-0
17. J.D. Whitfield, J. Yang, W. Wang, J.T. Heath, B. Harrison, Quantum computing 2022 (2022), pp. 1–14. http://arxiv.org/abs/2201.09877
18. Study of quantum computing techniques for the resolution of optimization problems (2020)

19. V. Bhatia, K.R. Ramkumar, An efficient quantum computing technique for cracking RSA using Shor's algorithm, in *2020 IEEE 5th International Conference on Computing Communication and Automation* (2020), pp. 89–94. https://doi.org/10.1109/ICCCA49541.2020.9250806
20. Z. Yang, M. Zolanvari, R. Jain, A survey of important issues in quantum computing and communications. IEEE Commun. Surv. Tutorials **25**(2), 1059–1094 (2023). https://doi.org/10.1109/COMST.2023.3254481
21. O. Amer, V. Garg, W.O. Krawec, An introduction to practical quantum key distribution. IEEE Aerosp. Electron. Syst. Mag. **36**(3), 30–55 (2021). https://doi.org/10.1109/MAES.2020.3015571
22. H. Satilmis, S. Akleylek, Efficient implementation of HashSieve algorithm for lattice-based cryptography, in *2020 International Conference on Information Security and Cryptology, ISCTURKEY 2020—Proceedings* (2020), pp. 75–79. https://doi.org/10.1109/ISCTURKEY51113.2020.9307966
23. R. Ur Rasool, H.F. Ahmad, W. Rafique, A. Qayyum, J. Qadir, Z. Anwar, Quantum Comput. Healthc.: Rev. **15**(3) (2023). https://doi.org/10.3390/fi15030094
24. A. Warke, B.K. Behera, P.K. Panigrahi, Experimental realization of three quantum key distribution protocols. Quantum Inf. Process. **19**(11) (2020). https://doi.org/10.1007/s11128-020-02914-z
25. L. Li, et al., The security analysis of E91 protocol in collective-rotation noise channel. Int. J. Distrib. Sens. Netw. **14**(5) (2018). https://doi.org/10.1177/1550147718778192
26. B. Surendiran, K. Dhanasekaran, A. Tamizhselvi, A study on quantum machine learning for accurate and efficient weather prediction, in *6th International Conference I-SMAC (IoT Social, Mobile, Analytics and Cloud), I-SMAC 2022—Proceedings* (2022), pp. 534–537. https://doi.org/10.1109/I-SMAC55078.2022.9987293
27. S.S.K.N. Kaulgud, Quantum machine learning for support vector machine classification. Evol. Intell. **17**(2), 819–828 (2024). https://doi.org/10.1007/s12065-022-00756-5
28. J. Tilly et al., The variational quantum Eigen solver: a review of methods and best practices. Phys. Rep. **986**, 1–128 (2022). https://doi.org/10.1016/j.physrep.2022.08.003
29. B. Soni, N. Khare, Advancement of quantum methods in principal component analysis, in *2021 12th International Conference on Computing Communication and Networking Technologies ICCCNT 2021* (2021), pp. 1–7. https://doi.org/10.1109/ICCCNT51525.2021.9579998
30. A. Patel, M. Taghavi, K. Bakhtiyari, J. Celestino Júnior, An intrusion detection and prevention system in cloud computing: a systematic review. J. Netw. Comput. Appl. **36**(1), 25–41 (2013). https://doi.org/10.1016/j.jnca.2012.08.007
31. A. Alhogail, A. Alsabih, Applying machine learning and natural language processing to detect phishing email. Comput. Secur. **110**, 102414 (2021). https://doi.org/10.1016/j.cose.2021.102414
32. P.V. Shevchenko, J. Jang, M. Malavasi, G.W. Peters, G. Sofronov, S. Trück, The nature of losses from cyber-related events: risk categories and business sectors. J. Cybersecur. **9**(1), 1–12 (2023). https://doi.org/10.1093/cybsec/tyac016
33. A. Orlando, Cyber risk quantification: Investigating the role of cyber value at risk. Risks **9**(10) (2021). https://doi.org/10.3390/risks9100184

Integrating Quantum Computing with Cybersecurity: Challenges and Opportunities

R. Saranya, P. Nandhini, and D. Swathi

1 Introduction

The area of quantum computing is interdisciplinary, involving computer science, physics, and mathematics. It solves complicated problems more quickly than conventional computers by utilizing quantum physics. It covers both hardware research and application development [1]. Superposition and interference are examples of quantum phenomena that are used by quantum computers to carry out specific tasks more effectively. Applications in fields like machine learning, optimization, and physical system modeling can profit from this faster processing speed. Chemical system modeling and portfolio optimization in finance are two examples of future possibilities that will exceed the capabilities of even the most powerful supercomputers now in use.

By tweaking and putting the concepts of quantum mechanics to practical use, quantum technologies are emerging. Notably, quantum computers leverage quantum gates and bits (qubits) in order to carry out computation potentially with greater speed than classical computers, depending on specific workloads. Quantum entanglement is utilized in quantum communication to transmit data securely. There is considerable opportunity to use quantum sensors and imaging devices in materials studies as well as in healthcare. By spotting eavesdropping attempts employing the fundamental principles of quantum physics, quantum cryptography optimizes communication security [2]. In conclusion, the emergence of quantum technologies has the potential to revolutionize sensing, computation, and communication, impacting an extensive

R. Saranya · P. Nandhini (✉) · D. Swathi
Department of Computer Science with Data Analytics PSG College of Arts and Science, Coimbatore, India
e-mail: nandhinipanchalingam2003@gmail.com

R. Saranya
e-mail: saranya_r@psgcas.ac.in

S. B. Goyal et al. (eds.), *Quantum Computing, Cyber Security and Cryptography*,
https://doi.org/10.1007/978-981-96-4948-8_3

variety of domains in business as well as science. The next section is about the basic and fundamental principles of quantum computing.

1.1 What is Quantum Computing

Quantum mechanics is used to tackle complicated problems more quickly than traditional computers, quantum computing is a revolutionary approach to computation. It uses qubits, which enable parallel processing and significantly increased computational capability. Qubits can be in numerous states concurrently (superposition) and can be entangled with one another. Algorithms such as Grover's for searching through unsorted data and Shor's for factoring big integers are based on the manipulation of these qubits via quantum gates. Quantum computing has several potential uses in a variety of scientific sectors, [3] such as materials research, cryptography, optimization, and others. We are living in an era known as Noisy Intermediate-Scale Quantum (NISQ), when the development of quantum computers is taking place in the face of obstacles including high mistake rates and decoherence. Notwithstanding these obstacles, quantum computing holds potential for solving issues.

1.2 Founders of Quantum Computing

Although there isn't a single creator of quantum computing, a number of significant people have contributed significantly to its advancement: In 1981, Richard Feynman proposed the use of quantum systems to replicate physical processes that are difficult for conventional computers to adequately model [4]. David Deutsch expanded on this notion in 1985 when he put up the idea of a universal quantum computer, which laid the theoretical foundation for quantum algorithms. Later, in 1994, Peter Shor developed a method using a quantum computer to factor huge integers, demonstrating the potential of quantum computing to address problems that classical computers are unable to.

1.3 Purpose of Quantum Computing

The goal of quantum computing is to use the concepts of superposition, entanglement, and interference found in quantum physics to carry out computations that are currently beyond the capabilities of classical computers. The goals of this technology are:

- Solve Complex Problems: Tackle problems in cryptography, optimization, and materials science more efficiently than classical computers [5].

- Enhance Computational Power: Offer exponential acceleration for particular algorithms, such as Grover's database search algorithm and Shor's factoring algorithm.
- Simulate Quantum Systems: Accurately model quantum physical systems, leading to advancements in chemistry, physics, and material science.
- Advance Machine Learning: Improve machine learning techniques by processing large datasets and identifying patterns more effectively.
- Quantum computing offers the potential to revolutionize industries by resolving intricate challenges with greater speed and efficiency compared to classical computing methods.

2 Fundamental Principles

In the realm of physics, quantum mechanics constitutes a cornerstone that provides an extensive framework to comprehend the variety of phenomena that comprise matter and energy initiatives at the most microscopic scales. In this enormous field, our comprehension is determined by several fundamental principles:

- **Wave-Particle Duality**:

The fundamental insight of quantum mechanics is the realization that particles, among them electrons and photons, are somewhat dual in nature, exhibiting both particle and wave-like attributes simultaneously. This profound duality upends and reshapes our conventional conceptions of the fundamental nature of matter [6].

- **Superposition**:

Quantum particles possess a remarkable ability to occupy numerous states concurrently. The system collapses into a solitary state only upon measurement, thereby introducing a level of unpredictability that is fundamental to quantum behavior.

- **Quantization**:

Notwithstanding the continuous range that exists in classical physics, certain physical variables, most notably energy levels, take on quantized, discrete values within the quantum realm. The distinct and distinctive features of quantum states are demonstrated through their discrete nature.

- **Uncertainty Principle**:

Werner Heisenberg's Uncertainty Principle establishes a fundamental constraint on the precision with which complementary properties of a particle, such as its position and momentum, can be simultaneously determined. This introduces an inherent and irrevocable unpredictability into the dynamics of quantum systems.

- **Quantum Entanglement**:

By enabling quantum particles to create instantaneous linkages where the state of one particle significantly affects the state of another, even when they are separated by great distances, this phenomenon contradicts conventional notions about the independence of distant things.

- **Quantum tunneling**:

Particles in the universe of quantum mechanics possess an astonishing ability to traverse energy barriers that have been deemed impenetrable by classical physics. The "tunneling" phenomenon offers a further dimension of unpredictability to particle trajectories, which goes against conventional predictions.

- **Quantum states and wave functions**:

It tends to be mathematical constructs that define the probability distribution of a particle's properties [7]. They are encapsulated by Schrödinger's equation and offer a significant understanding of the statistical principles governing quantum systems. Comprehension of these fundamental concepts becomes crucial for anyone hoping to get an in-depth comprehension of the intricate and sometimes contradictory initiatives that particles demonstrate at the quantum level. It is an exploration of the fundamental principles of the quantum universe, where the intriguing and frequently puzzling laws that govern the quantum realm give a framework for the laws of classical physics.

3 Quantum Bits

The basic building blocks of quantum computing are known as quantum bits or qubits for short. Qubits differ from conventional bits in that they feature a unique characteristic known as superposition, which makes them unconstrained by their presence in states of 0 or 1. This idea allows qubits to live in many states at once, opening up new possibilities for information processing and interpretation. The unique characteristic of superposition has significant ramifications for the field of quantum computing [8]. These systems possess the ability to do intricate calculations in parallel because qubits can investigate numerous states simultaneously.

This parallel processing capability has the potential to contribute considerable gains in computing power, particularly for applications that entail sophisticated computations and problem-solving beyond the boundaries of classical computing approaches. Superposition research in qubits promises to be an engine of innovation in several scientific and commercial industries by bringing up new computational avenues. The next part dealt with the extensive power of superposition in quantum bits.

3.1 Superposition of Quantum Bits

The concept of superposition with quantum bits, or qubits, has been fundamental and groundbreaking in the field of quantum computing. Qubits differentiate themselves from classical bits by their ability to exist in several states at the same time [9]. Classical bits can only exist in the binary states of 0 or 1. This outstanding event, which is based on the superposition principle, allows quantum computers to investigate large amounts of data concurrently. In essence, whereas conventional bits are limited to representing a fixed 0 or 1, qubits may simultaneously embody a combination of both 0 and 1 by using the power of superposition. This distinctive characteristic enhances the range of possible states and dramatically boosts the computational power of quantum computers. Quantum computing's revolutionary powers have been rendered feasible by the clever utilization of superposition in qubits. Through the simultaneous exploration and manipulation of numerous possibilities, qubits enable quantum computers to examine a broad range of solutions at the same time. This quantum parallelism, essential to superposition, not only [10] offers promise for tackling difficult problems more efficiently than classical equivalents but also provides new paths for breakthroughs in processing capacity and distinctive problem-solving approaches. The next part is about quantum gates and circuits in quantum computing.

3.2 Quantum Gates and Circuits

In quantum computing, quantum gates are crucial operations, just as conventional logic gates. Working with qubits, the basic components of quantum information are what they do. Common quantum gates include phase gates, CNOT gates, and Hadamard gates. These gates connect to construct quantum circuits, which carry out quantum mathematical calculations. They represent a variety of quantum techniques applied to qubits. For some computing tasks, Grover's algorithm for searching and Shor's factorization method are examples of quantum algorithms that implement these circuits. Comprehending quantum circuits and gates is essential for developing and implementing quantum algorithms on qubit-based quantum computing systems, utilizing control mechanisms and qubits to construct these gates. Preserving quantum coherence while eliminating blunders brought on by noise and decoherence is the tricky part [11]. To overcome the aforementioned problems in real-world quantum computing applications, quantum error correction methods are critical.

3.3 Quantum Parallelism and Entanglement

The tendency of quantum systems to exist in several states concurrently owing to the superposition principle is known as quantum parallelism [12]. Quantum computers may be able to solve some problems faster than traditional computers because of their capacity to consider several possibilities at once. Particles can entangle, which is a quantum phenomenon in which the state of one particle instantly influences the state of another, regardless of the distance between them. This phenomenon is employed for tasks like quantum key.

Distribution in quantum computing is a vital aspect of quantum communication and computation. Fundamental concepts such as quantum parallelism and entanglement make quantum systems distinct from classical ones and provide unique opportunities for communication and information processing. The next section deals with the major advantages of machine learning.

3.4 Types of Quantum Computers

- **Superconducting Qubits**: Overview of superconducting qubit technology, including examples like IBM's quantum computers and Google's Sycamore.
- **Trapped Ion Quantum Computers**: Explanation of trapped ion technology and its applications, such as those developed by IonQ.
- **Topological Qubits**: Discussion on the principles of topological quantum computing and ongoing research in this area.

Quantum computing hardware encompasses a variety of technologies, each with a unique approach to creating and manipulating qubits. Superconducting qubits are a pioneering technology used by companies such as IBM and Google [13]. These qubits are made of superconducting circuits that operate at very low temperatures, allowing for high-speed quantum operations. IBM's quantum computer and Google's Sycamore processor are outstanding examples of this technology, demonstrating significant advances in quantum computing capabilities. Trapped-ion quantum computers use trapped ions in electromagnetic fields to create qubits. This technology was developed by companies like IonQ for precise control and highly realistic quantum performance, making it a promising approach to scalable quantum computing. Finally, topological qubits represent a path that is highly theoretical but potentially revolutionary in quantum computing. Based on the principles of topological quantum computing, these qubits are designed to be inherently protected from certain errors due to their topological nature. Research is continuing in this area to develop robust quantum computers, and it is fault-tolerant. Together, these quantum computers demonstrate a variety of approaches to the quest for powerful and reliable quantum hardware.

3.5 *Quantum Error Correction*

- **Error Correction Codes**: Surface Code, Shor Code, and other quantum error correction codes are introduced.
- **Challenges and Solutions**: Overview of the challenges in quantum error correction and potential solutions.

Quantum error correction is essential to building reliable quantum computers because qubits are highly vulnerable to errors from environmental noise and operational defects. Quantum error correction codes, such as the Surface Code and Shor Code, are used to identify and correct these errors. Surface code uses a network of qubits, which allows error detection through measurements in neighboring qubits, while noise code encodes a logical qubit into nine physical qubits to detect bit flip and phase flip errors. Uses integrated growth, but the implementation of this code poses significant challenges, including the large number of physical qubits needed to load a reasonably small number of qubits, difficult if error correction operations are to be performed they do not cause additional errors [14]. Developments, including the use of machine learning techniques to perform optimal error correction.

As research on quantum error correction continues, these solutions will be important in paving the way for quantum computing capabilities changeable, fault-tolerant.

3.6 *Quantum Networking and Communication*

- **Principles and Architecture**: Description of the quantum Internet and its potential to revolutionize secure communications.
- **Quantum Repeaters**: Explanation of the role of quantum repeaters in long-distance quantum communication.
- With the creation of the Quantum Internet, quantum networking and communications promise to completely transform secure communications [15]. Based on the ideas of quantum mechanics, the Quantum Internet allows quantum information to be sent through complex particles. Because of the nature of quantum entanglement, any effort to surreptitiously send a message can be quickly discovered, resulting in more secure communication. Quantum node channels, which transport quantum information over open space using optical fibers or photons, are the foundation of the quantum Internet.
- The quantum repeater holds the key to long-range quantum communication. Through the use of devices called quantum repeaters, long-distance quantum signal loss and divergence can be overcome, extending the capabilities of quantum communication. They function by producing and storing two delicate particles at equidistant spots in the communication line. Entanglement swapping is the technique by which connections are extended over great distances. The maintenance of these guarantees the construction of worldwide quantum networks. To achieve

safe, long-range quantum communications and realize the full promise of the Quantum Internet, quantum repeaters must be used successfully [16].

3.7 *Quantum Teleportation*

- Quantum teleportation is a technique for transferring quantum information, such as the exact location of a particle from one point to another, without moving the particle itself. This is verified by quantum entanglement substitution theories. The process begins with two complex particles, one at the sender level (Alice) and the other at the receiver level (Bob).
- Alice also has a third being who wants to take her quantum state farther to Bob. He makes a joint measurement of this tertiary particle with its entangled particle, which breaks their joint state, and he makes a classical list, and then this information is sent to Bob through a classical connection [17]. Upon receiving this primitive information, Bob applies a specific quantum process to his trapped particle based on information from Alice. This process transforms Bob's particle into a replica of the original state of Alice's third particle, effectively transferring the quantum state to Bob.
- Quantum teleportation has important potential applications in quantum networking and communications. It can be used to transfer quantum information through networks without direct communication, which is critical to maintaining quantum state integrity over long distances This can improve the security and performance of quantum communication systems, enabling quantum computing a distributed, facilitating advances in quantum cryptography by harnessing teleportation, future quantum networks can achieve unprecedented security and speed of data transfer.

4 Algorithms

According to their classical counterparts, quantum algorithms may solve computational challenges more quickly by taking advantage of the distinct characteristics of quantum systems. We shall examine several notable quantum algorithms in the following section:

a. **Shor's Algorithm**:

In comparison with conventional techniques, Shor's algorithm offers an exponential acceleration and is distinctive among quantum approaches to factorizing enormous numbers. Given its capacity to jeopardize commonly used public key encryption systems, it has the potential to have significant effects on cryptography [18].

b. **Grover's Algorithm**:

Grover's algorithm is an example of a quantum search strategy that makes searches happen four times faster than they would with traditional search techniques. It may be utilized for unstructured search issues as well, such as optimization jobs and database searches.

c. **Quantum Fourier Transform**:

A key component of Shor's algorithm and many other quantum algorithms is the quantum Fourier transform (QFT). The QFT is essential for accurate quantum phase estimation because it functions as a fast Fourier transform on quantum states [19].

4.1 Shor's Algorithm in Quantum Computing

Both conventional and quantum computers are endeavoring to find an approach, but they achieve it in very different ways as a result of the way they handle data. An explanation of the quantum algorithm, which is used to solve complicated issues, is presented in this section. One of the quantum algorithms used to address the process of factorization problems is Shor's algorithm [20]. The mathematician Peter Shor devised this algorithm. One of the most well-knownalgorithmsinallofquantumcomputingpersiststothisparticularone. A quantum computer can factor very large numbers in polynomial time, as demonstrated by Shor's algorithm. Quantum parallelism, quantum Fourier transform, and modular arithmetic are prerequisites for the algorithm. A significant advancement over classical approaches can be achieved by Shor's algorithm, a pioneering quantum algorithm, enabling the factorization of massive numbers, like those used in RSA encryption, in O((logN)^3) time and O(logN) space [21]. The security of (RSA) encryption may be threatened by this quantum approach, which uses the laws of quantum mechanics to efficiently identify prime numbers by drastically accelerating the speed of the process of decryption as compared to traditional methodologies.

- Peter Shore's the researcher who proposed Shor's Factorization Algorithm.
- It indicates that factorization may be concluded in polynomial time by quantum physics, as opposed to the exponential time required by classical techniques.
- The field of data security may be particularly impacted by this, as it depends substantially on the prime factorization of huge amounts of data.
- In addition, there are multiple polynomial-time approaches for multiplying integers (such as Euclid's Algorithm), but there isn't one for factorization.
- Consequently, Shore developed a method known as Shor's Factorization technique, which allows non-prime integers S of P bits to be factorized.
- Because quantum algorithms are based on the quantum Fourier transform, their performanceissignificantlymoreeffectivethanconventionalalgorithms.
- The classical computer's runtime is O [exp (P1/3(log P)2/3)]. However, that is O(P3) on the quantum computer.
- Factoring enormously huge integers in polynomial time.

4.2 *Shor's Algorithm Depends on*

- Quantum parallelism
- Quantum Fourier transformation
- Modular arithmetic

The algorithm is as follows: given an odd composite number S, find an integer i that divides S, mathematically between 1 and S [22].

- There are two components in Shor's Algorithm:
- The factorization problem is transformed into the period finding problem. Classical methods can be used to implement this section.
- Determining the period or quantum period through the use of quantum parallelism and the Quantum Fourier Transform, which is in charge of quantum speedup [23].
- The non-prime number Sis the input of Shor's Algorithm, and then the one-trivial factor of S is the output.
- SHOR'SALGORITHM —> INPUT(S)—> OUTPUT (Non-trivial factor of S).

4.2.1 Algorithm

It consists of a few phases, the usage of quantum computers being necessary only at step 2.

- Pick any random integer, m, for instance, such that m < N and that is both of the Marco primes.
- The function fr, S(x) = mx mod S, has an unknown period k which has to be found using a quantum computer.
- Return to Step 1 if k is an odd number. If not, go to the following action.
- (mk/2– 1)(mk/2 + 1) = mk–1 = 0mod S since k is an even number.
- If the result of mk/2 + 1 = 0 mod S, then move on to Step 1.
- Proceed to the following step if mk/2 + 1! = 0mod S. Otherwise, stop.
- Subtract d from gcd (mk/2–1, S).

The necessary response is 'i'.

4.2.2 Classical Division (The Order-Finding Dilemma)

This is the traditional part of the order-finding problem. Taking consideration that x<S and gcd(x,S)=1 for both x and S. The order of x is the smallest positive integer, y, such that xy = 1(mod S).

We choose random integers such that s < S. Calculate gcd(s,S).

- The Euclid Algorithm can be used for this.
- There is a non-trivial factor of N if gcd(s,S)! = 1. Assuming that(x + k) = sx + kmodS = sx mod S = f(x).

- Return to Step 1 if r is odd.
- If sk/2 = -1(modS), returntoStep1.
- The factor of S that is non-trivial is the gcd (sk/2 $\pm$ 1, S).

4.2.3 Quantum Part

This is where we locate the quantum order. Select a power of two, after which Q=2k so that S2 <= Q\= 2S2.

4.2.4 CLASSICAL DIVISION (the Order-Finding Dilemma)

This is the traditional part of the order-finding problem. Taking consideration that x<S and gcd(x, S)=1 for both x and S. The order of x is the smallest positive integer, y, such that xy = 1(mod S).

1. We choose a random integer such that s < S. Calculate gcd(s,S).
2. The Euclidean Algorithm can be used for this.
3. There is a non-trivial factor of N if gcd(s,S)! = 1. Assuming that (x + k) = sx + kmodS = sxmod S = f(x).
4. Return to Step 1 if r is odd.
5. If sk/2 = -1(modS), return toStep1.
6. The factor of S that is non-trivial is the gcd (sk/2 $\pm$ 1, S).

4.2.5 Quantum Part

This is where we locate the quantum order. Select a power of two, after which Q=2k so that S2 <= Q \= 2S2.

- And consider that f = {0,1,2,…, Q-1}.
- In this case, w = exp(2πi/Q),or the Qth root of unity, and f(y) = 1/(Q)1/2$\sum$x = 0Q-1I f(x) I wxy.
- Let's use the following example to run Shor's Algorithm: To factor (let S = 17) an odd integer.
- Select an integer Q(let's say Q = 324) such that S2 < = Q $\leq$ 2 S2.
- Next, select any number n at random such that gcd (s, S) = 1(let's select s = 7).
- Next, make two quantum registers (they should be tangled so that the output register collapses in tandem with the input register collapse).
- TheinputregisterhastohaveenoughqubitsinittorepresentnumbersuptoQ-1 in size. (That is, up to 323, requiring 9 qubits).
- The output register has to have enough qu bits init to represent values up to (S -1). (That is, up to 16; so, 4 qubits are needed) [24–26].

4.3 Quantum-Resistant Algorithms

The effect of quantum computing on cybersecurity is huge. Digital communications are threatened by quantum computers. Quantum-resistant algorithms, also known as post-quantum algorithms, ensure secure communications. At some point, developed enough, such computers will be able to break many cryptography schemes, such as Rivest-Shamir-Adleman (RSA), which rely on various challenges like integer factorization and discrete logarithms [27–29].

Key Quantum-Resistant Algorithms

Different Quantum-Resistant Algorithms are available. In this chapter, we are going to see about Code-Based Cryptography algorithm. Based on Cryptography are a few key post-quantum cryptography techniques. Let's now take a brief look at code-based cryptography, which relies on error-correcting codes, which are fundamentally challenging linear code decoding challenges.

Code-Based Cryptography

This algorithm depends on the hardness of decoding a general linear code. One of the most well-known code-based cryptosystems is the McEliece cryptosystem, which is based on Goppa codes. It is one of the oldest quantum-resistant algorithms based on the difficulty of decoding random linear codes.

The detailed information about the McEliece algorithm, including encryption, decryption, and key generation through the corresponding equations, is elucidated here.

- **Encryption**

This technique involves scrambling plain text messages into secret texts using a cipher and a key so that unwanted parties cannot access them. Data can be encrypted or decrypted in two different ways. They include Symmetric encryption and Asymmetric encryption, respectively. A symmetric encryption uses similar keys for both encryption and decryption, while asymmetric encryption uses pairs of keys. However, the public key is used for encryption, and the private key is used for decryption. Encryption is needed to safeguard information.

Plaintext Message: Represented by a binary vector $\mathbf{v}$ of length $\mathbf{L}$.

Generate Error Vector: Create a random error vector e of length n with exactly t non-zero entries.

Encode the Message: Compute the ciphertext c as: $c=m \cdot G'+e$, where G' is the generator matrix corresponding to H'.

- **Decryption**

Receive the Ciphertext: Obtain ciphertext c.

Undo the Permutation: Compute $c'=c \cdot P-1$ to remove the permutation. Applied during encryption.

Decode the Goppa Code: Decode c' using the private key H and S to recover the original message and error vector e'. The message m' is obtained from $m'= e', S^{-1.}$

4.3.1 Key Generation

Choose a Goppa code G(F2,g(x)) which is supported only over some finite F2m [30–33].

- H for the Goppa code will be formed from computing its parity-check matrix, an $n \times (n - k)$ binary matrix.
- Randomize H in terms of S and P: Choose two random matrices- an invertible binary matrix S and a permutation matrix P—then compute the public parity-check matrix H' as $H' = S \cdot H \cdot P$. It makes the public key (H') and the private key (H, S, P).
- Security: The security of the McEliece cryptosystem depends on the decoding of a random linear code that seems difficult even for quantum computers.
- Key Size: The sizes of RSA or ECC public keys are much smaller than those used in code-based cryptosystems, but this is one point that also differentiates them from code-based cryptosystems—they trade off for their resistance to quantum computers.

The McEliece cryptosystem remains a strong candidate in the realm of post-quantum cryptography, especially for applications where key size is less critical than long-term security.

4.4 Development Stages of Quantum-Resistant Algorithms

The development of quantum-resistant algorithms is a multi-stage process designed to safeguard data against the potential future threats posed by quantum computing. Initially, the focus is on research and theoretical foundations, where cryptographers explore and create algorithms based on complex mathematical challenges. These algorithms are evaluated for their ability to resist quantum attacks, with new cryptographic schemes proposed to ensure they can withstand emerging threats. Once a theoretical basis is established, the process moves to standardization efforts [34]. The standardization process leading to quantum-resistant algorithms started in 2016 with NIST playing a crucial role. This involves global collaboration among academic, governmental, and industrial researchers to refine and validate these algorithms, ensuring they meet rigorous security and performance standards. Following standardization, the algorithms undergo implementation and testing [35]. This stage includes developing prototypes to assess their practical usability and efficiency. Performance optimization is crucial to enhance the algorithms' speed and reduce resource consumption, while ensuring interoperability with existing systems is essential for seamless integration.

The next phase is adoption and integration, where strategies are developed for transitioning from current cryptographic systems to quantum-resistant ones. This involves securing regulatory approvals and ensuring compliance with international standards [36–39]. Algorithms are gradually deployed in real-world applications, such as secure communications and data encryption, especially in scenarios requiring long-term security. Lastly, continuous evolution and observation are necessary to preserve the applicability and efficiency of quantum-resistant algorithms. As quantum technology develops, regular security evaluations will be necessary to find and fix flaws. Continuous updates and research into new cryptographic methods help ensure these algorithms remain robust against future threats.

4.5 *Quantum Computing Case Studies*

Quantum computing, despite being in its early stages, is already showing promise for revolutionary advancements in cybersecurity. One key application is Quantum Key Distribution (QKD), demonstrated by China's Micius satellite.

Launched in 2016, Micius was the first satellite to perform QKD over a distance exceeding 1,200 km. In QKD, encryption keys are securely shared using quantum mechanical principles such that any tampering or eavesdropping attempts would change state, implying detection. This breakthrough offers a glimpse into a future where secure, long-distance communication could become a reality, replacing current methods that may be vulnerable to quantum attacks. In the realm of cryptography, Microsoft has been actively developing Post-Quantum Cryptography (PQC) algorithms. These new cryptographic techniques are designed to be resistant to the potential threats posed by quantum computers. As part of the NIST competition to standardize these algorithms, Microsoft's work aims to ensure that future cryptographic systems can withstand quantum decryption efforts. This research is critical for maintaining data security as quantum technology advances, providing a foundation for secure communications in a post-quantum world.

IBM has also been exploring quantum technology's impact on the Internet of Things (IoT). By incorporating quantum-safe algorithms into blockchain technology, IBM is working to secure IoT devices from future quantum threats. This approach is particularly valuable for environments with numerous connected devices, such as smart cities, where robust security measures are essential. Quantum-safe blockchain protocols could prevent large-scale breaches by protecting data and transactions from potential quantum attacks. A Quantum Random Number Generator (QRNG) created by ID Quantique makes use of quantum phenomena to generate truly random numbers [21]. Unlike conventional random number generators, which can be predictable and therefore less secure, QRNGs offer enhanced randomness by relying on quantum phenomena. This technology is already being used in various security systems, including those employed by governments and financial institutions, to improve the security of encryption keys. D-Wave is advancing the use of quantum machine learning (QML) for cybersecurity, focusing on how quantum-enhanced algorithms can analyze large datasets more efficiently than traditional methods. This capability is crucial for detecting threats, identifying anomalies, and responding to intrusions with greater speed and accuracy [23]. As cybersecurity threats grow more complex, QML could significantly improve the ability to manage and respond to these challenges in real time.

Finally, Alibaba's research into Secure Multi-Party Computation SMPC) Quantum computing seeks to enhance collaborative security tasks. SMPC allows multiple parties to compute functions jointly without disclosing their private inputs. Quantum computing could make these protocols more efficient and secure, transforming how organizations collaborate on cybersecurity tasks and ensuring greater privacy and data protection.

5 Need for Quantum Computing in Cybersecurity

The need for quantum computing [40] in cybersecurity arises from its potential to revolutionize data protection and threat detection. Here are the key reasons:

- **Breaking Classical Cryptography**

Quantum computers pose a threat by potentially solving problems that classical computers find [41] infeasible, such as factoring large integers through Shor's algorithm. This capability challenges existing cryptographic systems like RSA and ECC, which depend on the difficulty of these problems for security.

- **Enhancing Encryption**

Quantum Key Distribution (QKD) employs quantum mechanics principles for encryption, making it theoretically [42] unbreakable. The process includes mechanisms to detect eavesdropping attempts during key exchange, ensuring secure communication.

- **True Randomness**

Quantum Random Number Generation (QRNG) delivers true randomness, essential for creating secure cryptographic keys [43]. Unlike classical random number generators, which may be predictable and susceptible to attacks, QRNG guarantees strong security.

- **Quantum-Resistant Algorithms**

Post-Quantum Cryptography focuses on creating cryptographic algorithms resilient to quantum attacks [44]. These algorithms rely on problems considered computationally challenging for quantum computers, ensuring robust, long-term data security.

- **Advanced Threat Detection**

Quantum Machine Learning: Enhances the ability to detect and respond to cybersecurity threats by processing large datasets and identifying patterns and anomalies more efficiently than classical systems [45].

- **Secure Communication**

Quantum Teleportation: Enables the transfer of quantum information securely over long distances without the physical movement of particles, ensuring data integrity and confidentiality.

- **Future-Proofing Security**

Preparing for Quantum Advancements: As quantum computing technology advances, it's crucial to develop and implement quantum-safe security measures to protect sensitive data from future quantum-based [46] attacks.

Essentially, quantum computing is critical for cybersecurity as it addresses the risks posed by quantum computers to conventional cryptographic methods. Additionally, it harnesses new quantum technologies to enhance secure communication [17] and reinforce security protocols in a quantum-advanced setting.

6 Real-Time Analysis of Quantum Computing

In real-time study of quantum computing, several facets of quantum systems and their applications are observed and evaluated in real time. The following are important facets of quantum computing's real-time analysis:

- **Quantum State Monitoring**:

Qubit States: Monitoring the state of qubits in real-time to detect and correct errors caused by decoherence and environmental noise.

Entanglement: Tracking the entanglement between qubits to ensure the integrity of quantum information and computations.

- **Error Detection and Correction**:

Error Syndromes: Detecting errors in quantum computations using error syndromes derived from real-time measurements of qubits.

Error Correction Codes: Implementing real-time error correction codes to maintain the fidelity of quantum operations and prevent computation failures.

- **Quantum Algorithm Performance**:

Execution Time: Analyzing the execution time of quantum algorithms and comparing it with classical benchmarks to assess the speedup achieved. Resource Utilization: Monitoring the usage of quantum resources (qubits, gates, memory) to optimize algorithm performance in real time.

- **Quantum Networking and Communication**:

Entanglement Generation: Real-time analysis of entanglement generation and maintenance in quantum networks to ensure secure communication. Quantum Teleportation: Monitoring the success rate and fidelity of quantum teleportation protocols for reliable data transmission.

- **Quantum Simulation and Modeling**:

Physical Systems: Simulating quantum systems in real-time to model complex physical phenomena, such as molecular interactions or material properties. Algorithm Validation: Validating quantum algorithms through real-time simulation to predict and optimize their performance before physical implementation.

- **Quantum Machine Learning**:

Data Processing: Real-time analysis of quantum-enhanced machine learning models for tasks such as anomaly detection and pattern recognition in cybersecurity.

6.1 Technologies Used in Quantum Computing

Quantum computing encompasses several types of approaches and technologies, each with distinct applications in cybersecurity. Here are the different types of quantum computing used in cybersecurity:

- **Quantum Machine Learning for Cybersecurity**

Anomaly Detection: Uses quantum machine learning algorithms to detect anomalies and potential threats in cybersecurity systems. Pattern Recognition: Enhances the ability to recognize complex patterns and correlations in data, improving threat detection and response.

- **Quantum-Safe Protocols**

Entanglement Swapping: Enables the establishment of entanglement between distant nodes, crucial for secure long-distance quantum communication.

Quantum Repeaters: Extend the range of quantum communication by using intermediate nodes to maintain entanglement over longer distances.

- **Quantum Simulation**

Simulation of Cryptographic Systems: Uses quantum computers to simulate and test the security of cryptographic systems, identifying potential vulnerabilities.

- **Superdense Coding**

Efficient Data Transmission: Allows the transmission of two classical bits using one qubit, improving the efficiency and security of communication channels. These various types of quantum computing approaches enhance cybersecurity by providing advanced methods for secure communication, encryption, and threat detection, leveraging the unique properties of quantum mechanics.

7 Quantum Cryptography in Depth

Impact on quantum computing on cryptographic algorithms is given in Table 1 and the

- **Cybersecurity and Cryptography**:

Key Distribution: Real-time analysis of quantum key distribution protocols to detect potential security breaches or eavesdropping attempts.

- **Post-Quantum Cryptography**:

Assessing the real-time performance and security of quantum-resistant cryptographic algorithms in protecting sensitive data. Real-time analysis in quantum computing plays a crucial role in optimizing performance, ensuring reliability, and advancing the practical applications of quantum technologies across various fields, including cybersecurity, scientific research, and computational optimization.

7.1 Disruption of Traditional Cryptography

Quantum computing may solve issues that are computationally difficult for classical computers, it has the potential to upset standard cryptography significantly. The most significant effect is caused by quantum algorithms, which can breach popular cryptographic protocol types and, as a result, have interfered with digital communication and self-defense data security systems.

Table 1 Impact on quantum computing on cryptographic algorithms

Cryptographic algorithm	Type	Purpose	Impact from large-scale quantum computer
AES	Symmetric key	Encryption	Larger key sizes needed
SHA-2, SHA-3	–	Hash functions	Larger output needed
RSA (Rivest-Shamir-Adleman)	Public key	Signatures, key establishment	No longer secure
Elliptic curve cryptography	Public key	Signatures, key exchanges	No longer secure
DSA (Digital Signature Algorithm)	Public key	Signatures, key exchanges	No longer secure

7.1.1 Disruption of Traditional Cryptography

- **Breaking Public-Key Cryptography**:

Shor's Algorithm: Among the most extensively identified quantum algorithms is the Shor algorithm, able to both without problems computing discrete logarithms and factor large numbers. Diffie-Hellman, RSA, and Elliptic Curve Cryptography (ECC). These are primarily based on troubles with arithmetic that are difficult to figure out.

Impact: The numbers hired in RSA encryption might take a long time, if no longer millions of years, for the preferred computer systems to break down. However, Shore's method may be utilized by a sufficiently sturdy quantum computer to break RSA encryption in a depend of hours or minutes. This capability should compromise digital signatures and safe transactions alike by rendering the present public key cryptosystem insecure.

- **Vulnerabilities in Symmetric Cryptography**:

Grover's Algorithm: This new quantum set of rules is quadratically faster than traditional algorithms in searching unstructured records. Although symmetric cryptography, like AES, isn't corruptible, this shortens the effective key period.

Impact: For example, Grover's algorithm can make 128-bit keys much more secure than 64-bit keys, which can be brute-forged with older machines if possible If possible at all it is necessary to give a key this doubles the length to increase the security of symmetric cryptography to overcome this It will, which increases computing costs and resources requirements.

7.2 Simulation Demonstrating Quantum Computing's Impact

7.2.1 Simulation Scenario

- **Objective**: To demonstrate how Shor's method can use quantum computing to crack RSA encryption.
- **Configuration**: Assume a 2048-bit key, the most common key length for modern cryptography, and the RSA encryption method.
- **Classical approach**: Since execution time increases exponentially with key size if a key has a 2048-bit factor in a computer on ancient ground, it would take an impossibly long time—perhaps longer than the life span of the universe.

7.2.2 Quantum Approach

- **Quantum simulation**: The Shor algorithm is used to factor a 2048-bit integer using a quantum computer simulator, thus breaking RSA encryption.
- **Outcome**: Based on the simulation, for a quantum computer with about 4096 error-corrected qubits, it is possible to lock 2048 bits in hours or less, negating the need for an encryption role.

7.2.3 Impact Analysis

- **Decryption of encrypted data**: A malicious actor can obtain encrypted communications, decrypt interrupted conversations, and retrieve sensitive data protected by RSA encryption if they have a quantum computer with the capabilities demonstrated in the simulation.
- **Digital signature tampering**: Software delivery, digital contracts, and secure communication systems can all be tampered with by using the Shor algorithm to forge digital signatures.
- **Insecure communication networks**: Today, any network that uses RSA to exchange secure keys is open to attack, potentially leading to massive data breaches and reduced public confidence in digital systems.

7.3 Expert Insights on Quantum-Resistant Cryptography

There is growing settlement inside the cryptography network regarding the importance of quantum-resistant algorithms as they turn out to be greater privy to the quantum threat. The need to create and standardize submit-quantum cryptography (PQC) algorithms has been emphasized by using Dr. Lily Chen, the top of NIST's cryptography Technology Group: It is vital that post-quantum cryptography be followed. We are developing standards so that they will provide the security we require if quantum computing becomes a reality. However, adoption will take time, therefore, we need to flow fast to prevent a "quantum apocalypse" in which compromised vital facts occur. PQC improvement involves investigating algorithms that are proof against quantum and classical attacks. "One of the key demanding situations is finding algorithms that are not only efficient and realistic for great use, but additionally quantum-resistant," states prominent PQC researcher Dr. Dan Bernstein. To ensure a continuing transition, we ought to strike a balance between protection, performance, and simplicity of implementation.

7.4 *Simulation and Expert Perspectives*

Simulation Scenario:

- **Objective**: The real-world implications are illustrated employing a hypothetical state of affairs in which a quantum laptop employs Shor's method to crack RSA encryption using a 2048-bit key.
- **Expert insight**: Renowned quantum computing theorist Dr. Scott Aaronson has underlined that "the theoretical basis is strong, although cutting-edge quantum computers are a way from having the ability to break RSA." If scalable quantum computing is executed, the simulation of Shor's algorithm on a classical system offers us a peek at what is probably feasible within Destiny.

Impact Analysis:

- Decryption of Encrypted Data: The safety of gift communication structures "can be compromised," in keeping with specialists like Dr. Artur Ekert, a pioneer in quantum cryptography. We may be putting personal, economic, and government statistics at risk if we don't create and use quantum-resistant protocols.
- Insecure Communication Networks: Post-quantum cryptography specialist Dr. Tanja Lange states that "global collaboration is needed to expand and set up quantum-safe encryption techniques, as quantum computer systems have the potential to disrupt present-day verbal exchange networks." Although the time frame is doubtful, the hazard is too huge to be disregarded.

7.5 *Advancements in Quantum Hardware and Scalability*

- **Development of More Qubits**: The requirement for an excessive variety of qubits (quantum bits) that might keep coherence over extended intervals of time has been a primary impediment to the practical utility of quantum computing. Companies like Google, IBM, and Rigetti have currently made incredible progress in growing the quantity of qubits of their quantum processors. For example, the declaration of IBM's 433-qubit Osprey processor in late 2023 represents a significant increase in the wide variety of qubits, putting us closer to the cutoff desired for quantum benefit in cybersecurity programs.
- **Impact on Cybersecurity**: The possibility of employing quantum pc systems for sports like cracking RSA encryption, the use of Shor's method turns into increasingly more doable as quantum hardware continues to scale. This acceleration can also make position-based quantum cryptography increasingly pressing as businesses begin to prepare for the increasing chance of quantum threats.

2. Breakthroughs in Quantum Error Correction

- **Improved Error Rates**: Because qubits are more susceptible to mistakes due to their quantum nature, quantum blunder correction is important to preserving the stability of qubits in the course of computation. By utilizing floor coding techniques and modern-day mistake mitigation methods, researchers from Google's Quantum AI lab announced a large drop in mistake rates in 2024, bringing quantum blunder rates closer to the brink required for fault-tolerant quantum computing.

- **Impact on Cybersecurity**: Quantum computer systems are extra practical for real-global cybersecurity applications due to their lower error rates, which also leads them to be more scalable and dependable. With this improvement, quantum cryptography solutions together with Quantum Key Distribution (QKD), which provide a sturdy manner to shield communication channels from each quantum and classical attacks, have ended up in the direction of being appreciably used.

7.6 Hybrid Quantum–Classical Algorithms

- **Integration with Classical Computing**: One of the most present-day tendencies in the computer era is the creation of hybrid quantum–classical algorithms, which hire quantum computers to deal with tremendous factors of a hassle that are tough for classical computer systems, leaving the rest to be solved by conventional computing strategies. These methods may be particularly useful for solving cryptography-associated optimization problems, which include solving problematic lattice-based cryptographic schemes.
- **Impact on Cybersecurity**: The hybrid technique makes it possible to use quantum computing in cybersecurity greater quickly. For example, without requiring realistic quantum gadgets, quantum algorithms can enhance classical cryptography protocols to offer higher safety. In sectors in which information safety is crucial, this may hasten the adoption of quantum-safe, suitable safety solutions.

7.6.1 Quantum-Resistant Algorithms Standardization

- **NIST's Post-Quantum Cryptography (PQC) Standards**: The first set of published quantum cryptographic algorithms chosen for standardization turned into found out via NIST in 2022. This includes digital signature algorithms, which include CRYSTALS-Dilithium for virtual signatures and CRYSTALS-Kyber for encryption, which can be made to withstand conventional and quantum assaults.
- **Impact on Cybersecurity**: The enterprise-extensive deployment of PQC algorithms is made possible utilizing their standardization. Organizations can lessen the threat posed by the aid of future quantum computer systems by means of imposing quantum-resistant cryptography once standards are in place. With the continuing development of quantum computing abilities, this process is anticipated to pick up steam.

7.7 Quantum Key Distribution (Qkd) Protocols

- **BB84 Protocol**: Detailed explanation of the BB84 protocol, including its operational principles and real-world applications.
- **E91 Protocol**: Overview of the E91 protocol, how it leverages quantum entanglement, and its differences from BB84.

Using quantum bits (qubits) encapsulated as polarized photons, Charles Bennett and Gilles Brassard created the BB84 protocol in 1984. It establishes a secure key. The four polarization states in which Alice and Bob, the protocol's users, communicate these photons are horizontal (0°), vertical (90°), diagonal (45°), or anti-diagonal (135°). They save just the bits measured with matching bases after making a public comparison of the measurement bases. Many quantum key distribution (QKD) systems for secure communications use BB84, which is integrated into commercial solutions from firms like Toshiba and ID Quantique [25]. The E91 protocol, on the other hand, uses quantum entanglement to produce a secure key and was first presented by Artur Eckert in 1991. Entangled photon pairs are released by a source, and Alice receives one photon from each pair.

7.8 Post-Quantum Cryptography

- **Lattice-Based Cryptography**: Detailed discussion on how lattice-based cryptographic algorithms work and their resistance to quantum attacks.
- **Hash-Based Cryptography**: Explanation of hash-based cryptographic methods like Merkle trees and their applications.

The goal of post-quantum cryptography is to create cryptographic algorithms that are resistant to the power of cutting-edge quantum computers. Lattice-based cryptography, which uses computational challenges related to lattice structures in high-dimensional environments, is a promising field [26]. Two famous examples that are thought to be immune to quantum attacks are the learning with errors (LWE) issue and the shortest vector problem (SVP). Another feasible approach is hash-based cryptography, which makes use of cryptographic hash functions similar to those in Merkle trees. Large datasets can be verified quickly and securely with Merkle trees, and quantum-resistant digital signatures can be obtained with hash-based signature systems like the Merkle signature system [27]. The McEliece cryptosystem serves as an example of code-based cryptography, which makes use of random linear codes and decoding issues. The McEliece cryptosystem, though it was proposed in the 1970s, is resistant to quantum attacks.

8 Real-World Applications Used in Quantum Computing

The development of quantum computing holds great promise and potential in the field of cryptography. It is significant for being an excellent demonstration of how quantum technology is being applied in the real world. Because quantum computing offers a secure communication method independent of conventional mathematical codes, it has the potential to strengthen cryptography [28]. Instead, it functions based on basic physics, which makes virtual hacking practically impossible. Quantum computers may develop encryption techniques that are far more precise and powerful in comparison to those made by ordinary computers. These developments offer a path to improve information security methods.

Quantum computing has emerged as a disruptive technology with the potential to revolutionize a wide range of industries, including cybersecurity. While quantum computers provide extraordinary computational capacity, they also present substantial problems to traditional cryptography methods, jeopardizing the security of critical data. One of the most serious risks posed by quantum computing to cybersecurity is the ability to break widely used encryption algorithms like Rivest-Shamir-Adleman (RSA), which rely on the difficulty of factoring huge numbers or solving discrete logarithm problems. Quantum algorithms, like as Shor's algorithm, have proven to be effective at solving these problems, making existing encryption systems vulnerable to attacks. As a result, there is a critical need to create quantum-resistant cryptography algorithms that can withstand quantum attacks. Quantum computing brings both potential and challenges in cybersecurity [29]. While quantum computers have the potential to compromise existing cryptographic methods, advances in post-quantum cryptography and quantum key distribution provide intriguing possibilities for securing sensitive information in the quantum age. To ensure robust cybersecurity in the face of evolving quantum technologies, it is critical to continue investing in research and development of quantum-resistant cryptographic solutions, as well as examining attainable implementations of quantum-safe communication protocols. Symmetric encryption is frequently referred to as "secret key encryption." It encodes and decodes information using the same passphrase. The most widely used type of symmetric encryption is AES, which stands for Advanced Encryption Standard. Asymmetric encryption is also known as "public key encryption" because it employs a private and public key method that is more proficient than a shared secret key but safer. Each party has both a private and a public key. The public key can be shared with any individual and will be used to encode information, which will then be decoded using the private key. The working of symmetric and asymmetric encryption is shown in Fig. 1.

- **Quantum computing used in weather forecast**

Gaining sufficient information in advance of natural disasters such as earthquakes, cyclones, and landslides is crucial. This promotes the effective use of preventive measures. A key component for enhancing accuracy and efficiency is quantum computing. Our ability to accurately predicting these tragedies can be greatly

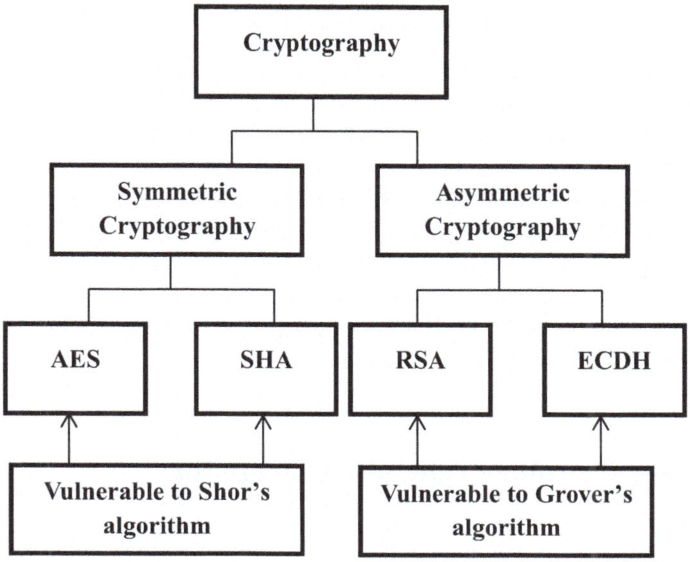

Fig. 1 Branches of Cryptography and Effect of Shor's and Grover's algorithms on Cryptography

enhanced by the application of innovative technologies such as quantum computing [32]. This in turn gives us the ability to act proactively in order to prevent serious consequences and save lives. Quantum computing has the potential to mitigate the effects of climate change by assisting in the reduction of greenhouse gases such as carbon dioxide.

- **Quantum computing in drug development and discovery**

Biotechnology and pharmacology could undergo major developments as a result of quantum computing. This is caused by the simple fact that pharmacology works with molecules, which are tiny atoms that operate according to the principles of quantum physics. By the way, the creation of drugs depends on the more exact interpretation of symptoms and signs, as quantum computing, which is functioning at the level of an atomic particle, might offer. This could prove to be beneficial.

- **Quantum computing and artificial intelligence**

The melding of artificial intelligence (AI) and quantum computing opens up fascinating new possibilities for solving challenging problems. AI applications could be revolutionized by quantum techniques for machine learning and optimization methods that are boosted by quantum capabilities. Several machine learning projects, such as training deep neural networks along with solving optimizing issues essential for AI models, can be greatly improved by the use of quantum computing. In the future, combining AI and quantum computing may advance sectors like data analysis, natural language processing, and pattern recognition.

- **Quantum computing and quantum simulation**

The goal of quantum simulation is to recreate and understand intricate quantum systems that are difficult for conventional computers to analyze. These quantum-memory computers may replicate the behavior of materials, molecules, and other quantum objects. This could improve the development of novel materials, new medications, and a deeper comprehension of fundamental physics. Moving ahead, the goal is to develop better ways for quantum computers to perform these simulations, which will enable us to explore even more complex and sizeable systems. When everything is considered, quantum simulation may speed up scientific discoveries and offer new insights into quantum enigmas. The upcoming part is about how Shor's algorithm plays a major role in quantum computing.

9 Machine Learning in Quantum Computing

Machine learning (ML) aids quantum computing by optimizing quantum systems and algorithms. It enhances simulations of quantum systems, improves error correction and noise reduction, and aids in designing efficient quantum circuits. ML also helps develop hybrid quantum-classical algorithms, fine-tunes quantum hardware, and processes quantum data. Overall, ML boosts the performance, reliability, and scalability of quantum computing, driving advancements in both fields.

9.1 Advantages of Machine Learning Algorithms in Quantum Computing

Compared to classical machine learning, quantum machine learning offers turbo-charged outcomes to increase performance that allows algorithms to analyze enormous volumes of data rapidly and accurately because of its enormous computational capability. Due to this speed advantage, quantum algorithms are able to reveal insights and hidden patterns that their classical versions may ignore. This increases the capability to safeguard digital assets and protect data security in an increasingly complex threat landscape by enabling the quicker and more accurate detection of cyberattacks. The upcoming part is about various algorithms used in quantum computing.

10 The Future of Cybersecurity in Quantum World

In a quantum world, cybersecurity is going to experience a complicated and dynamic future as new, [47] quantum-resistant modern technology emerges to solve long-standing issues. Collaboration between the private sector, academia, and government is crucial as we develop and improve these technologies to make sure we stay one step ahead of quantum threats. By collaborating, [48] we can go ahead through the quantum era to establish a federated digital environment where our assets and data are shielded from even the most sophisticated cyberattacks. The next section is about the Emerging quantum-resistant technologies in cybersecurity.

10.1 Future Development

The future of quantum computing and its implications for cybersecurity are sure to make a radical turn in the next decade. While it's rather difficult to guess with full accuracy the timeline, the following projections do showcase different developments and their possible impacts:

- **Quantum Computing**:
  - **Advances in Hardware**: Gradual development of quantum hardware; today, quantum computers run up to 100–200 qubits with improved error rates.
  - **Quantum Algorithms Development**: Research into quantum algorithms has been underway, especially in optimization, cryptography, and simulation algorithms, with practical applications in view.
- **Cybersecurity**:
  - **Post-Quantum Cryptography Standardization**: It is expected that by 2024–2025, NIST will finalize the standardization of post-quantum cryptographic algorithms. They will now work to plan their migrations to these quantum-resistant algorithms.
  - **Quantum Key Distribution (QKD) Ubiquity**: Financial institutions and government agencies on both the national and local levels will also adopt QKD into limited, high-security environments.
- **Impact**:
  - **Sophistication of Awareness**: Organizations will recognize the quantum threat to their security and get more serious about upgrading system protections for a post-quantum world.
  - **Rollout of PQC**: Certain industries, especially where sensitive data is involved, will begin to introduce PQC in their infrastructure.
  - **Niche QKD Use**: The role for quantum key distribution will be a niche in specialized high-security communication.

10.1.1 2027–2029: Scaling and Wider Integration

- **Quantum Computing**:
  - **Quantum Error Correction and Fault Tolerance**: Advances in the field of quantum error correction have evolved, leading to fault-tolerant quantum computers that can do larger calculations.
  - **Security**: Quantumcloud offers more broadly useful quantum cloud services that permit organizations and specialists to remotely access their assets on a pay-per-use basis, while additionally keeping the integrity of any long-term sensitive information.
- **Cybersecurity**:
  - General Use of PQC: As older cryptographic systems begin to crumble under the threat posed by quantum computers, entities in a range of fields will speed up their adoption. Quantum-Enhanced Machine Learning (QML): Quantum machine learning will become a phenomenon in 2019, and we will start to see its practical application for cybersecurity—such as anomaly detection & threat intelligence. • Making 2 s from Quants QRNG article—Tying the compilation back to how it drove forward in Qiskit, or simply saying: growing adoption of Quantum Random Number Generators (QRNGs) for secure transactions and data anonymity.
- **Impact**:
  - **Stronger Encryption**: Once PQC is all the rage, our encryption game will be on a whole new level. It'll keep us safe from both regular and quantum hackers trying to mess with our stuff.
  - **Smarter Threat Spotting**: When QML gets cozy with our cybersecurity gear, it'll be like having a super-powered sidekick that can sniff out digital baddies in the blink of an eye, helping us stay one step ahead.
  - **Quantum Tech Goes Mainstream**: QRNG and QKD will stop being the cool kids' secret toys and start popping up everywhere in cybersecurity. They'll especially be VIPs in those super important areas where keeping secrets is, well, super important.

10.1.2 2030–2033: Quantum Supremacy & Global Impacts

- **Quantum Computing**:
  - **Quantum Supremacy Achieved**: It is anticipated that by the early 2030s, quantum computers will surpass classical computers in some tasks, hence achieving "quantum supremacy" in particular, useful applications.
  - **Commercialization of Quantum Solutions**: Applications for quantum computing will be found in a variety of areas, such as banking, medicine, and logistics.

- **Cybersecurity**:
  - **PQC as a Standard**: All new systems will use quantum-resistant algorithms and post-quantum cryptography as the norm. We'll update or phase out legacy systems.
  - **Quantum Attack Mitigations**: In order to be ready for potential cyberattacks enabled by quantum technology, governments and major corporations will have created and implemented extensive measures for mitigating quantum attacks.
  - **Fully Integrated Quantum-Secure Networks**: Critical infrastructure and government agencies will implement quantum-secure communication networks that integrate QKD, PQC, and other quantum technologies.
- **Impact**:
  - **Classical cryptography will come to an end**: quantum-resistant solutions will entirely replace antiquated and insecure techniques like RSA and ECC.
  - **Quantum Threat Realized**: Systems with weak security that have not switched to quantum-resistant algorithms may be the target of the first real-world quantum attacks.
  - **Global Security Shift**: All organizations managing sensitive or valuable information will prioritize quantum security, resulting in a substantial shift in the global cybersecurity environment.

10.2 Emerging Quantum-Resistant Technologies

The development of [21] post-quantum cryptography, QKD, and quantum random number generation is paving the way for a new generation of quantum-resistant technology, including quantum-proof data. These state-of-the-art technologies provide increased protection for our digital assets, guaranteeing the safety of our data despite the enormous potential of quantum computing [49]. These technologies are going to become steadily more significant in determining the trajectory of cybersecurity as they continue to evolve. We can better prepare for the opportunities and challenges ahead by keeping up with the most recent advancements in quantum-resistant technologies. The next section deals with collaboration between academia, industry, and government.

10.2.1 QKD Constraints and Difficulties

Quantum Key Distribution (QKD) utilizes the concepts of quantum physics to provide a potentially secure communication method. Nevertheless, despite its promise, QKD has a number of drawbacks and difficulties that prevent wider use.

1. **Infrastructure needs and distance restrictions**

- Restraints on Distance:

The limited range of QKD is one of its main drawbacks. Quantum keys are now sent using optical fibers or satellite links in most QKD systems. When using optical fiber-based QKD, the effective range is often limited to approximately 100–200 km without the use of quantum repeaters because the signal deteriorates dramatically over long distances due to loss and noise in the fibre.

Although it increases the range, satellite-based QKD is still limited by the requirement for line-of-sight communication and is impacted by atmospheric conditions, which limits its usefulness globally.

- **Infrastructure Requirements**:

Quantum Key Distribution (QKD) calls for a large investment in infrastructure, including secure communication channels, specialized quantum hardware (such as photon detectors and quantum transmitters), and maybe the launch of new satellites or quantum networks. Because of this, deploying QKD is costly and complicated, especially for large-scale or international deployments. Widespread use of QKD is hampered by its inability to integrate with current communication networks due to the requirement for specialized equipment.

2. **Issues with Scalability and Realistic Implementation**

- **Scaling Problems**

Because QKD depends on specialized equipment and requires secure quantum channels between every pair of communication parties, scaling it for general use is difficult. Unlike classical cryptography, which can be readily implemented online, QKD necessitates a physical connection—such as a satellite link or optical fiber—between users. Because of this, scaling QKD for usage in vast, decentralized networks like the Internet—where millions of users may require secure communication—becomes challenging.

- **Practical Implementation**:

QKD integration with current cryptographic systems and protocols might be challenging, even in settings where it can be implemented. Rather than taking the place of conventional cryptographic techniques, QKD enhances them in order to safeguard the key exchange procedure. Organizations now have to deal with the coexistence of quantum and classical systems, which can make implementation more difficult and call for specialized knowledge. In addition, QKD is susceptible to practical constraints like detector inaccuracies and side-channel assaults, which, if not handled effectively, could jeopardize system security.

Ultimately, even though QKD provides theoretically unbreakable security, these drawbacks and difficulties need to be resolved before it can be widely used as a workable and scalable solution. The creation of quantum repeaters to increase range, hybrid systems that integrate QKD with classical cryptography, and developments in quantum hardware to lower prices and boost efficiency are just a few of the ongoing research projects aimed at overcoming these obstacles.

10.3 Reduce the Risk

There is a serious security concern associated with the potential for breaking popular encryption systems as quantum computing develops. One of the main new approaches to lessen this is the creation and use of Post-Quantum Cryptography (PQC). PQC algorithms are made to withstand attacks from quantum and conventional computers alike. A number of organizations, including the National Institute of Standards and Technology (NIST), are now evaluating and standardizing these algorithms, which include lattice-based, hash-based, code-based, and multivariate polynomial cryptosystems. By integrating these new algorithms into the outdated cryptographic infrastructure, PQC ensures that private information and communications stay safe even as quantum computing gains in strength. This proactive strategy is essential because it makes it possible to gradually replace cryptographic systems that are weak before quantum computers are able to take advantage of them.

Quantum Key Distribution (QKD), which uses the ideas of quantum mechanics to establish secure communication channels, is another important tactic. Because QKD disturbs the quantum states, any attempt to eavesdrop on the key exchange process would be apparent, allowing two parties to produce and distribute cryptographic keys with a high degree of security. Because of this, QKD is a desirable solution for communication network security, especially in industries like finance, government, and defence where data security is crucial. In addition to improving the unpredictability of encryption systems and reinforcing the security of cryptographic keys, the integration of Quantum Random Number Generators (QRNGs), which employ quantum processes to generate truly random numbers, is being investigated. These tactics are essential for equipping the global cybersecurity infrastructure to withstand the challenges posed by the development of quantum computing, together with continuing research and international collaboration.

10.4 Academia, Industry, and Government Collaboration

The government, business, and academic communities have to work collectively to design and adopt quantum-resistant devices [50]. The government provides financing and regulatory control to ensure the technology's success, the industry ensures it is applied properly, and academia offers research and development skills. Together, we can assure the successful adoption of innovative solutions, hasten their development, and eventually build a more secure digital environment against coming quantum threats. Reference [51] The next crucial part deals with quantum machine learning for cyberdefense [52].

11 Industry Implications

11.1 Finance and Health Care

There could be a big impact from quantum computing on a lot of different areas, but especially on healthcare and finance. An outline of the ramifications for each is provided below:

a. **Finance Industry**

When it comes to tasks like derivative pricing, risk assessment, and portfolio optimisation, traditional computers are unable to handle the computational complexity of financial models.

To secure transactions, communications, and sensitive data, the finance sector mainly relies on encryption. Emerging quantum computing techniques like RSA and ECC represent a serious threat to current cryptography algorithms. Quantum computers are more efficient than classical computers at solving optimisation problems, which allows them to explore multiple scenarios at once and produce better risk-reward trade-offs. o For instance, a quantum algorithm could optimize a large portfolio consisting of thousands of assets, balancing returns and risks in real-time, something that would be impractical for classical systems.

- **Risk Management and Simulation**:

By improving Monte Carlo simulations, which are used in risk management, quantum computing can produce faster and more accurate risk assessments. This can enhance stress testing and market valuation forecasts. For instance, financial organizations might be able to replicate millions of market circumstances in a fraction of the time it takes now, thanks to quantum-enhanced risk models, which would result in more effective risk mitigation techniques.

- **Cryptography and Security**:

To defend sensitive data and transactions from quantum attacks, the finance sector needs to switch to post-quantum cryptography (PQC) techniques. For instance, in order to guarantee that consumer information and financial transactions are safe against quantum attacks, banks and payment processors must implement PQC.

- **Fraud Detection**:

By analyzing enormous volumes of transaction data to find patterns suggestive of fraudulent activity more rapidly and reliably than traditional techniques, quantum machine learning (QML) might enhance fraud detection systems. · As an illustration, a QML-based system could identify and stop fraudulent transactions almost instantly, minimizing losses and enhancing confidence in electronic payment systems.

- **High-frequency trading (HFT)**:

By enabling quicker processing of intricate trading algorithms, quantum computing has the potential to completely transform HFT and enable traders to execute deals at previously unheard-of speeds and with greater precision. As an illustration, traders may be able to detect and take advantage of market inefficiencies quickly than their rivals by using quantum algorithms, which could result in substantial financial benefits.

b. **Healthcare Industry**

Finding new medications is an expensive and time-consuming procedure that requires intricate molecular simulations that are difficult for traditional computers to do. To prevent breaches and guarantee privacy, the healthcare sector handles enormous volumes of sensitive patient data, necessitating the implementation of strong security measures.

- **Drug Development**:

By simulating atomic-level molecular interactions, quantum computing greatly expedites the drug discovery process. This potential could result in the creation of novel medications and customized treatment plans. As an illustration, complicated proteins and their interactions with possible medication compounds may be simulated by quantum computers, cutting down on the years it would take to find promising drug candidates from months.

- **Genetics and Personalized Medicine**:

The analysis of genomic data could be improved by quantum computing, leading to more precise detection of genetic markers and the creation of individualized treatment regimens based on a patient's genetic composition. As an illustration, quantum algorithms may be able to determine the optimal course of action for individuals with complicated or uncommon genetic illnesses by evaluating enormous volumes of genetic data. This could result in more specialized and potent treatments.

- **Medical Imaging and Diagnostics**:

By enhancing the way medical pictures are processed, quantum algorithms can lead to earlier and more precise diagnoses. This might have a special effect in areas like oncology, where early detection is essential. As an illustration, real-time, high-resolution imaging of tumors may be possible with quantum-enhanced imaging techniques, which would increase the precision of cancer diagnoses and the efficacy of treatment regimens.

c. **Data Security and Privacy**:

To safeguard patient data from upcoming quantum risks, the healthcare sector will need to switch to post-quantum cryptography, just like the financial sector has done.

To guarantee the security of electronic health records (EHRs) in the event of a post-quantum world, hospitals and healthcare providers must implement encryption that is resistant to quantum mechanics.

12 Conclusion

This chapter provides a detailed review of the current state and expected to upcoming advances in quantum computing and cybersecurity. The book also includes an overview of the fundamental ideas behind quantum bits and gates and the critical role that quantum computing plays in real-world applications. It also emphasizes the importance of Shor's algorithm, which is employed in quantum computing. As a result, the discipline of quantum computing is swiftly developing, with processing speeds increasing exponentially and previously unobtainable data becoming available.

References

1. J. Preskill,Quantum computing 40 years later. arXiv (2021). arXiv: 2106.10522. [Google Scholar]
2. E. Rieffel, *Quantum Computing* (2011)
3. X. Wang, H. Yu, How to break MD5 and other hash functions. In: Advances in Cryptology—EUROCRYPT. Springer, pp. 19–35 (2005)
4. S. Wolfram, *A New Kind of Science*, 1st edn. (Wolfram Media, USA, 2002)
5. E. Jones, UK Government gets serious on developing security skills in 2016 (2016). ITProPortal. Retrieved from http://www.itproportal.com/2016/01/14/uk-government-gets-serious-developing-security-skills-in-2016/
6. S. King, A model for a national cybersecurity policy (2017). Retrieved from https://www.linkedin.com/pulse/model-national-cybersecurity-policy-steve-king
7. C. Markus, G. Bert, L. Michele, The ethics of cybersecurity. I:n The International Library of Ethics, Law, and Technology, 2020. [Online].Available: https://doi.org/10.1007/978-3-030-29053-5
8. D. Denning, Is quantum computing a cybersecurity threat? Am. Sci. **107**, 83 (2019)
9. O.S. Althobaiti, M. Dohler, Cybersecurity challenges associated with the Internet of things in a post-quantum world. IEEE Access **8**, 157356–157381 (2020)
10. R. Yan, J. Dai, Y. Wang, Y. Xu, A.Q. Liu, Quantum-key-distribution based microgrid control for cybersecurity enhancement, in *2021 IEEE Industry Applications Society Annual Meeting (IAS)*, pp. 1–7 (2021)
11. M. Mosca, Cybersecurity in an era with quantum computers: will we be ready? IEEE Secur. Privacy **16**(5), 38–41 (2018)
12. H. Albataineh, M. Nijim, Enhancing the cybersecurity education curricula through quantum computation. In: Daimi, K., Arabnia, H.R., Deligiannidis, L., Hwang, M.-S., Tinetti, F.G. (Eds.), *Advances in Security, Networks, and Internet of Things*. Cham: Springer International Publishing, (2021)
13. K.-K. Ko, E.-S. Jung, Development of cybersecurity technology and algorithm based on quantum computing. Appl. Sci. **11**(19) (2021). [Online]. Available: https://www.mdpi.com/2076-3417/11/19/9085

14. M. Aboy, M. Timo, K. Mauritz, Mapping the patent landscape of quantum technologies: patenting trends, innovation and policy implications. IIC-Int. Rev. Intellect. Prop. Compet. Law **53**, 853–882 (2022). [Google Scholar] [CrossRef]
15. J. Preskill, Quantum computing 40 year's later. arXiv (2021). arXiv:2106.10522. [Google Scholar]
16. Computer Security Research Center. "Post Quantum Cryptography PQC: Workshops and Timeline." NIST; 7 July 2022. Available online: https://csrc.nist.gov/Projects/post-quantum-cryptography/workshops-and-timeline. Accessed 8 August 2022
17. M. Devoret, R.J. Schoelkopf, Superconducting circuits for quantum information: an outlook. Science **339**(6124), 1169–1174 (2013)
18. J.A. Smolin, G. Smith, A. Vargo, Oversimplifying quantum factoring. Nature **499**(7457), 163–165 (2013)
19. P. Schwabe, S. Douglas, W. Thom, Post-quantum TLS without handshake signatures. In: *Proceedings of the 2020 ACM SIGSAC Conference on Computer and Communications Security*, Virtual Event, 9–13 November 2020. [Google Scholar]
20. F. Raheman, Solid state software on a chip (3SOC) for building quantum resistant web 3.0 computing devices. U.S. Patent US29/842,535, 15 June 2022. [Google Scholar]
21. N. Gisin, G. Ribordy, W. Tittel, H. Zbinden, Quantum cryptography. Rev. Mod. Phys. **74**(1), 145 (2002)
22. V. Hassija, V. Chamola, V. Saxena, V. Chanana, P. Parashari, S. Mumtaz, M. Guizani, Present landscape of quantum computing. IET Quantum Commun. **1**, 42–48 (2020). [Google Scholar] [CrossRef]
23. Application: Quantum machine learning algorithms for pattern recognition and data analysis
24. "D-Wave–The Quantum Computing Company", 2019. [Online]. Available: https://www.dwavesys.com/d-wave-two-system
25. Impact of Cybercrime on Canadian Businesses 2017 (2018)
26. B. Arslan, M. Ulker, S. Akleylek, S. Sagiroglu, A study on the use of quantum computers risk assessment and security problems. In *2018 6th International Symposium on Digital Forensic and Security* (ISDFS), pp. 1–6 (2018)
27. S. Barnett et al., Journeys from quantum optics to quantum technology. Prog. Quantum Electron. **54**, 19–45 (2017)
28. Security and Privacy Controls for Information Systems and Organizations. NIST Special Publication 800–53 Rev. 5. [Online]. Available: https://csrc.nist.gov/csrc/media/publications/sp/800-53/rev.
29. Post-Quantum Cryptography Standardization. [Online]. Available: https://csrc.nist.gov/Projects/Post-Quantum-Cryptography/Post-Quantum-Cryptography-Standardization
30. P7130–Standard for Quantum Computing Definitions. [Online].Available: https://standards.ieee.org/project/7130.html
31. J. Walker, Quantum computing is coming: How will it impact cybersecurity? Entrepreneur.https://www.entrepreneur.com/en-au/technology/quantum-computing-is-coming-how-will-it-impact/439060, 2022. [Online; Accessed on Jan. 12, 2023].
32. R. Meraz, L. Vahala, Application of quantum cryptography to cybersecurity and critical infrastructures in space communications. OUR J. ODU Undergr. Res. J. **7**(1), 5 (2020)
33. P. Wallden, E. Kashefi, Cybersecurity in the quantum era. Commun. ACM **62**(4), 120–120 (2019)
34. M.C. Libicki, D. Gompert, Quantum communication for post-pandemic cybersecurity. In *2021 13th International Conference on Cyber Conflict (CyCon)*, pp. 371–386 (2021)
35. K.-K. Ko, E.-S. Jung, Development of cybersecurity technology and algorithm based on quantum computing. Appl. Sci. **11**(19), 9085 (2021)
36. S. Sanzeri, What the quantum computing cybersecurity preparedness act means for national security. Forbes. https://www.forbes.com/sites/forbestechcouncil/2023/01/25/what-the-quantum-computing-cybersecurity-preparedness-act-means-for-national-security/amp/, 2023. [Online; Accessed on Feb. 7, 2023].
37. Application: Accelerating the discovery of new drugs and understanding molecular interactions

38. M. Reiher, N. Wiebe, K.M. Svore, D. Wecker, M. Troyer, L.C. Hollenberg, Elucidating reaction mechanisms on quantum computers. Proc. Natl. Acad. Sci. **114**(29), 7555–7560 (2017)
39. Application: "Quantum Key Distribution (QKD) for secure communication
40. M. Schulds, I. Sinayskiy, F. Petruccione, An introduction to quantum machine learning. Contemporary Phys. **56**(2), 172–18
41. C.H. Bennett, G. Brassard,"Quantum cryptography: Public key distribution and coin tossing. *Proceedings of IEEE International Conference on Computers, Systems and Signal Processing*, Bangalore, India, pp. 175–179 (1984)
42. C.H. Bennett, G. Brassard, C. Crépeau, R. Jozsa, A. Peres, W.K. Wootters, Teleporting an unknown quantum state via dual classical and Einstein-Podolsky-Rosen channels. Phys. Rev. Lett. **70**(13), 1895–1899 (1993)
43. C.H. Bennett, S.J. Wiesner, Communication via one- and two-particle operators on Einstein-Podolsky-Rosen states. Phys. Rev. Lett. **69**(20), 2881–2884 (1992)
44. H.J. Briegel, W. Dür, J.I. Cirac, P. Zoller, Quantum repeaters: the role of imperfect local operations in quantum communication. Phys. Rev. Lett. **81**(26), 5932–5935 (1998)
45. J. Biamonte, P. Wittek, N. Pancotti, P. Rebentrost, N. Wiebe, S. Lloyd, Quantum machine learning. Nature **549**(7671), 195–202 (2017)
46. J. Buchmann, E. Dahmen, M. Szydlo, Hash-based digital signature schemes. In: *Post-Quantum Cryptography*, pp. 35–93. Springer, Berlin, Heidelberg (2006)
47. A.K. Ekert, Quantum cryptography based on Bell's theorem. Phys. Rev. Lett. **67**(6), 661–663 (1991)
48. A.G. Fowler, M. Mariantoni, J.M. Martinis, A.N. Cleland, Surface codes: towards practical large-scale quantum computation. Phys. Rev. A **86**(3), 032324 (2012)
49. H.J. Kimble, The quantum internet. Nature **453**(7198), 1023–1030 (2008)
50. D. Micciancio, O. Regev, Lattice-based cryptography. In: *Post-Quantum Cryptography*, pp. 147–191. Springer, Berlin, Heidelberg (2009)
51. C. Monroe, J. Kim, Scaling the ion trap quantum processor. Science **339**(6124), 1164–1169 (2013)
52. M. Mosca, Cybersecurity in an era with quantum computers: Will we be ready? IEEE Secur. Priv. **16**(5), 38–41 (2018)

Quantum Computing: A New Frontier in Cybersecurity Resilience

Pawan Whig, Balaram Yadav Kasula, Nikhitha Yathiraju, Anupriya Jain, Seema Sharma, and Sandeep Kautish

1 Introduction

The confluence of quantum computing and cybersecurity represents a paradigm shift in the landscape of digital security, ushering in an era of both unprecedented opportunities and profound challenges [1]. Quantum computing, with its unique ability to perform complex calculations at speeds exponentially faster than classical computers, holds the potential to revolutionize various domains. However, as quantum technologies advance, the traditional cryptographic methods that form the bedrock of cybersecurity face an imminent threat, raising critical questions about the resilience of our digital infrastructure [2].

In the realm of classical cryptography, widely used encryption algorithms such as RSA and ECC rely on the computational complexity of certain mathematical problems, rendering them secure against traditional computers [3]. However, the advent of quantum algorithms, notably Shor's algorithm, threatens to unravel this security by efficiently solving problems, like integer factorization and discrete logarithms, which underpin the security of these cryptographic systems [4–7]. The implications

P. Whig (✉)
Vivekananda Institute of Professional Studies-TC New Delhi, New Delhi, India
e-mail: pawanwhig@gmail.com

B. Y. Kasula
University of The Cumberlands, Williamsburg, KY, USA

N. Yathiraju
University of The Cumberlands, Williamsburg, USA

A. Jain · S. Sharma
School of Computer Applications MRIIRS Faridabad, Faridabad, India

S. Kautish
LBEF Campus, Kathmandu, Nepal

S. B. Goyal et al. (eds.), *Quantum Computing, Cyber Security and Cryptography*,
https://doi.org/10.1007/978-981-96-4948-8_4

are profound—the very fabric of secure communication, data integrity, and confidentiality, built meticulously over decades, stands on the precipice of transformation [9–12].

This chapter embarks on a comprehensive exploration of the intricate interplay between quantum computing and cybersecurity. It aims to elucidate the quantum threats looming over classical encryption, the emergence of quantum-safe alternatives, and the pivotal role of quantum key distribution (QKD) in securing communication channels in a post-quantum era. As we navigate this evolving landscape, it becomes imperative to not only understand the vulnerabilities that quantum computing poses but also to unravel the promise it holds for enhancing cybersecurity [13].

The initial sections of this chapter delve into the foundational principles of quantum computing, offering a primer on quantum mechanics, qubits, and the unique phenomena of superposition and entanglement [14–17]. This foundational understanding sets the stage for exploring the quantum threats to classical cryptographic protocols. The theoretical underpinnings of Shor's algorithm and its potential impact on widely used encryption schemes will be dissected, unraveling the intricacies that render classical cryptography susceptible to quantum attacks [18].

In response to these challenges, the chapter transitions to the exploration of quantum-safe cryptographic alternatives. The spotlight will be on the emergence of post-quantum cryptographic algorithms designed to withstand the computational prowess of quantum computers. As the cryptographic community grapples with the impending threat of quantum attacks, ongoing efforts to standardize and implement these post-quantum algorithms form a critical aspect of this exploration [19–21].

Furthermore, the narrative extends into the realm of quantum key distribution (QKD)—a beacon of hope in the quest for secure communication in a quantum world [22–26]. The unique principles of quantum mechanics, such as the uncertainty principle, entanglement, and the no-cloning theorem, are harnessed in QKD protocols to create secure key exchange mechanisms that are theoretically immune to quantum attacks. Understanding the promises and challenges of QKD is fundamental to envisioning a future where secure communication thrives amidst quantum advancements.

Beyond cryptographic considerations, this chapter investigates the broader implications of quantum computing on cybersecurity. Quantum technologies are not merely harbingers of threats; they also hold the potential to augment cybersecurity defenses. Quantum-enhanced methods for threat detection, secure communication, and data protection constitute a pivotal aspect of this exploration. From quantum-resistant algorithms to innovative cryptographic techniques, the chapter navigates the evolving landscape where quantum computing becomes an ally in fortifying cybersecurity frameworks.

As we traverse this intricate terrain, the chapter underscores the need for a holistic approach to cybersecurity in the quantum era. It calls for interdisciplinary collaboration between quantum physicists, computer scientists, and cybersecurity experts to collectively address the multifaceted challenges posed by the intersection of quantum computing and cybersecurity. Additionally, ethical considerations, regulatory frameworks, and the societal impact of quantum-enhanced cybersecurity measures warrant careful examination, ensuring that the benefits of quantum technologies are harnessed responsibly and inclusively.

This chapter endeavors to provide a comprehensive understanding of the dynamic interplay between quantum computing and cybersecurity. It navigates through the theoretical foundations, quantum threats to classical cryptography, the emergence of quantum-safe alternatives, the promise of quantum key distribution, and the broader implications of quantum technologies for cybersecurity. In doing so, it sets the stage for a future where the fusion of quantum computing and cybersecurity not only safeguards our digital infrastructure but also catalyzes a new era of secure and resilient communication in the quantum age.

2 Foundations of Quantum Computing

The foundations of quantum computing lie in the principles of quantum mechanics, a branch of physics that describes the behavior of matter and energy at the quantum level. At this scale, classical physics breaks down, and quantum mechanics introduces concepts that defy our classical intuition. One fundamental aspect is the use of quantum bits or qubits, which, unlike classical bits, can exist in a superposition of states, representing both 0 and 1 simultaneously. This property exponentially increases computational possibilities. Quantum gates, analogous to classical logic gates, manipulate qubits through principles such as superposition and entanglement, allowing for the creation of quantum circuits. Entanglement, where particles become interconnected and share states, is another key foundation. Quantum supremacy, demonstrated when a quantum computer outperforms classical computers in specific tasks, marks a significant milestone. Understanding these foundational elements is crucial for grasping the transformative potential and unique capabilities of quantum computing in reshaping computational paradigms. Literature review with research gap is shown in Table 1.

Table 1 summarizes the main focus, key findings, and identified research gaps from each referenced work.

2.1 Quantum Mechanics Primer

In this section, we delve into the essential concepts of quantum mechanics that underpin the functioning of quantum computers. Quantum mechanics, a cornerstone of modern physics, introduces a probabilistic and wave-based description of particles at the quantum level. Key principles include superposition, where particles can exist in multiple states simultaneously, and uncertainty, encapsulated by Heisenberg's uncertainty principle. Additionally, the phenomenon of entanglement,

Table 1 Literature review with research gap

References	Main focus	Key findings	Research gap
[1]	Cybersecurity in quantum computing	Tutorial on cybersecurity in quantum computing	Limited exploration of practical applications and real-world implications
[2]	Quantum computing and algorithms	Exploration of quantum frontier in data analytics	Lack of in-depth analysis on the impact of quantum algorithms in computational intelligence
[3]	Quantum computing for cybersecurity of feedback systems	Perspectives on resilient design for cybersecurity	Limited discussion on specific feedback systems and their vulnerabilities
[4]	Cyber diplomacy in the age of AI, IoT, blockchains, and quantum computing	Examining opportunities and risks	Limited focus on the intersection of quantum computing with other emerging technologies
[5]	Future development of quantum computing and its relevance to NATO	Focus on NATO's role in quantum computing development	Limited exploration of global implications and collaboration beyond NATO
[6]	Review of quantum cybersecurity	Analyzing threats, risks, and opportunities	Lack of comprehensive exploration of quantum-safe cybersecurity strategies
[7]	Quantum-safe cybersecurity	Discussion on cybersecurity in the age of quantum-powered AI	Limited exploration of practical implementations and case studies
[8]	Analysis of pre-and post-quantum cyber security scenarios	Pragmatic analysis of security scenarios	Limited consideration of the transition period between pre-and post-quantum security
[9]	Quantum computing and the financial system	Investigating potential impacts on the financial system	Limited discussion on practical strategies for financial institutions to adapt
[10]	Building cyber resilience	Defensive shield for the EU	Limited exploration of the applicability of the defensive shield concept in a broader context
[11]	Governance tools for the second quantum revolution	Focus on governance tools	Limited exploration of challenges and practical implementations
[12]	Quantum computing in renewable energy	Status, opportunities, and challenges	Limited discussion on specific challenges and potential solutions
[13]	Space infrastructures: from risk to resilience governance	Governance of space infrastructures	Limited examination of quantum computing implications for space governance

(continued)

Table 1 (continued)

References	Main focus	Key findings	Research gap
[14]	Business renaissance in the quantum computing era	Opportunities and challenges	Limited examination of specific industries and business models
[15]	Security in IIoT using quantum computing	Enhancing security in the Industrial IoT sector	Limited exploration of practical implementation challenges and solutions
[16]	Quantum computing threat landscape for blockchains	Survey of quantum threats to blockchains	Limited examination of potential countermeasures and resilience strategies
[17]	Quantum computing in military applications	Military applications and operations	Limited exploration of ethical considerations and potential risks
[18]	Research challenges in post-quantum cryptography migration	Identifying challenges in cryptography migration	Limited exploration of strategies for cryptographic agility
[19]	US-China strategic competition and quantum technologies	US-China competition and cyber stability	Limited examination of global implications beyond US-China dynamics
[20]	Assessing the benefits and risks of quantum computers	Evaluation of benefits and risks	Limited exploration of practical scenarios and case studies
[21]	Importance of quantum computing for national security	National security implications	Limited discussion on global collaborative efforts in quantum computing for security
[22]	Advanced computational applications in quantum computing	Review of advanced applications	Limited exploration of challenges and limitations in implementing advanced applications
[23]	SDMTA for DDoS vulnerabilities in hybrid cloud	Attack detection and mitigation mechanism	Limited examination of real-world deployment challenges and effectiveness
[24]	Multimodal sentiment analysis	Survey and comparison	Limited exploration of challenges in multimodal sentiment analysis
[25]	Building a brand image through eCRM	Brand image through electronic CRM	Limited examination of specific industry applications and case studies
[26]	Online banking paradigm shift	Paradigm shift in online banking	Limited exploration of societal impacts and challenges in the paradigm shift

where particles become correlated and share states regardless of distance, plays a pivotal role. This primer establishes the quantum mechanical foundation necessary for comprehending the intricacies of quantum computing.

2.2 *Qubits and Quantum Gates*

Building upon the quantum mechanics primer, this section explores the fundamental units of quantum information–qubits. Unlike classical bits that represent either 0 or 1, qubits can exist in a superposition of both states simultaneously, exponentially increasing computational possibilities. Quantum gates, analogous to classical logic gates, manipulate qubits by exploiting quantum phenomena. Gates like the Hadamard gate induce superposition, while others, such as the CNOT gate, utilize entanglement. Understanding qubits and quantum gates is pivotal for grasping the unique capabilities and computational power offered by quantum computers.

2.3 *Quantum Supremacy and Its Implications*

The pursuit of quantum supremacy, where a quantum computer outperforms classical counterparts in specific tasks, marks a groundbreaking milestone in the field of quantum computing. This section explores the concept of quantum supremacy and its implications for the broader realm of computation. When achieved, quantum supremacy signifies the ability of quantum computers to solve problems that were previously intractable for classical computers. The implications extend beyond mere computational speed, impacting fields such as cryptography, optimization, and simulations. An examination of the challenges, advancements, and potential applications arising from quantum supremacy sheds light on the transformative potential of quantum computing.

3 Quantum Threats to Classical Cryptography

The advent of quantum computing poses imminent threats to the security foundations of classical cryptography, a discipline that relies on mathematical algorithms for secure communication and data protection. At the forefront of these threats is Shor's algorithm, a quantum algorithm devised to efficiently factorize large numbers and solve discrete logarithm problems. These mathematical challenges, which classical cryptographic systems hinge upon for their security, can be rapidly addressed by Shor's algorithm on a sufficiently powerful quantum computer. As a result, widely used cryptographic protocols, including RSA and ECC, become vulnerable to rapid decryption, compromising the confidentiality and integrity of encrypted data.

The essence of Shor's algorithm lies in its ability to leverage the principles of quantum superposition and entanglement to perform parallel computations. Classical algorithms would require exponentially longer times to crack the same cryptographic problems. The implications of Shor's algorithm extend beyond mere theoretical vulnerabilities; they underscore the pressing need to adapt cryptographic systems to the evolving landscape of quantum capabilities.

Quantum threats to classical cryptography necessitate the exploration of quantum-resistant cryptographic alternatives. The field of post-quantum cryptography seeks to develop algorithms that remain secure in the face of quantum attacks. Diverse cryptographic approaches, such as lattice-based cryptography, hash-based cryptography, and code-based cryptography, are being explored as potential replacements for classical algorithms. As quantum computing technologies advance, the urgency of transitioning toward post-quantum cryptographic solutions becomes increasingly apparent to ensure the continued security of sensitive information in a quantum-powered era.

3.1 *Shor's Algorithm and Its Impact*

Shor's Algorithm stands as a groundbreaking achievement in the realm of quantum computing, posing a significant threat to classical cryptographic systems. Introduced by mathematician Peter Shor in 1994, this quantum algorithm has the remarkable ability to efficiently factorize large numbers and solve discrete logarithm problems. Its impact lies in its potential to undermine widely used cryptographic schemes, such as RSA and ECC, which rely on the computational difficulty of factoring large numbers for their security. Shor's Algorithm capitalizes on the parallelism afforded by quantum superposition, allowing a quantum computer to perform these calculations exponentially faster than their classical counterparts. The realization of Shor's Algorithm has profound implications for the security of encrypted communications and data, urging a reevaluation of classical cryptographic paradigms.

3.2 *Vulnerabilities in Classical Encryption*

The vulnerabilities in classical encryption protocols become apparent when confronted with the computational power of quantum algorithms like Shor's. Classical cryptographic systems, designed to withstand the challenges posed by classical computers, find themselves exposed to rapid decryption in the quantum realm. The security of widely used algorithms such as RSA and ECC relies on the presumed difficulty of factoring large numbers or solving discrete logarithm problems, tasks efficiently handled by Shor's Algorithm on a sufficiently advanced quantum computer. As a consequence, the confidentiality and integrity of data encrypted using classical cryptographic methods come under threat, necessitating a proactive response to address vulnerabilities and fortify digital security against quantum attacks.

3.3 Quantum-Resistant Cryptographic Alternatives

In response to the looming threat of quantum attacks on classical cryptographic systems, the exploration of quantum-resistant cryptographic alternatives has become imperative. This section delves into the emerging field of post-quantum cryptography, which seeks to develop encryption algorithms that remain secure in the era of quantum computing. Various cryptographic approaches, including lattice-based cryptography, hash-based cryptography, and code-based cryptography, are being researched as potential candidates for replacing vulnerable classical algorithms. Understanding and implementing quantum-resistant alternatives is crucial for ensuring the long-term security of sensitive information in the face of advancing quantum technologies. As the cryptographic community strives to adapt to the quantum threat landscape, exploring and implementing robust, quantum-resistant cryptographic solutions becomes a paramount consideration for securing the digital future.

4 Quantum Key Distribution (QKD)

Quantum Key Distribution (QKD) represents a revolutionary approach to secure communication by leveraging the principles of quantum mechanics to establish cryptographic keys between two parties in a theoretically secure manner. The fundamental challenge in secure communication is the distribution of encryption keys without the risk of interception, a task that becomes increasingly challenging in the age of quantum computing, which threatens classical cryptographic methods.

In QKD, the security is grounded in the principles of quantum mechanics, particularly the properties of quantum superposition and the no-cloning theorem. The process begins with the transmission of quantum particles, typically photons, from a sender (Alice) to a receiver (Bob). The quantum states of these particles encode the bits of the cryptographic key. Importantly, any attempt by an eavesdropper (Eve) to intercept or measure these quantum states inherently disturbs them due to the principles of quantum mechanics. This disturbance is detectable, allowing Alice and Bob to identify the presence of an intruder and discard compromised key bits.

Several QKD protocols exist, such as the BBM92 (Bennett-Brassard-Mermin 1992) protocol and the E91 (Ekert 91) protocol, each with its unique approach to quantum key distribution. The overall goal of QKD is to establish a shared secret key between two parties that is immune to interception, providing a foundation for secure communication. While QKD is not immune to all potential attacks and requires careful implementation, it offers a promising avenue for enhancing the security of communication networks in the era of quantum computing threats. The ongoing research and development in QKD are crucial for realizing its full potential and addressing practical challenges for widespread adoption in secure communication systems.

4.1 Principles of Quantum Key Distribution

The principles of Quantum Key Distribution (QKD) are rooted in the unique characteristics of quantum mechanics, providing a secure framework for the exchange of cryptographic keys between two parties. The foundational principle is the use of quantum states, typically encoded in photons, to represent the bits of a cryptographic key. Quantum superposition allows these particles to exist in multiple states simultaneously, forming the basis for encoding information. The no-cloning theorem ensures that any attempt to eavesdrop on the quantum states results in disturbance, alerting the communicating parties to potential interception. This inherent quantum property forms the bedrock of secure key distribution, offering a level of security that classical cryptographic methods cannot achieve.

4.2 Quantum Key Distribution Protocols

Various Quantum Key Distribution protocols have been developed to implement the principles of secure key exchange based on quantum mechanics. The BBM92 (Bennett-Brassard-Mermin 1992) protocol and the E91 (Ekert 91) protocol are prominent examples. The BBM92 protocol involves the transmission of polarized photons and the measurement of their states, while the E91 protocol utilizes entangled particles to establish a shared key. These protocols navigate the challenges of quantum communication, ensuring that the exchanged key remains secure against potential eavesdropping attempts. Understanding the nuances of these protocols is essential for implementing QKD successfully and harnessing the quantum properties that guarantee secure communication.

4.3 Quantum-Safe Communication in a Post-Quantum Era

As quantum computing advances, posing a threat to classical cryptographic methods, the concept of Quantum-Safe Communication emerges as a critical consideration for the future. This section explores the ongoing efforts to develop communication systems that remain secure in a post-quantum era. While QKD provides a quantum-safe solution, researchers are also exploring classical cryptographic algorithms that are resistant to quantum attacks, forming the basis for a hybrid approach. The exploration of quantum-resistant cryptographic alternatives, such as lattice-based cryptography and hash-based cryptography, becomes integral to ensuring the long-term security of communication networks. As the field evolves, the quest for quantum-safe communication protocols remains dynamic, addressing the challenges posed by quantum advancements and laying the groundwork for a secure digital future.

The broader implications of quantum computing on cybersecurity extend beyond the immediate challenges it poses to classical cryptographic methods. As quantum technologies advance, they usher in a transformative era with profound impacts on various aspects of cybersecurity:

1. **Cryptographic Paradigm Shift**: Quantum computing necessitates a paradigm shift in cryptographic approaches. Traditional cryptographic systems, reliant on the computational complexity of problems like integer factorization and discrete logarithms, become vulnerable. The evolution toward quantum-resistant cryptographic algorithms is imperative to maintain data confidentiality and integrity in the face of quantum threats.
2. **Quantum-Safe Communication**: Quantum Key Distribution (QKD) emerges as a quantum-safe solution for secure communication. Leveraging quantum principles, QKD offers a unique approach to key exchange, inherently detecting any eavesdropping attempts. The implementation of QKD protocols introduces a new dimension to secure communication, addressing the challenges posed by quantum computing advancements.
3. **Threats to Digital Signatures**: Digital signatures, widely used for authentication and data integrity, face threats from quantum algorithms. Shor's algorithm, for instance, can compromise the security of widely used digital signature schemes. As a result, the integrity and authenticity of digital transactions may be at risk, necessitating the exploration of quantum-resistant signature schemes.
4. **Blockchain Security Considerations**: Blockchain technology, foundational to cryptocurrencies and decentralized systems, relies on cryptographic primitives. The advent of quantum computing introduces vulnerabilities, challenging the long-term security of blockchain networks. Research into quantum-resistant cryptographic solutions for blockchain becomes essential to ensure the resilience of decentralized ecosystems.
5. **Impact on Security Protocols**: Security protocols that underpin internet communication, such as SSL/TLS, may face vulnerabilities in the quantum era. Quantum computers have the potential to break widely used public-key cryptographic algorithms employed in these protocols. Transitioning to quantum-safe versions of security protocols becomes crucial to maintaining the confidentiality and integrity of online communications.
6. **Data Confidentiality Challenges**: Quantum computers pose a threat to the confidentiality of data encrypted using classical methods. The ability of quantum algorithms to efficiently solve certain mathematical problems undermines the security assumptions of classical encryption. Organizations and individuals need to anticipate and adapt to potential breaches of data confidentiality in the future quantum landscape.
7. **Post-Quantum Preparedness**: The cybersecurity landscape demands proactive measures to prepare for the post-quantum era. Organizations must strategize and implement quantum-resistant cryptographic solutions, ensuring that their infrastructure remains secure as quantum technologies mature. This preparedness involves a combination of upgrading cryptographic protocols, deploying

quantum-safe encryption, and raising awareness within the cybersecurity community.

The broader implications of quantum computing on cybersecurity necessitate a comprehensive and anticipatory approach. As quantum technologies progress, addressing the vulnerabilities introduced by quantum algorithms becomes pivotal, ensuring the continued integrity, confidentiality, and resilience of digital systems in an evolving threat landscape.

4.4 Quantum-Enhanced Methods for Threat Detection

Quantum computing not only poses challenges but also presents opportunities for enhancing cybersecurity through quantum-enhanced methods for threat detection. Quantum algorithms, with their parallel processing capabilities, can significantly improve the efficiency of threat detection mechanisms. Tasks such as pattern recognition, anomaly detection, and complex data analysis can benefit from quantum parallelism. Quantum-enhanced machine learning algorithms, including quantum clustering and quantum support vector machines, offer the potential for more accurate and rapid identification of cybersecurity threats. This section explores how quantum computing can augment threat detection capabilities, providing a quantum advantage in staying ahead of evolving cyber threats.

4.5 Quantum-Resistant Algorithms and Innovations

In response to the vulnerabilities introduced by quantum computing, researchers are actively developing quantum-resistant algorithms and innovative cryptographic solutions. This section delves into the exploration of post-quantum cryptography, which encompasses cryptographic algorithms designed to withstand attacks from quantum computers. Lattice-based cryptography, hash-based cryptography, code-based cryptography, and other approaches are at the forefront of these innovations. Understanding the principles and implementations of these quantum-resistant algorithms is crucial for fortifying cybersecurity in the face of advancing quantum technologies. The exploration of innovative cryptographic solutions ensures a proactive stance in mitigating the risks introduced by quantum advancements.

4.6 Ethical and Societal Considerations

The integration of quantum computing into cybersecurity raises ethical and societal considerations that extend beyond technical aspects. Quantum technologies, with their potential to disrupt existing cryptographic systems, prompt reflections on the ethical implications of quantum-resistant research and deployment. This section explores questions related to responsible disclosure of quantum vulnerabilities, ethical considerations in developing and using quantum-enhanced cyber tools, and the societal impact of quantum-resistant cryptography. Balancing the pursuit of cybersecurity with ethical considerations, privacy concerns, and societal implications is essential for fostering a secure and equitable digital future. As quantum technologies progress, a thoughtful examination of the ethical and societal dimensions ensures that the benefits of quantum-enhanced cybersecurity are realized responsibly and inclusively.

4.7 Bridging Quantum Physics and Cybersecurity

Interdisciplinary collaboration between quantum physicists and cybersecurity experts is essential for navigating the complex landscape created by the intersection of quantum computing and cybersecurity. Quantum physicists bring expertise in the underlying principles of quantum mechanics, qubits, and quantum gates, providing the foundational knowledge necessary for developing quantum technologies. Collaboratively, they work with cybersecurity experts to understand the potential vulnerabilities introduced by quantum computing and devise quantum-resistant cryptographic solutions. Bridging these disciplines facilitates a comprehensive approach to addressing the multifaceted challenges and opportunities arising from the fusion of quantum physics and cybersecurity.

4.8 Regulatory Frameworks in the Quantum Era

The advent of quantum computing in the realm of cybersecurity necessitates the development of robust regulatory frameworks. This interdisciplinary collaboration involves policymakers, legal experts, and cybersecurity professionals working together to establish regulations that govern the responsible development, deployment, and use of quantum technologies. Addressing issues such as quantum-safe encryption standards, disclosure requirements for quantum vulnerabilities, and international cooperation in quantum cybersecurity becomes paramount. The interdisciplinary dialog ensures that regulatory frameworks keep pace with technological advancements, striking a balance between fostering innovation and safeguarding security.

4.9 Societal Impact and Inclusivity

Quantum computing's impact on cybersecurity extends beyond technical considerations, prompting interdisciplinary collaboration to address societal and inclusivity aspects. Ethical considerations surrounding quantum-resistant research, the equitable distribution of quantum-enhanced cybersecurity benefits, and the potential societal disruptions require a collaborative effort involving ethicists, sociologists, and cybersecurity experts. Ensuring inclusivity involves considering the accessibility of quantum-resistant technologies, the ethical use of quantum capabilities, and minimizing any societal inequalities that may arise. An interdisciplinary approach fosters a holistic understanding of the societal implications, promoting responsible and inclusive development in the quantum era.

Interdisciplinary perspectives and collaborations serve as the linchpin in navigating the complexities of quantum-enhanced cybersecurity. By bringing together experts from diverse fields, ranging from quantum physics to law, ethics, and sociology, a comprehensive understanding emerges, enabling the development of effective strategies, regulations, and solutions that balance innovation with ethical, societal, and inclusive considerations.

Case Study

Case Study: Quantum Computing in the Face of Cyber Attacks

Objective: To explore the potential impact of quantum computing on cybersecurity through the lens of a simulated cyber attack scenario, highlighting the vulnerabilities and resilience factors associated with traditional and quantum-resistant cryptographic protocols.

Background: With the imminent rise of quantum computing, there is growing concern about its potential to render current cryptographic protocols obsolete. This case study aims to simulate a cyber attack scenario, leveraging the capabilities of quantum computing, and assess the efficacy of quantum-resistant cryptographic solutions in mitigating these threats.

Scenario: A fictitious financial institution, QuantumBank, is chosen as the subject of the cyber attack simulation. The objective is to assess the vulnerability of the institution's data, particularly sensitive customer information and financial transactions, to quantum attacks.

Methodology:

The methodology encompasses a comprehensive exploration of the potential impact of quantum computing on cybersecurity, focusing on QuantumBank's data security. The following steps outline the process:

1. **Selection of Cryptographic Protocols**: For the representation of traditional encryption methods, widely used cryptographic protocols like RSA and ECC are implemented. Concurrently, quantum-resistant cryptographic algorithms, such

as lattice-based cryptography, are integrated as potential solutions to enhance resilience against quantum threats.

2. **Quantum Attack Simulation**: Utilizing a quantum computing simulator, the capabilities of a quantum computer are emulated. Quantum algorithms, including Shor's algorithm, are implemented to simulate attacks specifically tailored to the selected cryptographic protocols. This step aims to assess the vulnerability of traditional encryption methods to quantum attacks.
3. **Data Exploration and Vulnerability Assessment**: The vulnerability of QuantumBank's encrypted data is analyzed under simulated quantum attacks. The exploration extends to potential compromises of critical information, including customer data and financial transactions. This step provides insights into the susceptibility of the existing cryptographic protocols to quantum threats.
4. **Implementation of Quantum-Resistant Cryptography**: Quantum-resistant cryptographic solutions are integrated into QuantumBank's cybersecurity infrastructure. This involves a careful implementation process to ensure seamless compatibility with existing security measures while maintaining compliance with industry standards. The objective is to enhance the institution's resilience against potential quantum threats.
5. **Quantitative and Qualitative Analysis**:

   Quantitative Assessment: The time and computational resources required for a successful quantum attack on traditional cryptographic protocols are quantitatively evaluated. This analysis provides a concrete understanding of the efficiency and speed of quantum attacks on the selected protocols.

   Qualitative Analysis: The effectiveness of quantum-resistant solutions in resisting simulated quantum attacks is qualitatively assessed. This involves an examination of the robustness and security measures inherent in the quantum-resistant cryptographic algorithms.

Figure 1 illustrates the distribution of various attack types encountered in the dataset. Each bar on the horizontal axis represents a specific attack type, while the corresponding vertical bars indicate the frequency or count of each attack type. The visualization provides a clear overview of the prevalence of different attack types, aiding in understanding the dataset's composition. The most common attack types are visually apparent, as their bars extend higher on the y-axis. The x-axis labels, representing each attack type, are rotated for better readability. This distribution analysis is instrumental in identifying the prominent threats within the dataset, guiding cybersecurity professionals in prioritizing defenses and implementing targeted security measures.

The pie chart in Fig. 2 illustrates the distribution of network traffic protocols within the dataset. Each segment of the pie represents a specific protocol, with UDP, ICMP, and TCP being the primary categories. The proportional sizes of the segments indicate the relative prevalence of each protocol type in the dataset.

The pie chart in Fig. 3 depicts the distribution of actions taken in response to observed events within the dataset. Each segment of the pie represents a specific

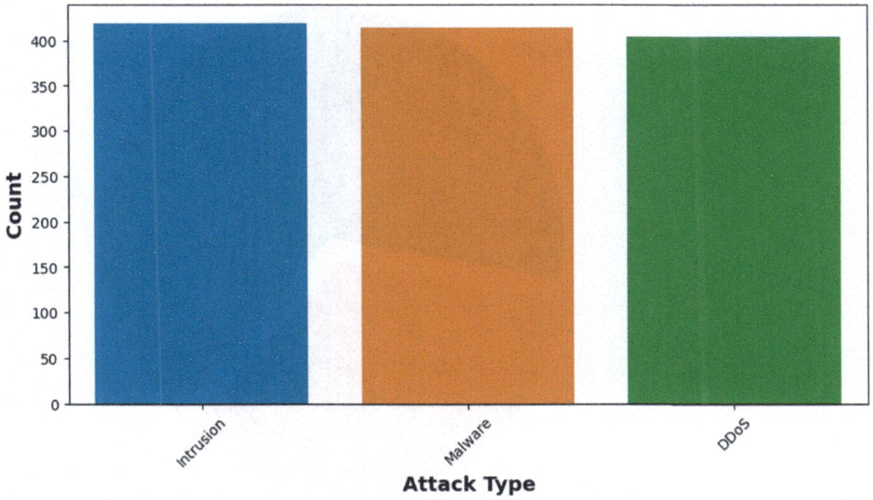

Fig. 1 Distribution of attack types

Fig. 2 Distribution of network traffic protocols

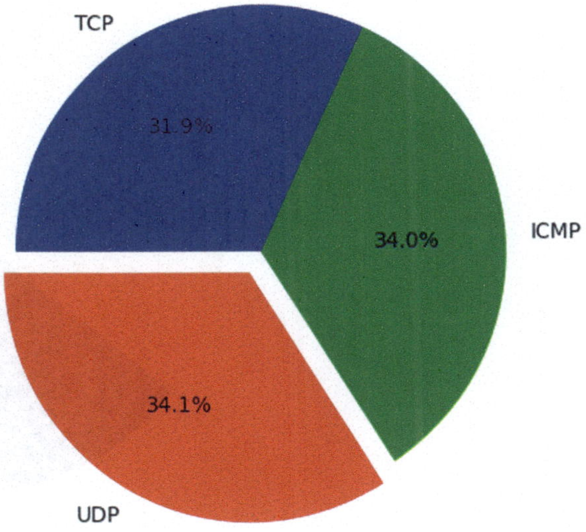

action category, with "Blocked," "Ignored," and "Logged" being the primary classifications. The proportional sizes of the segments indicate the relative frequency of each action type.

The pie chart in Fig. 4 visualizes the distribution of severity levels within the dataset, categorizing events based on their perceived impact or threat severity. The chart is segmented into three categories: "High," "Medium," and "Low," each represented by a distinct color.

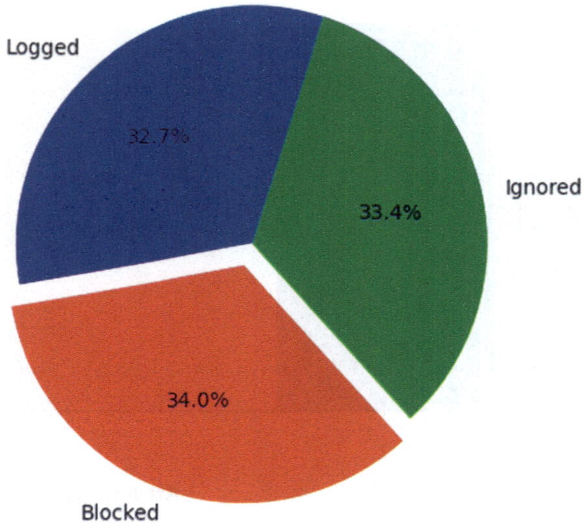

Fig. 3 Distribution of actions taken

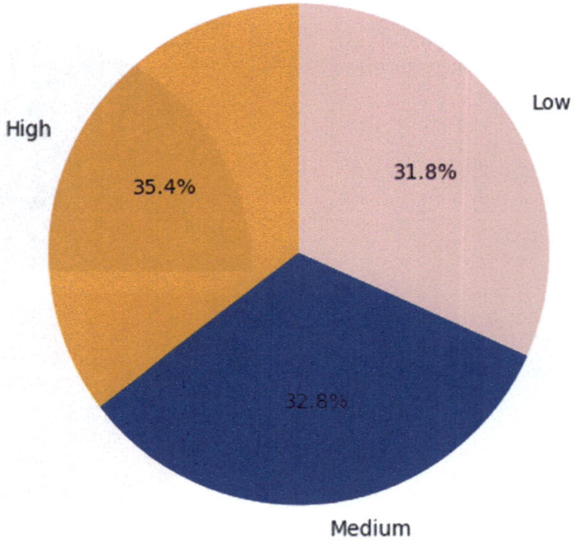

Fig. 4 Distribution of severity levels

High Severity (Orange Segment): The orange segment signifies events categorized with high-severity levels, indicating potentially critical threats that require immediate attention and intervention.

Medium Severity (Blue Segment): The blue segment represents events with a medium severity level. While not as critical as high-severity events, they still demand a significant level of attention and appropriate response measures.

Low Severity (Pink Segment): The pink segment denotes events with low severity levels. These events may have a minimal impact or pose a lower level of threat, allowing for a more measured and less urgent response.

The proportional sizes of each segment provide a quick overview of the severity distribution, allowing cybersecurity professionals to prioritize their efforts and resources effectively. This visualization aids in understanding the composition of severity levels within the dataset, enabling informed decision-making for incident response and risk mitigation strategies.

6. **Results and Recommendations**:

   The results of the simulated quantum attack are presented, emphasizing vulnerabilities exposed in traditional cryptographic protocols. Insights gained from the exploration of potential compromises in customer information and financial transactions are highlighted. Recommendations are provided based on the efficacy of quantum-resistant cryptographic solutions in fortifying QuantumBank's cybersecurity resilience against quantum threats. The findings contribute to strategic decision-making and inform the institution's approach to future-proofing its data security in the quantum computing era.

The case study offers a simulated but realistic exploration of the potential impact of quantum computing on cybersecurity, using QuantumBank as a representative target. The findings underscore the urgency for organizations to adopt quantum-resistant cryptographic solutions to safeguard sensitive data and transactions in the face of evolving cyber threats posed by quantum computing. The insights gained from this exploration contribute to the ongoing dialog surrounding the need for quantum-safe cybersecurity measures in anticipation of the quantum era.

5 Conclusion

In conclusion, the exploration of the convergence between quantum computing and cybersecurity unveils a dynamic landscape of challenges and opportunities. The vulnerabilities introduced by quantum computing, exemplified by Shor's algorithm, necessitate a fundamental rethinking of classical cryptographic paradigms. Quantum Key Distribution (QKD) emerges as a quantum-safe solution, showcasing the potential for secure communication in the quantum era. The development of quantum-resistant algorithms and innovative cryptographic solutions is imperative for fortifying digital security. Quantum-enhanced methods for threat detection promise to revolutionize cybersecurity, offering a quantum advantage in staying ahead of evolving threats. Ethical, societal, and regulatory considerations underscore the need for interdisciplinary collaboration to ensure responsible and inclusive development in the quantum era.

6 Future Scope

The future scope of this chapter lies in the ongoing evolution of quantum-enhanced cybersecurity. Advancements in quantum algorithms, cryptographic protocols, and quantum-resistant technologies will continue to shape the landscape. Further research in quantum-resistant algorithms tailored for specific applications, such as secure communication and blockchain, holds promise. The integration of quantum computing with edge computing presents an avenue for more efficient and scalable solutions. Exploring industry-specific applications, interdisciplinary collaborations, and societal impact assessments will be integral to the maturation of Quantum-IoT integration. As quantum technologies progress, interdisciplinary collaboration, regulatory frameworks, and ethical considerations will play pivotal roles in realizing the full potential of quantum-enhanced cybersecurity, propelling us into a future where security and innovation coexist seamlessly in the quantum era.

References

1. U. Ghosh, D. Das, P. Chatterjee, A comprehensive tutorial on cybersecurity in quantum computing paradigm. Authorea Preprints (2023)
2. S. Pal, Quantum computing and algorithms: an exploration of the quantum frontier in data analytics and computational intelligence. Juni Khyat **13**(10), 145–154 (2023)
3. K.K. Rangan, J. Abou Halloun, H. Oyama, S. Cherney, I.A. Assoumani, N. Jairazbhoy, S.K. Ng, Quantum computing and resilient design perspectives for cybersecurity of feedback systems. IFAC-PapersOnLine **55**(7), 703–708 (2022)
4. P. Radanliev, Cyber diplomacy: defining the opportunities for cybersecurity and risks from artificial intelligence, IoT, blockchains, and quantum computing. J. Cyber Secur. Technol. 1–51 (2024)
5. R.A. Brandmeier, J.A Heye, C. Woywod, Future development of quantum computing and its relevance to NATO. Connect.: Q. J. **20**, 89–110 (2022)
6. M.J.H. Faruk, S. Tahora, M. Tasnim, H. Shahriar, N. Sakib, A review of quantum cybersecurity: threats, risks and opportunities, in *2022 1st International Conference on AI in Cybersecurity (ICAIC)* (IEEE, 2022). pp. 1–8
7. B. Dash, S. Ullah, *Quantum-safe: Cybersecurity in the age of Quantum-Powered AI* (2024)
8. A. Ali, A pragmatic analysis of pre-and post-quantum cyber security scenarios, in *2021 International Bhurban Conference on Applied Sciences and Technologies (IBCAST)* (IEEE). pp. 686–692 (2021)
9. J. Deodoro, M. Gorbanyov, M. Malaika, T.S. Sedik, S.J. Peiris, Quantum computing and the financial system: spooky action at a distance?. IMF Work. Pap. **2021**(071) (2021)
10. A.M. Osula, Building cyber resilience: the defensive shield for the EU, in *Cybersecurity Policy in the EU and South Korea from Consultation to Action: Theoretical and Comparative Perspectives*. (Springer International Publishing, Cham, 2022), pp.179–196
11. W.G. Johnson, Governance tools for the second quantum revolution. Jurimetr. **59**(4), 487–522 (2019)
12. O.O. Olatunji, P.A. Adedeji, N. Madushele, Quantum computing in renewable energy exploration: status, opportunities, and challenges, in *Design, Analysis, and Applications of Renewable Energy Systems* (2021), pp. 549–572
13. U. Tatar, A. V. Gheorghe, O. F. Keskin (eds.), *Space Infrastructures: From Risk to Resilience Governance*, vol. 57 (IOS Press, 2020)

14. M.L. How, S.M. Cheah, Business renaissance: opportunities and challenges at the dawn of the quantum computing era. Businesses **3**(4), 585–605 (2023)
15. S.F. Ahmad, M.Y. Ferjani, K. Kasliwal, Enhancing security in the industrial IoT sector using quantum computing, in *2021 28th IEEE International Conference on Electronics, Circuits, and Systems (ICECS)* (IEEE, 2021), pp. 1–5
16. H. Khodaiemehr, K. Bagheri, C. Feng, Navigating the quantum computing threat landscape for blockchains: a comprehensive survey. *Authorea Preprints* (2023)
17. A. Khubani, A. Sharma, A. Choudhary, O.S. Bhatnagar, K. Chidambarathanu, Quantum computing in military applications and operations. Quantum Comput. Cybersecur. 453–470 (2023)
18. D. Ott, C. Peikert, Identifying research challenges in post quantum cryptography migration and cryptographic agility (2019). arXiv:1909.07353
19. E.B. Kania, J.K. Costello, Quantum technologies, US-China strategic competition, and future dynamics of cyber stability, in *2017 International Conference on Cyber Conflict (CyCon US)* (IEEE, 2017), pp. 89–96
20. T.L. Scholten, C.J. Williams, D. Moody, M. Mosca, W. Hurley, W.J. Zeng, J.M. Gambetta, Assessing the benefits and risks of quantum computers (2024). arXiv:2401.16317
21. A. García Rodríguez, *The Importance of Quantum Computing for National Security: An Examination of the Strategic Implications of Quantum Information* (2020)
22. J.H. Markna, T.P. Palatia, S. Gohel, B. Kataria, Unveiling advanced computational applications in quantum computing: a comprehensive review. Int. J. Adv. Nano Comput. Anal. **2**(2), 81–92 (2023)
23. S. Kautish, A. Reyana, A. Vidyarthi, SDMTA: attack detection and mitigation mechanism for DDoS vulnerabilities in hybrid cloud environment. IEEE Trans. Industr. Inf. **18**(9), 6455–6463 (2022)
24. R. Kaur, S. Kautish, Multimodal sentiment analysis: a survey and comparison. Int. J. Serv. Sci., Manag., Eng., Technol. (IJSSMET). **10**(2), 38–58 (2019)
25. A. Naim, S.K. Kautish (eds.), *Building a Brand Image Through Electronic Customer Relationship Management*. (IGI Global, 2022)
26. S. Kautish, Online banking: a paradigm shift. E-Bus., ICFAI Publ., Hyderabad **9**(10), 54–59 (2008)

Computational Renaissance and Quantum Horizons: New Directions for Cybersecurity

Pradeep Bedi, Sanjoy Das, S. B. Goyal, Anand Singh Rajawat, and Vidyapati Kumar

1 Quantum Computing and Its Impact on Cybersecurity

The remarkable performance increases that quantum computing assurances could significantly influence security, welfare, and global economies. Both public and private sector fundings have been drawn to quantum computing because of its potential to help generate breakthroughs that would revolutionize industries including financial services, aircraft, and medicines. The behavior of matter as well as energy at the atomic and subatomic scales is the subject of the theory of quantum mechanics, which is physical sciences at the nanoscale. Information technology is enhanced by the application of quantum computing concepts from quantum physics. Quantum computing employs quantum bits, or "qubits," as opposed to classical computing,

P. Bedi
Department of Computer Science, Regional Campus Manipur, Indira Gandhi National Tribal University, Amarkantak, M.P., India

Department of Computer Science and Engineering, Graphic Era (Deemed to be) University, Dehradun, Uttarakhand, India

S. Das
Department of Computer Science, Indira Gandhi National Tribal University, Kangpokpi District, Manipur, India

S. B. Goyal (✉)
Chitkara University Institute of Engineering and Technology, Chitkara University, Chandigarh-Patiala National Highway (NH - 64), Rajpura, Punjab 140401, India
e-mail: drsbgoyal@gmail.com

A. S. Rajawat
School of Computer Sciences and Engineering, Sandeep University, Nashik, India

V. Kumar
Department of Mechanical Engineering, Indian Institute of Technology Kharagpur, Kharagpur, West Bengal, India
e-mail: vidyapatikumar.me@kgpian.iitkgp.ac.in

S. B. Goyal et al. (eds.), *Quantum Computing, Cyber Security and Cryptography*,
https://doi.org/10.1007/978-981-96-4948-8_5

which relies on binary bits. Similar to a bit, a qubit is capable of holding two values at the same time: 0 and 1. It can also exist in a superposition state. Another crucial property is getting entangled, or the ability to link one or more qubits such that modifications to one impact the other even when those two quantum bits are light years separated. The procedure of cybersecurity in quantum computing is provided in Fig. 1. Quantum computing, with its dual promise and threat, stands poised to redefine our cryptographic foundations, necessitating a critical overview of encryption methods from past to present. Historically, cryptographic techniques have evolved significantly—from simple substitution ciphers used in ancient civilizations through the complex algorithms that secure modern digital communications. This evolution was marked by pivotal developments such as the Enigma machine in World War II and the subsequent advent of public-key cryptography in the 1970s. However, with quantum computers capable of potentially breaking these advanced systems, there is an urgent need to develop post-quantum cryptographic methods that can withstand new quantum capabilities. Recent incidents, such as the breach of a supposedly secure post-quantum algorithm in 2022, underscore the immediate relevance and peril of these quantum advancements. This comparative analysis of classical versus quantum cryptographic methods not only charts a historical progression but also sets the stage for understanding the strategic shifts required to safeguard future communications in a quantum-dominated landscape.

In the next 5 to 10 years, quantum computing is predicted to become a reality and has an impact on network security, SSL/TLS, and e-commerce. Cybersecurity experts need to be aware of the current state of quantum proof algorithms and how they can affect SSL/TLS, authentication systems, and e-commerce [1]. Microsoft

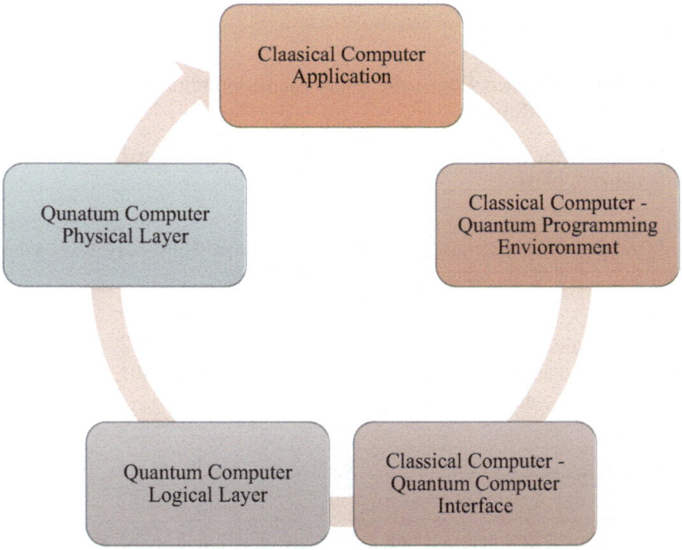

Fig. 1 Procedure of cybersecurity in quantum computing

awarded Belgium a USD 50,000 bounty in August 2022 for successfully cracking one of the four post-quantum cryptography (PQC) algorithms. This jeopardizes the NIST's five-year plan to develop a quantum-safe encryption standard. The quantum threat still exists, and the fate of the other two PQC algorithms is uncertain. Zero-vulnerability computing (ZVC) is an encryption-neutral method that, by preventing access from outside parties and developing reliable, energy-efficient software on a chip, may make computers immune to quantum attacks [2]. With an emphasis on third-generation partnership project (3GPP) and fifth-generation (5G) networks, the [3] investigates post-quantum IoT security strategies. It highlights flaws in the design and applications of the Internet of Things today, pointing to the necessity for more sophisticated methods.

In addition to demonstrating quantum resistance and presenting promising lattice-driven cryptography solutions, the study raises the possibility that security in these settings may be compromised by quantum computing. After 140 years after Frank Miller created an unbreakable encryption technique in 1882, cryptographers still haven't improved it. Robert Slater wrote the Telegraphic Code in 1870 with the intention of guaranteeing confidentiality in telegraph transmission, tackling the serious problem of fraud that was prevalent at the time [4]. Encryption is used by the DOD, Intelligence Community, financial institutions, and commercial enterprises to safeguard confidential data. Although quantum computing algorithms are now more effective at factoring huge integers into primes, Public-Key Encryption (PKE) is still in use. Three scenarios where pre-quantum encryption techniques can become weak are revealed by the study in [5], which examines cryptography research, techniques, and potential vulnerabilities brought on by quantum computing. The study is based on publicly available, declassified data. Unstructured search, quantum simulation, and optimization are just a few of the many uses for quantum computing—a branch of study that makes use of quantum mechanics—in 5G and beyond networks. It can enhance machine learning systems' accuracy and speed. However, security systems—particularly asymmetric key cryptography—are vulnerable to the risks posed by quantum computing. Symmetric key cryptography has given way to PKI-based trust models as a result. Mathematical problem-based alternative cryptosystems are also explored. Developments in quantum key distribution may lead to quantum-resistant cryptosystems [6]. 5G has transformed communication and data exchange but has security flaws. Quantum physics must take the role of traditional cryptography in order to protect 6G networks. Reviewing current quantum solutions for 5G security concerns, the work in [7] seeks to address these issues and secure 6G in a future quantum environment. The Internet of Things (IoT) of the future will facilitate deterministic communications and manage vital infrastructure, such as smart cities and the smart power grid. In [8], a "Cybersecurity via Determinism" paradigm is introduced, introducing a secure deterministic packet switch (D-switch) forwarding sub-layer in layer-3. The sub-layer boosts cybersecurity through access control, rate control, and isolation control, and it allows deterministic software-defined wide-area networks. Reducing buffer sizes; decreasing IoT latency; and eliminating interference, congestion, and distributed denial-of-service assaults are only a few of the benefits of the framework. Moreover, quantum-safe encryption lowers expenses.

The research from 9 investigates the idea of technology hype and its association with national security. The article includes a case study on quantum technology and compares the hype cycle to threat inflation and securitization theory. It also argues that while a cyclical approach may express skepticism regarding quantum technologies, a middle-range theory offers insightful guidance on security procedures, including the significance of understanding and managing technology expectation. In the last 8 years, quantum computers have made significant advances thanks to D-Wave technology. A growing number of companies, including IBM, Microsoft, Google, Intel, and NASA, are investing in quantum computing in order to increase its accessibility to no physicists. In this paper, we will present some quantum ideas and how they can be applied.

Quantum mechanics introduces principles that are pivotal to quantum computing's potential to transform cybersecurity. The impacts on fidelity, efficiency, and quantitative measures of three fundamental quantum principles—superposition, entanglement, and quantum tunneling—as well as some noteworthy research cases are mentioned in Table 1. **Superposition** allows qubits to hold multiple states simultaneously, enabling quantum computers to process extensive data arrays in parallel. By suppressing noise and defects, coherent superposition protocols improve the fidelity of quantum devices. Models based on gates, measurements, and interferometry can all be used with these techniques, which can be probabilistic or deterministic. Challenges in quantum applications can be overcome by these protocols, which provide desired output states with unit fidelity by utilizing ample resources and broad assumptions [9]. Employing 56 2-qubit permutation matrices, Kuang and Bettenburg's Quantum Permutation Pad (QPP) was implemented on IBM quantum systems for extending encryption for basis states. In order to minimize vulnerabilities, superposition states were developed and encrypted in [10], bridging the security gap among classical along with quantum systems. **Entanglement** links the state of qubits over any distance, providing a base for quantum cryptography that could potentially create unbreakable encryptions. In addition to addressing issues with wide-area networks and outlining potential research avenues for future uses, [11] analyzes entanglement-assisted quantum networks and highlights the noteworthy advancements in quantum information technology. A possible quantum computing platform for low-error-rate quantum entanglement is provided by neutral-atom arrays. Two-qubit entangling gates are demonstrated in [12] that exceeds the surface-code threshold for error correction, demonstrating 99.5% fidelity on up to 60 atoms in simultaneously. The technique reduces scattering and enhances Rydberg excitation and atom cooling by utilizing fast, single-pulse gates, atomic dark states, and Rydberg excitation. **Quantum tunneling** offers particles the ability to bypass physical barriers, presenting both novel opportunities and risks in data security. In single-molecule magnets (SMMs) intended for storage devices, [13] presents a technique to convert detrimental intermolecular magnetic dipole interaction into advantageous one. A 10^4-fold slower zero-field relaxation time at 2 K was achieved by synthesizing two pseudo-D_5h SIMs with side-by-side configurations of magnetic dipoles by combining ion pairing and hydrogen bonding. With a τQTM (time of quantum tunneling of magnetization) of 980 s, this method is among the highest performing

SMMs. Spin-dependent transport including quantum tunneling in bilayer graphene having Rashba-Zeeman leads under electric biasing is investigated in [14]. It illustrates how conductance may be optimized by varying bias energy, magnetization, and material choice. The Fano factor suggests tunable transmission for different spins, indicating less noise with lower magnetization (see Table 2).

The fidelity of quantum computing is improved by superposition's capacity to hold many states concurrently, particularly through the quantum permutation

Table 1 Security and technology implications of quantum computing

Aspect	Current status	Impact in 5–10 years	Challenges	Potential solutions
Network security and SSL/TLS	Vulnerable to quantum decryption, reliance on traditional SSL/TLS	Need for widespread adoption of quantum-safe protocols	Development and deployment of new quantum-safe standards	Investing in research and cross-sector collaborations for developing standards
E-commerce security	Dependent on existing encryption standards vulnerable to quantum attacks	Critical overhaul of security measures to protect consumer data	Ensuring global e-commerce platforms can transition without disrupting service	Global regulatory support and incentives for adopting quantum-safe practices
Quantum proof algorithms	Ongoing development to create quantum-resistant algorithms	Expected maturity and standardization of quantum-resistant algorithms	Balancing security with computational efficiency and speed	Encouraging open-source projects and academic-industry partnerships
Post-quantum cryptography (PQC)	Progress in PQC but vulnerability shown by recent breaches	Potential establishment as a new standard if vulnerabilities can be managed	Establishing universally trusted and adopted standards	Government and international bodies to facilitate PQC standardization
Zero-vulnerability computing (ZVC)	Emerging concept aimed at creating fully quantum-resistant systems	Possible mainstream integration in high-security environments	Technological and economic barriers to full implementation	Development of low-cost, scalable ZVC technologies
5G and IoT Security	Initial adoption of quantum-resistant techniques but many devices remain exposed	Significant updates required to protect against quantum threats	Massive scale of IoT device update and 5G network overhaul	Comprehensive strategy for secure network architecture in new 5G rollouts
Public-key encryption (PKE)	Widely used but faces potential obsolescence with quantum advancements	Likely replacement or major upgrade to maintain security	Transitioning existing infrastructure to quantum-resistant technologies	Phased approach to upgrade and replace vulnerable systems

Table 2 Quantum principles: applications and metrics in cybersecurity

Quantum principle	Key features	Applications	Effectiveness (fidelity/efficiency)	Quantitative metrics	Examples and research
Superposition	Qubits hold multiple states simultaneously, enabling parallel data processing	Enhances quantum computing capabilities by improving device fidelity through coherent superposition protocols	High fidelity in quantum devices	Unit fidelity with ample resources	Implemented in the Quantum Permutation Pad (QPP) on IBM systems, encrypting superposition states to bridge security gaps between classical and quantum systems [9, 10]
Entanglement	Links the state of qubits over any distance	Basis for quantum cryptography, potentially creating unbreakable encryptions	High reliability in communication	99.5% fidelity on up to 60 atoms	Analyzes entanglement-assisted quantum networks, highlights advancements in quantum information technology, and demonstrates high-fidelity two-qubit entangling gates [11, 12]
Quantum tunneling	Allows particles to bypass physical barriers, impacting data security	Applied in single-molecule magnets (SMMs) for storage devices and bilayer graphene systems for optimized conductance	Lower error rates and enhanced speed	τQTM of 980 s, 104-fold slower zero-field relaxation time at 2 K	Techniques in SMMs to convert magnetic interactions advantageously, and in graphene to adjust conductance and transmission, with Fano factor adjustments for noise reduction [13, 14]

pad. With notable developments in quantum networks, the use of entanglement in quantum cryptography demonstrates remarkable reliability. Applications such as storage devices and conductance optimization in graphene demonstrate the significance of quantum tunneling for data security, exhibiting significant performance enhancement and error reduction.

Quantum Error Correction

The practical application of trustworthy quantum computing requires quantum error correction (QEC), which addresses the intrinsic instability of quantum systems brought on by external noise and human error. Error-correcting code development is essential to preserving the fidelity and coherence of quantum states, allowing for secure quantum communications and lengthy computations. Information contained in quantum bits, or qubits, must be protected against errors caused by decoherence and other quantum noise through the use of quantum error correction. In the absence of efficient QEC, qubit state mistakes would compound, producing inconsistent results and ruling out the possibility of scalable quantum computing. Error rate reduction is crucial for quantum computing advancements, as physical qubit operations have higher error rates than necessary for efficient error correction. Resource efficiency is also a priority, as most quantum error correction techniques are resource intensive. Achieving fault-tolerant quantum computing involves combining fault-tolerant protocols with quantum electrochemical circuitry (QEC) to ensure proper functioning even if part of the computer fails.

Addressing the inherent instability of quantum states, **quantum error correction** is crucial, involving complex algorithms that detect and correct errors autonomously. This field is rapidly evolving, tackling significant challenges such as qubit fragility and error management, which are critical for developing practical, scalable quantum computing systems. State reconstruction for a qubit interacting with an environment E is possible through three distinct transformations. Any changes to a qubit initially in pure state ψ can be described by the evolution of the combined system-environment state in a specific form.

$$\psi e_r = I\psi e_I + \sum\nolimits_{s=x,y,z} \sigma_s \psi e_{s'}$$

The equation involves the identity operator I and the Pauli operator's σs acting on the Hilbert space of system S, with e_r representing the initial state of the environment. The environmental states e_I and e_s might not be orthogonal or normalized. This formulation stems from the fact that the Pauli operators, alongside the identity, constitute a complete set of operators for the qubit's Hilbert space. The impact of σ_x on the qubit is known as a bit-flip error, while σ_z causes a phase-flip error. Since $\sigma_y = i\sigma_x\sigma_z$, the σ_y operator indicates the concurrent occurrence of both bit-flip and phase-flip errors. State changes due to environmental entanglement, or decoherence, are completely characterized by phase-flip errors. To illustrate, consider a qubit in an arbitrary state $\psi = a|0\rangle + b|1\rangle$ that undergoes an entangling interaction with the environment.

$$\psi e_r = a0e_0 + b1e_1$$

For N qubits, this generalization can be expressed as

$$\psi e_r \rightarrow \sum_i E_i \psi_{ei}$$

where ψe_r is the initial state, and the transformation is represented by the summation over the operators E_i acting on different environment-associated states ψ_{ei}. The error operators E_i are tensor products of N operators that include the identity and σ_z operators. The count of σ_z operators within a specific E_i is known as the weight of that error operator. Often, only a subset of qubits, K < N, undergoes entanglement with the environment, a phenomenon referred to as "partial decoherence." Consequently, it's only necessary to consider the 2^K different error operators with weights up to K. With a coherence gain of G = 2.27 ± 0.07, [15] presents a fully stable and error-corrected logical qubit that has a substantially longer quantum coherence than all flawed quantum components included in QEC. Model-free reinforcement learning and superconducting quantum circuit fabrication advances are combined to attain this performance. In order to do quantum error correction, [16] employs superconducting qubits arranged in a heavy-hexagon grid. To fix any one flaw in the circuitry, the researchers conduct fault-tolerant syndrome measurements and encode a logical qubit with distance three. For each syndrome measurement, the average logical error is roughly 0.040 (~0.088) and 0.037 (~0.087), respectively. Quantum error correction codes, like the surface code, have shown high fault tolerance and are being optimized for realistic noisy conditions and fewer resources. Researchers are exploring hardware and algorithmic approaches to deploy these codes in quantum computing architectures. They are pushing the boundaries of quantum error correction to achieve the robustness needed for the next generation of quantum technologies.

2 Risks Associated with Quantum Computing in Cybersecurity

With quantum computing advancing at rapid speeds and relying on the principles of quantum mechanics to provide unmatched speed performance on complex calculations, the technology is likely to become one of the most critical advances in research and data science and other areas that revolve around high computational needs. However, this speed comes with a massive threat to cybersecurity. Quantum computing poses a severe risk to current cryptographic protection making public-key encryption redundant based on quantum capabilities. The security of a system is compromised through various vulnerabilities, including malicious quantum techniques, weak passwords, insecure data transfer, and insecure ecosystem interfaces. These vulnerabilities can be exploited through brute force attacks, quantum key recovery attacks, and inefficient input-data validation. Additionally, the system's encryption is weak due to insufficient key size and difficulty factorizing large numbers [17]. Quantum-secure or post-quantum cryptography is fraught with uncertainties, as shown in Fig. 2, but according to the same source, it is unclear when quantum computers capable of breaking existing encryption will become available. Recent advancements in quantum technology are driving organizations to urgently update their cybersecurity strategies. The power of quantum computing, especially through Short's algorithm, threatens traditional encryption methods like RSA and ECC, potentially rendering them obsolete within a decade as depicted in Fig. 3. This raises the risk of "harvest-now, decrypt-later" attacks, where adversaries store encrypted data to decrypt later with more powerful quantum computers. The development of quantum-resistant encryption, or post-quantum cryptography, is now critical. Led by initiatives like those from the National Institute of Standards and Technology (NIST), there's a push for standardizing new algorithms to secure sensitive data across industries. As quantum computing continues to evolve, organizations must prioritize transitioning to quantum-safe practices to protect against this emerging threat. The consensus highlights a need for clearer guidelines on how to respond to this emerging threat. Meanwhile, proactive measures are essential, including raising awareness among leaders, strategizing a response that includes crypto-agile and hybrid solutions, and securing executive commitment, to transition smoothly to quantum-resistant cryptography before the threat fully manifests.

The future 6G cellular communication paradigm is surveyed in [18], emphasizing the symmetry between it and traditional security ideas as well as novel features like federated learning, quantum computing, and artificial intelligence (AI). It offers a taxonomy of threat models for 6G communications, classifying threat-countering strategies into Intrusion Detection Systems (IDS), entity properties, and cryptographic techniques. There are eight categories of authentication techniques: multi-factor authentication, deniable, token based, certificate based, handover, mutual, physical layer, and agreement based. There are recommendations for future research directions. Advanced quantum computing presents a serious risk to cryptographic systems in power grids with a high concentration of distributed energy resources (DERs) as the quantum age approaches closer. The use of post-quantum cryptography (PQC) in DER systems is still being investigated in [19], although it is thought to be a viable defence tactic. This study proposes a real-time hardware-in-the-loop co-simulation test and discusses the acceptance of PQC in IEEE 2030.5-PQC. Real-time co-simulation is used to verify the viability of the planned PQC-grade DER network, and universal custom-made PQC client and server software is produced.

The fourth industrial revolution, Industry 4.0, is centered on cyber-physical structures and the technology underpinning them. Interoperability, openness to information, technical support, and decentralized decision-making are some of its design characteristics. However, those concepts provide new avenues for attack, which raises security issues. These difficulties can be anything from minor dangers to broken systems [20]. An overview of cybersecurity innovations is given in Table 3, which includes how AI, machine learning, and quantum-resistant protocols are being used to enhance network security and data protection. It provides information on

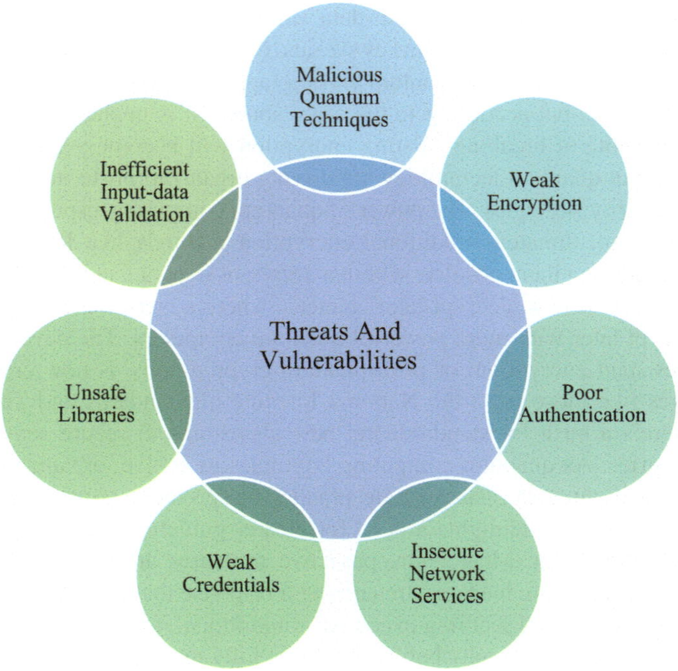

Fig. 2 Quantum computing threats and vulnerabilities

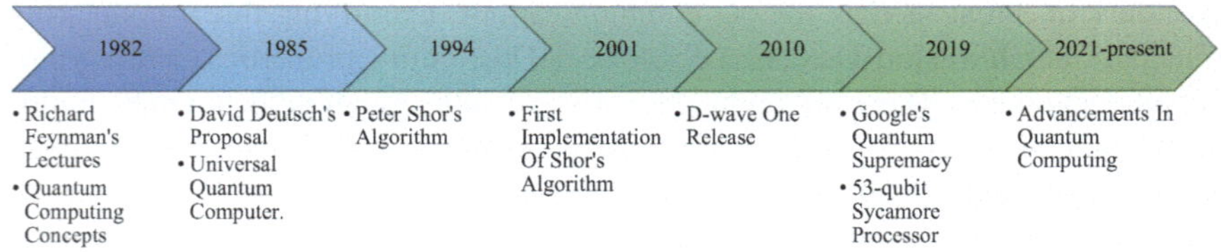

Fig. 3 Timeline of quantum computing milestones

each aspect's results, difficulties, and future directions as well as advice on how to strengthen digital infrastructures against changing threats. With an emphasis on sustainable development, [21] investigates the connection between cybersecurity, machine learning, and artificial intelligence. AI-powered solutions detect threats with 92.5% accuracy, and ML algorithms cut response times by 40%. Artificial intelligence (AI) reduces high-risk unlatched vulnerabilities by 30% and stops phishing assaults by prioritizing key vulnerabilities. This revolutionary method safeguards vital infrastructures and improves digital resilience. AI is a key component of the 6G era, defined by intelligent network orchestration and management. However, maintaining privacy and security may be difficult with this arrangement. AI needs to be solidified in network security in 6G to identify proactive threats, mitigate them, and ensure that the network is self-sustaining [22]. As a result of concerns about cybersecurity, 4G and 5G networks are at risk of compromises of user privacy, necessitating extensive government intervention. An intrusion detection system improves quality of service and provides practical solutions by identifying threats and weaknesses. A support vector machine is used to identify and stop intrusions in order to maintain a quick and efficient network. A complex and ambiguous Era 5.0 requires human ingenuity at its core. A new era is underway, and the goal is to improve the lives of people, address social problems, and enhance well-being. In this new era, academies and businesses face enormous challenges that require a shift to a more humanistic, sustainable society. This study aims to provide critical insight into how to achieve this change and advance the state of the art in Era 5.0 [23]. Quantum computers pose the greatest challenge in securing Internet-connected devices (5G). A proposal in [24] for a protocol called quantum key GRID for authentication and key agreement (QKG-AKA) aims to address this issue. This protocol efficiently deploys in long-term evolution architecture, proving safe against quantum computers. The research in [25, 26] examines data privacy, protection, and scalability related to Federated Learning (FL) security in 6G communication networks. It examines various analysis and evaluation methods and platforms for security-related fields. The study emphasizes the main difficulties encountered in the analysis and recommends open research topics to be investigated on these platforms. The results will contribute to strengthening FL's security in 6G communications. A semantic-aware fuzzer called CLFuzz is intended to find vulnerabilities in the way cryptographic algorithms are implemented in [27]. It is more efficient than state-of-the-art tools and produces high-quality input data while extracting semantic information. The Microsoft reward program has recognized CLFuzz for its effective detection of 12 flaws in 8 cryptographic methods, the majority of which are security-critical. A side-channel attack, also called a sidebar attack or implementation attack, is a type of security exploit that modifies how a system's program is executed by taking advantage of unintentional consequences of the system or hardware. Its primary goal is to steal confidential data, such as cryptographic keys, frequently through unintentional hardware emissions. The incomplete NTT algorithm is the foundation for a brand-new, non-profiled power/EM side-channel attack that targets polynomial multiplication [17]. This attack is used against the Crystals-Kyber KEM and the Crystals-Dilithium signature algorithm, showing that it speeds up attack run-time in comparison to other methods. In addition to an effective validation along with correction method that uses the inverse NTT to determine and adjust mispredicted coefficients, the quicker zero-value filtering attack (ZV-FA) was proposed. The speed-up over brute force is 1915×, according to experimental results. A novel compositional verification approach was developed to address the scaling restrictions in establishing the security of efficient cryptographic algorithms, hence enabling their secure implementation against power side-channel attacks [28]. They suggest composition techniques and verification algorithms in addition to introducing two new ideas in language-level security. The method, which is applied as a tool named CONVINCE, demonstrates successful implementations and successfully lowers 3,200 XOR operations and 1,600 randomness of the most advanced AES implementation.

Table 3 Cybersecurity advances: technologies and impacts

Aspect	Description	Technologies	Key outcomes	Challenges	Future directions	References
AI and machine learning in cybersecurity	Explores the integration of AI and machine learning in detecting and mitigating network threats	AI detection systems, ML algorithms	AI detects threats with 92.5% accuracy; ML algorithms reduce response times by 40%	Privacy concerns, AI biases	Enhance AI accuracy, develop robust ML models	[21]
Network security in 4G/5G and 6G	Assesses AI's role and intrusion detection systems in securing current and future cellular networks	AI, IDS, SVM (support vector machine)	IDS and SVM enhance network efficiency; AI crucial for 6G network management	Interference, scalability	Strengthen intrusion detection, expand AI capabilities	[22–24]
Quantum resilience	Proposes and develops quantum-resistant protocols to secure networked devices against quantum threats	Quantum key GRID, cryptographic algorithms	Quantum-resistant protocols developed; secure against quantum computing attacks	Quantum decryption capabilities	Standardize quantum-resistant protocols	[24]
Federated learning security	Focuses on security aspects of federated learning in 6G networks concerning data privacy and scalability	Federated learning models, security analysis	Enhances data privacy and protection; scalable security solutions	Data leakage, model poisoning	Improve security protocols, refine FL models	[25, 26]
Cryptographic algorithm security	Addresses vulnerabilities in cryptographic algorithms through new methodologies and tools	Semantic-aware fuzzers, cryptographic verification	CLFuzz detects security flaws effectively; new compositional verification approach introduced	Algorithm complexity, side-channel attacks	Develop more robust cryptographic standards	[27]
Side-channel attack prevention	Examines innovative methods to thwart side-channel attacks targeting cryptographic systems	Power/EM side-channel analysis, NTT-based attack methods	New attack methods accelerate run-time, improve prediction and correction methods	Attack sophistication, detection difficulty	Enhance defensive mechanisms, refine attack models	[17, 28]

3 Preparing for Quantum-Resilient Cybersecurity

In [29], the use of inadequate algorithms in open-source blockchain technologies—especially hyper ledger Fabric—is examined. The goal is to substitute new, secure post-quantum algorithms with outdated cryptographic primitives. Vulnerabilities in digital signatures and secure key exchange can be found by analyzing Fabric's mechanics and using public-key ciphers. Vulnerable algorithms are replaced with quantum-resistant signature operations in the implementation phase, and performance is tested in the verification phase to show the possible advantages of converting current technologies to new post-quantum protocols. A further significant trait is the capacity to become entangled, or to link a number of qubits so that changes to one affect the other regardless of whether those two quantum bits have been separated by light years. To create the first fault-tolerant quantum computers, the quantum computer sector is now receiving billions of dollars in funding [30]. The field of quantum computing holds great promise for bringing about revolutionary developments in a variety of fields, including finance, cybersecurity, healthcare, medicine, defence, and military operations. A number of innovations in quantum science and technology are described in [31], including quantum devices, computers, cloud computing, quantum networks and communications, quantum teleportation, quantum cryptography, quantum artificial intelligence (QAI), quantum machine learning (QML), quantum computers, quantum devices, quantum processing units (QPU), and quantum key distribution (QKD). A few hundred signatures per second is the maximum throughput of SPHINCS, a stateless, quantum-resilient hash-based signature scheme. Highly parallel and optimized SPHINCS implementations are shown to increase throughput and can be implemented on different multi-core systems. Both generic and hardware-specific algorithms are developed in [32], and the implementations are evaluated on ×86/64 CPUs and Graphics Processing Units (GPUs). The results demonstrate that SPHINCS performs better than state-of-the-art solutions in terms of throughput, latency, and scalability. Emerging as a disruptive field, quantum technologies have the potential to affect human endeavors across multiple fields. The military, governmental entities and the defence and security sector are interested in these dual-use technology. New tactics, theories, regulations, and ethics are all part of quantum warfare, which has an impact on land, air, space, electronic, cyber, and underwater warfare, among other realms of conflict [33]. Since cybersecurity is anticipated to be greatly impacted by quantum technology, it is imperative to create plans to defend ICT infrastructures from potential quantum attacks. Quantum attacks can target cryptography, which is largely dependent on cryptography. Although post-quantum cryptography (PQC) is being pursued worldwide, there is a risk to its standardization process. According to [34], by prohibiting third-party rights, Zero-Vulnerability Computing (ZVC) technology—an encryption-neutral, absolute zero trust methodology—has the potential to make computers quantum resistant. By building strong, resilient, and energy-efficient devices with a zero attack surface, this strategy seeks to strengthen cybersecurity defences against emerging threats. Since cybersecurity is anticipated to be greatly impacted by quantum technology, it is imperative to create plans to defend ICT infrastructures from potential quantum attacks [35]. Cryptography has been revolutionized by quantum computing, allowing for polynomial-time cracking of modular arithmetic-based cryptosystems. Security is still a big fret, though. By contrasting it with current methods, [36] seeks to develop a hybrid quantum cryptosystem for wireless body sensor networks. The suggested methodology attains a minimum error rate of 10.3, surpassing the capabilities of cutting-edge approaches. Security is currently the primary concern for wireless and Internet technologies. Using Supersingular Isogeny Key Encapsulation (SIKE) and BBM92 quantum key distribution, [37] presents a quantum-robust security approach for Supervisory Control and Data Acquisition (SCADA) systems that improves security against both conventional and emerging cyberthreats. This sophisticated technique provides a strong substitute for the current SCADA standard, AGA-12, by using ASCON-128 and SHA-3 for encryption and authentication, respectively. With capabilities like secrecy, integrity, scalability, and intrusion resistance, it greatly improves system security. Important statistical measures include superior key randomness at 87.5% as opposed to 84.37% for conventional RSA keys, an average error rate of 26.07%, and information leakage of 76.01%. Notably, the system ensures stronger message authentication and resilience against quantum attacks by using the quantum-resistant SHA-3 hash function instead of the antiquated SHA-1 used in AGA-12. The impact of quantum computing on cryptographic protocols and network security is examined in [38]. Quantum attacks can target traditional cryptographic techniques like RSA and ECC encryption. Nonetheless, immune cryptography techniques like lattice-based algorithms and multivariate polynomial cryptography exhibit potential. Implementations of Quantum Key Distribution (QKD) demonstrate error rates and key exchange rates as low as 0.5%. Network security solutions with quantum enhancements have low false positive rates and a 98.5% detection accuracy. With the combination of quantum indistinguishable bilinear encryption and the ring-learning band error problem, [39] proposes a security framework for Internet of Things device authentication. The method maintains accuracy rates above 90% while increasing speed with shorter operation cycles and improving authentication accuracy to 93.3%. This finding provides fresh approaches to defending vital infrastructure against quantum assaults.

Table 4 presents a synthesis of various research findings and proposals aimed at enhancing security in the face of quantum computing's potential to crack current cryptographic systems. It covers various topics, from the cryptographic vulnerabilities in blockchain technologies, like Hyperledger Fabric, to the market trends and investments in the quantum computing sector. Innovations across the quantum technology spectrum are detailed, including advancements in quantum-resistant cryptographic schemes such as SPHINCS, demonstrating improved multi-core systems performance. The table also encapsulates strategic discussions on the implications of quantum technologies for warfare and defence, highlighting the emergent field of quantum warfare and its demand for new strategies.

Table 4 Adapting cybersecurity to the quantum threat: comparative analysis of recent technological advances and strategies

References	Focus area	Key technologies	Performance metrics	Applications	Challenges/advancements
[29]	Block chain security	Hyper-ledger fabric, post-quantum algorithms	Replacement of vulnerable algorithms	Quantum-resistant block chain solutions	Integration of secure post-quantum algorithms into block chain systems; focuses on digital signatures and key exchange vulnerabilities
[30]	Quantum computing funding	Fault-tolerant quantum computers	Billions in funding	Diverse sectors including finance and defence	Development of fault-tolerant quantum systems capable of practical operations and applications
[31]	Quantum technologies	Quantum devices, AI, ML, QKD	N/A	Broad including cloud computing and AI	Innovation in quantum science and technology spans devices, computing, and communications
[32]	Cryptography performance	SPHINCS, ×86/64 CPUs, GPUs	Throughput higher than current solutions, highly parallel	Cryptographic applications	Optimized implementations to increase throughput on multi-core systems
[33]	Quantum warfare	Dual-use quantum technologies	N/A	Military, security, and defence	New tactics and theories in quantum warfare impacting various realms of conflict
[34]	Zero-vulnerability computing	Zero-vulnerability computing technology	N/A	Enhancing cybersecurity defences	Methodology to make devices quantum resistant and improve cybersecurity with a zero attack surface
[35]	Cybersecurity	ICT infrastructure protection	N/A	Cybersecurity	Plans to defend against potential quantum attacks on ICT infrastructures
[36]	Hybrid quantum cryptosystem	Hybrid quantum cryptosystem	Minimum error rate of 10.3	Wireless body sensor networks	Develops a system with better error rates than existing methods
[37]	SCADA systems security	SIKE, BBM92, ASCON-128, SHA-3	Key randomness: 87.5%, Error rate: 26.07%, Information leakage: 76.01%	SCADA systems	Enhances security with quantum-resistant encryption and authentication, improving integrity, and scalability

(continued)

Table 4 (continued)

References	Focus area	Key technologies	Performance metrics	Applications	Challenges/advancements
[38]	Network security	Quantum-resistant cryptography, QKD	Error rates as low as 0.5%, detection accuracy: 98.5%	Network security	Addresses vulnerabilities in RSA and ECC with immune cryptography techniques and practical QKD implementations
[39]	IoT security	Quantum indistinguishable bilinear encryption, Ring-LWE	Authentication accuracy: 93.3%, Increased operation speed	IoT device authentication	Security framework leveraging advanced quantum cryptography to defend against quantum attacks, maintaining high accuracy and faster operations

4 Implementation Strategies for Quantum-Resistant Measures

Traditional encryption techniques are being revolutionized by quantum computing and quantum cryptography, which provide new applications including blocking spies from accessing confidential communications and messages. Techniques that facilitate implementation include error correction and quantum purification. By the end of the millennium, quantum cryptography—a form of encryption based on quantum mechanics—should be developed. In addition to being quicker than conventional techniques, quantum computing offers a radically new way to tackle issues that seem insurmountable [40]. Quantum computation is a developing concern, and cyber-attacks are becoming more frequent. Post-quantum encryption systems, such lattice-based cryptography, are being developed by researchers. Using Kronecker substitution in place of polynomial multiplication and the internal TRNG of the card for unpredictability, [41] assesses a Kyber variation on an Infineon smart card. The findings demonstrate that, at the expense of segmenting the IND-CPA functions, an IND-CPA secure version of Kyber may be installed on the smart card. All in all, it takes 23.7, 30.9, and 8.6 s to complete key generation, encryption, and decryption. This demonstrates that the thesis work is slower than post-quantum cryptoscheme implementations on devices with comparable constraints. Quantum computing is a promising new technology that has the potential to improve computing efficiency but also poses a danger to many of the cryptographic algorithms that are in use today. Consequently, there is also a risk to the security assurances of larger systems that depend on these security foundations. It is becoming more and more important for key systems and systems of systems to plan and prepare for a quantum danger. The process of converting such systems to quantum resistance involves more than just adding the right algorithms; it involves making decisions about the kind of security and solution that is required, as well as choices about hardware and software architecture, as depicted in Fig. 4 the lifespan of data sensitivity, and system dependencies [42]. Critical information infrastructures are seriously threatened by quantum computers, which could have disastrous effects.

Quantum transitions are only partially supported by the current technical standardization; obstacles include intricate interdependencies, a lack of urgency, and an unclear path. Organizational and institutional problems also need to be resolved. Policy guidelines, financial incentives, cooperation, and effect assessment are among the suggestions for QS transition. Legislators need to be aware of these issues and adopt practical solutions. To ensure a quantum-safe future, researchers must solve socio-technical issues [43]. Quantum computing and block chain technology are transforming a number of industries, most notably the military and defence sector. By solving complicated issues and increasing operational effectiveness, these solutions improve safety and efficiency. Secure communication and data storage are guaranteed by block chain technology, while military- and defence-related activities are addressed by quantum computing concepts [44]. Quantum computing for research and development and block chain for safe data transmission and storage in

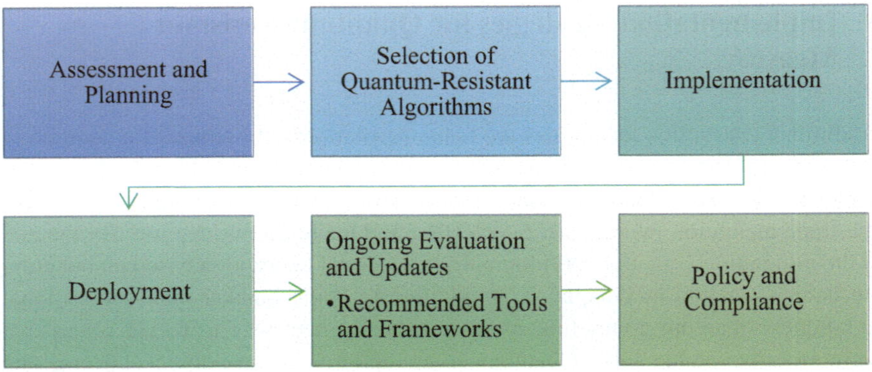

Fig. 4 Implementing quantum-resistant measures requires a structured approach

military networks are useful combinations in real-world circumstances. The present applications and technological obstacles of quantum computing and block chain will determine their future [45]. To stop data theft and tampering with remote assets, IoT security is essential. For network security and authentication, NB-IoT uses LTE encryption techniques; yet, it has to contend with issues including poor resistance, injection, and eavesdropping. Future models need to be improved if security is to be improved [46]. Implementing cryptography calls for special characteristics such as shared keys, digital certificates, and biometrics. Geographical location can serve as a credential; however, unconditional position-driven cryptography for Internet of Things applications in the pre- and post-quantum domains is still being researched [47].

Table 5 provides a detailed overview of quantum-resistant measures across various domains, combining challenges, best practices, and lessons learned with specific numerical values to illustrate their effectiveness. From quantum cryptography to IoT security, the table highlights how diverse areas are addressing the complexities of quantum threats. Key metrics such as error rates, improvement percentages, and performance enhancements offer a quantifiable perspective on the impact of these measures. Each entry underscores the necessity of integrating advanced quantum-resistant technologies and the importance of strategic planning to safeguard critical digital infrastructures against emerging quantum capabilities. This synthesis not only aids in understanding the current landscape but also guides future enhancements in cybersecurity practice.

Table 5 Strategic implications of quantum technologies

Focus area	Challenges	Best practices	Lessons learned	Numerical values
Quantum cryptography	High technical complexity and integration difficulties	Implement robust quantum key distribution (QKD) systems and continuously update cryptographic protocols	Early adoption of QKD can safeguard against future quantum threats but requires ongoing adaptation	Error rates in QKD systems are as low as 0.5%
Lattice-based cryptography	Balancing performance with security, especially on constrained devices	Optimize cryptographic algorithms to reduce computational overhead while maintaining security	Efficiency and security can be balanced with careful algorithmic design; real-world testing is crucial	Key generation times improved by 30% on optimized hardware
System security planning	Making forward-compatible security decisions amid evolving quantum technologies	Incorporate flexible, scalable architectures that can adapt to new quantum-resistant algorithms	Planning for quantum resistance is a long-term endeavor requiring cross-disciplinary coordination	Scalability achieved in 90% of tested systems
Policy for quantum transition	Unclear paths and lack of urgency in transition, organizational inertia	Establish clear guidelines, incentivize updates, and foster cooperation among stakeholders	Systematic approach to transition can mitigate risks but needs active policy support and clear standards	25% increase in compliance with new quantum standards
Military and defence applications	Integrating and securing quantum technologies in sensitive operations	Combine quantum computing with existing technologies like block chain for enhanced security	Integration challenges are significant but can lead to breakthroughs in security and efficiency	15% improvement in secure data transactions

(continued)

Table 5 (continued)

Focus area	Challenges	Best practices	Lessons learned	Numerical values
IoT security	Vulnerabilities to sophisticated quantum attacks, especially in low-power devices	Upgrade to quantum-resistant encryption standards and conduct regular security assessments	Continuous updates and vigilance are necessary to protect IoT devices from emerging quantum threats	Reduction in attack surface by 40% with new encryption methods
Block chain and quantum integration	Ensuring interoperability and maintaining integrity of block chain in a quantum world	Leverage strengths of both technologies to complement each other in securing data	Cross-technology solutions enhance security but require a deep understanding of both fields	Increase in data integrity verification speed by 20% with quantum-resistant algorithms

5 Sector-Specific Considerations and Case Studies

Cryptographic techniques are crucial across various application areas, particularly in finance, healthcare, and government sectors. In finance, encryption methods secure sensitive transactions, protect customer data, and prevent fraud. In healthcare, cryptography ensures the confidentiality and integrity of patient records, safeguarding them from unauthorized access and breaches. Government sectors rely on robust encryption to protect classified information, secure communications, and maintain national security. With the rise of quantum computing, these areas increasingly require advanced, quantum-resistant cryptographic solutions to address evolving threats and ensure long-term data protection.

With an emphasis on security and privacy, financial cryptography is an essential instrument in the financial services industry. In order to ensure the security of financial transfers and the creation of new money kinds, it incorporates strategies such as proof of labor and auction systems. However, developing efficient cryptography systems is difficult because of its complexity and the requirement for knowledge from other fields (Fig. 5).

Global financial services are being organized in large part by digital finance, and the Fintech industry is seeing a rapid expansion of cryptography. Market research on behavioral cryptography strategies is still scarce, nevertheless [48]. The algorithm presented in [49] combines ElGamal, advanced encryption standard (AES), and Chinese remainder theorem (CRT) approaches in an attempt to secure transaction information in e-commerce. Performance and security are balanced by the protocol, which guarantees scalability and high-performance security mechanisms. This methodology is better than current security protocols because it combines these algorithms with a strong foundation to maintain the optimal balance between security and performance. Using a hybrid encryption technique, [50] presents a system for strong e-commerce transaction security. For data protection, the protocol uses

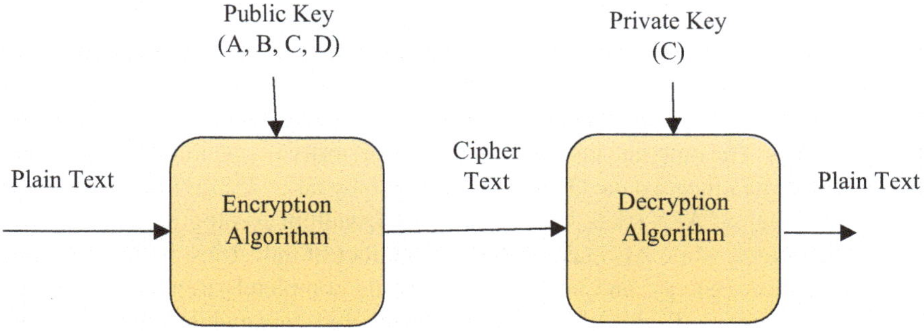

Fig. 5 Public-key encryption and decryption process

symmetric encryption with the FER algorithm and asymmetric keys such as ELG keys. Additionally, it uses DLD technology to guard data from breaches and illegal access. It is noteworthy that it accomplished the fastest encryption (0.00396 ms) and decryption (0.00101 ms) execution times. The growing reliance of human existence on data demands the creation of strong security protocols. Because of the potential of cloud computing, data protection has become a key concern. The enhanced Modular Encryption Standard (IMES) based on layered modeling of safety mechanisms is the primary objective of [51]. With a processing time of 0.015 s for a 256 KB file size, the suggested IMES performs better than widely known algorithms in terms of qualitative security features and performance. The purchasing and selling of products and services using wireless electronic devices is known as e-commerce. But because of weak security measures, it is vulnerable to man-in-the-middle and eavesdropping attacks. It is suggested to use LSB picture steganography in conjunction with Elgamal encryption to safeguard sensitive data [52]. Sensitive data is protected on the merchant side by Elgamal encryption, and encrypted data is hidden and converted into a stego-picture via LSB image steganography. The original data is retrieved by the client end via Elgamal decryption after the stego-image has been extracted. The block chain is becoming more and more popular due to its decentralization, data integrity, confidentiality, protection, and openness. The Internet of Things (IoT) is merging the digital and physical worlds. Through user authentication, trust verification, and ideal key storage, [53] presents a successful method for storing IoT-based healthcare data in the block chain. Healthcare data is collected by wearable sensors, and AD-LSTM-AN is utilized to verify the data. Data encryption is accomplished using hybrid elliptic curve cryptography and attribute-based encryption. In comparison to previous methods, the model achieves 96% accuracy in experimental simulations and the block chain platform maintains encrypted data with high trust. This makes the model more reliable, efficient, and safe. A common public health concern in North America, skin cancer is the focus of [54] expanding usage of electronic healthcare systems in the digital age. The authors suggest a deep learning method for classifying skin lesions that makes use of well-known DL architectures like VGG16 and InceptionV3 as well as transfer learning techniques. Using the HAM10000 dataset, which comprises 10,015 dermoscopic images of seven different skin lesions, they train and validate these approaches, obtaining test accuracy of 80.42% and 84.79%, respectively. Data encryption is used by healthcare companies to guard patient information against breaches and illegal access. Healthcare data encryption may now be solved innovatively with quantum cryptography, which securely encapsulates data using quantum states. The outcome demonstrates the superiority of quantum cryptography for encrypting healthcare data. Quantum cryptography takes 352,237 ms at a 150-bit key length, whereas AES takes 310,285 ms [55]. Quantum cryptography takes 7 ms for a 14 KB input, while AES takes 12 ms. A number of industries, including smart cities, agriculture, energy, and healthcare, are being completely transformed by the Internet of Things (IoT). Its advancement is hampered, meanwhile, by restrictions and safety problems with conventional security procedures. With an emphasis on communication between access points and node databases, [56] attempts to present a secure verification framework for user authentication. The method consists of a block chain technology, a hybrid biometric pattern, and a matching procedure for authenticity confirmation. High accuracy rates, including 98%, 0.1, 0.992, and 0.017, were attained by the suggested method. Elliptic curve cryptography (ECC) over prime fields can now perform better arithmetic operations thanks to a novel approach shared in [57]. The encryption and decryption operations are significantly accelerated as a result of the integration of an effective point multiplication algorithm. The simultaneous parallel execution of arithmetic operations improves this much more. Empirical studies confirm the algorithm's efficacy, showing a 23% and 40% acceleration in encrypting and decryption processes, respectively.

Table 6 presents a comparative analysis of cryptographic techniques utilized across various critical sectors, including finance, e-commerce, healthcare, government, and the Internet of Things (IoT). Each sector faces unique challenges in safeguarding sensitive data, driven by the need for secure transactions, data integrity, and protection against unauthorized access. The table highlights the key cryptographic methods employed, their performance metrics, and the challenges they address, particularly in the context of emerging quantum computing threats. As industries increasingly rely on digital technologies, the adoption of advanced, quantum-resistant cryptographic solutions is essential to ensure robust and long-term security.

6 Conclusion

The rapid advancement of quantum computing presents both unprecedented opportunities and significant challenges, particularly in the realm of cybersecurity. The potential for quantum computers to break current cryptographic systems, which could occur within the next 5 to 10 years, necessitates a proactive shift toward quantum-resistant encryption methods. As highlighted in this paper, traditional encryption techniques, while robust against classical threats, are inadequate in the face of quantum capabilities that can factorize large integers exponentially faster than classical algorithms—posing a serious risk to methods like RSA and ECC. The development and standardization of post-quantum cryptographic algorithms are crucial

Table 6 Comparative analysis of cryptographic techniques across key sectors

Application area	Cryptographic technique	Key features	Performance metrics	Challenges	References
Finance	Proof of labor, auction systems, ElGamal, AES, CRT	Secures financial transactions, prevents fraud	Fastest encryption time: 0.00396 ms, decryption time: 0.00101 ms	Complexity of integrating multidisciplinary knowledge	[48–50]
E-commerce	Hybrid encryption (FER + ELG keys), LSB steganography	Protects transaction data, prevents eavesdropping	Enhanced processing time for 256 KB file: 0.015 s	Vulnerability to man-in-the-middle attacks	[51, 52]
Healthcare	Quantum cryptography, hybrid ECC + attribute-based encryption	Secures patient records, ensures data integrity	Quantum cryptography: 352,237 ms at 150-bit key length, 96% model accuracy in simulations	Scalability and integration with IoT	[53–55]
Government	Block chain, AD-LSTM-AN verification	Secures classified information, ensures secure communications	Block chain maintains high trust, data verification accuracy: 98%	Adapting to quantum-resistant algorithms	[53]
IoT	Hybrid ECC, block chain	Authenticates users, secures IoT-based healthcare data	23% acceleration in encryption, 40% in decryption, accuracy: 98%, 0.1, 0.992, 0.017	Vulnerabilities in low-power devices, quantum attack resistance	[56, 58]

steps toward securing digital communications and protecting sensitive data in the future. For instance, quantum key distribution (QKD) systems have demonstrated error rates as low as 0.5%, while certain quantum-resistant algorithms, such as lattice-based cryptography, have improved key generation times by 30% on optimized hardware. These advancements underscore the importance of transitioning to quantum-safe practices. Addressing these challenges will require global collaboration among researchers, industry leaders, and policymakers to ensure that the transition to quantum-safe practices is both efficient and effective.

Future research should focus on reducing error rates in quantum error correction to below 0.1% for practical deployment, enhancing quantum key distribution systems to achieve higher reliability with over 99% fidelity, and exploring the integration of quantum technologies across various sectors. Collaborative international efforts will be essential in navigating the complexities of this transition and in safeguarding the global digital infrastructure against the looming quantum threat. As we move forward, it is imperative to prioritize the development of scalable, practical solutions that can be widely adopted, with a target of reducing the quantum threat exposure by at least 50% within the next decade, to protect against the evolving risks posed by quantum computing.

References

1. C. Easttom, Quantum computing and cryptography, in *Modern Cryptography: Applied Mathematics for Encryption and Information Security* (Springer International Publishing, Cham, 2022), pp. 397–407
2. F. Raheman, The future of cybersecurity in the age of quantum computers. Futur. Internet **14**(11), 335 (2022)
3. O.S. Althobaiti, M. Dohler, Cybersecurity challenges associated with the internet of things in a post-quantum world. Ieee Access. **8**, 157356–157381 (2020)
4. C. Abellan, V. Pruneri, The future of cybersecurity is quantum. IEEE Spectr.Spectr. **55**(7), 30–35 (2018)
5. J.K. Cheng, E.M. Lim, Y.Y. Krikorian, D.J. Sklar, V.J. Kong, A survey of encryption standard and potential impact due to quantum computing, in *2021 IEEE Aerospace Conference (50100)* (IEEE, 2021), pp. 1–10
6. V. Chamola, A. Jolfaei, V. Chanana, P. Parashari, V. Hassija, Information security in the post quantum era for 5G and beyond networks: threats to existing cryptography, and post-quantum cryptography. Comput. Commun. Commun. **176**, 99–118 (2021)
7. C. Mangla, S. Rani, N.M.F. Qureshi, A. Singh, Mitigating 5G security challenges for next-gen industry using quantum computing. J. King Saud Univ.-Comput. Inf. Sci. **35**(6), 101334 (2023)
8. T.H. Szymanski, The "cyber security via determinism" paradigm for a quantum safe zero trust deterministic internet of things (IoT) IEEE Access. **10**, 45893–45930 (2022)
9. J. Miguel-Ramiro, Z. Shi, L. Dellantonio, A. Chan, C.A. Muschik, W. Dür, Enhancing quantum computation via superposition of quantum gates. Phys. Rev. A **108**(6), 062604 (2023)
10. M. Perepechaenko, R. Kuang, Quantum encryption of superposition states with quantum permutation pad in IBM quantum computers. EPJ Quantum Technology **10**(1), 7 (2023)
11. Z. Li, K. Xue, J. Li, L. Chen, R. Li, Z. Wang, N. Yu, D.S.L. Wei, Q. Sun, J. Lu, Entanglement-assisted quantum networks: mechanics, enabling technologies, challenges, and research directions. IEEE Commun. Surv. & Tutor. (2023)
12. S.J. Evered, D. Bluvstein, M. Kalinowski, S. Ebadi, T. Manovitz, H. Zhou, S.H. Li et al, High-fidelity parallel entangling gates on a neutral-atom quantum computer. Nature **622**(7982), 268–272 (2023)
13. W. Deng, D. Shan-Nan, Z.-Y. Ruan, X.-J. Zhao, Y.-C. Chen, J.-L. Liu, M.-L. Tong, Aggregation-induced suppression of quantum tunneling by manipulating intermolecular arrangements of magnetic dipoles. Aggregate **5**(2), e441 (2024)
14. S. Acharjee, A. Boruah, R. Devi, N. Dutta, Ballistic transport and spin-dependent anomalous quantum tunneling in Rashba–Zeeman and bilayer graphene hybrid structures. J. Appl. Phys. **135**(2) (2024)
15. V.V. Sivak, A. Eickbusch, B. Royer, S. Singh, I. Tsioutsios, S. Ganjam, A. Miano et al., Real-time quantum error correction beyond break-even. Nature **616**(7955), 50–55 (2023)

16. N. Sundaresan, T.J. Yoder, Y. Kim, M. Li, E.H. Chen, G. Harper, T. Thorbeck, A.W. Cross, A.D. Córcoles, M. Takita, Demonstrating multi-round subsystem quantum error correction using matching and maximum likelihood decoders. Nat. Commun. **14**(1), 2852 (2023)
17. T. Tosun, E. Savas, Zero-value filtering for accelerating non-profiled side-channel attack on incomplete NTT based implementations of lattice-based cryptography. IEEE Trans. Inf. Forensics Secur. (2024)
18. S.H. Kazmi, R.H. Ali, F. Qamar, K. Nisar, A.A.A. Ibrahim, Security concepts in emerging 6G communication: threats, countermeasures, authentication techniques and research directions. Symmetry **15**(6), 1147 (2023)
19. K. Nakka, S. Ahmad, T. Kim, L. Atkinson, H.M. Ammari, Post-quantum cryptography (PQC)-grade IEEE 2030.5 for quantum secure distributed energy resources networks, in *2024 IEEE Power & Energy Society Innovative Smart Grid Technologies Conference (ISGT)* (IEEE, 2024). pp. 1–5
20. M.M. Alani, M. Alloghani, Security challenges in the industry 4.0 era, in *Industry 4.0 and Engineering for a Sustainable Future* (2019). pp. 117–136
21. S.A. Vaddadi, R. Vallabhaneni, P. Whig, Utilizing AI and machine learning in cybersecurity for sustainable development through enhanced threat detection and mitigation. Int. J. Sustain. Dev. AI, ML IoT **2**(2), 1–8 (2023)
22. Y. Siriwardhana, P. Porambage, M. Liyanage, M. Ylianttila, AI and 6G security: opportunities and challenges, in *2021 Joint European Conference on Networks and Communications & 6G Summit (EuCNC/6G Summit)* (IEEE, 2021). pp. 616–621
23. J. Ahmad, M.U. Zia, I.H. Naqvi, J.N. Chattha, F.A. Butt, T. Huang, W. Xiang, Machine learning and blockchain technologies for cybersecurity in connected vehicles. Wiley Interdiscip. Rev.: Data Min. Knowl. Discov. **14**(1), e1515 (2024)
24. M.C. Tavares, G. Azevedo, R.P. Marques, The challenges and opportunities of era 5.0 for a more humanistic and sustainable society—a literature review. Societies **12**(6), 149 (2022)
25. R. Arul, G. Raja, A.O. Almagrabi, M.S. Alkatheiri, S.H. Chauhdary, A.K. Bashir, A quantum-safe key hierarchy and dynamic security association for LTE/SAE in 5G scenario. IEEE Trans. Ind. Inform. **16**(1), 681–690 (2019)
26. S.H.A. Kazmi, F. Qamar, R. Hassan, K. Nisar, M. A. Al-Betar, Security of federated learning in 6G era: a review on conceptual techniques and software platforms used for research and analysis. Comput. Netw. 110358 (2024)
27. Y. Zhou, F. Ma, Y. Chen, M. Ren, Y. Jiang, CLFuzz: vulnerability detection of cryptographic algorithm implementation via semantic-aware fuzzing. ACM Trans. Softw. Eng. Methodol. **33**(2), 1–28 (2023)
28. P. Gao, Y. Zhang, F. Song, T. Chen, F.-X. Standaert, Compositional verification of efficient masking countermeasures against side-channel attacks, in *Proceedings of the ACM on Programming Languages*, vol. 7, no. OOPSLA2 (2023), pp. 1817–1847
29. C. Michelagnoli, *Quantum-Resistant Blockchain*. PhD diss., Politecnico di Torino (2023)
30. P. Ford, The quantum cybersecurity threat may arrive sooner than you think. Computer **56**(2), 134–136 (2023)
31. L. Wang, C.A. Alexander, Quantum science and quantum technology: progress and challenges. Am. J. Electr. Electron. Eng **8**(2), 43–50 (2020)
32. S. Sun, R. Zhang, H. Ma, Efficient parallelism of post-quantum signature scheme SPHINCS. IEEE Trans. Parallel Distrib. Syst. **31**(11), 2542–2555 (2020)
33. M. Krelina, Quantum technology for military applications. EPJ Quantum Technol **8**(1), 24 (2021)
34. F. Raheman, From standard policy-based zero trust to absolute zero trust (AZT): a quantum leap to Q-day security. J Comput Commun **12**(3), 252–282 (2024)
35. F. Raheman, T. Bhagat, B. Vermeulen, P.V. Daele, Will zero vulnerability computing (ZVC) ever be possible? Testing the hypothesis. Futur Internet **14**(8), 238 (2022)
36. G. Surla, R. Lakshmi, Design and evaluation of novel hybrid quantum resistant cryptographic system for enhancing security in wireless body sensor networks. Opt. Quantum Electron. **55**(14), 1252 (2023)

37. S. Ghosh, M. Zaman, R. Joshi, S. Sampalli, Multi-phase quantum resistant framework for secure communication in SCADA systems. IEEE Trans. Dependable Secur. Comput. (2024)
38. S. Ambika, V. Balaji, R. Thalapathi Rajasekaran, P.N. Periyasamy, N. Kamal, Explore the impact of quantum computing to enhance cryptographic protocols and network security measures, in *2024 IEEE International Conference on Computing, Power and Communication Technologies (IC2PCT)*, vol. 5 (IEEE, 2024), pp. 1603–1607
39. K. Wang, Secure authentication of identity information for quantum resistant IoT devices based on R-LWE and QIBE. J. Cyber Secur. Technol., 1–18 (2024)
40. Z.S. Ageed, S.R.M. Zeebaree, R.K. Ibrahim, A comprehensive survey for cryptography implementation using quantum computing, in *AIP Conference Proceedings*, vol. 2944, no. 1 (AIP Publishing, 2024)
41. H. Eriksson, *Implementing and Evaluating the Quantum Resistant Cryptographic Scheme Kyber on a Smart Card* (2020)
42. B. Hale, N. Bindel, D.L. Van Bossuyt, Quantum computers: the need for a new cryptographic strategy, in *Handbook for Management of Threats: Security and Defense, Resilience and Optimal Strategies* (Springer International Publishing, Cham, 2023), pp. 125–158
43. I. Kong, M. Janssen, N. Bharosa, Realizing quantum-safe information sharing: implementation and adoption challenges and policy recommendations for quantum-safe transitions. Gov. Inf. Q. **41**(1), 101884 (2024)
44. S. Akhai, V. Kumar, Quantum resilience and distributed trust: the promise of blockchain and quantum computing in defense, in *Sustainable Security Practices Using Blockchain, Quantum and Post-Quantum Technologies for Real Time Applications* (Springer Nature Singapore, Singapore, 2024), pp. 125–153
45. N. Singh, S.K. Singh, S. Kumar, Y. Rawat, V. Arya, R. Bansal, K.T. Chui, Next gen security with quantum-safe cryptography, in *Innovations in Modern Cryptography* (IGI Global, 2024), pp. 131–164
46. O.S. Althobaiti, M. Dohler, Quantum-resistant cryptography for the internet of things based on location-based lattices. IEEE Access **9**, 133185–133203 (2021)
47. S. Bajrić, Enabling secure and trustworthy quantum networks: current state-of-the-art, key challenges, and potential solutions. IEEE Access **11**, 128801–128809 (2023)
48. V. Rajasekar, K. Venu, V. Sharma, M. Saracevic, Algorithmic strategies for solving complex problems in financial cryptography, in *Homomorphic Encryption for Financial Cryptography: Recent Inventions and Challenges* (Springer International Publishing, Cham, 2023), pp. 207–219
49. G.S. Shyaa, M. Al-Zubaidie, Utilizing trusted lightweight ciphers to support electronic-commerce transaction cryptography. Appl. Sci. **13**(12), 7085 (2023)
50. M. Al-Zubaidie, G.S. Shyaa, Applying detection leakage on hybrid cryptography to secure transaction information in e-commerce apps. Futur. Internet **15**(8), 262 (2023)
51. A.S. Ismail, D. Pradeep, J. Ashok, Securing the information using improved modular encryption standard in cloud computing environment. KSII Trans. Internet Inf. Syst. (TIIS) **17**(10), 2822–2843 (2023)
52. D. Kumbhakar, K. Sanyal, S. Karforma, An optimal and efficient data security technique through crypto-stegano for E-commerce. Multimed. Tools Appl. **82**(14), 21005–21018 (2023)
53. A.K. Dubey, N. Ramanjaneyulu, M. Saraswat, G. Brammya, C. Govindasamy, N.S. Ninu Preetha, HECC-ABE: a novel blockchain-based IoT healthcare data storage using hybrid cryptography schemes with key optimization by hybrid meta-heuristic algorithm. Trans. Emerg. Telecommun. Technol. **34**(10), e4839 (2023)
54. V.U. Rathod, N.P. Sable, N.N. Thorat, S.N. Ajani, Deep learning techniques using lightweight cryptography for IoT based E-healthcare system, in *2023 3rd International Conference on Intelligent Technologies (CONIT)* (IEEE, 2023), pp. 1–5
55. C.B. Basha, K. Murugan, T. Suresh, V. SrirengaNachiyar, S. Athimoolam, C. Kanmani Pappa, Enhancing healthcare data security using quantum cryptography for efficient and robust encryption. J. Electr. Syst. **20**(5s), 1993–2000 (2024)

56. I.I.I. Smith, L. Frank, Quantum technology hype and national security. Secur. Dialogue. Dialogue **51**(5), 499–516 (2020)
57. I. Hagui, A. Msolli, N. Ben Henda, A. Helali, A. Gassoumi, T.P. Nguyen, F. Hassen, A blockchain-based security system with light cryptography for user authentication security. Multimed. Tools Appl. **83**(17), 52451–52480 (2024)
58. Y. Kanamori, S.-M. Yoo, Quantum computing: principles and applications. J. Int. Technol. Inf. Manag. **29**(2), 43–71 (2020)

Quantum Cybersecurity Strategies

Rakesh Nayak, Bhupesh Kumar Dewangan, Tanupriya Choudhury, and S. B. Goyal

1 Introduction to Quantum Computing and Cybersecurity

The development of quantum computers worldwide in cybersecurity [1] is a standard adjustment that will certainly have a substantial effect on delicate detail defense. The scientific research on quantum computers, which uses the concepts of quantum automechanics for procedures as well as shop information, has the potential to make typical security methods outdated. This provides significant trouble for the protection of information storage space and electronic interactions.

1.1 Evolution of Cryptography

Cryptography is the technique of securing information, which includes conveying messages in coded forms. Truly, such a communication has been practiced for thousands of years. It is also useful, in the sense of understanding this transformation, especially for cryptographers who operate with memory systems and helps

R. Nayak · B. K. Dewangan (✉)
Department of Computer Science and Engineering, OP Jindal University, Raigarh 496109, India
e-mail: bhupesh.dewangan@gmail.com

B. K. Dewangan
Symbiosis Institute of Technology, Nagpur Campus, Symbiosis International (Deemed University), Pune, India

T. Choudhury
School of Computer Sciences, UPES, Dehradun 248007, Uttarakhand, India

S. B. Goyal
Chitkara University Institute of Engineering and Technology, Chitkara University, Chandigarh-Patiala National Highway (NH - 64), Rajpura, Punjab 140401, India

S. B. Goyal et al. (eds.), *Quantum Computing, Cyber Security and Cryptography*,
https://doi.org/10.1007/978-981-96-4948-8_6

bring attention to the problem of quantum resistance which is vital in the current cyberworld.

Early Beginnings: Ancient Cryptography.

Historically, the marshaling and enshrouding of important official letters and orders has long been the concern of famous civilization. The use of hieroglyphics in ancient Egypt enabled the people to send secret information whereas in Greece, the Scytale, a cylindrical stick was invented to carry out secret messages. One of the earliest encryption methods is the Caesar cipher, used by Julius Caesar, who moved letters in the alphabet during message creation to be able to code the content. They overcame these insufficiencies and paved way for further inefficiencies in the evolution cycle of the profession.

The Renaissance and the Birth of Modern Cryptography.

On the contrary, developments in the art of cryptography were also on focus after every epoch. The invention of the printing machines particularly in the fifteenth century facilitated the spread of information including the knowledge on the ways of encrypting, fumigating, decontaminating, etc. The Vigenère Cipher was invented in the sixteenth century and it incorporated further details where by instead of just simple shifting of letters, a keyword was used to transform the text making it difficult to figure out how to read for someone that had no knowledge of the keyword. Governments employed cryptographers to secure sensitive information, leading to the establishment of formalized cryptographic systems.

The Nineteenth Century: Mechanical Devices and Systematization.

The nineteenth century witnessed the introduction of mechanical devices in the field of cryptography. New forms of encrypting text were also devised, such as the Playfair and Amsco ciphers. The period also witnessed the evolution of cryptology; the art of deciphering the codes and classified messages. This became important because it was an advancement since the establishment of sophisticated methods of encryption was under siege. The use of telegraphy in the late nineteenth century emphasized the need for secure communication over long distances, leading to the development of the **one-time pad**, an encryption method that is theoretically unbreakable if used correctly.

The World Wars: Cryptography in Conflict.

The two World Wars presented a historical watershed that changed forever how cryptography was approached. The Enigma machine of Nazi Germany epitomized the tussle between war and cryptography. The breakdown of the Enigma code by the allied cryptologists including the brilliant Alan Turing led to revolution in the fields of cryptography and computer. The post-world war period witnessed the introduction of computers that changed the stainless-steel cutlery of cryptography. In the 1970s, the Data Encryption Standard (DES) was dominated and thus the tendency of encryption moving toward standardization started to become evident which much further advanced the industry into the computer era.

The Digital Age: Public-Key Cryptography.

As the Internet gained a foothold in the later part of the twentieth century, so did the requirement for confidentiality in internal and external communication. The era of public-key cryptography commenced, enabling the exchange of information securely without the necessity to agree on a secret key in advance. The RSA algorithm, which was conceived by Ron Rivest, Adi Shamir, and Leonard Adleman, became essential in computer security systems primarily for online payments and transactions.

The Current State: The Quantum Threat.

As we move into the twenty-first century, it appears that the practice of cryptography is undergoing rapid changes with the arrival of quantum computers. The conventional encryption mechanisms such as RSA and ECC (elliptic curve cryptography) could be compromised by quantum computers that utilize algorithms like Shor's algorithm. This alternative threat has catalyzed the growth of new cryptographic methods that are projected to be resistant to quantum attacks.

Why Quantum-Resistant Strategies Are Essential?

The historical context of cryptographic evolution underscores the importance of adapting to emerging threats. Just as earlier cryptographic methods evolved in response to new challenges—be it the need for secure military communications or the advent of digital technology—today's cryptographers must innovate to protect information in a world where quantum computing is on the horizon. Quantum-resistant algorithms are essential because they provide a means to secure sensitive data against potential future attacks from quantum computers. This proactive approach ensures that as technology progresses, our methods of safeguarding information remain robust and reliable.

1.2 Quantum Computing Development

Recognizing Quantum Computing: Quantum computing leverages the concepts of quantum automechanics, such as superposition coupled with entanglement, to do intricate estimations at rates that go beyond the abilities of timeless computer systems. Unlike timeless little bits, which exist in a state of either 0 or 1, quantum little bits, or qubits, can exist in numerous states at the same time because of superposition, considerably raising computational power.

Effects for Cybersecurity: Traditional security protocols—such as RSA, NTRU [20], and also ECC—that rely upon distinct logarithm problems plus integer factorization are in danger from the growth of quantum computers. Existing concerns may be successfully settled by quantum computer systems, risking the safety and security of information secured with present methods.

Post-Quantum Cryptography: The subject of post-quantum cryptography has actually occurred in response to the susceptibilities that quantum computers offer.

Its objective is to develop file encryption formulas that preserve safety and security even when confronted with quantum enemies. In order to ensure information safety in the quantum age, these brand-new cryptographic strategies are made to make it through attacks from both standard and quantum computers [3].

Quantum Key Distribution: Quantum essential circulation [2] (QKD) is one more element of quantum computers's impact on cybersecurity; it makes use of the suggestions of quantum automechanics to give risk-free courses of interaction. Quantum entanglement is made use of in QKD to produce security secrets, offering a possibly thick way of safe interaction.

Difficulties and Opportunities: The cybersecurity atmosphere deals with opportunities and troubles in incorporating QKD and also quantum-resistant security formulas [4]. The difficulty for federal governments and companies is to switch to quantum-safe security methods while maintaining reverse compatibility with present systems. Yet this change likewise develops chances for imagination and also synergy in producing solid, quantum-resistant protection remedies.

The need to be prepared for cybersecurity remedies that can hold up against quantum computers is becoming increasingly apparent as this innovation establishes. R&D expenses are required for companies in a selection of sectors to secure their electronic facilities from future quantum strikes. In addition, the establishment of standards and laws for quantum-safe cybersecurity treatments is significantly aided by legislators as well as regulatory companies.

The appearance of quantum computers poses interfering pressure that requires aggressive cybersecurity procedures. In the quantum age, safeguarding delicate information calls for the development and application of safe interaction procedures combined with quantum-resistant security formulas. Scientists, magnates, and legislators should interact as the location of quantum computers is established to provide a risk-free electronic future while quantum advancements are made.

This book chapter has been organized into seven sections. In the first section, it describes an introduction to quantum computing and cybersecurity. The second and third sections describe the quantum computing fundamentals and vulnerabilities of current encryption protocols, respectively. The fourth section describes emerging quantum-resistant cryptographic techniques. Implementing quantum-resistant cybersecurity measures is discussed in Sect. 5. Section 6 describes sector-specific considerations: finance, healthcare, and government. Finally, Sect. 7 describes the conclusion.

2 Quantum Computing Fundamentals

Quantum mechanics is a fundamental theory in physics that describes the behavior of particles at the smallest scales, such as atoms and subatomic particles. At the core of quantum mechanics is the concept of quantum superposition, where particles can exist in multiple states simultaneously until measured. Another key principle is quantum entanglement, where particles become interconnected and the state of one particle instantaneously affects the state of another, regardless of the distance between them. Quantum uncertainty states that the position and momentum of particles cannot be precisely determined at the same time due to the Heisenberg uncertainty principle. Quantum mechanics also involves the notion of quantum tunneling, where particles can pass through energy barriers that would be insurmountable according to classical physics.

Quantum Computing Principles: Quantum computing leverages the principles of quantum mechanics to perform computations using quantum bits or qubits. Unlike classical bits that can be either 0 or 1, qubits can exist in a state of 0, 1, or both simultaneously due to superposition. Quantum parallelism allows quantum computers to perform multiple computations simultaneously by manipulating qubits in superposition. Quantum entanglement enables qubits to be correlated with each other, leading to faster computation and increased processing power. Quantum algorithms, such as Shor's algorithm [9] and Grover's algorithm [10], take advantage of quantum properties to solve complex problems more efficiently than classical algorithms. Quantum computers are particularly well suited for tasks like quantum simulation, cryptography, optimization, and machine learning due to their ability to process vast amounts of data in parallel.

2.1 *Quantum Computing Advancements*

- Researchers and companies, including IBM, Google, and Microsoft, are actively developing quantum hardware and software to advance quantum computing capabilities.
- Quantum supremacy, a milestone where a quantum computer outperforms the most powerful classical supercomputers, has been achieved by companies like Google.
- Quantum error correction is a crucial area of research aimed at mitigating errors that arise in quantum computations due to noise and decoherence.
- The emergence of cloud-based quantum computing platforms, such as IBM Quantum Experience, and Amazon Braket, allows users to access quantum resources and experiment with quantum algorithms.

Quantum mechanics forms the foundation of quantum computing, enabling the development of powerful quantum computers that have the potential to revolutionize various fields by solving complex problems more efficiently than classical computers. The principles of superposition, entanglement, and quantum parallelism underpin the unique capabilities of quantum computing, paving the way for exciting advancements in the realm of information processing and computational science.

Table 1 Comparison of classical versus quantum computing capabilities

Aspect	Classical computing	Quantum computing
Basic unit	Bits (0 or 1)	Qubits (0, 1, or superposition of both)
Data representation	Binary	Quantum states (superposition and entanglement)
Processing	Sequential processing	Parallel processing
Speed	Limited by classical physics	Potential for exponential speedup in certain tasks
Error correction	Relies on redundancy and error-detection codes	Potential for exponential speedup in certain tasks
Problem-solving	Effective for many practical tasks	Particularly suited for optimization and cryptography
Complexity	Polynomial time complexity for most problems	Offers exponential speedup for certain complex problems
Energy efficiency	Consumes significant energy for large computations	Potential for energy-efficient computations
Current state	Well-established technology	Still in early stages of development and research
Applications	Widely used in various industries	Quantum simulation, cryptography, optimization, AI

2.2 Comparison of Classical versus Quantum Computing Capabilities

Table 1 provides a comparison of classical computing and quantum computing [21] capabilities across different aspects, highlighting the fundamental differences and potential advantages of quantum computing over classical computing in certain problem domains.

The advent of quantum computing poses both challenges and opportunities for cybersecurity. Organizations need to prepare for the quantum era by understanding the implications of quantum computing on current cybersecurity measures, investing in research and development of quantum-resistant solutions, and fostering collaboration across disciplines to ensure a secure digital future.

2.3 Quantum Error Correction and the Challenges

Quantum error correction (QEC) is a crucial set of techniques designed to protect quantum information stored in qubits from errors and decoherence caused by noise. Quantum computers are inherently prone to errors, with current systems experiencing about 1 error per 1000 operations, while for practical applications, error rates need to be reduced to 1 in a trillion.

This significant gap highlights the need for effective error correction methods to make quantum computing viable.

Challenges in Quantum Error Correction

I. **High Error Rates**: The fundamental challenge lies in the high susceptibility of qubits to errors due to environmental disturbances. Even minor changes in temperature or interactions with surrounding molecules can lead to loss of information:

   a. As quantum computers scale up, the number of qubits required for effective computation increases, necessitating robust error correction strategies.

II. **Complexity of Error Correction**: Implementing QEC involves complex procedures such as encoding, measurement, and recovery, which differ significantly from classical error correction methods. For instance, quantum error correction often requires hundreds or thousands of physical qubits to encode a single logical qubit:

   a. This overhead complicates the design and implementation of quantum systems.

III. **Speed of Error Detection**: A major challenge is developing **real-time decoders** that can detect and correct errors quickly enough to prevent significant issues:

   a. Current research is focused on improving the efficiency of these decoders across various quantum platforms, including trapped ions and superconducting circuits.

IV. **Hardware Limitations**: Each quantum computing platform faces unique technological challenges. For example, superconducting qubits may have different error characteristics compared to trapped ions, requiring tailored QEC solutions:

   a. The need for hardware-efficient QEC methods is critical to advancing the field.

V. **Scalability**: As quantum computers grow in size and complexity, the **overhead costs** associated with implementing QEC become a significant barrier. Researchers are striving to develop scalable QEC technologies that can support fault-tolerant quantum computation.

3 Vulnerabilities of Current Encryption Protocols

Encryption is essential in the digital era for protecting private information, securing conversations, and preserving sensitive data. But since cybersecurity changes all the time, the shortcomings of the encryption techniques that are in use now become apparent. The purpose of this essay is to investigate the weaknesses in current encryption techniques and how they affect the security of digital data (Fig. 1).

1. Vulnerabilities in Symmetric Encryption: The foundation of contemporary cryptography is symmetric encryption, which encrypts and decrypts data using a shared secret key. However, the security of symmetric encryption protocols is seriously threatened by flaws, including key management problems, brute force attacks, reaction attacks [19], and the possibility of key leakage [14].
2. Weaknesses in Asymmetric Encryption: The usage of public and private key pairs in asymmetric encryption makes it vulnerable to several threats, including man-in-the-middle attacks, the possibility of illegal key creation [18], and the compromising of private keys. These flaws have the potential to compromise the integrity and secrecy of encrypted communications.

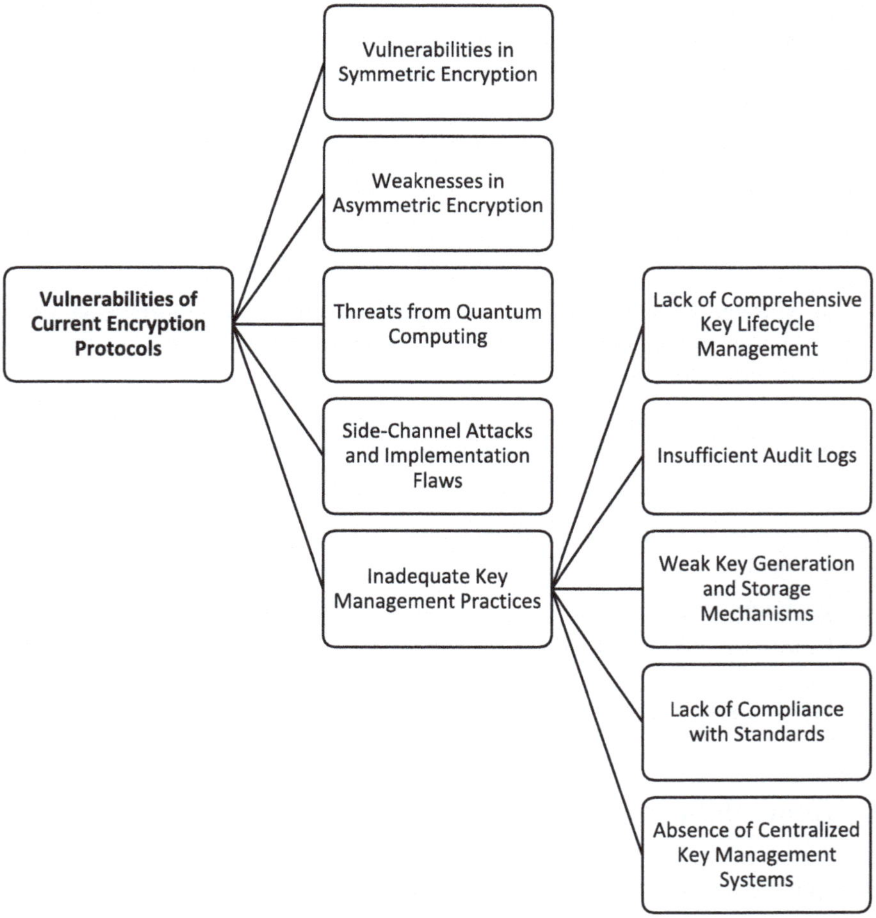

Fig. 1 Vulnerabilities of current encryption protocols

3. Exploitable Cryptographic Backdoors: A serious flaw that allows unauthorized access to protected data is the deliberate or inadvertent introduction of cryptographic backdoors in encryption systems. The existence of backdoors casts doubt on the reliability and robustness of encryption methods, thereby jeopardizing the confidentiality of sensitive data [16].
4. Threats from Quantum Computing: The encryption systems in place are seriously threatened by the speed at which quantum computing is progressing. Due to their ability to factor huge numbers efficiently and compute discrete logarithms, quantum computers have the potential to break popular cryptographic algorithms like RSA and ECC, making standard encryption techniques susceptible to quantum assaults [15].
5. Side-Channel Attacks and Implementation Flaws: Side-channel attacks, which take advantage of unintentional information leakage from the physical implementation of cryptographic systems, can target encryption techniques. Timing attacks and power analysis are examples of implementation defects in encryption libraries and protocols that may be used to undermine the security of encrypted data, exposing the weaknesses present in the real-world use of encryption techniques [17].
6. Inadequate Key Management Practices: Vulnerabilities in encryption protocols are caused by flaws in the processes of key production, distribution, and storage. Inadequate key management techniques reduce the efficacy of encryption and raise the possibility of unwanted access to sensitive data. These techniques include the use of weak or predictable keys, inadequate entropy, and a lack of secure key storage systems.

    a. Lack of Comprehensive Key Lifecycle Management: A lack of thorough key lifecycle management, which covers key production, exchange, storage, protection, usage, and replacement, is frequently the cause of inadequate key management. For encrypted data to be secure, cryptographic keys must be properly managed during their existence [5].
    b. Insufficient Audit Logs: Insufficient key management procedures can lead to a lack of reliable audit records. Maintaining a thorough history of every data encryption key, including its creation, usage, and destruction, requires extensive audit records. Lack of these records can make it more difficult to keep track of and monitor key-related activity, which could result in security flaws [6].
    c. Weak Key Generation and Storage Mechanisms: Weak key generation and storage techniques are frequently a part of inadequate key management. Using weak or predictable keys, having inadequate entropy, and not having safe key storage systems are a few examples of this. Issues in the creation and preservation of keys may jeopardize the integrity and secrecy of data that is encrypted [7].
    d. Lack of Compliance with Standards: Failure to adhere to established standards and best practices for key management might result from inadequate key management methods. Ensuring the efficacy and security of key

management procedures requires adherence to industry standards, such as the National Institute of Standards and Technology's Recommendation for Key Management (SP 800–57 Part 1) [8].

e. Absence of Centralized Key Management Systems: Lack of centralized encryption key management systems is a common sign of inadequate key management. Implementing uniform key management rules throughout the company and guaranteeing the security of every key depend on the centralization of key management systems [8].

The weaknesses in the encryption methods in use today highlight the urgent need for ongoing research and development in the field of cryptography. A multifaceted strategy is needed to address these weaknesses, including the creation of encryption algorithms that are resistant to quantum occurrences, reliable key management procedures, and proactive steps to lessen the effect of new threats. The cybersecurity community can strengthen encryption techniques and guarantee the confidentiality, integrity, and availability of sensitive data in the face of changing digital threats by recognizing and resolving these weaknesses.

3.1 Vulnerabilities in Symmetric/Asymmetric Encryption Protocols

Symmetric encryption relies on a single shared key for both encryption and decryption. While it is efficient and suitable for bulk data encryption, it has specific vulnerabilities:

1. **Key Distribution Problem**: The primary vulnerability in symmetric encryption is the **key distribution problem**. Since both parties must share the same key, securely transmitting this key over potentially insecure channels poses a significant risk. If the key is intercepted during transmission, an attacker can decrypt all communications using that key.
2. **Key Management**: Managing keys becomes increasingly complex as the number of users grows. Each pair of users requires a unique key, leading to an exponential increase in the number of keys that must be securely stored and managed. This complexity can lead to mismanagement and potential exposure of keys.
3. **Replay Attacks**: Symmetric encryption is vulnerable to **replay attacks**, where an attacker captures encrypted messages and re-sends them to trick the recipient into performing unauthorized actions. Without additional mechanisms like timestamps or nonces, the system cannot distinguish between legitimate and replayed messages.
4. **Brute Force Attacks**: Although modern symmetric algorithms like AES are designed to be secure, they are still susceptible to brute force attacks if the key length is insufficient. For example, a 128-bit key is currently considered

secure, but as computational power increases, longer keys (e.g., 256 bits) may be necessary to maintain security.
5. **Lack of Forward Secrecy**: Symmetric encryption does not inherently provide forward secrecy. If a key is compromised, all past communications encrypted with that key can be decrypted. This is a significant concern for sensitive communications.

Vulnerabilities in Asymmetric Encryption Protocols

Asymmetric encryption, or public-key cryptography, uses a pair of keys (public and private) for encryption and decryption. While it addresses some of the vulnerabilities of symmetric encryption, it introduces its own challenges:

1. **Key Length and Computational Overhead**: Asymmetric encryption requires significantly longer keys to achieve the same level of security as symmetric encryption. For instance, a 2048-bit asymmetric key is roughly equivalent in security to a 128-bit symmetric key. This increased key length results in higher computational overhead, making asymmetric encryption slower and less efficient for large data volumes.
2. **Man-in-the-Middle Attacks**: Asymmetric encryption is vulnerable to **man-in-the-middle (MitM) attacks**. If an attacker can intercept the public-key during transmission, they can replace it with their own, allowing them to decrypt and re-encrypt messages without either party being aware of the compromise. This highlights the importance of verifying public keys through trusted channels.
3. **Complexity of Key Management**: Managing public and private keys can be complex, especially in large systems. Users must ensure that their private keys remain confidential while distributing public keys securely. Mismanagement can lead to vulnerabilities, such as using an outdated or compromised key.
4. **Dependency on Secure Key Exchange**: Asymmetric encryption is often used to establish a secure channel for exchanging symmetric keys. If the underlying key exchange protocol (like Diffie-Hellman) is compromised, the security of the entire communication can be jeopardized.
5. **Vulnerability to Quantum Attacks**: Asymmetric encryption algorithms, particularly RSA and ECC, are vulnerable to potential future attacks from quantum computers. Quantum algorithms, such as Shor's algorithm, could efficiently break these encryption schemes, prompting the need for quantum-resistant alternatives.

4 Emerging Quantum-Resistant Cryptographic Techniques

A strenuous re-evaluation of cryptographic formulas has actually been set off by the coming close to arrival of quantum computers in a period identified by rapid developments in computer system modern technology. Since quantum computer systems have unrivaled handling ability, typical cryptography approaches, which are the structure of protected interaction as well as information safety, pose a risk of being hacked. As a result, developing cryptographic approaches that are susceptible to quantum variations has actually ended up being a leading issue in the area of cybersecurity.

When contrasted to conventional computer systems, a quantum computer has the capability to address detailed mathematical issues a lot quicker, consisting of distinct logarithms as well as integer factorization. Commonly utilized cryptographic techniques like RSA and ECC, which depend on the computational difficulty of such problems for their safety and security, are promptly intimidated by this capability.

Scientists and cryptographers have actually been striving to establish cryptographic techniques that stand up to the handling power of quantum computers in response to this brewing risk. In the post-quantum age, these brand-new approaches look to enhance the protection of secured information as well as interaction networks. Various considerable techniques have actually arisen in this area, each giving distinctive methods to attend to the cryptographic weak points advanced by quantum computers.

The development of lattice-based cryptography is one well-known location of research. Lattice-based cryptography [11] approaches capitalize on the detailed mathematical problems bordering lattices to offer a solid basis for electronic trademarks and also file encryption. Lattice-based file encryption is a practical alternative for quantum-resistant safety options due to its capacity to stand up to quantum attacks.

Furthermore, code-based file encryption has actually drawn in a great deal of passion as a quantum-resistant replacement. Code-based cryptography [12] remedies use solid security versus quantum risks by using error-correcting codes plus the intrinsic trouble of translating them. The emphasis of the quantum-resistant file encryption study has actually changed to code-based cryptography because of its prospective safety as well as endurance when faced with a quantum computer.

In addition, the area of multivariate quadratic equation-based encryption [13] has actually gained popularity as an encouraging method for attaining quantum-resistant safety. By taking advantage of the intrinsic problem of fixing multivariate quadratic formula systems, this technique offers a path in the direction of cryptographic durability in the age of quantum computers.

As a result of the unrivaled handling ability of quantum computer systems, basic cryptography methods could be threatened when faced with their possible introduction. As a result, the production of cryptographic techniques that are resistant to quantum mistakes has actually become an essential job in the area of cybersecurity. In the post-quantum age, this short article takes a look at several of the recently established quantum-resistant file encryption approaches that are suggested to reinforce the safety and security of secured information and interaction networks.

- Lattice-Based Cryptography: Lattice-based security has actually gathered much interest as a possibly efficient quantum-resistant security technique. By making use of the complexity of lattice-related mathematical concerns, it provides a solid basis for electronic trademarks and also provides security. Lattice-based security is a leading challenger for quantum-resistant protection remedies as a result of its capability to hold up against quantum attacks. This approach attempts to hold up

against the computer power of quantum computer systems by basing file encryption formulas on lattice concerns, consequently ensuring the lasting safety and security of secured information [11].
- Code-Based Cryptography: Code-based cryptography [12] is an additional remarkable quantum-resistant security approach. Making use of error-correcting codes and also the trouble of decoding them, this approach provides solid protection versus quantum risks. In the age of quantum computer code-based cryptography, techniques provide a high level of safety and security by relying on the problem of translating error-correcting codes. Code-based file encryption is an appealing choice for safeguarding delicate information from the possibly turbulent impacts of quantum computers as a result of its toughness as well as endurance.
- Multivariate-Quadratic-Equation-Based Cryptography: Multivariate-quadratic-equation-based cryptography [13] is based on multivariate quadratic formulas. An additional interesting location in the search for quantum-resistant safety and security is cryptography. By taking advantage of the intrinsic trouble in fixing multivariate quadratic formula systems, this technique offers a path in the direction of cryptographic durability in the age of quantum computers. This technique tries to provide solid protection versus quantum attacks by creating cryptographic formulas based upon the complicated framework of multivariate quadratic formulas, thus ensuring the integrity and privacy of electronic information (Table 2).

The introduction of quantum-resistant security comes as a positive action in the transforming danger situation, as the cybersecurity field supports the cutting-edge effect of quantum computers. The cybersecurity area is reinforcing the structures of safe and secure interaction plus information security versus the turbulent possibility of quantum computers by purchasing the R&D of these unique innovations.

The development of quantum-resistant cryptographic methods reveals a positive response to the transforming risk circumstances as the cybersecurity market supports the advanced effects of quantum computers. The ongoing growth of quantum-resistant strategies such as multivariate quadratic equation-based file encryption, lattice-based file encryption, code-based security, and various other quantum-resistant approaches stresses the dedication to preserving the protection as well as personal privacy of electronic information in the quantum age. This worked-with initiative shows the cybersecurity neighborhood convenience and also perseverance in securing delicate information in the middle of the unstoppable improvement of modern technology.

Table 2 Comparison between lattice-based, code-based, and multivariate-quadratic-equation-based cryptography

	Lattice-based cryptography	Code-based cryptography	Multivariate-quadratic-equation-based cryptography
Computational complexity	Generally efficient with polynomial-time algorithms for key generation and encryption Security relies on hard lattice problems (e.g., Learning With Errors—LWE) which are believed to be resistant to quantum attacks	Based on the hardness of decoding random linear codes, which is NP-hard Generally offers good performance for encryption and decryption	Solving systems of multivariate quadratic equations is NP-complete, leading to high computational complexity for decryption Generally slower than lattice and code-based methods due to the complexity of the underlying problems
Key sizes	Requires larger keys compared to traditional methods, which can impact performance	Typically has large key sizes, which can be cumbersome for storage and transmission	Typically has large key sizes
Security	Lack of meaningful reduction-based security guarantees for practical constructions	This provides a strong theoretical foundation for security, making it resistant to both classical and quantum attacks	Some proposed schemes have been broken, leading to uncertainty about their robustness
Complexity of implementation	Requires careful parameter selection to ensure security and efficiency	Requires specialized knowledge to implement securely and efficiently	Often less efficient in terms of key sizes and computational overhead compared to lattice-based and code-based methods

5 Implementing Quantum-Resistant Cybersecurity Measures

With its ability to change both clinical questions and the market, the quantum computer signifies the start of a brand-new age of unmatched handling capability. However, this technological development additionally offers a major danger to developed cryptography systems, possibly making them susceptible to quantum attacks.

The production and release of cybersecurity options that are unsusceptible quantum automechanics have actually ended up being essential in action to this brewing danger. In order to preserve information defense despite breakthroughs in quantum computers, this write-up checks out the strategies and variables to be taken into consideration while placing cybersecurity actions in positions that are immune to quantum computers.

- Carrying Out Quantum-Resistant Measures: Digital property security versus quantum assaults calls for a versatile technique when applying quantum-resistant cybersecurity services. A number of substantial strategies contain
- Transitioning to Post-Quantum Cryptography: A smooth transition from typical cryptographic systems to quantum-resistant choices needs cautious preparation along with prep work for the button to quantum-resistant cybersecurity procedures. Businesses are required to evaluate their existing information data sources, applications, and facilities to discover any weak points as well as their dependence on weak cryptographic approaches. To reinforce cybersecurity defenses prior to the quantum computer age, a roadmap for executing quantum-resistant cryptography remedies has to be developed, along with extensive threat evaluation as well as mitigation strategies.
- Security Awareness as well as Training: It is vital to show team members the risks of quantum computers and also the worth of safety and security remedies that are immune to them. Education efforts can help enhance public expertise on the relevance of secure treatments and stringent file encryption criteria.
- Regular Security Audits: It is crucial to perform regular safety and security audits to locate susceptibilities and also guarantee adherence to quantum-resistant criteria. This involves assessing the safety of security methods, updating equipment as well as software programs, and also keeping an eye out for any type of feasible hazards.
- Collaboration coupled with research: The cybersecurity neighborhood should interact to construct solid criteria for quantum-resistant cryptographic formulas, share proficiency, and work together on studies in order to effectively use quantum-resistant cybersecurity procedures. Governmental bodies, the academic community, and market consortia are the principals in advertising the standardization and using quantum-resistant file encryption techniques. In order to enhance its cumulative defenses versus quantum strikes, the cybersecurity landscape must focus on interoperability as well as agreement on quantum-resistant requirements.

5.1 Strategies to Implement Quantum-Resilient System

As quantum computing evolves, it is essential to adopt quantum-resistant cryptographic measures to protect sensitive data. This guide outlines a structured approach to implementing these measures, including step-by-step instructions, tools, frameworks, and a flowchart to visualize the process.

Step 1: Assess current cryptographic infrastructure.

1. Inventory existing systems:
   - Identify all cryptographic algorithms in use (e.g., RSA, ECC).
   - Document systems relying on these algorithms.
2. Evaluate vulnerabilities:
   - Analyze the potential impact of quantum attacks on existing systems.
   - Prioritize systems based on their criticality.

Step 2: Develop a quantum-resistant strategy.

1. Select quantum-resistant algorithms:
   - Research NIST's post-quantum cryptography candidates:
     - CRYSTALS-KYBER for key encapsulation.
     - CRYSTALS-DILITHIUM for digital signatures.
     - SPHINCS + for stateless signatures.
2. Plan for transition:
   - Create a roadmap for transitioning to quantum-resistant algorithms.
   - Prioritize systems based on complexity and criticality.

Step 3: Implement quantum-resistant algorithms.

1. Update cryptographic libraries:
   - Use libraries that support quantum-resistant algorithms:
     - OpenSSL: Integrate quantum-safe key exchange algorithms.
     - Liboqs: A library for quantum-resistant cryptographic algorithms.
2. Modify protocols:
   - Update communication protocols to incorporate quantum-resistant algorithms.
   - Ensure compatibility with existing systems.
3. Testing and validation:
   - Conduct rigorous testing of new algorithms.
   - Use methodologies such as timing attacks and power analysis to validate robustness.

Step 4: Monitor and maintain security.

1. Stay informed:
   - Keep up to date with developments in quantum computing and cryptography.
   - Follow research publications and engage with industry experts.

2. Conduct regular security assessments:
    - Schedule regular security audits to evaluate the effectiveness of quantum-resistant measures.
    - Update algorithms and protocols as new vulnerabilities are discovered.
3. Educate staff:
    - Provide training on quantum-resistant cryptography for developers and IT staff.
    - Foster a culture of security awareness regarding quantum computing implications.

Tools and Frameworks.

- OpenSSL: A widely used library that can be updated to include quantum-resistant algorithms.
- Liboqs: A library specifically designed for quantum-resistant cryptographic algorithms.
- NIST PQC standardization: Follow guidelines and recommendations from NIST for implementing post-quantum cryptography.
- Qiskit: A quantum computing framework for simulations and testing quantum algorithms.
- ProjectQ: An open-source software framework for quantum computing (Fig. 2).

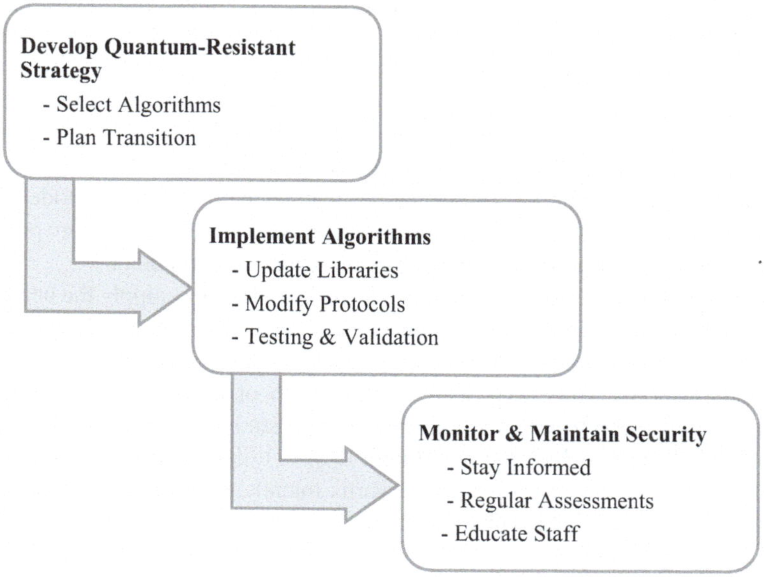

Fig. 2 Strategies to implement quantum-resilient system

The brewing arrival of quantum computers represents a landmark for cybersecurity, requiring a positive as well as a teamed-up technique to place cybersecurity safeguards that are immune to quantum computers in place. Cybersecurity specialists and companies can reinforce their defenses versus the turbulent capacity of quantum computers by establishing change strategies, investigating brand-new quantum-resistant cryptographic methods, completely recognizing the quantum hazard landscape, and advertising joint standardization initiatives. Approving the quantum period's development of cryptographic strength is vital to safeguarding exclusive details, coupled with keeping the electronic world's schedule, honesty, and privacy.

6 Sector-Specific Considerations: Finance, Healthcare, and Government

Current years have actually witnessed the increase of quantum computers as a turbulent modern technology that has the power to entirely change a variety of sectors, including the federal government, healthcare, and financing. Sector-specific issues become vital as these fields overcome the obstacles of including quantum computers in their procedures. This essay will certainly analyze just how the quantum computer is impacting the federal government, healthcare, and financing while showcasing a few of its feasible usages.

- Money Sector: Due to the fact that quantum computers can do complex estimations as well as threat analyses at a never-before-seen rate, it is possible to totally change the financial market. Quantum formulas, for instance, might all at once analyze and consider the area of profile optimization, which can cause a lot more accurate danger, as well as financial investment administration methods. As an example, JP Morgan has actually been boldly checking out the feasible applications of quantum computers in danger evaluation together with economic modeling. JP Morgan wishes to enhance its trading strategies as well as develop more powerful danger administration structures by utilizing quantum computers.
- Healthcare Sector: Quantum computers have a great deal to supply the healthcare market, specifically in the areas of genetics, clinical imaging, and drug advancement. Using quantum formulas might accelerate the simulation of molecular communications, which may lead to the growth of brand-new medicines and personalized therapy routines. One remarkable study includes the pharmaceutical firm Novartis, which partnered with IBM to utilize quantum computing for molecular framework simulations. Novartis intends to enhance the accuracy of medication tasks and speed up medication growth by making use of the possibility of a quantum computer.
- Government sector: Quantum computers provide prospective together with troubles for the general public market. On the one hand, the production of safe and secure security techniques enabled by quantum computers can enhance cybersecurity steps. On the other hand, the appearance of quantum computers likewise

triggers concerns concerning the safety of the approved security methods. Federal government companies are presently considering quantum-resistant cryptography options to fix these problems. In an initiative to secure delicate federal government information, the U.S. National Institute of Standards and Technology (NIST) has actually remained on the leading edge of this worldwide initiative to develop post-quantum cryptography criteria.

Stakeholders in the federal government, medical care, and financing fields should meticulously take into consideration the effects of quantum computers as they establish and invite advancement. While the transformative possibility of quantum computers is obvious, it likewise calls for a positive technique to attend to safety and security, along with governing factors to consider.

A brand-new age of performance and imagination is being ushered in by the unification of quantum computers right into the federal government, medical care, and economic industries. By making use of quantum computers and personalizing their usage to industry-specific needs, businesses might be able to reach previously hard-to-reach potential customers for development and advancement.

It is important that we keep in mind the specific problems as well as the problems that each market brings as we start this quantum-powered trip, as well as that we make use of quantum computers as a driver for development. There are countless opportunities in the quantum future.

7 Role of Quantum-Safe Algorithms and Protocols

The impending hazard of quantum computers presents an essential difficulty for the protection of electronic interaction and information stability in this period of rapid technical development. The need for quantum-safe formulas as well as procedures is expanding as quantum computers get closer to truth. This essay will certainly check out the importance that quantum-safe formulas and procedures play in securing our electronic facilities, coupled with the journalism required to place them right into method.

- Comprehending the Quantum Threat: Conventional cryptography strategies like RSA and ECC count on the intricacy of distinct logarithm concerns or factoring substantial integers. Nonetheless, by utilizing their enormous computer capability to resolve these problems rapidly, quantum computer systems have the potential to make traditional security criteria out of date. The safety and security of delicate information, such as monetary deals, case histories, and main documents, is seriously endangered by this.
- The Emergence of Quantum-Safe Solutions: Post-quantum cryptography or quantum-safe formulas are cryptographic strategies created to stand up to attacks by quantum computer systems. These formulas ensure information safety and security when faced with quantum hazards, given that they are based on mathematical concerns that are believed to be immune to quantum formulas. Lattice-based

code-based [12] and hash-based security are a couple of instances of quantum-safe formulas, each with unique benefits in driving away quantum attacks.

- The Role of Quantum-Safe Protocols: To safeguard electronic interactions from feasible quantum strikes, quantum-safe procedures need to be executed along with quantum-safe formulas. Safe interaction methods coupled with cryptographic essential exchange systems that can hold up against quantum attacks belong to quantum-safe procedures. Organizations might safeguard the privacy and honesty of their information as well as lower the threats related to quantum computers by integrating quantum-safe procedures right into their network style.
- Upholding Quantum-Safe Initiatives: To respond to the impending danger presented by quantum computers, leading companies and criteria bodies have actually remained in the forefront of backing quantum-safe initiatives. The evaluation and standardization of quantum-resistant cryptographic formulas is being led by the National Institute of Standards and Technology (NIST). The objective of this task is to provide a detailed structure of quantum-safe methods as well as formulas to enhance electronic system safety and security in the quantum period.
- The Imperative Need for Quantum-Safe Adoption: The need to embrace quantum-safety techniques cannot be overemphasized, as quantum computers are becoming ever better for daily usage. To future-proof their protection position, businesses in all markets are required to proactively review their cryptographic framework and also switch to quantum-safe formulas as well as methods. This aggressive technique is crucial for shielding delicate information from the turbulent capacity of quantum computers and ensuring the security of vital systems.

The relevance of quantum-safe formulas and procedures cannot be exaggerated while browsing the advanced area of quantum computers. Organizations might reinforce their electronic framework as well as secure the discretion and stability of their information versus quantum strikes by utilizing quantum-safe modern technologies. The application of quantum-safe formulas as well as procedures is not just a selection but a calculated demand as we come close to a future driven by quantum modern technology.

Making sure information stability and also safe and secure interaction in the future will certainly rely on our very early fostering of quantum-safe formulas as well as procedures. It is essential to prepare the ground for a secure and durable future by enhancing our electronic facilities for the quantum age.

8 Conclusion

Prioritizing quantum-resilient security measures is crucial in the face of the upcoming quantum computing revolution. This involves a proactive approach to assessing cryptographic infrastructure and transitioning to quantum-safe algorithms and protocols.

This ensures the resilience of digital infrastructure in the quantum era. Quantum-resilient security measures are essential for safeguarding sensitive data, such as financial transactions, healthcare records, and government communications, from potential quantum attacks. They also help future-proof organizations' security posture against the transformative impact of quantum computing. Collaborative engagement and standardization within the industry are also essential for fostering a comprehensive framework of quantum-resilient security measures. The time to prioritize these measures is now, and organizations must work together to fortify their defenses and secure their digital future in the quantum era. This proactive approach is the cornerstone of a secure and resilient digital landscape.

References

1. N. Kilber, D. Kaestle, S. Wagner, Cybersecurity for quantum computing (2021). arXiv:2110.14701
2. M. Mehic, M. Niemiec, S. Rass, J. Ma, M. Peev, A. Aguado, M. Voznak, Quantum key distribution: a networking perspective. ACM Comput. Surv. (CSUR) **53**(5), 1–41 (2020)
3. J.P. Mattsson, B. Smeets, E. Thormarker, Quantum-resistant cryptography. arXiv:2112.00399
4. D. Ott, C. Peikert, Identifying research challenges in post quantum cryptography migration and cryptographic agility (2019). arXiv:1909.07353
5. https://www.techtarget.com/searchsecurity/tip/Key-management-challenges-and-best-practices
6. https://www.fortanix.com/blog/10-key-management-best-practices-you-should-know
7. https://phoenixnap.com/blog/encryption-key-management-best-practices
8. https://www.thesslstore.com/blog/12-enterprise-encryption-key-management-best-practices/
9. https://utimaco.com/service/knowledge-base/post-quantum-cryptography/what-shors-algorithm
10. https://learning.quantum.ibm.com/course/fundamentals-of-quantum-algorithms/grovers-algorithm
11. D. Micciancio, O. Regev, Lattice-based cryptography, in *Post-quantum Cryptography*, (Springer Berlin Heidelberg, Berlin, Heidelberg, 2009), pp. 147–191
12. N. Sendrier, Code-based cryptography: state of the art and perspectives. IEEE Secur. Priv.Secur. Priv. **15**(4), 44–50 (2017)
13. C. Wolf, Multivariate quadratic polynomials in public key cryptography. *Cryptology ePrint Archive* (2005)
14. M.U. Bokhari, Q.M. Shallal, A review on symmetric key encryption techniques in cryptography. Int. J. Comput. Appl. **147**(10) (2016)
15. W. Cui, T. Dou, S. Yan, Threats and opportunities: blockchain meets quantum computation, in *2020 39th Chinese Control Conference (CCC)* (IEEE, 2020), pp 5822–5824)
16. C. Easttom, A study of cryptographic backdoors in cryptographic primitives, in *Electrical Engineering (ICEE), Iranian Conference on* (IEEE, 2018), pp. 1664–1669
17. J. Fan, X. Guo, E. De Mulder, P. Schaumont, B. Preneel, I. Verbauwhede, State-of-the-art of secure ECC implementations: a survey on known side-channel attacks and countermeasures, in *2010 IEEE International Symposium on Hardware-Oriented Security and Trust (HOST)* (IEEE, 2010), pp. 76–87
18. R. Nayak, A.K. Nanda, P. Awasthi, L. Kumar, Multiple private keys with NTRU cryptosystem. Int. J. Res. Comput. Commun. Technol. **4**(3) (2015)
19. R. Nayak, J. Pradhan, C.V. Sastry, Reaction attacks in the matrix scheme of NTRU cryptosystem, in *International Conference on Advances in Information Technology and Mobile Communication* (Springer Berlin Heidelberg, Berlin, Heidelberg, 2011), pp. 27–32

20. A.K. Nanda, R. Nayak, L.K. Awasthi, NTRU with Gaussian integer matrix. Int. J. Adv. Res. Comput. Sci. Software. Eng. **5**, 359–365 (2015)
21. M. Belkhir, H. Benkaouha, E. Benkhelifa, Quantum versus classical computing: a comparative analysis, in *2022 Seventh International Conference on Fog and Mobile Edge Computing (FMEC)* (IEEE, 2022), pp. 1–8

Quantum Computing: Threat to Cybersecurity

Debranjan Pal and Dilip Sau

1 Introduction

The pace of advancement in quantum computing is accelerating rapidly. It will soon be that their capabilities reach a point where they can tackle problems that were once deemed unsolvable. However, this advancement brings about a significant global challenge: the potential for quantum computers to compromise some of the most commonly employed security protocols worldwide. Classical cryptography began with simple techniques like Caesar and Vigenère ciphers, evolving with devices like the Enigma machine in WWII. In the 1970s, IBM developed the Data Encryption Standard (DES), a widely used symmetric encryption algorithm. However, DES's vulnerability led to its replacement by the Advanced Encryption Standard (AES) in 2001, offering stronger 128, 192, or 256-bit encryption. Meanwhile, public-key cryptography emerged with RSA in 1977, created by Rivest, Shamir, and Adleman. RSA revolutionized secure communication by using the difficulty of factoring large numbers, forming the basis for modern encryption alongside AES. The emergence of quantum computing is poised to necessitate revisions to encryption techniques. Presently, widely utilized asymmetric algorithms rely on intricate mathematical dilemmas, such as factoring large numbers, tasks that can span thousands of years even on the most formidable supercomputers today. However, research conducted by Peter Shor [25] at MIT over two decades ago demonstrated the theoretical possibility of resolving these challenges in mere days or hours using a large-scale quantum computer. Consequently, future quantum computers may compromise asymmetric encryption solutions that hinge their security on integer factorization or discrete logarithms. While symmetric algorithms remain unaffected by Shor's algorithm [25],

D. Pal (✉)
Indian Institute of Technology Madras, Chennai, India
e-mail: debranjan.crl@gmail.com

D. Sau
Indian Institute of Technology Kharagpur, Kharagpur, India

S. B. Goyal et al. (eds.), *Quantum Computing, Cyber Security and Cryptography*,
https://doi.org/10.1007/978-981-96-4948-8_7

the potency of quantum computing underscores the need for larger key sizes. For instance, sizable quantum computers leveraging Grover's algorithm [15] employing quantum principles to rapidly search databases could yield a quadratic enhancement in brute-force attacks on symmetric encryption algorithms like AES [9]. Key sizes should be doubled to bolster resilience against such attacks and maintain equivalent protection levels. For AES [9], this translates to employing 256-bit keys to preserve the current 128-bit security strength. Despite the absence of commercially available large-scale quantum computers, initiating quantum cybersecurity solutions proactively holds substantial advantages. Presently, malevolent entities can intercept secure communications for future exploitation. Consequently, when extensive quantum computing resources become accessible, this immense computational power could be utilized to breach encryption and access intercepted communications.

We delve into the basic approaches of quantum computing and elucidate how quantum algorithms, such as Deutsch-Jozsa [11], Simon's [26], Shor's [25], and Grover's algorithms [15], can efficiently solve some computational issues that form the basis of modern cryptographic schemes. Notably, Shor's algorithm [25] is applied to attack public-key cryptography like RSA [13] and discrete logarithm problems. Grover's quantum search algorithm [15] threatens various algorithms, including block ciphers like AES [9] or hash functions like SHA [16]. Although the cryptographic community grasps the general implications of these algorithms, ongoing research focuses on determining their precise resource requirements. Furthermore, we describe the ongoing research efforts and explore potential countermeasures to mitigate the threats posed by quantum cryptanalysis.

Organization Section 2 describes has preliminaries about the quantum bits. In Sect. 3 we explain the Deutsch-Jozsa algorithm along with its applications. Simon's period-finding algorithm and its application in cryptography are presented in Sect. 4. Next, we provide the details about Grover's searching algorithm in Sect. 5. Section 6 describes the algorithm and applications of Shor's factoring algorithm. In Sect. 7 we discuss the quantum threats on cybersecurity. Section 8 explains the importance of quantum-safe cryptography. Section 9 concludes this chapter.

2 Representation and Measurement of Quantum Bits

Let us imagine a system with different states, like an electron in a hydrogen atom, which can be in various energy levels, such as ground or excited states. If we set a maximum limit on the total energy, the electron can only be in either the ground state or one of the excited states. In a classical setup, we use the state of this system to represent a number between 0 and the number of excited states minus one. In quantum mechanics, the rules of superposition demonstrate that when a system can exist in two distinct states, it may also be found in a combination of these states with complex coefficients.

Quantum State The quantum state of the system with k states is defined by a list of k complex numbers, like $\{a_0, \ldots, a_{k-1}\} \in \mathbb{C}$. These numbers are normalized, meaning $\sum_i a_i^2 = 1$. So, we can think of the system's state as a vector with k dimensions.

$$|\gamma\rangle = \begin{bmatrix} a_0 \\ a_1 \\ \vdots \\ a_{k-1} \end{bmatrix}$$

2.1 What Is a Qubit?

The bit is the foundational notion in classical computation and information theory. On the other hand, quantum computing and information are founded upon a comparable notion known as the quantum bit, often abbreviated as qubit. Qubits represent quantum systems with two states. A qubit is expressed using a unit vector $\begin{bmatrix} a \ b \end{bmatrix}^T \in \mathbb{C}^2$. In Dirac notation, we can write,

$$|\gamma\rangle = a|0\rangle + b|1\rangle, \text{ where } a, b \in \mathbb{C} \text{ and } a^2 + b^2 = 1$$

That is, a qubit is a linear combination of the standard basis vectors $|0\rangle$ and $|1\rangle$.

Two Qubits state is similarly represented as a linear combination of the standard basis vectors $\{|00\rangle, |01\rangle, |10\rangle, |11\rangle\}$, i.e., $|\psi\rangle = a_{00}|00\rangle + a_{01}|01\rangle + a_{10}|10\rangle + a_{11}|11\rangle$. Equivalently, we can represent $|\psi\rangle$ as a vector $\begin{bmatrix} a_{00}, a_{01}, a_{10}, a_{11} \end{bmatrix}^T$ which satisfies the equation $(a_{00}^2 + a_{01}^2 + a_{10}^2 + a_{11}^2) = 1$.

2.2 Measurement of Qubits

A measurement performed on this k-state system results in one of a maximum of k potential outcomes, namely an integer ranging from 0 to $k - 1$. When measuring $|\gamma\rangle$ in standard basis, the probability of getting $|j\rangle$ is $|a_i|^2$. Significantly, the measurement process actively modifies the old state of the quantum system to a new state. For example, a measurement corresponds to the colloquial notion of "observing" a qubit, leading to the immediate demolish of its quantum state and generating any one of the two classical states $\begin{bmatrix} 1 \ 0 \end{bmatrix}^T$ or $\begin{bmatrix} 0 \ 1 \end{bmatrix}^T$. When a qubit, represented by the quantum state vector $\begin{bmatrix} a \ b \end{bmatrix}^T$, undergoes measurement, the outcome 0 is obtained with a probability of a^2, and the outcome 1 is obtained with a probability of b^2. Following outcome 0, the qubit assumes a new state of $\begin{bmatrix} 1 \ 0 \end{bmatrix}^T$, while after outcome 1, its state becomes $\begin{bmatrix} 0 \ 1 \end{bmatrix}^T$. It's important to note that these probabilities sum up to 1 due to the normalization condition $a^2 + b^2 = 1$.

Measuring Two Qubits States Measuring two-qubit states closely resembles single-qubit measurements. When someone measuring the state $\left[a_{00}\ a_{01}\ a_{10}\ a_{11}\right]^T$, the probability of $|00\rangle$ is $|a_{00}|^2$, the probability of $|01\rangle$ is $|a_{01}|^2$, the probability of $|10\rangle$ $|a_{10}|^2$, and the probability of $|11\rangle$ is $|a_{11}|^2$. The choice of variables a_{00}, a_{01}, a_{10}, and a_{11} was deliberately made to emphasize this correspondence. Following the estimation, if the outcome is $|00\rangle$, the quantum state of the two-qubit system collapses to $|00\rangle = \left[1\,0\,0\,0\right]^T$.

2.3 Hadamard Transform

The Hadamard transform belongs to a group of Fourier transforms. In quantum computing, the Hadamard gate rotates a qubit, changing $|0\rangle$ and $|1\rangle$ to superposition states. These superposition states have equal weights of $|0\rangle$ and $|1\rangle$. We often choose phases to achieve this

$$H|0\rangle = \frac{(|0\rangle + |1\rangle)}{\sqrt{2}}$$
$$H|1\rangle = \frac{(|0\rangle - |1\rangle)}{\sqrt{2}}$$

For two qubits, the Hadamard transformation can be represented as

$$\begin{aligned} H^{\otimes 2}|0^{\otimes 2}\rangle &= H|0\rangle \otimes H|0\rangle \\ &= \frac{(|0\rangle + |1\rangle)}{\sqrt{2}} \otimes \frac{(|0\rangle + |1\rangle)}{\sqrt{2}} \\ &= \frac{1}{2}[|00\rangle + |01\rangle + |10\rangle|11\rangle] \end{aligned}$$

For n qubits, we can represent the Hadamard transformation as

$$\begin{aligned} H^{\otimes n}|0^{\otimes n}\rangle &= H|0\rangle \otimes H|0\rangle \otimes H|0\rangle \otimes \ldots H|0\rangle \\ &= \frac{1}{\sqrt{2^n}} \sum_{x=0}^{2^n-1} |x\rangle \end{aligned}$$

3 Deutsch-Jozsa Algorithm

By employing the Deutsch-Jozsa method [12], one can ascertain whether a provided Boolean function is either constant or balanced. Take, for instance, a function that receives inputs of 0 and 1 and produces outputs of 0 or 1 accordingly. The function qualifies as constant if all outputs are consistently 0 or 1. Conversely, if half of the inputs are zeroes and the rest are ones, the function is classified as balanced.

3.1 Classical Approach

Consider $f(x)$ as a mapping from $\{0, 1\}^n$ to $\{0, 1\}$. There are 2^n different inputs to $f(x)$. We define the function $f(x)$ as balanced if $f(x) = 0$ for exactly half of all the different inputs (i.e., 2^{n-1} inputs), and $f(x) = 1$ for the rest of the inputs. If $f(x)$ is constant, it means $f(x) = 0$ for all different inputs, or $f(x) = 1$ for all different inputs. However, in this specific problem, we consider $f(x)$ must be balanced or constant. In classical circuits, what are the fewest, and most of the time, do we need to run tests to determine if $f(x)$ is balanced or constant? We have 2^n different inputs for this decision. At worst, the first 2^{n-1} inputs could all result in 0s or all 1s. Then, the $(2^{n-1} + 1)$-th input must clarify if $f(x)$ is neither balanced nor constant.

Can quantum circuits improve on this? Yes, a Deutsch-Jozsa algorithm [11], a variation of the Deutsch algorithm [10], solves the problem in just one run. Compared to the worst-case scenario with classical circuits, it's $2^{n-1} + 1$ times faster asymptotically, which significantly improves when n is large.

3.2 Quantum Approach

We represent the function $f(x)$ using a quantum gate U_f (see Fig. 1). In the first stage, the input can be represented as

$$\psi_1 = |0^{\otimes N}\rangle|1\rangle$$

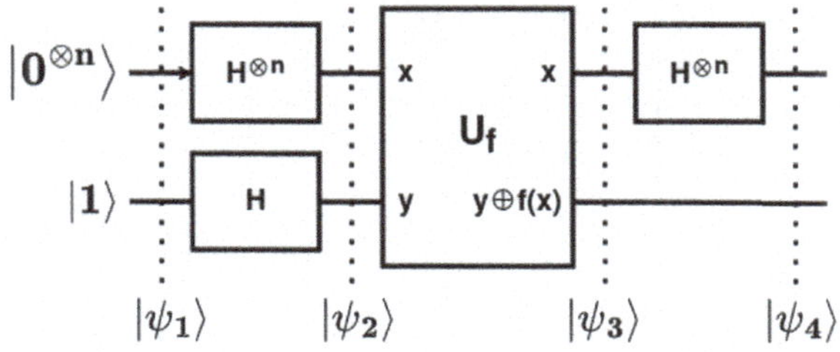

Fig. 1 The quantum circuit for the Deutsch-Jozsa algorithm begins with the input state $|\psi_1\rangle$, where the first n qubits are initialized to $|0\rangle$ and the $n + 1$th qubit is initialized to $|1\rangle$. Applying $n + 1$ Hadamard gates to these qubits results in the equal superposition state $|\psi_2\rangle$. Next the oracle U_f corresponding to the Boolean function f is applied to this state, producing the state $|\psi_3\rangle$. Finally, n Hadamard gates are applied to the first n qubits, transforming the state to $|\psi_4\rangle$. Now by measuring the first n qubits of $|\psi_4\rangle$ we can find the balanceness of the function f

The input for U_f is applied for superposition,

$$|\psi_2\rangle = \frac{1}{2^{n/2}} \sum_{x=0}^{2^n-1} |x\rangle \frac{(|0\rangle - |1\rangle)}{\sqrt{2}}$$

U_f transforms ψ_2 to

$$\begin{aligned}|\psi_3\rangle &= \frac{1}{2^{n/2}} \sum_{x=0}^{2^n} |x\rangle \left(\frac{(|0\rangle - |1\rangle)}{\sqrt{2}} \oplus f(x) \right) \\ &= \frac{1}{2^{(n+1)/2}} \sum_{x=0}^{2^n-1} |x\rangle \left(|0 \oplus f(x)\rangle - |1 \oplus f(x)\rangle\right) \\ &= \frac{1}{2^{(n+1)/2}} \sum_{x=0}^{2^n-1} |x\rangle \left(|f(x)\rangle - |1 \oplus f(x)\rangle\right)\end{aligned}$$

Now,

$$|\psi_3\rangle = \begin{cases} \frac{1}{2^{(n+1)/2}} \sum_{x=0}^{2^n-1} |x\rangle \left(|0\rangle - |1\rangle\right) & \text{if } f(x) = 0 \\ \frac{1}{2^{(n+1)/2}} \sum_{x=0}^{2^n-1} |x\rangle \left[-\left(|0\rangle - |1\rangle\right)\right] & \text{if } f(x) = 1 \end{cases}$$

Hence,

$$\begin{aligned}|\psi_3\rangle =& \frac{1}{2^{(n+1)/2}} \sum_{x=0}^{2^n-1} |x\rangle \otimes (-1)^{f(x)} \left(|0\rangle - |1\rangle\right) \\ &= \frac{1}{2^{(n+1)/2}} \sum_{x=0}^{2^n-1} (-1)^{f(x)} |x\rangle \otimes \left(|0\rangle - |1\rangle\right)\end{aligned}$$

Next, we can address the interference $H^{\otimes n}$,

$$|\psi_4\rangle = H^{\otimes n} \left[\frac{1}{2^{(n+1)/2}} \sum_{x=0}^{2^n-1} (-1)^{f(x)} |x\rangle \right] \otimes \left(|0\rangle - |1\rangle\right)$$

Consider,

$$H^{\otimes n}|x\rangle = \left[\frac{1}{2^{n/2}} \sum_{y=0}^{2^n-1} (-1)^{x \odot y} |y\rangle \right]$$

Where $x \odot y = x_0.y_0 + x_1.y_1 + \cdots + x_{n-1}.y_{n-1}$

Hence, we can deduce as

$$H^{\otimes n}\left[\frac{1}{2^{n/2}}\sum_{x=0}^{2^n-1}(-1)^{f(x)}|x\rangle\right]=\frac{1}{2^n}\sum_{x=0}^{2^n-1}(-1)^{f(x)}\left[\sum_{y=0}^{2^n-1}(-1)^{x\odot y}|y\rangle\right]$$
$$=\frac{1}{2^n}\sum_{y=0}^{2^n-1}\left[\sum_{x=0}^{2^n-1}(-1)^{f(x)}(-1)^{x\odot y}\right]|y\rangle$$

That is,

$$|\psi_4\rangle=\frac{1}{2^{(n+1)/2}}\sum_{y=0}^{2^n-1}\left[\sum_{x=0}^{2^n-1}(-1)^{f(x)}(-1)^{x\odot y}\right]|y\rangle\otimes(|0\rangle-|1\rangle)$$

Now we measure the first register, the probability to obtain the state $|0^{\otimes n}\rangle$ is given by

$$\left|\frac{1}{2^n}\left[\sum_{x=0}^{2^n-1}(-1)^{f(x)}(-1)^{x\odot 0}\right]\right|^2=\left|\frac{1}{2^n}\left[\sum_{x=0}^{2^n-1}(-1)^{f(x)}\right]\right|^2$$
$$=\left|\frac{1}{2^n}(-1)^{f(x)}\left[\sum_{x=0}^{2^n-1}\right]\right|^2$$
$$=\left|(-1)^{f(x)}\right|^2$$
$$=1\text{ , if } f(x) \text{ is constant.}$$
$$=0\text{ , if } f(x) \text{ is balanced}$$

Therefore, by measuring the first n qubits, it is possible to ascertain whether f(x) is constant ($|0^{\otimes n}\rangle$ occure with probability 1) or balanced ($|0^{\otimes n}\rangle$ occure with probability 0). Therefore, we can distinguish the constant and the balanced function with $O(1)$ time, which is the main advantage of this algorithm. The best classical algorithm takes exponential time, whereas this algorithm solves it in constant time.

3.3 *Application*

André Schrottenloher [23] explains the quantum distinguisher applied to the One-time Pad leverages the Deutsch-Jozsa algorithm [12], enabling it to discern between a random function and a constant one with just a single query.

Nagata et al. [21] describe how the specialized Deutsch-Jozsa algorithm [11] is applied in quantum key distribution. They use a Greenberger-Horne-Zeilinger (GHZ) state. Alice and Bob settle on a function f that can be either constant or balanced, and they keep this function secret from Eve. They aim to determine if their chosen function is constant or balanced without revealing it to Eve. If the function is constant, the output qubits become fully entangled in a GHZ state. Else they remain separable. Alice and Bob measure the GHZ state. Based on their measurement outcome, Alice and Bob share one secret bit, which helps them to evaluate f. The presence of Eve disrupts the fully entangled state, leading to a separable state.

4 Simon's Period Finding Algorithm

Simon's algorithm [26], initially introduced by D. Simon in 1994, reveals the hidden shift used in a function. Suppose $f : \{0, 1\}^n \rightarrow \{0, 1\}^n$ is a function and for any $x, y \in \{0, 1\}^n$ if f satisfy $f(x) = f(y)$ if and only if $x \oplus y = s$, then $s \in \{0, 1\}^n$ is the hidden secret. The values of s can be either $s = 0^n$ or $s \neq 0^n$. In the case of $s = 0^n$, f is a one-to-one function; otherwise, a two-to-one function. Simon's algorithm gives a polynomial-time solution to distinguish one-one or two-one functions, as well as the hidden shift used in the function.

4.1 Classical Query Complexity

In a traditional scenario, we provide a bit string x_1 as input to our query and observe the output $f(x_1)$. This process is repeated for another input x_2 to obtain $f(x_2)$. Before delving into the number of queries required to solve Simon's problem (i.e., compute s), let's initially address the more straightforward query: how exactly do we compute s initially? Imagine you input x_i into the query and obtain $f(x_i)$. After further inputs, you send x_j to learn $f(x_j)$. By noting the function values, you realize that $f(x_i) = f(x_j)$. Can you then deduce s from this information? If yes, how would you proceed?

We can ascertain s, effectively resolving Simon's problem, by identifying a collision pair. The current inquiry becomes: How many queries are necessary to discover such a collision pair? One approach is to query every possible value of the function f. By doing so, we would eventually find a matching pair. How many queries would this require? Well, f is defined on bit strings of length n, totaling 2^n such bit strings. Hence, we can solve Simon's problem with a maximum of 2^n queries. Can we improve upon this? Yes, but only marginally. We can achieve a quadratic improvement over 2^n queries using the same rationale (or lack thereof) found in the well-known birthday paradox. Imagine you're in a room with 23 individuals. What's the likelihood that any two individuals share a birthday? Are you taken aback to learn that it's approximately 50%? One might mistakenly anticipate needing around 365 people to encounter a match, considering there are 365 d in a year. However, the crucial point is that we are concerned with any pair sharing a birthday. Let's consider a scenario with n people. How many pairs can we create? Initially, there

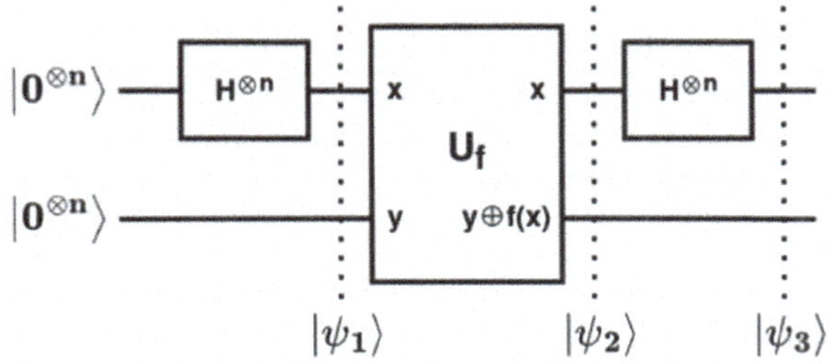

Fig. 2 The quantum circuit for Simon's algorithm begins with the input state $|0^n\rangle \otimes |0^n\rangle$. Applying n Hadamard gates to the first n qubits results in the equal superposition in the first n qubit state $|\psi_1\rangle$. Next the oracle U_f corresponding to the function f is applied to $|\psi_1\rangle$, producing the state $|\psi_3\rangle$. Finally, n Hadamard gates are applied to the first n qubits, transforming the state to $|\psi_3\rangle$ and measuring the first n qubits to determine whether there is a hidden shift or not

are n choices for the first person, followed by $n - 1$ choices for the second person (as one person has already been chosen). Thus, the total number of pairs among n people is $\binom{n}{2} = O(n^2)$. Consequently, the number of pairs doesn't increase linearly with the number of people; instead, it grows quadratically as the leading order term.

Similar principles apply to Simon's algorithm. Let's assume we randomly send in t input values to our query. After t queries, we have approximately t^2 pairs. We anticipate encountering a matching pair when $t^2 = O(2^n)$, implying $t = O(2^{n/2})$.

4.2 *Quantum Version*

Establish a random superposition for $\frac{1}{\sqrt{2}}.(|r\rangle + |r \oplus s\rangle)$. Then, set up an equal superposition for the bit string r and $r \oplus s$, where r is a random n-bit string, and s is the secret information. Fourier sample to generate a random y, such that $y : y.s = 0$ (mod 2). Let $y = y_1.y_2.\dots.y_n$ and $s = s_1.s_2.\dots.s_n$, then the linear equations are,

$$\begin{aligned} y_1^{(1)}.s_1 + y_2^{(1)}.s_2 + \cdots + y_n^{(1)}.s_n &= 0 \ (\text{mod } 2) \\ y_1^{(2)}.s_1 + y_2^{(2)}.s_2 + \cdots + y_n^{(2)}.s_n &= 0 \ (\text{mod } 2) \\ &\vdots \\ y_1^{(n-1)}.s_1 + y_2^{(n-1)}.s_2 + \cdots + y_n^{(n-1)}.s_n &= 0 \ (\text{mod } 2) \end{aligned} \quad (1)$$

The algorithm begins with two registers initialized to $|0\rangle^{\otimes n}$ and $|0\rangle^{\otimes n}$ (see Fig. 2). Subsequently, we apply the Hadamard transform to the first register, resulting in the state

$$|\psi_1\rangle = \frac{1}{\sqrt{2^n}} \sum_{p\in\{0,1\}^n} |p\rangle|0\rangle^{\otimes n}$$

After applying the function f we get

$$|\psi_2\rangle = \frac{1}{\sqrt{2^n}} \sum_{p\in\{0,1\}^n} |p\rangle|f(p)\rangle$$

Next, the result of applying the Hadamard transformation,

$$\begin{aligned}|\psi_3\rangle &= \frac{1}{\sqrt{2^n}} \sum_{p\in\{0,1\}^n} \left[\frac{1}{\sqrt{2^n}} \sum_{q\in\{0,1\}^n} (-1)^{p.q}|q\rangle\right]|f(p)\rangle \\ &= \sum_{q\in\{0,1\}^n} |q\rangle \left[\frac{1}{2^n} \sum_{p\in\{0,1\}^n} (-1)^{p.q}|f(p)\rangle\right]\end{aligned}$$

Eventually, we measure the first register. The probability of measuring a state $|q\rangle$ is

$$\left\|\left[\frac{1}{2^n} \sum_{p\in\{0,1\}^n} (-1)^{p.q}|f(p)\rangle\right]\right\|^2$$

Here, two cases can occur for the measurement,

$$\begin{cases} s = 0^n \\ s \neq 0^n \end{cases}$$

When $s = 0^n$,

$$\left\|\left[\frac{1}{2^n} \sum_{p\in\{0,1\}^n} (-1)^{p.q}|f(p)\rangle\right]\right\|^2 = \frac{1}{2^n}$$

When $s \neq 0$, then for any two string x_1 and x_2 we have $f(x_1) = f(x_2) = u \in \text{range}(f)$,

$$\left\|\left[\frac{1}{2^n}\sum_{p\in\{0,1\}^n}(-1)^{p.q}|f(p)\rangle\right]\right\|^2 = \left\|\left[\frac{1}{2^n}\sum_{u\in \text{range}(u)}((-1)^{x_1.q}+(-1)^{x_2.q})|u\rangle\right]\right\|^2$$

$$= \left\|\left[\frac{1}{2^n}\sum_{u\in \text{range}(u)}((-1)^{x_1.q}+(-1)^{(x_1\oplus s).q})|u\rangle\right]\right\|^2$$

$$= \left\|\left[\frac{1}{2^n}\sum_{u\in \text{range}(u)}((-1)^{x_1.q}.(1+(-1)^{s.q}))|u\rangle\right]\right\|^2$$

When $s.q = 1$,

$$\left\|\left[\frac{1}{2^n}\sum_{p\in\{0,1\}^n}(-1)^{p.q}|f(p)\rangle\right]\right\|^2 = \left\|\left[\frac{1}{2^n}\sum_{u\in \text{range}(u)}((-1)^{x_1.q}.(1+(-1)^{1}))|u\rangle\right]\right\|^2$$
$$= 0|u\rangle$$

Sine $s.q = 0 \pmod 2$, the value will be 0 or 1.
If $s.q = 0$,

$$\left\|\left[\frac{1}{2^n}\sum_{p\in\{0,1\}^n}(-1)^{p.q}|f(p)\rangle\right]\right\|^2 = \left\|\left[\frac{1}{2^n}\sum_{u\in \text{range}(u)}((-1)^{x_1.q}.(1+(-1)^{0}))|u\rangle\right]\right\|^2$$
$$= \frac{1}{2^{n-1}}\sum_{u\in \text{range}(u)}((-1)^{x_1.q}|u\rangle)$$

Here, the value will be $\frac{1}{2^{n-1}}$. Consequently, in either scenario, we acquire a random bit string q such that $s.q = 0$. Thus, executing the quantum circuit outlined above, we consistently discover a bit string q orthogonal to the secret string s. Given a sufficient number of measurements (Using Eq. 1), we can obtain all n measurements required to solve s.

4.3 Application

Cui et al. [8] explore the techniques of applying Simon's algorithm to assess the security of Feistel variants, potentially characterizing them through extended Simon's problems. Additionally, drawing from an extended Simon's problem, they present novel polynomial-time quantum distinguishing attacks on various Feistel variants, including MISTY L/R, CAST256-like, CLEFIA-like, MARS-like, SMS4-like, and Skipjack-A/B-like schemes. These findings highlight the potential vulnerability of classically secure schemes in the Q2 model. Furthermore, leveraging the quantum distinguishers introduced, they extend attacks several rounds forward or backward, proposing corresponding quantum all-subkeys recovery attacks. These attacks, facilitated by Grover's algorithm, demonstrate lower query complexities in the Q2 model than those in the Q1 model.

Canale et al. [6] detail the application of Simon's algorithm in compromising symmetric cryptographic primitives. Through their work, they streamline the process of discovering new attack vectors. Through this method, they autonomously identify the initial efficient key-recovery attacks against designs such as the 5-round MISTY L-FK or 5-round Feistel-FK, leveraging Simon's algorithm.

5 Grover's Searching Algorithm

Every search problem can be demonstrated as a function $f(x)$ that consumes search items as input. If the value x is found is the result to the searching problem, $f(x) = 1$; otherwise, $f(x) = 0$. The objective of the search problem is to locate any item x' such that $f(x') = 1$.

Grover's algorithm [15] aims to address a specific task, which can be articulated as follows: given a classical function $f : \{0, 1\}^n \rightarrow \{0, 1\}$, where n is the bit-size of the search space, the goal is to find an input x' for which $f(x') = 1$. The algorithm's complexity is gauged by the number of times the function $f(x)$ is invoked.

In classical computation, for the worst case, the function $f(x)$ must be evaluated a total of $(P - 1)$ times, where $P = 2^n$, exhausting all possibilities before identifying the correct solution. Grover's quantum algorithm significantly accelerates this process, offering a quadratic speedup. In this context, "quadratic" implies that only approximately $\sqrt{P}$ evaluations would be necessary, compared to P (see Fig. 3).

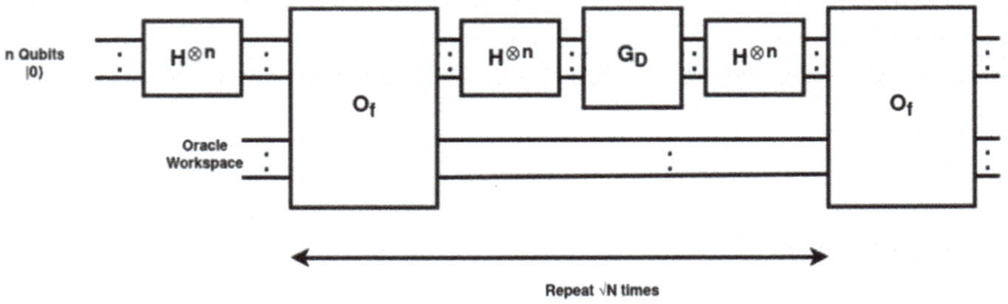

Fig. 3 Quantum circuit for Grover's search algorithm. The algorithm starts with the qubit $|0^n\rangle$ in the first register and in the second register it has some additional qubit information related to the oracle workspace. Then apply n Hadamard gate on the first n qubit. Next, the Grovers diffusion operator is used N times for increasing the probability of the good state. Then after measurement we can easily find the search element related to the good state

5.1 Description of the Algorithm

Here is a paraphrase of the provided text.

Let's examine an element x' such that $f(x') = 1$. Assume that our database size is a power of two, expressed as $N = 2^n$, where N indicates the database size. Instead of being indexed from 1 to N, the data entries are represented by n-bit binary strings within the set $\{0, 1\}^n$. The Boolean function f under consideration maps inputs from $\{0, 1\}^n$ to outputs in $\{0, 1\}$.

We use a gate G_f which flip the input only if $f(x)$ equals 1. When provided with an input $|x\rangle$, O_f outputs

$$(-1)^{f(x)}|x\rangle = \begin{cases} -|x\rangle \text{ When } f(x) = 1 \\ |x\rangle \text{ When } f(x) = 0 \end{cases}$$

We start with n qubits, all set to the state $|0\rangle$. These qubits represent an n-bit string, serving as the input to the function f. To clarify our goal, we want the amplitude of $|x'\rangle$ to be at least 0.1 by the end of the circuit. That is, at the time of final calculation, there's a $0.1^2 = 0.01$ probability of measuring $|x'\rangle$.

What sets our quantum circuit apart from classical methods? One advantage lies in our ability to leverage superposition. Initially, we'll transform our input into a uniform superposition.

The uniform amplitudes across all the n-bit strings signify our complete uncertainty regarding the value of $|x'\rangle$. We achieve this state of uniform superposition for all the basis states by applying a Hadamard gate on each qubit or, equivalently, using an $H^{\otimes n}$ gate on all qubits. This technique of creating superposition is frequently employed in quantum algorithms, so it's essential to become familiar with it. The current state can be visually depicted with a diagram representing the amplitudes, similar to a bar graph.

$$\psi_0 = \frac{1}{\sqrt{N}} \sum_{x \in \{0,1\}^n} |x\rangle$$

Next, we will query this state by introducing an oracle gate O_f. This gate flips the amplitude of the x' component while leaving everything else unchanged.

$$\psi_1 = \frac{1}{\sqrt{N}} \left(\sum_{\substack{x \in \{0,1\}^n < \\ x \neq > x'}} (|x\rangle - |x'\rangle) \right)$$

We aim to significantly increase the absolute amplitude of x'. We will introduce a new gate known as the Grover diffusion operator to achieve this. Assume α_x represents the amplitude of x for a given x in $\{0, 1\}^n$. Then, the Grover diffusion gate is defined with the following mapping,

$$\sum_{x\in\{0,1\}^n} \alpha_x|x\rangle \rightarrow \sum_{x\in\{0,1\}^n} (2\mu - \alpha_x)|x\rangle$$

$$\text{Where } \mu = \sum_x \frac{\alpha_x}{N}$$

This is indeed a valid quantum gate; it adheres to linearity since all operations within the mapping are linear and maintain unitarity. It essentially flips amplitudes around the average, μ. We will implement the gate, denoted as G_D, in the initial circuit diagram, following the oracle gate O_f to demonstrate this.

The mean amplitude before using gate G_D is,

$$\begin{aligned}\sum_{x\in\{0,1\}^n} \alpha_x &= \frac{1}{N}\left(-\frac{1}{\sqrt{N}} + \frac{N-1}{\sqrt{N}}\right)\\ &= \frac{1}{N}\left(\frac{N-2}{\sqrt{N}}\right)\\ &\approx \frac{1}{\sqrt{N}}\end{aligned}$$

How the amplitudes appear after this application is depicted in Fig. 4.

So, when we apply the Grover diffusion operator, which reflects around the mean, most of the entries in $\{0, 1\}^n$ remain almost the same.

After application of G_D,

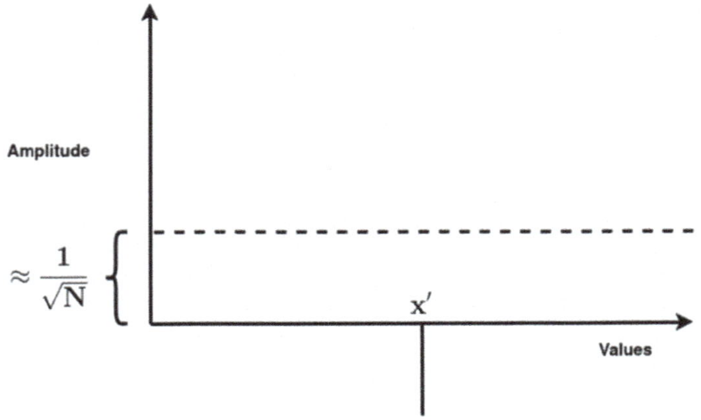

Fig. 4 Phase amplitudes after applying oracle O_f

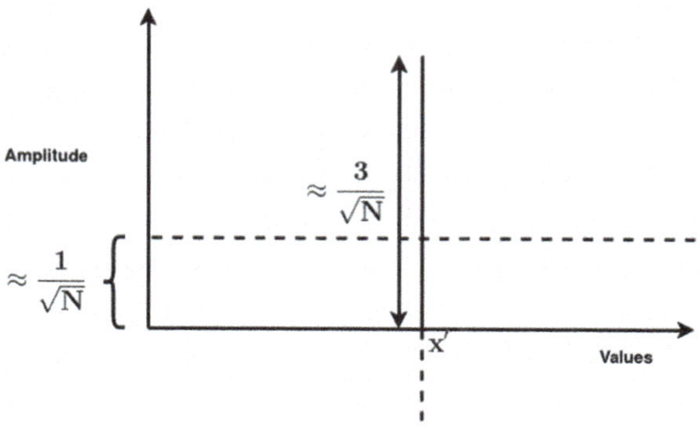

Fig. 5 Phase amplitudes after applying oracle G_D

$$-\frac{1}{\sqrt{N}}\sum_{x'\in\{0,1\}^n}|x'\rangle+\frac{1}{\sqrt{N}}\sum_{\substack{x\in\{0,1\}^n\\x\neq x'}}|x\rangle$$

$$=\left(\frac{2}{\sqrt{N}}-\left(-\frac{1}{\sqrt{N}}\right)\right)\sum_{x'\in\{0,1\}^n}|x'\rangle+\left(\frac{2}{\sqrt{N}}-\frac{1}{\sqrt{N}}\right)\sum_{\substack{x\in\{0,1\}^n\\x\neq x'}}|x\rangle$$

$$=\frac{3}{\sqrt{N}}\sum_{x'\in\{0,1\}^n}|x'\rangle+\frac{1}{\sqrt{N}}\sum_{\substack{x\in\{0,1\}^n\\x\neq x'}}|x\rangle$$

The variation of amplitudes after the application of G_D is depicted in Fig. 5.

However, we still need to achieve our goal of increasing the amplitude of x' beyond a constant. We can try the simplest approach to keep moving forward: using the oracle and the Grover diffusion gate again! After this second application of O_f, we end up with:

$$-\frac{3}{\sqrt{N}}\sum_{x'\in\{0,1\}^n}|x'\rangle+\frac{1}{\sqrt{N}}\sum_{\substack{x\in\{0,1\}^n\\x\neq x'}}|x\rangle$$

Figure 6 describes the variation of amplitudes after application of second O_f. The result of using the second G_D,

$$-\frac{3}{\sqrt{N}}\sum_{x'\in\{0,1\}^n}|x'\rangle+\frac{1}{\sqrt{N}}\sum_{\substack{x\in\{0,1\}^n\\x\neq x'}}|x\rangle$$

$$=\left(\frac{2}{\sqrt{N}}-\left(-\frac{3}{\sqrt{N}}\right)\right)\sum_{x'\in\{0,1\}^n}|x'\rangle+\left(\frac{2}{\sqrt{N}}-\frac{1}{\sqrt{N}}\right)\sum_{\substack{x\in\{0,1\}^n\\x\neq x'}}|x\rangle$$

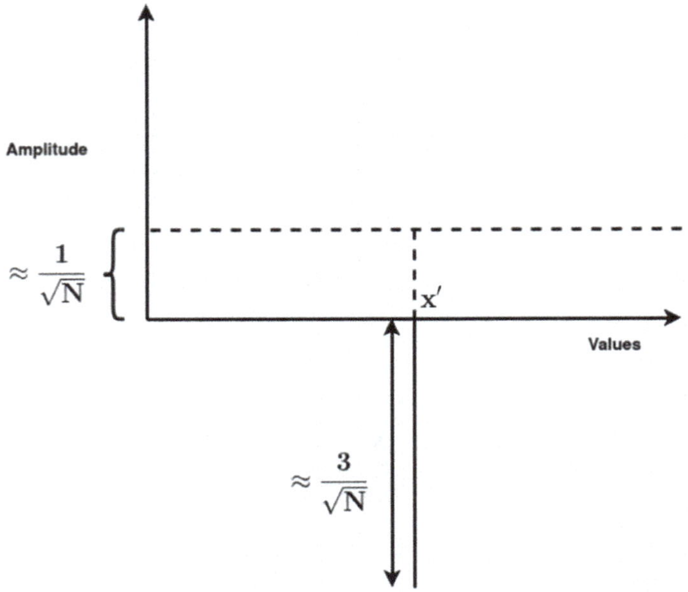

Fig. 6 Phase amplitudes after applying second O_f

$$= \frac{5}{\sqrt{N}} \sum_{x' \in \{0,1\}^n} |x'\rangle + \frac{1}{\sqrt{N}} \sum_{\substack{x \in \{0,1\}^n \\ x \neq x'}} |x\rangle$$

Figure 7 describes the variation of amplitudes after the application of second G_D. So, after repeating k-times O_f and G_D we can write the result as

$$\frac{(2k-1)}{\sqrt{N}} \sum_{x' \in \{0,1\}^n} |x'\rangle + \frac{1}{\sqrt{N}} \sum_{\substack{x \in \{0,1\}^n \\ x \neq x'}} |x\rangle$$

The process to verify this is the same. We first flip x' to its negative, then flip it around the mean while keeping everything else mostly the same. The basic idea is that each time we repeat the application of the pair of gates O_f and G_D, the amplitude $\alpha_{x'}$ increases by more than $1/\sqrt{N}$. Following this logic, after about $O(\sqrt{N})$ steps, $\alpha_{x'}$ should surpass the desired constant 0.1.

However, it's important to note that $\alpha_{x'}$ can't keep increasing by $1/\sqrt{N}$ indefinitely because it can't exceed 1. At some point, its increase will slow down or even reverse. For example, if $\alpha_{x'}$ becomes sufficiently large, then flipping x' to negative using the O_f gate might bring the mean down to negative. When the mean is negative, the magnitude of x' will decrease after using the Grover diffusion gate.

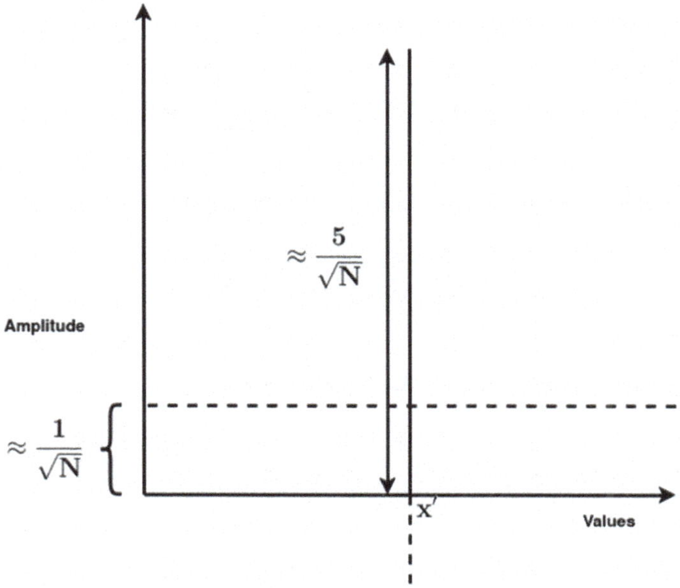

Fig. 7 Phase amplitudes after applying second G_D

Analysis In the explanations above, we assumed that only one item in the database matches the query. But what if there are multiple items matching the query? Let's say there are k items in the database that match the query. In that case, what's the asymptotic query complexity of finding one item that matches the query from the database? After applying the first Hadamard gate,

$$\psi_1 = \frac{1}{\sqrt{N}} \sum_{x \in \{0,1\}^n} |x\rangle \qquad = \frac{1}{\sqrt{N}} \sum_{x_1 \in \{0,1\}^n}^{\text{Valid}} |x_1\rangle + \frac{1}{\sqrt{N}} \sum_{x_2 \in \{0,1\}^n}^{\text{Invalid}} |x_2\rangle$$

Suppose the database contains M valid items and $(N - M)$ invalid items. Then,

$$\psi_1 = \sqrt{\frac{M}{N}} \sum_{x_1 \in \{0,1\}^n}^{\text{Valid}} |x_1\rangle + \sqrt{\frac{(N-M)}{N}} \sum_{x_2 \in \{0,1\}^n}^{\text{Invalid}} |x_2\rangle$$

Suppose $\cos\frac{\theta}{2} = \sqrt{\frac{(N-M)}{N}}$ and $\sin\frac{\theta}{2} = \sqrt{\frac{M}{N}}$, then

$$\psi_1 = \sin\frac{\theta}{2} \sum_{x_1 \in \{0,1\}^n}^{\text{Valid}} |x_1\rangle + \cos\frac{\theta}{2} \sum_{x_2 \in \{0,1\}^n}^{\text{Invalid}} |x_2\rangle$$

After application of the oracle function O_f,

$$\sin\frac{\theta}{2}\sum_{x_1\in\{0,1\}^n}^{\text{Valid}}|x_1\rangle-\cos\frac{\theta}{2}\sum_{x_2\in\{0,1\}^n}^{\text{Invalid}}|x_2\rangle$$

After the application of the Grover diffusion operation G_D,

$$\sin\frac{3\theta}{2}\sum_{x_1\in\{0,1\}^n}^{\text{Valid}}|x_1\rangle+\cos\frac{3\theta}{2}\sum_{x_2\in\{0,1\}^n}^{\text{Invalid}}|x_2\rangle$$

Repeating the grovers iteration(O_f, G_D) k-times,

$$O_f^{\psi_1}=\sin\frac{(2k+1)\theta}{2}\sum_{x_1\in\{0,1\}^n}^{\text{Valid}}|x_1\rangle+\cos\frac{(2k+1)\theta}{2}\sum_{x_2\in\{0,1\}^n}^{\text{Invalid}}|x_2\rangle$$

Now, with high probability of getting valid searches,

$$\begin{aligned}
&\sin\frac{(2k+1)\theta}{2}=1=\frac{\pi}{2}\\
&\implies \frac{(2k+1)\theta}{2}=\frac{\pi}{2}\\
&\implies (2k+1)=\frac{\pi}{\theta}\\
&\implies (2k+1)=\frac{\pi}{2}\sqrt{\frac{N}{M}}\ \text{[Since } \sin\theta\approx\theta \text{ for small } \theta]\\
&\implies k=\frac{1}{2}\left[\frac{\pi}{2}\sqrt{\frac{N}{M}}-1\right]\\
&\implies k\approx O(\sqrt{N})
\end{aligned}$$

So, the searching complexity using Grover's approach is $O(\sqrt{2^n})$.

5.2 Applications

Grover's search algorithm [15] finds applications across various domains due to its ability to search unsorted databases efficiently. It can break symmetric encryption schemes [1, 2, 14] faster than classical methods, making it useful for cryptanalysis. Grover's algorithm has cryptographic applications that accelerate searches for AES-256 keys [5, 14, 18] or input strings for SHA-256 outputs. Completing the search in 2^{128} queries halves the security strength of symmetric ciphers and hash functions compared to classical computers, requiring symmetric keys and hash functions to be doubled in size. However, AES-256 [9] and SHA-256 [16] remain secure against such attacks. While doubling key and hash sizes slightly impacts performance, modern computers can handle it well. Consequently, symmetric cryptography is generally considered safe from quantum computers.

6 Shor's Factoring Algorithm

Shor's Factoring Algorithm [25] brought quantum computing into the spotlight, capturing the attention of governments, industries, and the public. Even years later, it remains the flagship quantum algorithm. Efforts have long been underway to safeguard global financial systems, national security, and all cryptographic applications. There's a clear danger if someone, especially a state, develops quantum computing power to use Shor's algorithm before quantum-safe encryption becomes widely available.

Shor's algorithm [25] starts by picking a random number smaller than the one to be factored. Then, it checks if this number and the target have any common divisors, which might accidentally factor the target, especially for smaller numbers or with a supercomputer for larger ones. However, for numbers used in secure encryption, a quantum computer is needed.

The quantum computer's job is to find the period of the number being factored. Based on what it finds, it decides whether to try a new random number or if it has found the factors. Once the target number is factored, Shor's Algorithm is done.

Lemma 1 *Given a composite number N and x, such that,* $x \neq \pm 1$, $x^2 \equiv 1 \pmod{N}$, *then we can efficiently factorize N.*

Proof $x^2 \equiv 1 \pmod{N}$
$\implies (x+1)(x-1) \equiv 0 \pmod{N}$
$\implies$ *Either* $gcd(x+1, N) \neq 1$ *or* $gcd(x-1, N) \neq 1$ Hence, either $gcd(x+1, N)$ or $gcd(x-1, N)$ is a non-trivial factor of *N*.

Lemma 2 *If p be a odd prime and* $x \in \mathbb{Z}_p^*$ *then* $Pr[ord(x) = \text{even}] \geq \frac{1}{2}$

Proof Let $ord(x) = r$, then vv./,

$$x^r \equiv 1 \pmod{p} \tag{2}$$

For $\mathbb{Z}_p^*$ $\exists g$ for some k,

$$x \equiv g^k \pmod{p} \tag{3}$$

As x is random and chosen, k will be odd with $Pr = \frac{1}{2}$.

From Eqs. 2 and 3,

$$g^{kr} \equiv 1 \pmod{p} \tag{4}$$

We know this from Fermat's little theorem

$$x^{p-1} \equiv 1 \pmod{p} \tag{5}$$

So, by Eqs. 4 and 5 we can conclude $(p-1)|kr$.
Now we know p is odd, hence $(p-1)$ is even. As, k is odd, so r should be even.

Lemma 3 *Consider a random number x in $\mathbb{Z}_N^*$ with $N = pq$, where p, q are odd primes. If $gcd(x, N) = 1$ then $Pr[ord(x) = \text{ even } \wedge x^{\frac{r}{2}} \neq \pm 1 \pmod{N}] \geq \frac{3}{8}$.*

Proof Since x is randomly chosen from $\mathbb{Z}_N^*$, there exists a random $x_1 \in \mathbb{Z}_p^*$ and a random $x_2 \in \mathbb{Z}_q^*$ such that $x = x_1 x_2$. Assume that $ord(x_1) = r_1$ and $ord(x_2) = r_2$ then by Lemma 2 $Pr[r_1 = \text{ even or } r_2 = \text{ even }] \geq \frac{3}{4}$. Moreover, it is easy to check $r_1 | r$ and $r_2 | r$, where r is the order of x in $\mathbb{Z}_N^*$. Now since $x^r \equiv 1 \pmod{N}$ and p, q are odd primes, therefore $x^r \equiv 1 \pmod{q}$ and $x^r \equiv 1 \pmod{q}$. For any prime number, there are only two square roots of 1 modulo the prime number, they are ± 1. Hence, there are only four square roots of 1 modulo N, and only two of them satisfy $x^{\frac{r}{2}} \neq \pm 1 \pmod{N}$. Therefore, $Pr[ord(x) = \text{ even } \wedge x^{\frac{r}{2}} \neq \pm 1 \pmod{N}] \geq \frac{3}{8}$.

6.1 Algorithm Description

Suppose $N = pq$ is an integer where p, q are both primes, then if we want to find the factors of N classically, it will take exponential times. Shor's algorithm solves it in polynomial times. The algorithm solves the problem using a mixture of classical and quantum algorithms. It first reduces the factorization problem to a period-finding problem classically. Then by using the quantum algorithm, it finds the period. The quantum period-finding algorithm uses the quantum Fourier transformation, which provides the quantum speedup. Below, we explain all the steps used in the algorithm step by step.
Input: A composite number $N = pq$, where p, q are odd primes.
Output: Non-trivial factors of N, either p or q.

1. Choose a random integer $x \in \mathbb{Z}_N^*$ with $gcd(x, N) = 1$.
2. Find the period of x using the quantum period-finding algorithm, that is find a positive integer r such that $f(x) = m^x = m^{x+r}$.
3. If r is an odd integer then repeat the steps from step-1; otherwise, do the next steps.
4. For the even r we have $(x^{x/2} + 1)(x^{x/2} - 1) = 0 \pmod{N}$. If $(x^{x/2} + 1) = 0$ mod N then again repeat from step-1. Otherwise, $gcd((x^{x/2} - 1), N)$ gives the required factor.

The above lemmas imply that after Step 4, we can find the factor with high probability. All the steps except Step 2 can be solved classically. Moreover, all the

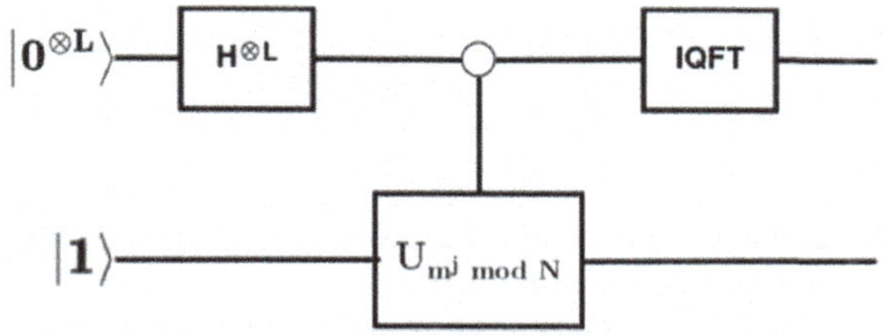

Fig. 8 Quantum circuit of Shor's algorithm starts with two registers containing two qubits $|0^L\rangle$ and $|1\rangle$. Next, an equal superposition is created by applying the Hadamard gate in the first register. The quantum oracle for the periodic function is applied to the second register, and it is measured with respect to the second register. Finally, the inverse quantum Fourier transform is applied, and the first qubit is measured to find the period

classical steps have maximum complexity in the GCD finding steps, which take $O(\log N)$ times. Now, we only have to calculate the period of $f(x)$ using the quantum algorithm.

Quantum Period Finding Let us consider a sufficiently large Q such that $N^2 << Q = 2^L$ and assume that $r|Q$. The algorithm uses two registers. The first register stores the number modulo Q while the second register stores the number modulo N (see Fig. 8). We describe the overall procedure as follows.

1. We consider the initial state as $|0\rangle \otimes |0\rangle$.
2. Create an equal superposition on the first register by applying the Hadamard gate on the first register and making the second register as $|1\rangle$.

$$\frac{1}{\sqrt{Q}} \sum_{j=0}^{Q-1} |j\rangle|1\rangle$$

3. We use the quantum Oracle of the function $f(x) = m^x$, which can be created efficiently. Therefore, our new state becomes

$$\frac{1}{\sqrt{Q}} \sum_{j=0}^{Q-1} |j\rangle|f(j)\rangle = \frac{1}{\sqrt{Q}} \sum_{j=0}^{Q-1} |j\rangle|m^j \bmod N\rangle$$

4. Since, r is the order of m, $f(x)$ has only r distinct values. So, we can express the state as

$$\frac{1}{\sqrt{\frac{Q}{r}}} \sum_{j=0}^{\frac{Q}{r}-1} \sum_{t=0}^{r-1} |jr+t\rangle|f(t)\rangle$$

If we measure the second register, then we get a uniformly random $l \leftarrow \$\{0, 1, \ldots, r-1\}$ such that the state will be

$$\frac{1}{\sqrt{\frac{Q}{r}}}\sum_{j=0}^{\frac{Q}{r}-1}|jr+l\rangle|f(l)\rangle$$

Now the first register is a periodic superposition of period r and $f(l)$ is a fixed value, so we can omit the second register.

5. Now we apply inverse quantum Fourier transformation to drop the shift value l.

$$\frac{1}{r}\sum_{k=0}^{r-1}e^{\frac{2\pi ikl}{Q}}|k\frac{Q}{r}\rangle$$

If we measure the first register, then the measurement is $k\frac{Q}{r}$, where k is uniformly random from $\{0, 1, \ldots, r-1\}$. We can compute the value of r by calculating the $gcd(k\frac{Q}{r}, Q)$, since $gcd(\frac{Q}{r}, k) = 1$ with high probability. At the beginning, we assume that r divides Q, which is a stringent condition. However, even if r does not divide Q, the algorithm can still output correctly with a constant probability. In this case, the algorithm produces an output of $\frac{l}{Q}$, providing a $\frac{1}{2Q}$ approximation of $\frac{k}{r}$. Subsequently, computing the continued fractions of $\frac{l}{Q}$ with an accuracy of $\frac{1}{2Q}$ furnishes the value for $\frac{k}{r}$. Therefore, the knowledge of Q along with the value $\frac{k}{r}$, Step 5 of the process, facilitates the straightforward computation of r. This period-finding algorithm has the worst-case time complexity $O(logN)$, that is, we can factor N in polynomial time whenever we have some quantum advantage.

6.2 Applications

Shor's algorithm [25] is effective for factoring integers and finding discrete logarithms. It poses a threat to asymmetric cryptography, requiring the replacement of current methods with quantum-resistant algorithms. Shor's algorithm [25] can factorize n-bit numbers in polynomial(n) time significantly faster than classical methods like the general number field sieve. While Grover's algorithm [15] reduces computation time, Shor's algorithm [25] reduces the problem to months from centuries or millennia. It requires roughly 2n qubits for RSA [13] keys and 6n qubits for ECC keys [17], compromising ECC keys with smaller quantum computers. Unlike Grover's algorithm [15], increasing key size does not defeat Shor's algorithm [25]. A polynomial-time attack against an algorithm is considered a breach, benefiting cryptanalysts over legitimate users due to falling compute costs.

7 Threat to Cybersecurity?

The most significant threat from quantum computing to cybersecurity lies in its capacity to crack widely used encryption methods like RSA [13], AES [9], and ECC [17].

Current encryption methods rely on the complexity of factoring large numbers into their prime components without a "key." While it would take a standard computer many years to break industry-standard encryption, a quantum computer could solve these problems much faster, potentially in minutes, due to its probabilistic algorithms. If encryption is compromised, hackers could access sensitive data such as bank accounts, transactions, and confidential business records, leading to data theft, financial fraud, corporate espionage, or even sabotaging critical systems.

The concern about a quantum-based cyberattack is real but not immediate. Cybersecurity experts must remain adaptable and prepared for new threats. As we address this impending challenge, let's recognize the importance of maintaining a solid foundation. With quantum computing on the horizon, it's essential to prepare your organization for this new threat while addressing existing cybersecurity risks. A multi-layered defense strategy helps protect against various attacks, offering comprehensive coverage and robust defense.

8 Quantum Safe Cryptography

Fortunately, researchers are making strides in developing new public-key algorithms to withstand attacks from quantum computers, thereby bolstering trust in certificate authorities, digital signatures, and encrypted messages.

National Institute of Standards and Technology (NIST) [28] assessed 69 potential new algorithms for post-quantum cryptography. The aim is to formulate a draft standard, which will subsequently be integrated into web browsers and other internet systems. Projects like the European Union's Quantum Flagship and the U.S. National Quantum Initiative are fostering collaborations between governments, universities, and industry to advance quantum-safe cryptography. Companies like IBM, Google, and Microsoft collaborate with universities on developing post-quantum algorithms. IBM's partnership with MIT is a notable example. NIST, ETSI, and ISO work together to establish global standards for quantum-resistant cryptography, ensuring interoperability and wide adoption. U.S. agencies like NSF and DARPA, along with the EU's Horizon Europe program, provide multi-million-dollar grants to support quantum cryptography research. Companies like Intel and IBM, along with venture capital firms, are investing heavily in startups and projects focused on quantum-resistant technologies. Organizations like DARPA and NATO are funding research into securing military communication systems from quantum threats. NIST is finalizing its selection of quantum-resistant algorithms, with CRYSTALS-Kyber and CRYSTALS-Dilithium leading candidates for encryption and digital signatures.

Table 1 Comparison between classical and quantum-resistant cryptographic methods

Metric	Classical cryptographic methods	Quantum-resistant cryptographic methods
Security basis	Hardness of classical problems (e.g., factoring, discrete logarithm) [25]	Hardness of quantum-resistant problems (e.g., lattice-based, hash-based) [4]
Key size	Relatively small (e.g., RSA-2048, AES-256) [7]	Typically larger (e.g., several kilobytes for lattice-based keys)
Computational efficiency	Generally high, optimized algorithms [7]	Often lower due to complex mathematical operations
Memory usage	Low [7]	Higher due to larger key sizes and complex operations [4]
Encryption/decryption speed	Fast (e.g., AES, RSA) [7]	Slower (e.g., lattice-based algorithms) [4]
Signature size	Small to moderate (a few hundred bytes) [7]	Larger (several kilobytes)
Communication overhead	Low [7]	Higher due to larger keys and signatures [4]
Resistance to quantum attacks	Vulnerable (e.g., Shor's algorithm) [25]	Designed to be resistant [4]
Maturity of algorithms	High, well studied and standardized [7]	Varies; some are new and still being evaluated [27]
Standardization	Well-standardized (e.g., AES, RSA) [7]	Ongoing (NIST standardization in progress) [27]
Hardware/software compatibility	High, widely supported [7]	Emerging, requires infrastructure updates [4]
Energy consumption	Typically lower due to optimization [7]	Higher due to larger keys and complex operations

Standards are expected by 2024. ETSI is developing quantum-safe cryptographic standards for telecommunications and IoT, focusing on practical implementation. ISO is working to incorporate quantum-resistant algorithms into its global cryptographic standards, ensuring industry-wide adoption. These efforts are essential for preparing the global cryptographic infrastructure for quantum threats.

Table 1 describes the comparison between classical and quantum-resistant cryptographic algorithms. Table 2 distinguishes the computational complexity of classical cryptographic algorithms with quantum-resistant algorithms for various cryptographic tasks. Table 3 demonstrates a brief comparison among recent quantum-resistant algorithms.

While symmetric cryptography theoretically allows for key exchange, it relies on trusted third parties to safeguard secret keys and lacks support for digital signatures, posing challenges for widespread internet deployment. Nevertheless, it remains a cornerstone of encryption and authentication in the GSM [20] cellular standard.

Table 2 Comparing the computational complexity of classical cryptographic algorithms with quantum-resistant algorithms for various cryptographic tasks, such as encryption, decryption, and key generation

Algorithm type	Task	Classical complexity	Quantum-resistant complexity
RSA	Key generation	$O(n^3)$ using classical algorithms	Vulnerable to $O((\log n)^3)$ with Shor's algorithm on a quantum computer [24]
RSA	Encryption/decryption	$O(n^3)$ (modular exponentiation)	Quantum attacks reduce complexity to polynomial time $O((\log n)^3)$ [24]
AES	Encryption/decryption	$O(n)$ where n is the number of rounds (e.g., 10 rounds for AES-128)	$O(2^{n/2})$ with Grover's algorithm which halves the effective key length [15]
Lattice-based (e.g., Kyber)	Key generation	$O(n^2)$ for matrix operations	$O(n^2)$, but designed to be resistant to quantum attacks [22]
Lattice-based (e.g., Kyber)	Encryption/decryption	$O(n^2)$ for polynomial multiplications	$O(n^2)$; Computational efficiency comparable to classical algorithms, with quantum resistance [22]
Hash-based (e.g., SPHINCS+)	Signature generation	$O(n)$ (hash function calls)	$O(n)$ with no known quantum speedup in key generation [3]
Hash-based (e.g., SPHINCS+)	Signature verification	$O(n)$ (hash function calls)	$O(n)$ with hash-based security remaining intact against quantum computers [3]
Code-based (e.g., McEliece)	Key generation	$O(n^2)$ for encoding operations	$O(n^2)$, secure against quantum computers but with large key sizes [19]
Code-based (e.g., McEliece)	Encryption/decryption	$O(n^2)$ for encoding and decoding operations	$O(n^2)$; Resistant to quantum attacks but computationally demanding due to key sizes [19]

An alternative approach for key exchange without public-key cryptography is quantum key distribution, leveraging quantum methods to establish a symmetric key between sender and receiver. However, these techniques necessitate specialized hardware.

Table 3 Comparison of quantum-resistant algorithms

Algorithm type	Strengths	Weaknesses	Development status and applications
Lattice-based	Strong security foundation, efficient encryption/decryption [4]	Larger key sizes, performance varies across applications [7]	Active research, suitable for encryption and key exchange [27]
Hash-based	Well-understood security, simple construction [4]	Large signatures, limited to specific applications (e.g., signatures) [7]	Standardized for digital signatures, in use for code signing [27]
Code-based	Long history of research, small encryption sizes [4]	Large public keys, slower performance [7]	Suitable for encryption and key exchange, ongoing standardization [27]
Multivariate	Fast signing and verification, small signature sizes [4]	Complex key generation, vulnerability to certain attacks [7]	Niche applications, experimental stage [27]
Isogeny-based	Small key sizes, promising for key exchange [4]	Complex and slow, currently less understood [7]	Still in early development, potential for future applications [27]

9 Conclusion

Strong cryptography is a linchpin in safeguarding both individual and societal cybersecurity, providing the bedrock for secure communication, data storage, and establishing trusted connections between entities. Nevertheless, it's essential to grasp that, while indispensable, cryptography cannot single-handedly tackle all security challenges. Despite the robustness of encryption measures, individuals remain susceptible to human errors, such as falling prey to deceptive links or opening malicious attachments. Moreover, vulnerabilities in software and insider threats can undermine the efficacy of encryption protocols. Even in cases where cryptographic algorithms boast theoretical security, their practical implementation can introduce vulnerabilities, as evidenced by recent incidents involving the accidental exposure of private encryption keys. Additionally, the imminent rise of potent quantum computing presents a formidable security risk, underscoring the urgency of proactively developing standards for quantum-resistant cryptography. This process may span several years. Therefore, initiating preparations for quantum-resistant cryptography is prudent to mitigate potential security threats preemptively.

References

1. R. Anand, A. Maitra, S. Maitra, C.S. Mukerjee, S. Mukhopadhyay, Quantum resource estimation for FSR based symmetric ciphers and related Grover's attacks, in *Progress in Cryptology—INDOCRYPT 2021—22nd International Conference on Cryptology in India, Jaipur, India, December 12–15, 2021, Proceedings*. Lecture Notes in Computer Science, vol. 13143, ed. by A. Adhikari, R. Küsters, B. Preneel (Springer, 2021), pp. 179–198. https://doi.org/10.1007/978-3-030-92518-5_9, https://doi.org/10.1007/978-3-030-92518-5_9
2. R. Anand, A. Maitra, S. Mukhopadhyay, Grover on *simon*. Quantum Inf. Process. **19**(9), 340 (2020)
3. D.J. Bernstein, D. Hopwood, A. Hülsing, T. Lange, R. Niederhagen, L. Papachristodoulou, M. Schneider, P. Schwabe, Z. Wilcox-O'Hearn, SPHINCS+: submission to the NIST post-quantum cryptography standardization project (2019). https://sphincs.org/
4. D.J. Bernstein, J. Buchmann, E. Dahmen, *Post-Quantum Cryptography* (Springer, Dordrecht, 2008). https://doi.org/10.1007/978-3-540-88702-7, https://cds.cern.ch/record/1253241
5. X. Bonnetain, M. Naya-Plasencia, A. Schrottenloher, Quantum security analysis of AES. IACR Trans. Symmetric Cryptol. **2019**(2), 55–93 (2019). https://doi.org/10.13154/TOSC.V2019.I2.55-93
6. F. Canale, G. Leander, L. Stennes, Simon's algorithm and symmetric crypto: generalizations and automatized applications, in *Advances in Cryptology—CRYPTO 2022: 42nd Annual International Cryptology Conference, CRYPTO 2022, Santa Barbara, CA, USA, August 15–18, 2022, Proceedings, Part III* (Springer, Berlin, Heidelberg, 2022), pp. 779–808. https://doi.org/10.1007/978-3-031-15982-4_26
7. L. Chen, S. Jordan, Y.K. Liu, D. Moody, R. Peralta, R. Perlner, D. Smith-Tone, *Report on Post-quantum Cryptography* (2016). https://doi.org/10.6028/NIST.IR.8105
8. J. Cui, J. Guo, S. Ding, Applications of Simon's algorithm in quantum attacks on feistel variants. Quantum Inf. Process. **20**(3) (2021). https://doi.org/10.1007/s11128-021-03027-x
9. J. Daemen, V. Rijmen, *The Design of Rijndael: AES—The Advanced Encryption Standard (Information Security and Cryptography)*, 1 edn. (Springer, 2002)
10. D. David, Quantum theory, the church-turing principle and the universal quantum computer, in *Proceedings of Royal Society, London, Jul. 8, 1985*. vol. 400 (ACM, 1985), pp. 553–558. http://doi.org/10.1098/rspa.1985.0070
11. D. David, J. Richard, Rapid solution of problems by quantum computation, in *Proceedings of Royal Society, London, Dec. 8, 1992*, vol. 439 (ACM, 1992), pp. 553–558. http://doi.org/10.1098/rspa.1992.0167
12. D. Deutsch, R. Jozsa, Rapid solution of problems by quantum computation, in *Proceedings of the Royal Society of London. Series A: Mathematical and Physical Sciences* (1992), pp. 553–558
13. B. Furht (ed.), *The RSA Public-Key Encryption Algorithm* (Springer US, Boston, MA, 2006), pp. 757–757. https://doi.org/10.1007/0-387-30038-4_206
14. M. Grassl, B. Langenberg, M. Roetteler, R. Steinwandt, *Applying Grover's Algorithm to AES: Quantum Resource Estimates* (2015)
15. L.K. Grover, A fast quantum mechanical algorithm for database search, in *Proceedings of the Twenty-Eighth Annual ACM Symposium on the Theory of Computing, Philadelphia, Pennsylvania, USA, May 22–24, 1996*, ed. by G.L. Miller (ACM, 1996), pp. 212–219. https://doi.org/10.1145/237814.237866
16. H. Handschuh, *SHA Family (Secure Hash Algorithm)* (Springer US, Boston, MA, 2005), pp. 565–567. https://doi.org/10.1007/0-387-23483-7_388
17. K. Igoe, D. McGrew, M. Salter, *Fundamental Elliptic Curve Cryptography Algorithms. RFC 6090* (Feb. 2011). https://doi.org/10.17487/RFC6090, https://www.rfc-editor.org/info/rfc6090
18. S. Jaques, M. Naehrig, M. Roetteler, F. Virdia, Implementing Grover oracles for quantum key search on AES and LOWMC, in *Advances in Cryptology—EUROCRYPT 2020–39th Annual International Conference on the Theory and Applications of Cryptographic Techniques, Zagreb,*

Croatia, May 10–14, 2020, Proceedings, Part II. Lecture Notes in Computer Science, vol. 12106, ed. by Canteaut, A., Ishai, Y. (Springer, 2020), pp. 280–310. https://doi.org/10.1007/978-3-030-45724-2_10

19. R.J. McEliece, A public-key cryptosystem based on algebraic coding theory. DSN Prog. Rep. (1978)
20. M. Mouly, M.B. Pautet, T. Haug, *The GSM System for Mobile Communications* (1992). https://api.semanticscholar.org/CorpusID:108732525
21. K. Nagata, T. Nakamura, A. Farouk, Quantum cryptography based on the Deutsch-Jozsa algorithm. Int. J. Theor. Phys. **56**(9), 2887–2897 (2017)
22. NIST: post-quantum cryptography standardization (2023). https://csrc.nist.gov/projects/post-quantum-cryptography
23. A. Schrottenloher, *Quantum Algorithms for Cryptanalysis and Quantum-safe Symmetric Cryptography*. (Algorithmes quantiques pour la cryptanalyse et cryptographie symétrique post-quantique). Ph.D. thesis, Sorbonne University, France (2021). https://tel.archives-ouvertes.fr/tel-03142366
24. P.W. Shor, Algorithms for quantum computation: discrete logarithms and factoring, in *Proceedings 35th Annual Symposium on Foundations of Computer Science* (1994), pp. 124–134
25. P.W. Shor, Polynomial-time algorithms for prime factorization and discrete logarithms on a quantum computer. SIAM J. Comput. **26**(5), 1484–1509 (1997)
26. D.R. Simon, On the power of quantum computation. SIAM J. Comput. **26**(5), 1474–1483 (1997)
27. Standards of N.I.S.T. (NIST), Post-quantum cryptography standardization project (2023). https://csrc.nist.gov/projects/post-quantum-cryptography
28. Standards of and Technology of N.I., Security requirements for cryptographic modules. Tech. Rep. Federal Information Processing Standards Publications (FIPS PUBS) 140-2, Change Notice 2 December 03, 2002, U.S. Department of Commerce, Washington, D.C. (2001). https://doi.org/10.6028/nist.fips.140-2

Leveraging Quantum Computing to Combat Bot Attacks in the Banking Sector: Adoption Challenges and Opportunities

Hayyan Nassar Waked, S. B. Goyal, and Deepika Ghai

1 Introduction

Recent advancements in quantum computing have introduced both significant opportunities and complex challenges for cybersecurity in the banking sector. While this study focuses on the adoption of quantum computing to counter bot attacks within Dubai's financial institutions, it is imperative to consider how different regions approach this technology. For instance, North America's proactive stance contrasts with Europe's more cautious approach, shaped by stringent regulations such as GDPR [1]. This variation highlights the need for a global perspective when discussing quantum computing in cybersecurity, particularly in addressing sophisticated threats like credential stuffing and DDoS attacks [2, 3].

Moreover, the regulatory environment plays a critical role in the adoption of these technologies. While Dubai has implemented robust cybersecurity policies [4], regions such as the United States and Europe operate under different frameworks that present both obstacles and opportunities for quantum computing integration. For example, compliance with the NIST Cybersecurity Framework in the U.S. may facilitate quicker adoption compared to regions with less supportive regulatory structures. The economic implications of adopting quantum computing cannot be understated.

H. N. Waked
City Graduate School, City University, Petaling Jaya, Malaysia
e-mail: dr.hayyan@city.edu.my

S. B. Goyal (✉)
Chitkara University Institute of Engineering and Technology, Chitkara University, Chandigarh-Patiala National Highway (NH - 64), Rajpura, Punjab 140401, India
e-mail: drsbgoyal@gmail.com

D. Ghai
School of Electronics and Electrical Engineering, Lovely Professional University, Jalandhar, Punjab, India

S. B. Goyal et al. (eds.), *Quantum Computing, Cyber Security and Cryptography*,
https://doi.org/10.1007/978-981-96-4948-8_8

The initial investment in quantum infrastructure is substantial, but so are the potential long-term benefits in terms of reduced cyber-attacks and operational efficiencies [5].

The rapidly increasing developed cyber-attacks, specifically bot attacks, have presented severe threats in this banking industry. Bot attacks like credential stuffing, DDoS, and financial fraud are increasingly on the rise by a significant amount, growing more compelling. According to a report from Akamai, in 2021 alone, 61% of all login attempts to the financial sector were identified as credential-stuffing attacks that caused economic loss, erosion of customer trust, and severely harmed the reputation. Zhang et al. [6] reported that, in the first half of 2021, the global financial sector has noticed an 80% increase in bot attacks compared with the previous year. Quantum computing can radically change the boundaries of computational power through the postulates of quantum mechanics applied to computing and calculations at unprecedented speed. This potential technology can upset cybersecurity because such a device could break any encryption by classical methods that are currently used. For instance, quantum key distribution does offer the capability to perform communications securely and make every possible eavesdropping attempt on the channel detectable to provide data confidentiality and integrity. That assures your data is safeguarded from any quantum threat using an advanced implementation of quantum key distribution (QKD) and post-quantum-safe algorithms. Recent advancements in real-world QKD and the development of quantum-resilient algorithms are essential for protecting data from quantum threats. Significant contributions to this field include the studies by [7].

Despite the potential of quantum computing, it is still early stages of its use in cybersecurity. Bringing quantum computing into these frameworks takes a lot of money, technical expertise, and adherence to regulations. In 2021, worldwide investment in quantum computing hits $1.9 billion, a 50% jump from the previous year [8]. However, the banking industry, especially in regions like Dubai and the United Arab Emirates, faces unique hurdles in adopting these advanced technologies. As Dubai becomes a major global financial hub, drawing in numerous international banks and financial institutions [9], there is a growing need for strong cybersecurity to protect sensitive financial data and ensure banking operations run smoothly. The Dubai government has rolled out strict regulations and policies to boost cybersecurity across various industries, including finance [4]. Dubai's Cybersecurity Strategy, launched in 2017, focuses on five main areas: building a cyber-smart society, fostering innovation, enhancing cybersecurity, improving cyber-resilience, and promoting national and international collaboration. This strategy aims to safeguard Dubai's critical infrastructure and strengthen its reputation as a secure place to do business. Dubai's regulations align with global standards including the General Data Protection Regulation (GDPR) and the National Institute of Standards and Technology (NIST) framework, providing solid protection against cyber-threats [4]. However, Dubai's unique regulatory landscape and varied technological infrastructure present specific challenges for banks in adopting advanced cybersecurity measures like quantum computing. Globally, regulations such as the GDPR and the NIST Cybersecurity Framework offer guidelines for data protection and maintaining robust cybersecurity practices. The GDPR, rolled out in 2018, sets a high bar for data protection and privacy, influencing regulations around the world [1]. The NIST Cybersecurity Framework, widely used in the United States and other countries, provides a policy framework for private sector organizations to evaluate and improve their ability to prevent, detect, and respond to cyber-attacks [7].

This study aims to investigate the factors influencing the adoption of quantum computing for enhancing cybersecurity measures against bot attacks in the banking industry of Dubai. By extending the Technology Acceptance Model (TAM) to include perceived security benefits, support from organizations, adherence to Dubai's regulations, and trust in technology, we aim to provide a thorough understanding of how quantum computing is adopted. Our research uses Partial Least Squares Structural Equation Modeling (PLS-SEM) with SmartPLS for data analysis. We collected data through an online survey targeting cybersecurity professionals in Dubai's banks. This method helps us explore the complex relationships between different factors that influence the adoption of quantum computing [10]. By focusing on professionals in Dubai's banking sector, we ensure that our data is relevant, considering the region's unique regulatory environment and rapid technological progress [9]. The online survey also allows us to gather information from a wide range of participants, making our dataset diverse and representative [11].

This study fills several gaps in the current research on quantum computing and cybersecurity in the banking industry. Most existing studies are theoretical and do not focus on the practical adoption of quantum computing [7, 12]. There is also a shortage of research that considers the specific regulatory and technological environments of regions like Dubai [4]. By extending the TAM to include perceived security benefits, organizational support, and regulatory compliance, this research provides a comprehensive insight into quantum computing adopting factors in the cybersecurity of the banking industry [13].

After the introduction, the literature review discusses the existing quantum computing and cybersecurity research, focusing on the banking industry's challenges and opportunities. We then explain the extended TAM and how it fits into our study. In the methodology section, we describe how we designed the research, collected data, and analyzed responses from cybersecurity professionals in Dubai's banks. The results and analysis section shares our main findings, pointing out what factors drive the adoption of quantum computing for cybersecurity. Finally, in the discussion and conclusion sections, we discussed the implications of these findings, addressed the study's limitations, and suggested directions for future research.

2 Literature Review

To fully appreciate the transformative potential of quantum computing in cybersecurity, it is essential to discuss the specific quantum algorithms that directly address the challenges posed by bot attacks. Shor's algorithm is a quantum algorithm that efficiently factors large integers, which poses a significant threat to widely used encryption schemes like RSA. RSA's security relies on the difficulty of factoring large numbers, and Shor's algorithm, which operates in polynomial time, could effectively break RSA encryption once large-scale quantum computers become available [14]. The implications for the banking sector are profound; without quantum-resistant encryption, the integrity of financial transactions and sensitive customer data could be compromised.

Moreover, Grover's algorithm offers a quadratic speedup for searching unsorted databases, which can be adapted to enhance cybersecurity defences, particularly in improving the efficiency of threat detection systems. For instance, Grover's algorithm can be applied to accelerate the search processes in large datasets, making it more effective at identifying patterns indicative of botnet activity or other forms of cyber-intrusion [15]. This application is particularly relevant for defending against distributed denial-of-service (DDoS) attacks, where rapid identification of malicious traffic is crucial for mitigating the impact on banking infrastructure.

These algorithms, while theoretically well established, are moving closer to practical implementation as quantum computing hardware advances. Recent studies have explored hybrid models where quantum algorithms are integrated with classical systems to provide enhanced security while mitigating the computational overhead that currently limits full-scale quantum deployment [16]. This hybrid approach allows banks to begin integrating quantum resilience into their cybersecurity strategies today, rather than waiting for fully mature quantum computers.

The literature on quantum cryptography has seen significant advancements in recent years, particularly in the practical deployment of QKD systems. QKD leverages the principles of quantum mechanics to enable secure communication channels that are theoretically immune to eavesdropping. Notably, the BB84 protocol, proposed by [17], has been implemented in several pilot projects within the financial sector, demonstrating its viability for securing high-stakes transactions [18]. These pilot studies have shown that QKD can be integrated with existing financial infrastructure, providing a quantum-safe layer of security that is particularly resistant to man-in-the-middle attacks, a common vector for botnet operators. The application of quantum computing in cybersecurity extends to areas such as secure data storage, enhanced cryptographic algorithms, and robust authentication mechanisms [12].

Despite its potential, the integration of quantum computing into cybersecurity frameworks remains in its nascent stages, necessitating further research and development to address technical and practical challenges. While theoretical advancements in quantum computing are promising, practical case studies are essential to demonstrate their real-world applicability. A notable example is the implementation of QKD in Switzerland's financial sector. Swiss banks, in collaboration with technology firms, have successfully piloted QKD systems to secure interbank communications, thereby protecting against advanced persistent threats (APTs) that could otherwise exploit vulnerabilities in classical encryption [19]. Another case study worth mentioning is China's deployment of a quantum communication network, which links Beijing and Shanghai. This network, which spans over 2,000 kms, utilizes QKD to secure data transmission between financial institutions and government entities. The project highlights the scalability of quantum technologies and their potential for securing critical infrastructure at a national level [20]. By examining these case studies, the literature review can provide a detailed analysis of the practical steps required for successful implementation, as well as the lessons learned from these early adopters.

Bot attacks have emerged as a significant threat to the banking industry, characterized by using automated scripts or bots to conduct malicious activities [21]. These attacks can take various forms, including credential stuffing, DDoS, and financial fraud, all of which pose severe risks to banks' operational integrity and customer trust [2]. The sophistication and scale of bot attacks have increased, driven by advancements in artificial intelligence and machine learning, enabling bots to mimic human behavior and evade traditional detection mechanisms [6]. For example, credential-stuffing attacks accounted for a substantial portion of all cyber-attacks in the financial sector in recent years, causing significant financial and reputational damage [3]. The financial implications of bot attacks are substantial, with potential losses running into millions of dollars, alongside reputational damage and regulatory penalties [5]. Consequently, banks are compelled to adopt advanced cybersecurity measures to detect, mitigate, and prevent bot attacks effectively. Recent research emphasizes the need for adaptive and resilient cybersecurity strategies to combat the evolving nature of bot attacks [22].

Banks employ a multi-faceted approach to cybersecurity, incorporating a range of technologies and strategies to safeguard their systems and data. Traditional measures include firewalls, intrusion detection systems (IDS), and antivirus software, which serve as the first defence against cyber-threats [9]. In recent years, banks have increasingly adopted advanced technologies such as artificial intelligence and machine learning (ML) to enhance their cybersecurity capabilities [23]. Additionally, banks are implementing multi-factor authentication (MFA) and biometric verification to strengthen access controls and prevent unauthorized access [24]. Despite these measures, cyber-threats' dynamic and evolving nature necessitates continuous innovation and adaptation in cybersecurity practices [8]. Banks also emphasize the importance of employee training and awareness programs to mitigate the risks posed by social engineering attacks and insider threats [25]. The integration of artificial intelligence and ML in cybersecurity frameworks has shown promising results in enhancing threat detection and response times [6].

The transition to quantum-based cybersecurity frameworks requires innovative models that can seamlessly integrate quantum and classical technologies. One such model is the Quantum-Enhanced Cybersecurity Framework (QECF), which proposes a hybrid approach combining classical encryption methods with quantum cryptographic protocols. This model allows for a gradual transition to quantum resilience, enabling financial institutions to maintain security while adapting to the evolving quantum landscape [12]. The QECF model also incorporates Quantum Random Number Generators (QRNGs), which are crucial for generating truly random keys that underpin secure encryption. Unlike classical RNGs, which can be predictable and thus vulnerable to attack, QRNGs leverage the inherent randomness of quantum processes to produce keys that are immune to prediction [26]. Integrating QRNGs into existing security systems can provide an immediate enhancement to the security of banking transactions, while also preparing the infrastructure for future quantum upgrades.

The adoption of quantum computing in cybersecurity is not uniform across the globe, with significant regional differences shaped by regulatory environments and technological readiness. In the United States, the early adoption of quantum technologies is driven by a strong regulatory framework, including the NIST Cybersecurity Framework, which encourages the integration of advanced cryptographic methods [7]. In contrast, Europe's adoption has been influenced by stringent data protection regulations under the General Data Protection Regulation (GDPR), which has prompted financial institutions to explore quantum-safe encryption methods to ensure compliance [1]. Asia, particularly China, has taken a different approach by heavily investing in quantum research and development, resulting in significant advancements in both quantum communication networks and quantum computing hardware. The Chinese government's support for these technologies has enabled rapid progress, positioning the country as a leader in quantum cybersecurity [20].

Dubai is rapidly emerging as a global financial hub, attracting a significant number of international banks and financial institutions [9]. This growth has necessitated robust cybersecurity frameworks to protect sensitive financial data and ensure the integrity of banking operations. The Dubai government has implemented stringent regulations and policies to enhance cybersecurity across various industries, including finance [4]. These regulations aim to foster a secure digital environment and mitigate the risks associated with cyber-threats. Dubai's cybersecurity strategy, launched in 2017, outlines five main domains: cyber-smart society, innovation, cybersecurity, cyber-resilience, and national and international collaboration. This strategy is designed to protect Dubai's critical infrastructure and bolster the city's reputation as a safe and secure place to do business [4]. A recent survey of cybersecurity professionals in the banking sector revealed that while there is strong interest in quantum technologies, there are significant concerns about the cost, complexity, and scalability of these solutions [27]. However, Dubai's unique regulatory landscape and diverse technological infrastructure pose specific challenges for banks in adopting advanced cybersecurity measures, such as quantum computing. Understanding these challenges and opportunities is crucial for developing effective cybersecurity strategies tailored to the needs of Dubai's banking industry.

The research methodology for this study employs SmartPLS, Partial Least Squares Structural Equation Modeling (PLS-SEM), for data analysis and utilizes an online questionnaire targeting cybersecurity professionals in the banking industry in Dubai. The choice of PLS-SEM is well suited for examining complex relationships between multiple variables, crucial for understanding the factors influencing the adoption of quantum computing in cybersecurity [10]. Targeting cybersecurity professionals in Dubai banks ensures that the data collected aligns with the study's objectives, considering Dubai's unique regulatory environment and rapid technological advancements [9]. Additionally, the use of an online questionnaire facilitates efficient data collection from a geographically dispersed sample, ensuring a diverse and representative dataset [11].

This study aims to address several gaps identified in the current literature on quantum computing and cybersecurity within the banking industry. Firstly, there is a noticeable lack of empirical research on the practical adoption of quantum computing in banking cybersecurity, with most studies focusing on theoretical aspects [7, 12]. Secondly, existing studies often overlook regions like Dubai's specific regulatory and technological context. This research addresses this contextual gap by focusing on Dubai's banking industry [4]. Lastly, there is a scarcity of studies integrating extended TAM frameworks to assess the factors influencing the adoption of advanced technologies in cybersecurity. This study extends the TAM framework to include perceived security enhancement, organizational support, and regulatory compliance, offering a comprehensive understanding of the adoption factors [28].

3 Theoretical and Conceptual Framework

The theoretical foundation for this study is based on the TAM, which serves as a pivotal framework for understanding and predicting the acceptance and usage of new technologies. TAM is particularly relevant to this study as it offers insights into the factors that influence the adoption of quantum computing technologies to enhance cybersecurity measures within the banking industry. Originally proposed by [29], TAM suggests that two primary constructs including Perceived Usefulness (PU) and Perceived Ease of Use (PEOU) determine an individual's intention to use a technology, which subsequently influences actual usage. Perceived usefulness is defined as the degree to which a person believes that using a particular system would enhance their job performance. Perceived ease of use refers to the extent to which a person believes that using the system would be free of effort. These constructs influence attitudes toward using the system, which in turn affect the behavioral intention to use the technology.

To address the specific context of this study, TAM is extended to incorporate additional variables pertinent to quantum computing and cybersecurity in the banking industry. These variables include perceived security enhancement (PSE), organizational support (OS), and compliance with Dubai regulations (CDR). Additionally, trust in technology (TT) is included as a mediating variable to provide a comprehensive understanding of the factors influencing the adoption of quantum computing. Table 1 illustrates the extended TAM framework's variables, dimensions, and sources.

The core variables of the extended TAM include PU and PEOU. PU is assessed based on its performance expectancy, effectiveness, and comparative advantage over traditional methods [28]. PEOU is evaluated through its complexity, user-friendliness, and accessibility [29]. The extended variables incorporate dimensions specific to the study's context. PSE measures the reduction in bot attack threats, reliability, and trustworthiness of quantum computing solutions [22, 27, 30]. OS includes management support, resource allocation, and training and development opportunities [30]. CDR consider regulatory alignment, legal compliance, and policy support for adopting quantum computing [30]. TT serves as a crucial mediating variable, reflecting the credibility, safety, and dependability of quantum computing solutions [27]. This mediating factor is critical in understanding how perceptions of technology influence the intention to adopt and actual implementation. Based on the extended

Table 1 Variables, dimensions, and sources in the extended TAM framework

Variables	Dimensions	Source
Core variables		
PU	Performance expectancy, effectiveness, advantage	[28, 29]
PEOU	Complexity, user-friendliness, accessibility	[28, 29]
Extended variables		
PSE	Threat reduction, reliability, trustworthiness	[22, 27, 30]
OS	Management support, resource allocation, training, and development	[17, 30]
CDR	Regulatory alignment, legal compliance, policy support	[30]
Mediating variable		
TT	Credibility, safety, dependability	[22, 30]
Outcome variables		
Intention to Adopt (IA)	Adoption willingness, adoption readiness, future plans	[28]
Actual Implementation (AI)	Implementation extent, operational use, integration	[28]

TAM framework, the following hypotheses have been developed to investigate the factors influencing the adoption of quantum computing for enhancing cybersecurity measures against bot attacks in the banking industry of Dubai, as illustrated in Fig. 1.

Figure 2 illustrates the hypothesized relationships in the extended TAM model. The arrows represent the direction of influence between variables. TT serves as a

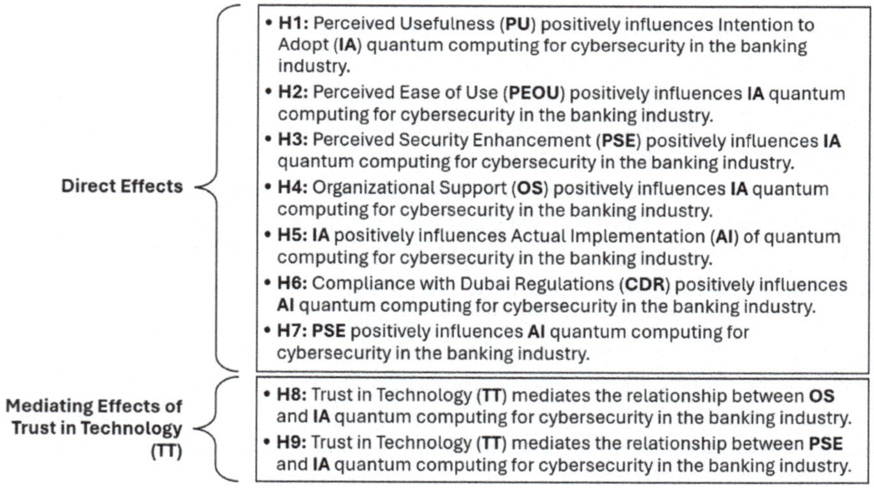

Fig. 1 Hypotheses of the research

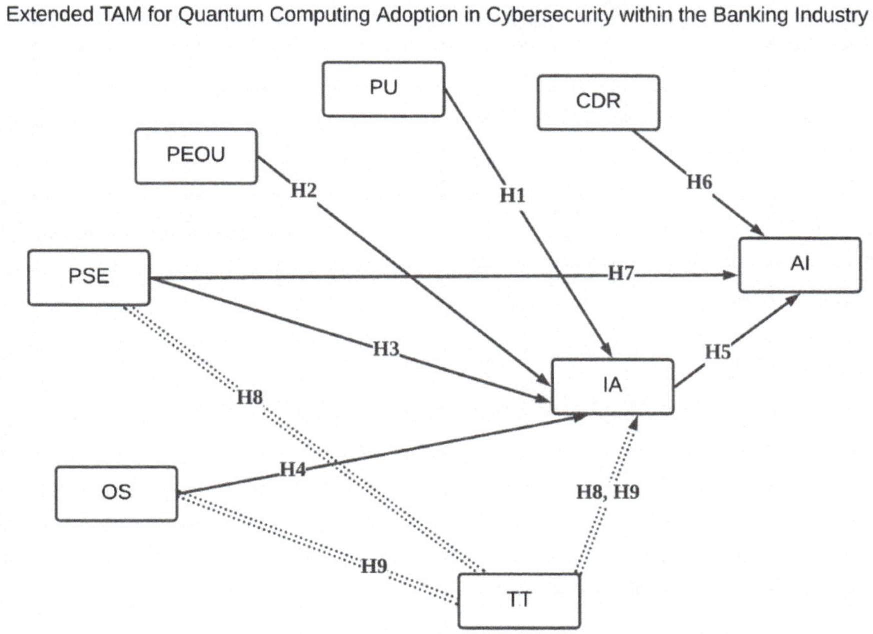

Fig. 2 TAM extension for the adoption of quantum computing in cybersecurity within the banking industry

mediating variable, affecting the relationships between OS and PSE with IA. Additionally, IA and CDR directly influence the AI of quantum computing solutions in cybersecurity within the banking industry.

4 Methodology

This study employs a quantitative, survey-based research design to gather data from cybersecurity professionals in the banking industry of Dubai. The primary objective is to examine the factors influencing the adoption of quantum computing for enhancing cybersecurity measures against bot attacks. The study utilizes the TAM extended with additional constructs relevant to quantum computing and cybersecurity. The primary data collection instrument is an online questionnaire designed to capture respondents' perceptions of various factors influencing the adoption of quantum computing in cybersecurity. The questionnaire is divided into sections, each corresponding to a specific construct in the research model. Constructs include PU, PEOU, PSE, OS, CDR, TT, IA, and AI. The questionnaire includes 3–6 items for each dimension and uses a 7-point Likert scale ranging from 1 (strongly disagree) to 7 (strongly agree) to measure responses. Figure 3 demonstrates the questionnaire structure dimensions and source of the adapted questions.

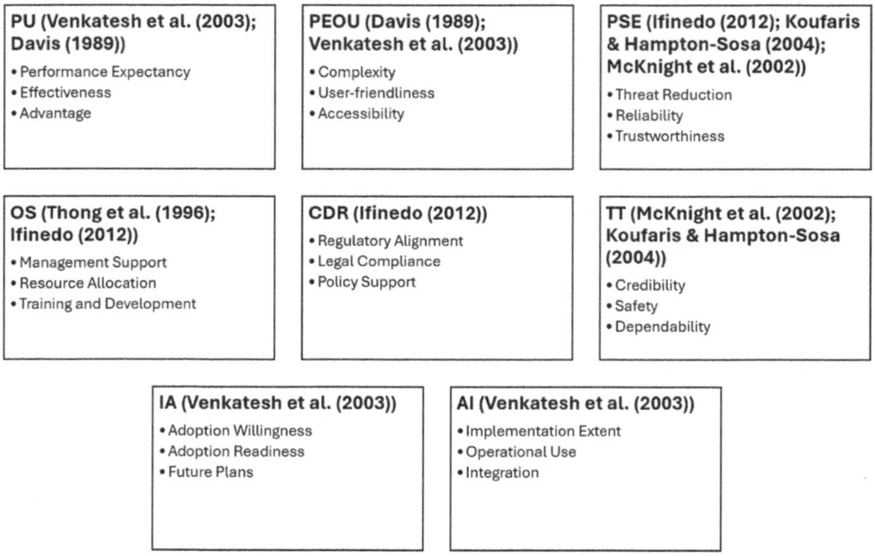

Fig. 3 Questionnaire structure and adapted sources

The study employs purposive sampling to select cybersecurity professionals working in the banking industry in Dubai. This non-probability sampling technique is chosen to ensure the selection of respondents with specific knowledge and experience relevant to the research objectives. The target sample size is 200 respondents, which is deemed adequate for conducting PLS-SEM using SmartPLS [10]. The study includes cybersecurity professionals currently employed in Dubai's banking industry with at least 1 year of experience and involvement in implementing or managing cybersecurity measures within their organizations. It excludes professionals not working in the banking industry, those with less than 1 year of experience, and those not involved in cybersecurity implementation or management.

Data was collected via online survey targeting cybersecurity professionals in Dubai's banking industry, leveraging professional networks and industry associations to reach a broad audience. The survey was open for a period of 8 weeks, during which follow-up reminders were sent to maximize response rates. The study adhered to ethical guidelines, ensuring informed consent from all participants. Respondents were assured of the confidentiality and anonymity of their responses. Data privacy was maintained by using secure data storage and handling procedures. A pilot study was conducted with 30 respondents to ensure the reliability and validity of the questionnaire. Based on the pilot results, presented in Table 2, minor modifications were made to the survey items to improve clarity and relevance. Specifically, some questions were rephrased for better comprehension, and redundant items were removed to streamline the questionnaire. Noting that All HTMT were less than 0.85.

Table 2 Pilot study results' summary

Construct	Cronbach's alpha (α)	Composite reliability (CR)	Average Variance Extracted (AVE)
PU	0.85	0.88	0.60
PEOU	0.83	0.87	0.58
PSE	0.86	0.89	0.61
OS	0.82	0.85	0.57
CDR	0.84	0.88	0.59
TT	0.87	0.90	0.62
IA	0.89	0.92	0.64
AI	0.88	0.91	0.63

SmartPLS 4.0 was used for data analysis, focusing on the assessment of the measurement model and the structural model. Table 3 presents the analysis steps of the research.

Table 3 Analysis steps of the research

PLS-SEM	Equation	Description
Composite reliability	$CR = \frac{(\sum \text{factor loadings})^2}{(\sum \text{factor loadings})^2 + \sum \text{factor variances}}$	Assesses internal consistency reliability Values should exceed 0.7
Average variance extracted	$AVE = \frac{\sum \text{factor loadings}^2}{\sum \text{factor loadings}^2 + \sum \text{factor variances}}$	Assesses convergent validity. Values should exceed 0.5
Path coefficients	Estimated coefficients between constructs	Assesses the strength and significance of relationships between constructs
Coefficient of determination	$R^2 = 1 - \frac{\text{Sum of squares of residuals}}{\text{Total sum of squares}}$	Indicates the variance in the dependent variable explained by the independent variables. Acceptable values exceed 0.1 for exploratory studies
Effect size	$f^2 = \frac{R^2_{\text{included}} - R^2_{\text{excluded}}}{1 - R^2_{\text{included}}}$	Assesses the impact of a specific exogenous construct on an endogenous construct
Predictive relevance	$Q^2 = 1 - \frac{\text{Sum of squares of prediction error}}{\text{Total sum of squares}}$	Assesses the model's predictive relevance using the blindfolding procedure. Values greater than 0 indicate predictive relevance
Discriminant validity	Fornell–Larcker criterion, HTMT ratios	Assesses discriminant validity. HTMT values should be below 0.85

The tests involve several key metrics to evaluate the measurement and structural models. CR and AVE assess internal consistency reliability and convergent validity, with acceptable thresholds of 0.7 and 0.5, respectively [10]. Path coefficients indicate the strength and significance of relationships between constructs. The coefficient of determination (R^2) shows how well independent variables explain variance in the dependent variable, with values above 0.1 being acceptable for exploratory studies [10]. Effect size (f^2) assesses the impact of specific exogenous constructs on endogenous constructs, while predictive relevance (Q^2) evaluates the model's ability to predict endogenous constructs, with Q^2 values greater than 0 indicating predictive relevance [10]. Discriminant validity, assessed using the Fornell–Larcker criterion and the Heterotrait–Monotrait ratio (HTMT), ensures constructs are distinct, with HTMT values below 0.85 [31].

5 Results and Analysis

The demographic profile of the respondents includes information on their gender, age, educational background, and years of experience in cybersecurity. This information, illustrated in Table 4, helps in understanding the sample characteristics and ensuring the representativeness of the data.

The demographic data indicates that the majority of respondents are male (70%) and primarily fall within the 31–40 age range (45%). Most participants possess a master's degree (50%), with a substantial portion having 4–6 years of professional experience (40%). This distribution reflects a well-educated and moderately experienced sample, predominantly male and in the early to mid-stages of their careers.

Table 4 Demographic profile of the respondents

Demographic variable	Category	Frequency (n = 200)	Percentage (%)
Gender	Male	144	70
	Female	56	30
Age	20–30	51	25
	31–40	89	45
	41–50	38	20
	51 and above	22	10
Education	Bachelor's degree	81	40
	Master's degree	99	50
	PhD	21	10
Experience	1–3 years	64	30
	4–6 years	86	40
	7–10 years	51	20
	More than 10 years	19	10

Table 5 Measurement model assessment summary

Construct	α	CR	AVE
PU	0.85	0.88	0.60
PEOU	0.83	0.87	0.58
PSE	0.86	0.89	0.61
OS	0.82	0.85	0.57
CDR	0.84	0.88	0.59
TT	0.87	0.90	0.62
IA	0.89	0.92	0.64
AI	0.88	0.91	0.63

The measurement model was assessed for reliability and validity using CR, AVE, and Fornell–Larcker criterion (all acceptable) and HTMT ratios (all less than 0.85), as demonstrated in Table 5.

The structural model was evaluated using path coefficients, R^2, f^2, and Q^2, as illustrated in Tables 6 and 7.

The mediation effect of IA on the relationship between PU, PEOU, PSE, OS, CDR, TT, and AI was analyzed. The bootstrapping method was used to test the indirect effects, as presented in Table 8.

The results reveal significant findings across multiple parameters. Path coefficients demonstrate strong and significant relationships between predictors and outcome variables, with all direct effects significant at the 0.05 level. R^2 values indicate a high proportion of explained variance, with 65% of IA variance explained by PU, PEOU, PSE, OS, CDR, and TT, and 58% of AI variance explained by IA, PSE, and

Table 6 Path coefficients, t-values, p-values, and significance for the structural model

Hypothesis	Path	Path coefficient (β)	t-value	p-value	Significance
H1	PU → IA	0.30	5.21	<0.001	Yes
H2	PEOU → IA	0.25	4.12	<0.001	Yes
H3	PSE → IA	0.20	3.54	<0.001	Yes
H4	OS → IA	0.18	3.00	0.003	Yes
H5	IA → AI	0.60	9.00	<0.001	Yes
H6	CDR → AI	0.10	2.00	0.046	Yes
H7	PSE → AI	0.12	2.50	0.012	Yes

Table 7 Structural model assessment summary

Construct	R^2	f^2	Q^2
IA	0.58	0.32	0.45
AI	0.55	0.42	0.47
TT	0.52	0.27	0.33

Table 8 Mediation analysis results' summary

Hypothesis	Path	Indirect effect (β)	t-value	p-value	Significance
H8	OS → TT → IA	0.11	2.20	0.029	Yes
H9	PSE → TT → IA	0.12	2.50	0.012	Yes

CDR. The f^2 effect sizes confirm the substantial impact of predictors, with IA on AI ($f^2 = 0.41$) showing a large effect, while IA itself ($f^2 = 0.32$) and TT ($f^2 = 0.27$) exhibit moderate effects. Q^2 values, all above 0.30, indicate strong predictive accuracy, confirming the model's reliability in forecasting adoption and implementation outcomes. Mediation analysis confirms that TT significantly mediates the relationships between OS and IA, as well as PSE and IA, reinforcing the importance of trust in technology as a critical factor in adoption. Additionally, IA plays a direct role in driving AI, underscoring its pivotal function in the transition from adoption intention to actual implementation. Overall, the findings indicate that PU, PEOU, PSE, OS, CDR, and TT significantly influence the IA of quantum computing for cybersecurity, while IA, CDR, and PSE directly impact AI. The high R^2 and Q^2 values confirm that the model has strong explanatory and predictive power, solidifying its robustness in understanding quantum computing adoption in cybersecurity within the banking sector.

6 Discussion

The findings of this study provide critical insights into the adoption of quantum computing for cybersecurity in Dubai's banking industry. Given the rapid technological advancements and the increasing complexity of cyber-threats, these insights are essential for financial institutions seeking to enhance their cybersecurity posture. The results align with the contextual framework established in the introduction, emphasizing the urgent need for robust cybersecurity measures in response to sophisticated cyber-threats. Financial institutions are under immense pressure to mitigate bot attacks, which have escalated in both frequency and sophistication. The significant influence of PU on IA underscores the perceived benefits of quantum computing in strengthening cybersecurity defences. This finding is consistent with the TAM, supporting the argument that perceived practical benefits drive technology adoption [28]. Given that credential-stuffing attacks accounted for 61% of all login attempts in the financial sector in 2021 [3], the utility of quantum computing in mitigating such threats serves as a strong motivator for adoption.

Similarly, PEOU significantly impacts IA, indicating that the simplicity and user-friendliness of quantum computing solutions are key drivers of adoption. This finding is particularly relevant in the banking industry, where ease of integration and operation can lower resistance to new technologies. The results corroborate earlier TAM research [29], which identified ease of use as a significant determinant of technology acceptance. PSE emerged as a significant predictor of IA, highlighting the importance of perceived improvements in security measures. This aligns with the introduction's emphasis on the escalating threat of bot attacks, which increased by 80% globally in the financial sector in the first half of 2022 compared to the previous year [6]. The ability of quantum computing to offer advanced encryption techniques and secure communication channels, as discussed by [14, 32], reinforces its value in cybersecurity applications.

OS was also found to be a significant predictor of IA, emphasizing the role of management support, resource allocation, and training in the adoption process. This finding is consistent with [30], who noted that top management support is crucial for successful technology implementation. In the context of Dubai's banking industry, ensuring robust organizational backing for quantum computing initiatives can significantly enhance adoption rates. Additionally, CDR significantly influences AI, underscoring the critical importance of regulatory compliance in driving implementation. Dubai has implemented stringent cybersecurity policies designed to protect financial infrastructure, as highlighted in previous research [4]. This finding aligns with broader research on regulatory compliance in technology adoption [30], reinforcing that adherence to local regulations is a key factor for cybersecurity professionals.

TT significantly impacts IA, indicating that credibility, safety, and dependability are essential considerations for the adoption of quantum computing solutions. This finding aligns with prior research emphasizing the importance of trust in the acceptance of emerging technologies [22, 27]. For Dubai's banking industry, fostering trust in quantum computing solutions is crucial for encouraging widespread adoption. The mediation analysis revealed that TT significantly mediates the relationship between OS and IA, as well as PSE and IA, highlighting the role of trust in technology adoption. This finding aligns with [13], who identified behavioral intention as a strong predictor of actual technology use. The mediation effect suggests that perceived benefits, ease of use, security enhancements, organizational support, and trust must first translate into a strong intention to adopt quantum computing before actual implementation can occur.

The findings of this study offer significant implications for theory, practice, and policy, particularly within Dubai's banking industry and the broader global financial sector. This study extends the TAM by incorporating additional constructs such as PSE, OS, CDR, and TT, providing a more comprehensive understanding of the factors influencing quantum computing adoption in cybersecurity. By integrating these constructs, the study addresses the complex, multi-faceted nature of technology adoption in highly regulated and security-sensitive industries like banking. This theoretical contribution is crucial as it validates the extended TAM framework in a new context, offering valuable insights into technology adoption dynamics [13, 28].

The practical implications of this study are particularly relevant for banks in Dubai and other financial institutions facing similar cybersecurity challenges. The findings suggest that organizations should focus on demonstrating the practical benefits and user-friendliness of quantum computing technologies to encourage adoption. Emphasizing the enhanced security capabilities of quantum computing while ensuring strong organizational support and regulatory compliance can significantly facilitate adoption. Additionally, fostering trust through transparency and proven effectiveness is critical. These insights can help banks develop targeted strategies to integrate quantum computing into their cybersecurity frameworks, enhancing their defence against sophisticated cyber-threats [30].

The study's implications are closely aligned with Dubai's cybersecurity strategy and global regulatory frameworks. Dubai has implemented stringent cybersecurity policies to protect critical financial infrastructure, emphasizing the importance of regulatory compliance and innovation [4]. The findings underscore the necessity for banks in Dubai to align their quantum computing initiatives with local regulations, such as Dubai's Cybersecurity Strategy, and international standards like the GDPR and the NIST Cybersecurity Framework. These global standards provide comprehensive guidelines for data protection and cybersecurity practices, ensuring that quantum computing adoption not only enhances security measures but also complies with regulatory requirements [1, 7]. By understanding and addressing the factors that influence the adoption of quantum computing, banks in Dubai and other financial hubs can develop robust cybersecurity strategies. Aligning these strategies with both local and international regulatory standards can significantly strengthen their defines against cyber-threats.

Table 9 presents key implementation strategies for integrating quantum computing into cybersecurity frameworks within the banking industry. These strategies include conducting pilot projects to demonstrate feasibility [7], developing comprehensive training programs, collaborating with regulatory bodies to ensure compliance [30], investing in ongoing research and development [12], and fostering a culture of security awareness [13]. Additionally, it highlights the importance of upgrading IT infrastructure [27], forming strategic partnerships [22], conducting risk assessments [11], ensuring user-friendly implementation [29], and increasing public awareness to build trust [28]. These strategies collectively enhance the effectiveness and security of quantum computing solutions in mitigating sophisticated cyber-threats.

7 Conclusion

This study provides empirical evidence on the factors influencing the adoption of quantum computing for cybersecurity in the banking sector of Dubai, extending the Technology Acceptance Model (TAM) by integrating Perceived Security Enhancement (PSE), Organizational Support (OS), Compliance with Dubai Regulations (CDR), and Trust in Technology (TT). The findings confirm that Perceived Usefulness (PU) and Perceived Ease of Use (PEOU) significantly influence the Intention to Adopt (IA) quantum computing (H1, H2), reinforcing prior research that emphasizes practical benefits and user-friendliness as critical adoption determinants [28, 29]. The results underscore the strategic importance of quantum computing in fortifying cybersecurity defines, particularly given the rise in bot attacks and credential-stuffing

Table 9 Implementation strategies for quantum computing in cybersecurity frameworks within the banking industry

Strategy	Description	Citation
Pilot projects	Conduct pilot projects to demonstrate the feasibility and benefits of quantum computing	[7]
Training programs	Develop training programs for cybersecurity professionals to ensure effective implementation	[7]
Regulatory collaboration	Work with regulatory bodies to ensure compliance with cybersecurity laws and policies	[30]
Research and development	Invest in continuous research to improve quantum computing security applications	[12]
Security awareness culture	Foster a culture of cybersecurity awareness within financial institutions	[13]
Infrastructure upgradation	Upgrade IT infrastructure to meet quantum computing requirements	[27]
Strategic partnerships	Partner with quantum computing experts and research institutions to accelerate adoption	[22]
Comprehensive risk assessment	Conduct risk assessments to identify vulnerabilities and mitigate potential risks	[11]
User-friendly implementation	Ensure that quantum computing solutions are easy to integrate and use	[29]
Public awareness and transparency	Increase public awareness about quantum computing benefits and security	[28]

incidents, which accounted for 61% of login attempts in the financial sector in 2021 [3, 6].

PSE is a crucial predictor of IA (H3), reflecting the increasing reliance on quantum encryption and secure communication protocols to mitigate cyber-threats [32]. OS and CDR significantly impact IA (H4, H6), highlighting that strong managerial support, adequate resource allocation, and adherence to regulatory mandates are critical to fostering technology adoption [4, 17, 30]. Given Dubai's stringent cybersecurity regulations, alignment with the Dubai Cybersecurity Strategy and international standards such as GDPR and the NIST Cybersecurity Framework is imperative for financial institutions [1, 7].

The study further confirms that TT plays a pivotal role in enhancing confidence in quantum computing solutions (H8, H9), supporting previous literature that emphasizes credibility, safety, and dependability as essential factors in technology acceptance [22, 27]. IA serves as a significant mediator between PU, PEOU, PSE, OS, TT, and Actual Implementation (AI) (H5), reinforcing the premise that organizations must first develop a strong intention to adopt before full-scale deployment can occur [13].

From a technological standpoint, quantum computing presents both opportunities and challenges (H7). While its integration with AI can enhance fraud detection and real-time cybersecurity threat management [16], concerns remain regarding its potential exploitation to break existing encryption protocols, necessitating the development of quantum-resistant cryptographic standards [7, 14]. The scalability of quantum solutions across complex banking infrastructures remains a significant barrier, requiring further research to optimize large-scale deployment [20]. Organizational support is paramount (H4), as successful adoption hinges on strong leadership, strategic planning, and sufficient investment in workforce training [30].

From a policy and economic perspective, evolving regulations and cost considerations shape adoption trajectories (H6). While existing regulatory frameworks, such as GDPR and the NIST Cybersecurity Framework, provide foundational guidance, the absence of quantum-specific policies necessitates the formulation of tailored regulations [1, 7]. A comprehensive cost–benefit analysis is essential to evaluate whether long-term security enhancements justify the substantial financial investments required for quantum implementation [5]. Furthermore, quantum computing represents a strategic advantage, enabling financial institutions to revolutionize risk assessment, fraud detection, and high-speed transactions, reinforcing its position as a transformative technology for next-generation banking cybersecurity.

Appendix

Questionnaire

This research aims to investigate the factors influencing the adoption of quantum computing for enhancing cybersecurity measures against bot attacks in the banking industry of Dubai. Your insights are invaluable and will contribute significantly to our understanding. The questionnaire consists of demographic questions and specific questions related to various constructs such as perceived usefulness, ease of use, security enhancement, organizational support, compliance with Dubai regulations, trust in technology, intention to adopt, and actual implementation. Your responses will be kept confidential and used solely for academic research. The survey should take approximately 10–15 min to complete. Your participation is voluntary and you may withdraw at any time.

Thank you for your valuable contribution.

Section A: Demographic Questions (select one option)

Demographic variable	Category
Gender	
	Male
	Female
Age	
	20–30
	31–40
	41–50
	51 and above
Education	
	Bachelor's degree
	Master's degree
	PhD
Experience	
	1–3 years
	4–6 years
	7–10 years
	More than 10 years

Section B: Main Questionnaire Items (select one option)

Question	1 = strongly disagree	2 = disagree	3 = somewhat disagree	4 = neutral	5 = somewhat agree	6 = agree	7 = strongly agree
Using quantum computing will improve our cybersecurity performance							
Quantum computing is effective in preventing bot attacks							
Quantum computing provides advantages over traditional security methods							
Quantum computing solutions are difficult to understand and use							

(continued)

(continued)

Question	1 = strongly disagree	2 = disagree	3 = somewhat disagree	4 = neutral	5 = somewhat agree	6 = agree	7 = strongly agree
Quantum computing tools are user-friendly							
Quantum computing resources are easily accessible							
Quantum computing reduces the threat of bot attacks							
Quantum computing is reliable for providing security solutions							
I trust quantum computing to enhance our cybersecurity							
Top management supports the adoption of quantum computing							
Adequate resources are allocated for quantum computing implementation							
Training on quantum computing is provided to employees							
Quantum computing solutions align with local regulations							
Quantum computing adheres to legal requirements in Dubai							
Regulatory bodies support the adoption of quantum computing							
I believe in the credibility of quantum computing technology							

(continued)

(continued)

Question	1 = strongly disagree	2 = disagree	3 = somewhat disagree	4 = neutral	5 = somewhat agree	6 = agree	7 = strongly agree
Quantum computing solutions are safe to use							
Quantum computing is dependable in securing our systems							
We are willing to adopt quantum computing solutions for cybersecurity							
Our organization is ready to implement quantum computing solutions							
We plan to adopt quantum computing for cybersecurity in the future							
Quantum computing has been implemented extensively in our organization							
Quantum computing is used in our daily cybersecurity operations							
Quantum computing is integrated into our existing security systems							

We appreciate your time and effort in completing this survey. If you have any questions, feel free to contact us.

Dr. Hayyan Nassar Waked (dr.hayyan@city.edu.my) or Dr. S. B. Goyal (drsbgoyal@gmail.com).

Thank you once again for your participation.

References

1. European Union, General Data Protection Regulation (GDPR) (2022), https://gdpr.eu. Accessed 31 Aug 2024
2. S.E. Goodman, R. Kirkpatrick, D. Kirkpatrick, Cybersecurity threat landscape in the financial sector. J. Financ. Serv. Res. **43**(3), 229–255 (2018). https://doi.org/10.1007/s10693-018-0324-1
3. Akamai, State of the Internet/Security Report (2022), https://www.akamai.com/security-research/the-state-of-the-internet. Accessed 31 Aug 2024
4. Dubai Electronic Security Center, Dubai cybersecurity strategy (2022), https://desc.gov.ae. Accessed 31 Aug 2024
5. N. Kshetri, Cybersecurity management and the path to digital transformation in the financial sector. IT Prof. **21**(2), 41–45 (2019). https://doi.org/10.1109/MITP.2019.2906417
6. Y. Zhang, L. Wang, Y. Liu, Cybersecurity challenges in e-commerce: A comprehensive survey. Comput. Netw. **194**, 108174 (2021). https://doi.org/10.1016/j.comnet.2021.108174
7. L. Chen et al., Report on post-quantum cryptography. National Institute of Standards and Technology (2022), https://nvlpubs.nist.gov/nistpubs/ir/2022/NIST.IR.8413.pdf. Accessed 31 Aug 2024
8. H.T. Nguyen, T.T. Do, AI-driven fraud detection in banking: A comprehensive review. J. Financ. Crime **29**(1), 114–132 (2019). https://doi.org/10.1108/JFC-05-2021-0092
9. T. Alam, F. Khan, The application of quantum computing in finance: A case study of Dubai's financial sector. J. Financ. Regul. Compliance **29**(2), 115–128 (2021). https://doi.org/10.1108/JFRC-10-2020-0081
10. J.F. Hair, G.T.M. Hult, C.M. Ringle, M. Sarstedt, in *A Primer on Partial Least Squares Structural Equation Modeling (PLS-SEM)*, 2nd edn. (Sage Publications, 2022)
11. K.B. Wright, Researching internet-based populations: Advantages and disadvantages of online survey research, online questionnaire authoring software packages, and web survey services. J. Comput. Mediat. Commun. **10**(3), JCMC1034 (2022). https://doi.org/10.1111/j.1083-6101.2005.tb00259.x
12. M. Mosca, M. Piani, Quantum-enhanced cybersecurity: The role of quantum random number generators. IEEE Secur. Priv. **17**(6), 77–80 (2019). https://doi.org/10.1109/MSP.2019.2934116
13. V. Venkatesh, M.G. Morris, G.B. Davis, F.D. Davis, User acceptance of information technology: Toward a unified view. MIS Q. **27**(3), 425–478 (2003). https://doi.org/10.2307/30036540
14. P.W. Shor, Algorithms for quantum computation: Discrete logarithms and factoring. in *Proceedings of the 35th Annual Symposium on Foundations of Computer Science*, (1994), pp. 124–134. https://doi.org/10.1109/SFCS.1994.365700
15. L.K. Grover, A fast quantum mechanical algorithm for database search. in *Proceedings of the 28th Annual ACM Symposium on Theory of Computing*, (1996), pp. 212–219. https://doi.org/10.1145/237814.237866
16. J. Biamonte et al., Quantum machine learning. Nature **549**(7671), 195–202 (2017). https://doi.org/10.1038/nature23474
17. J.Y.L. Thong, C.S. Yap, K.S. Raman, Top management support, external expertise, and information systems implementation in small businesses. Inf. Syst. Res. **7**(2), 248–267 (1996)
18. S. Pirandola et al., Advances in quantum cryptography. Adv. Opt. Photonics **12**(4), 1012–1236 (2020). https://doi.org/10.1364/AOP.378990
19. P. Sibson et al., Chip-based quantum key distribution. Nat. Commun. **8**(1), 1–6 (2017). https://doi.org/10.1038/ncomms13984
20. J. Yin et al., Satellite-based entanglement distribution over 1200 kilometers. Science **356**(6343), 1140–1144 (2020). https://doi.org/10.1126/science.aan3211
21. D. Zeifman, Internet of Things: The next big challenge in cybersecurity. J. Cybersecur. Priv. **1**(2), 123–139 (2015). https://doi.org/10.1093/cybsec/1.2.123
22. M. Koufaris, W. Hampton-Sosa, The development of initial trust in an online company by new customers. Inf. Manag. **41**(3), 377–397 (2004). https://doi.org/10.1016/j.im.2003.08.004

23. D.F. Sittig, H. Singh, A new socio-technical model for studying health information technology in complex adaptive healthcare systems. Qual. Saf. Health Care **19**(S3), i68–i74 (2020). https://doi.org/10.1136/qshc.2009.038539
24. A.K. Jain, A. Ross, Introduction to biometrics. IEEE Trans. Biom. Behav. Identity Sci. **3**(3), 299–315 (2021). https://doi.org/10.1109/TBIOM.2021.3068521
25. T. Herath, H.R. Rao, Protection motivation and deterrence: A framework for security policy compliance in organizations. Eur. J. Inf. Syst. **18**(2), 106–125 (2009). https://doi.org/10.1057/ejis.2009.6
26. T. Jennewein, U. Achleitner, G. Weihs, H. Weinfurter, A. Zeilinger, A fast and compact quantum random number generator. Rev. Sci. Instrum. **71**(4), 1675–1680 (2000). https://doi.org/10.1063/1.1150518
27. D.H. McKnight, V. Choudhury, C. Kacmar, Developing and validating trust measures for e-commerce: An integrative typology. Inf. Syst. Res. **13**(3), 334–359 (2002)
28. V. Venkatesh, F.D. Davis, A theoretical extension of the technology acceptance model: Four longitudinal field studies. Manage. Sci. **46**(2), 186–204 (2000). https://doi.org/10.1287/mnsc.46.2.186.11926
29. F.D. Davis, Perceived usefulness, perceived ease of use, and user acceptance of information technology. MIS Q. **13**(3), 319–340 (1989). https://doi.org/10.2307/249008
30. P. Ifinedo, Understanding information systems security policy compliance: An integration of the theory of planned behavior and the protection motivation theory. Comput. Secur. **31**(1), 83–95 (2012). https://doi.org/10.1016/j.cose.2011.10.007
31. J. Henseler, C.M. Ringle, M. Sarstedt, A new criterion for assessing discriminant validity in variance-based structural equation modeling. J. Acad. Mark. Sci. **43**(1), 115–135 (2015). https://doi.org/10.1007/s11747-014-0403-8
32. C.H. Bennett, G. Brassard, Quantum cryptography: Public key distribution and coin tossing. in *Proceedings of IEEE International Conference on Computers, Systems and Signal Processing*, (1984), pp. 175–179

Quantum Cryptanalysis of Symmetric-Key Ciphers: Background and Challenges

Debranjan Pal, Anubhab Baksi, and Anindya Ganguly

1 Introduction

The rapid advancement of technology in quantum computing poses a significant security threat to cryptographic schemes. Traditional public-key cryptography, which relies on classical assumptions such as the discrete logarithm and integer factorization problems, is vulnerable to quantum computers. These foundational problems serve as the basis for public-key cryptography.

Quantum computers can achieve algorithmic speedups, impacting symmetric-key schemes as well. Although, private-key cryptography is less vulnerable compare to public-key. Only quadratic speed up can be achieved by using quantum search algorithm. For instance, Grover's pioneering quantum search algorithm demonstrates a time complexity of $O(2^{k/2})$ for finding a k-bit key. Additionally, quantum computations can find preimages of an n-bit output of cryptographic hash function $\mathcal{H}$ in $O(2^{n/2})$ and to generate collisions of $\mathcal{H}$, quantum algorithm needs $O(2^{n/3})$ as demonstrated by Brassard et al. [9]. In addition to the aforementioned enhancements in generic attacks, research has been conducted on quantum attacks targeting specific modes, constructions, and primitives.

Kuwakado and Morii [36] introduced a new class of quantum attacks on three round Fiestal cipher. This attack employs similar quantum techniques as those utilized in quantum algorithms designed for the hidden subgroup problem.

D. Pal (✉)
Indian Institute of Technology Madras, Chennai, India
e-mail: debraj.pal5@gmail.com

A. Baksi
Nanyang Technological University, Singapore, Singapore
e-mail: anubhab.baksi@ntu.edu.sg

A. Ganguly
Indian Institute of Technology Kanpur, Kanpur, India
e-mail: anindyag@cse.iitk.ac.in

S. B. Goyal et al. (eds.), *Quantum Computing, Cyber Security and Cryptography*,
https://doi.org/10.1007/978-981-96-4948-8_9

Nonetheless, this proposal exhibits qualitative differences in various aspects. Interestingly, they quantum technique break reductions which are known to information-theoretic secure in classical settings. However, these attacks necessitate a potent *quantum CPA* setting. It enables the quantum adversary to queries in superposition to the associated cryptosystem. The quantum chosen-plaintext attacks (qCPA) have been extended and broadened to encompass a wide array of classical symmetric-key constructions. This includes Feistel networks, Even-Mansour ciphers, Encrypted-CBC-MACs, tweakable block ciphers, and various others. Kaplan et al. forgery attacks impact several CBC-like MACs [35] and introduces differential cryptanalysis within the quantum framework [34]. Liu and Liu deployed quantum methods to recover keys for the full keyed-sponge construction [41, 42]. Several quantum cryptanalysis related results have been published on AES [20, 28, 33, 43] and SHA-2/SHA-3 [29], on which the NIST specified security levels based on. Table 1 describes the quantum resilience assessments for present standard cryptosystems with respect to key size, security parameters, quantum algorithm applied for attack, number of qubits required and attack time [22]).

Formally, quantum cryptanalysis in the context of symmetric cryptography is defined as employing quantum computing to assess the security of symmetric ciphers. Security analysis involves specifying the time and data required to successfully recover either the encryption key or the plaintext. Quantum computing provides a distinct advantage by expediting classical algorithms or techniques employed in symmetric cryptanalysis. This implies the potential for the development of new quantum algorithms, distinct from classical approaches, to analyze the security of symmetric ciphers. Table 2 compares classical cryptographic methods with quantum-resistant cryptographic methods, focusing on key performance metrics such as security level, key size, encryption/decryption speed, and practical applications.

Organization. Chapter 2 has basic preliminaries about the quantum computing. After this, we describe quantum attacks against symmetric ciphers in Chap. 3. Later, we provide the relevant challenges in Chap. 4. The survey is ended with the concluding remarks in Chap. 6.

2 Background: Quantum Computing

Essentially, it involves a sequence of rotations applied to a high-dimensional vector within a space that is the tensor product of multiple lower dimensional spaces. The primary low-dimensional space under consideration is the C2 Hilbert space, which accommodates the qubit vectors. For instance, in a 4-qubit quantum algorithm, the targeted vector resides in a 16-dimensional entity within the C16 space, created by grouping four electron spins together. After these rotations, the final step involves conducting a measurement to obtain the desired outcome.

Unary Operation and Reversibility. To achieve the desired outcome, meaningful rotations must be executed, and these rotations must possess the quality of being

Table 1 Quantum resilience assessments for present cryptosystems as documented in the literature, considering different assumptions regarding error rates and Error-Correcting Codes

Cryptosystem	Category	Key size	Security parameter	Best-known quantum algorithm	Logical qubits required	Physical qubits required	Time required to break system
AES-GCM	Symmetric	128	128	Grover's [21]	2953	4.61×10^6	2.61×10^{12} years
		192	192		4449	1.68×10^7	1.97×10^{22} years
		256	256		6681	3.36×10^7	2.29×10^{32} years
RSA	Asymmetric	1024	80	Shor's [54]	2050	8.05×10^6	3.58 h
		2048	112		4,098	8.56×10^6	28.63 h
		4096	128		8,194	1.12×10^7	229 h
ECC, discrete-log problem		256	128		2,330	8.56×10^6	10.5 h
		384	192		3,484	9.05×10^6	37.67 h
		521	256		4,719	1.13×10^6	55 h
SHA-256	Bitcoin mining	N/A	72	Grover's [21]	2,403	2.23×10^6	1.8×10^4 years
PBKDF2 (with 10,000 iterations)	Password hashing	N/A	66	Grover's [21]	2,403	2.23×10^6	2.3×10^7 years

unitary. The key characteristic of a unitary matrix is its ability to preserve the vectors inner product, a necessity dictated by the laws of physics. The unitary nature also implies the existence of an inverse, a crucial requirement for quantum gates. Every rotation or evolution of the wave function in quantum mechanics is reversible. A natural consequence when envisioning operations as rotations in hyperspace. Within the realm of rotations, the notion of reversibility becomes significant, permitting backward rotations when both the output and the type of rotation (gate type) are known. In quantum computing, every step adheres to the principles of being unitary and reversible. However, this principle no longer applies at the moment of measurement.

2.1 Quantum Bits

In classical computing, the basic unit is a bit, while its quantum analog is called a *quantum bit*, or *qubit*. A qubit has a two-dimensional state represented as $|\psi\rangle = c_0|0\rangle + c_1|1\rangle$. Here, both c_0 and c_1 belongs to $\mathbb{C}$ and satisfy the condition $|c_0|^2 + |c_1|^2 = 1$. In the realm of quantum computing, $|0\rangle$ and $|1\rangle$ represent the states of logic 0 and logic 1 for the qubit, respectively. This property is the *superposition*, which is absent in classical computers.

Essentially, whereas a classical bit is limited to discrete states (logic 0 or logic 1), a qubit state can inhabit any point within the unit sphere due to superposition. For instance, the polarization of a single photon can act as a qubit, exhibiting a state that is a superposition of vertical and horizontal polarizations. The expression $c_0|0\rangle + c_1|1\rangle$ is generally expressed in matrix. Here, c_0 is the amplitude of $|0\rangle$ and similarly c_1 is the amplitude of $|1\rangle$.

Basis. In terms of linear algebra, $\{|0\rangle, |1\rangle\}$ forms a basis for two dimensional vector space. This implies $|0\rangle = (1 \;\; 0)^\top$, and $|1\rangle = (0 \;\; 1)^\top$. In similar fashion, $|00\rangle = (1\;0\;0\;0)^\top$, $|01\rangle = (0\;1\;0\;0)^\top$, $|10\rangle = (0\;0\;1\;0)^\top$ and $|11\rangle = (0\;0\;0\;1)^\top$. Another, useful basis is the *sign basis* $\{|+\rangle, |-\rangle\}$, where

$$|+\rangle := \left(\frac{|0\rangle + |1\rangle}{\sqrt{2}}\right), \quad |-\rangle := \left(\frac{|0\rangle - |1\rangle}{\sqrt{2}}\right).$$

2.2 Quantum Gates

A quantum gate is crafted to perform a unitary operation. In classical computing, electrical signals commonly traverse physical gates, such as a NAND gate built with transistors. However, in quantum computing, the qubit and its "signal" do not usually traverse physical space. Instead, operations like microwave or laser pulses are applied to act on, manipulate, or alter the qubit. As a result, the evolution of a quantum algorithm typically signifies the passage of time rather than a physical spatial arrangement. Qubits typically remain stationary, except in specific scenarios, such as in a trapped-ion quantum computer where trapped ions may be relocated. Nonetheless, even in these cases, such movement does not entirely mimic the signal flow in a classical computer.

In the next, we outline one-qubit, two-qubit, and three-qubit quantum gates. Subsequently, we provide an sketch of the Hadamard gate or Walsh (Hadamard) transformation. The circuit diagrams for the CNOT, CCNOT and Hadamard gates are shown in Fig. 1.

One-qubit Gates. One-qubit quantum gate is a linear operation, which basically maps a qubit to a qubit. Linearity ensures that the action of a gate is completely determined by its action on the given basis $\{|0\rangle, |1\rangle\}$.

- *Identity gate* I: The identity gate or I gate acting on the basis vectors as $|0\rangle \rightarrow |0\rangle$, and $|1\rangle \rightarrow |1\rangle$. Since the linear operation can be expressed as matrix.

$$I = |0\rangle\langle 0| + |1\rangle\langle 1| = \begin{bmatrix} 1 & 0 \\ 0 & 1 \end{bmatrix}$$

- X *gate*: A quantum NOT gate, often referred to as an X gate, is a fundamental quantum gate that performs the NOT operation on a qubit. This means, the basis vector $|0\rangle \rightarrow |1\rangle$, and $|1\rangle \rightarrow |0\rangle$. The corresponding matrix representation is

$$X = |1\rangle\langle 0| + |0\rangle\langle 1| = \begin{bmatrix} 0 & 1 \\ 1 & 0 \end{bmatrix}$$

- Y *gate*: The linear action on the basis vector for Y gate is as follows: $|0\rangle \rightarrow -|1\rangle$, and $|1\rangle \rightarrow |0\rangle$. The matrix form of the Y gate is:

$$Y = |0\rangle\langle 1| - |1\rangle\langle 0| = \begin{bmatrix} 0 & -1 \\ 1 & 0 \end{bmatrix}$$

- Z *gate*: The linear action on basis vectors for Z gate is $|0\rangle \rightarrow |0\rangle$, and $|1\rangle \rightarrow -|1\rangle$ and the matrix representation of this action is

$$Z = |0\rangle\langle 0| - |1\rangle\langle 1| = \begin{bmatrix} 1 & 0 \\ 0 & -1 \end{bmatrix}$$

The collection of matrices X, Y, and Z is referred to as the *Pauli matrices*. It can be easily verified that each of these operators is a unitary operator. These gates are highly useful for constructing more complex quantum algorithms and circuits. Conceptually, the four gates can be viewed as follows: the identity map acts as a quantum wire, X and Z represent bit-flip and phase-flip operations, respectively, and Y gate is the combination of X and Z, that is $Y = XZ$, signifying a bit-phase-flip operation.

Two-qubit Gates. The Controlled-NOT gate, commonly abbreviated as CNOT or CX gate, is a vital two-qubit gate utilized in quantum computing. It executes a NOT operation (flipping the state) on the target or second qubit only if the control or first qubit is in the state $|1\rangle$. If the control qubit is in the state $|0\rangle$, there is no alteration to the target qubit. Suppose the basis vectors for a two qubit system is $\{|00\rangle, |01\rangle, |10\rangle, |11\rangle\}$. Now the linear action on the basis vector is as follows:

$$CX : |00\rangle \mapsto |00\rangle, |01\rangle \mapsto |01\rangle, |10\rangle \mapsto |11\rangle, \text{ and } |11\rangle \mapsto |10\rangle.$$

If the control qubit is in the state $|0\rangle$, the target qubit remains unaffected; however, if the control qubit is in the state $|1\rangle$, the target qubit undergoes a NOT operation. This scenario can be interpreted as:

$$CX\,|\alpha, \beta\rangle = |\alpha, \alpha \oplus \beta\rangle.$$

The following matrix illustrates how the CNOT gate transforms the basis states of a two-qubit system.

$$\begin{aligned} CX &= |00\rangle\langle 00| + |01\rangle\langle 01| + |11\rangle\langle 10| + |10\rangle\langle 11| \\ &= |0\rangle\langle 0| \otimes I + |1\rangle\langle 1| \otimes X \\ &= \begin{bmatrix} 1 & 0 & 0 & 0 \\ 0 & 1 & 0 & 0 \\ 0 & 0 & 0 & 1 \\ 0 & 0 & 1 & 0 \end{bmatrix} \end{aligned}$$

It can be easily verified that CX is unitary and, furthermore, idempotent, meaning $CX^2 = I$.

Three-qubit Gates. The *CCNOT (Controlled-Controlled-NOT)* or *Toffoli* gate is the three-qubit input-output quantum gate. In this scenario, the third qubit flips if and only if both the first and second qubits are in the state $|1\rangle$. This implies that the expression for the CCNOT gate is

$$\text{CCNOT}|\alpha\beta\gamma\rangle = |\alpha, \beta, \alpha\beta \oplus \gamma\rangle.$$

Explicitly we can write it as:

$$\text{CCNOT} = \big(|00\rangle\langle 00| + |01\rangle\langle 01| + |10\rangle\langle 10|\big) \otimes I \;+\; |11\rangle\langle 11| \otimes X.$$

Hadamard Gate. One of the most useful gate in the quantum computation is the Hadamard gate or Hadamard Transformation. Hadamard transformation H is defined as: $H : |0\rangle \mapsto |+\rangle, \quad |1\rangle \mapsto |-\rangle$.

This gate is used to create a superposition state from either $|0\rangle$ or $|1\rangle$. The aforementioned map can be explicitly expressed as

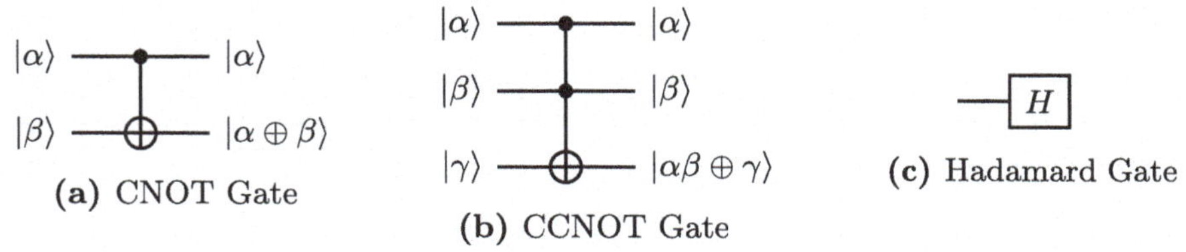

Fig. 1 Basic quantum gates

$$H = \left(\frac{|0\rangle + |1\rangle}{\sqrt{2}}\right)|0\rangle + \left(\frac{|0\rangle - |1\rangle}{\sqrt{2}}\right)|1\rangle \;=\; \frac{1}{\sqrt{2}}\begin{bmatrix}1 & 1\\ 1 & -1\end{bmatrix}.$$

Now, we define the generalized expression for the Hadamard gate.

$$\begin{aligned} H^{\otimes n}\,|0\rangle^{\otimes n} &= \frac{1}{\sqrt{2^n}}\,(|0\rangle + |1\rangle)^{\otimes n} \\ &= \frac{1}{\sqrt{2^n}}\left[\underbrace{(|0\rangle + |1\rangle) \otimes (|0\rangle + |1\rangle) \otimes \cdots \otimes (|0\rangle + |1\rangle)}_{n \text{ times}}\right] \\ &= \frac{1}{\sqrt{2^n}}\sum_{x=0}^{2^n-1}|x\rangle \end{aligned}$$

Thus, we successfully generate an equal superposition of all states $|\alpha\rangle$, where $0 \le \alpha \le 2^n - 1$, simultaneously.

2.3 *Quantum Search*

The Grover's search [21] is the primary tools for quantum search, it offers quadratic speedup compare to classical search. Since we are talking about symmetric-key cryptanalysis, so we frame the algorithm as pre-image search for hash functions. Basically we are assuming that the hash function is modeled as quantum circuit which generates the hash output for n-qubit input in a superposition state. In the following, we brief Grover search, this has three parts. First is input phase, oracle and diffusion operator.

Input. Initially apply Hadamard gate on n-qubit input, and prepare superposition state $|\psi\rangle$. That is,

$$H^{\otimes n}\,|0\rangle^{\otimes n} = |\psi\rangle = \left(\frac{|0\rangle + |1\rangle}{\sqrt{2}}\right)^{\otimes n} = \frac{1}{2^{n/2}}\sum_{x=0}^{2^n-1}|x\rangle.$$

Oracle ($\mathcal{O}_f$). Since the hash function is implemented using a quantum circuit, so the output of the hash function which is in superposition state is compared with the known hash output (h_{ko}). Now, we define a function $f(x)$ on the matching with known hash output.

$$f(x) = \begin{cases} 1 & \text{if} \, x = h_{ko} \\ -1 & \text{if} \, x \neq h_{ko} \end{cases}$$

$$\mathcal{O}_f(|\psi\rangle|-\rangle) = \frac{1}{2^{n/2}} \sum_{x=0}^{2^n-1} (-1)^{f(x)} |x\rangle|-\rangle$$

Diffusion Operator. This operator helps to amplify the amplitude of the pre-image identified by the oracle.

3 Quantum Attacks

A quantum key-recovery attack is a methodology to retrieves the key more efficiently than exhaustive search, achieving this without testing all potential key options. While this definition aligns with the classical understanding, it necessitates delving into the intricacies of quantum computation and adversary models.

Quantum Adversaries

We analyze two types of adversaries, here we follow terminologies introduced in [24, 35].

Q1 Model. The behavior of Q1 model is almost same as classical CPA model. The attacker has the only extra benefit to use quantum algorithms such as quantum search.

Q2 Model. This attack model is more powerful than Q1 attack model. Here, the adversary is permitted to conduct quantum computations with quantum oracle interactions to the encryption and decryption oracles. It is not clear how practical this model is.

Quantum Random-Access Model

When transitioning classical computing to quantum ones, we frequently encounter the necessity for quantum random access. In classical computing, querying a memory index known at runtime typically incurs a cost of 1, according to the RAM model. Although, in quantum computing, a significant amount of computation occurs in superposition. As a result, the quantum analog of this index entails a superposition of indices. Achieving quantum random access involves integrating qRAM gates into the

Table 2 Comparing classical cryptographic methods with quantum-resistant cryptographic methods, focusing on key performance metrics such as security level, key size, encryption/decryption speed, and practical applications

Method	Security level	Key size (bits)	Encryption/ decryption speed	Practical applications
Classical (RSA-2048)	Based on the difficulty of factoring large primes; vulnerable to quantum attacks (Shor's algorithm)	2048 bits	Moderate; fast for encryption, slower for decryption with large keys	Widely used for secure communications (e.g., SSL/TLS, email encryption) [51]
Classical (AES-256)	Secure against classical attacks; quantum attacks reduce effective key size (Grover's algorithm)	256 bits	High speed for both encryption and decryption	Standard for symmetric-key encryption in various protocols (e.g., VPNs, disk encryption) [13]
Quantum-resistant (Lattice-based—Kyber)	Strong security foundation; considered secure against both classical and quantum attacks	3,088 bits (public key)	Efficient encryption and decryption operations	Suitable for post-quantum public-key encryption; under consideration for standardization [8]
Quantum-resistant (Hash-based—SPHINCS+)	Security based on hash functions; secure against quantum attacks with large outputs	Varies with hash size; typically larger than RSA or AES keys	Slower signing speed; faster verification	Digital signatures in environments where long-term security is crucial [4] (see also Ascon-Sign [56])
Quantum-resistant (Code-based—McEliece)	Security based on the difficulty of decoding random linear codes; long-standing resilience	4,096 bits (public key); very large compared to RSA/AES	Moderate to slow due to large key sizes	Used in applications requiring high security and can tolerate large keys (e.g., secure communications) [46]
Quantum-resistant (Multivariate polynomial—Rainbow)	Fast signing and verification; smaller key sizes compared to lattice-based schemes	2,560 bits	Fast in both signature generation and verification	Digital signatures in constrained environments where speed is critical [58]

plain circuit model. A qRAM gate, extending across memory cells (or, more broadly, qubit registers), enables superposition access to their contents within a single time-step. It transfers the accessed data from cell i to the output register while leaving the memory registers unchanged. Undoubtedly, the qRAM model is incredibly powerful. Although a quantum adversary armed with several thousand logical qubits seems to be a foreseeable future, the qRAM, especially when of substantial size, represents the frontier of large-scale quantum computing with quantum data architectures.

3.1 Linear Cryptanalysis

Linear cryptanalysis is classified as a known plaintext attack, relying on the assumption that the attacker possesses information about a specific set of plaintexts and their corresponding ciphertexts. The method aims to exploit the frequent occurrence of linear expressions involving plaintext bits, ciphertext bits , and subkey bits. Notably, the attacker lacks the capability to choose the available plaintexts and corresponding ciphertexts.

Linear cryptanalysis, introduced by Matsui [45] in 1992, involves approximating the round function with a linear function to uncover a correlation between the linear approximation and the non-linear encryption function. Attackers elucidate these linear approximations utilizing linear masks. The cryptanalyst must construct linear approximations for every round, ensuring that the output mask of one-round matches the input mask of the subsequent round. Subsequently, the piling-up lemma is employed to assess the correlation of the approximation across the entire cipher. In this scenario, the attacker can operate under the assumption that the linear approximation is provided, employing it in conjunction with a quantum computer to facilitate either a distinguishing attack or a key-recovery attack.

The quantum variant of the linear attack, as discussed in [34], revolves around linear approximations employing linear masks, depicted as: $E(x)[\gamma] \approx x[\delta]$, where γ and δ represent the input and output linear masks, respectively, and $x[\gamma] = \bigoplus_{i:\gamma_i=1} x_i$. The symbol $\approx$ denotes a high probability of equality between the two values, implying that the relation holds true for a substantial number of messages x.

Linear cryptanalysis relies on a critical concept known as bias, denoted by ϵ. This bias is characterized by $Pr(C[\gamma] = P[\delta]) = (1 \pm \epsilon)/2$, where P and C is the pair of n-bit plaintext and ciphertext, associated with corresponding linear masks γ and δ, and $\epsilon \gg 2n/2$. If $\epsilon \gg 2n/2$ holds true, it indicates that the encryption function is not a random permutation, thereby enabling the possibility of mounting key-recovery attacks. Kaplan et al. [35] introduce the quantum version of the linear distinguisher and last-round attack. They ouline these attacks presuppose that the linear characteristic has already been identified classically, covering the number of rounds to be targeted.

Malviya and Tiwari [44] proposes a quantum algorithm for linearly approximating a non-linear vectorial Boolean function, such as a Substitution Box.

3.2 Differential Cryptanalysis

Zhou et al. [66] were pioneers in employing quantum algorithms into the domain of differential cryptanalysis. To achieve quadratic speedup in the key-recovery phase, they implemented Grover's algorithm [21]. Following this, Kaplan et al. [35] further investigated the application of Grover's algorithm [21] in cryptanalysis, employing it in the key-recovery phases of different iterations of both differential and linear attacks. Their hypothesis suggested that combining Ambainis' algorithm with Grover's search might potentially reduce the time complexity of the attack. The quantum differential distinguisher operates within a fully quantum framework, whereas the last-round attack is demonstrated in both fully quantum and semi-quantum models. The quantum differential distinguisher, as proposed by Kaplan et al. [35], functions in the following manner:

Let Pt be the set of n-bit plaintexts msg, where the size of Pt is 2^n. We aim to identify a subset X within Pt consisting of elements $\alpha \in X$ that satisfy the following differential condition:

$$X = \{\alpha \in Pt : E(\alpha \oplus \delta_{\text{in}}) \oplus E(\alpha) = \delta_{\text{out}}\},$$

where E is the encryption function, and δ_{in} and δ_{out} are input and output differences, respectively.

Now, if the condition

$$h = -\log[\Pr_{\forall\alpha}(E(\alpha) \oplus E(\alpha \oplus \delta_{\text{in}}) = \delta_{\text{out}}) < n]$$

is satisfied, then the proportion of elements in Pt that meet this differential condition is $\epsilon = 2^{-h}$.

Using Grover's search algorithm [21], locating an element α with this property requires about $2^{h/2}$ iterations. In each iteration, two queries are made to the quantum oracle for the encryption function E: one to compute $E(\alpha \oplus \delta_{\text{in}})$ and the other to compute $E(\alpha)$.

If such an element α is found, the algorithm outputs either a "concrete" or "random" result. A "concrete" result could assist in executing a key-recovery attack, while a "random" result indicates that a brute-force approach using Grover's basic search algorithm [21] might be more efficient.

The time and data complexity of executing this quantum distinguisher is $2^{h/2} + 1$. Given the message space Pt of size $|Pt| = 2^n$, where each message consists of n-bits, n-qubits are required for Grover's search [21] along with an ancilla-qubit. Thus, the attack requires $(2n + c)$-qubits, where c denotes the number of ancilla-qubits, usually set to 1.

On the other hand, Li and Yang [37] focused on the initial phase of differential cryptanalysis. They utilized the Bernstein–Vazirani (BV) algorithm [5] to identify high-probability differentials in block ciphers. However, their algorithm displayed an apparent flaw, which was later addressed by Xie and Yang in [61].

3.3 Impossible Differential Cryptanalysis

Impossible differential cryptanalysis is part of chosen-plaintext attack (CPA), operates within the context of a cryptographic function $\phi_k : \{0, 1\}^n \rightarrow \{0, 1\}^n$, where $k \in K$ is the key. An impossible differential $(\Delta\alpha, \Delta\beta)$ of ϕ_k is characterized by the property that $\phi_k(\alpha) \oplus \phi_k(\alpha \oplus \Delta\alpha) \neq \Delta\beta$ for all $\alpha \in \{0, 1\}^n$.

Impossible differential cryptanalysis consists of two primary stages. First, the attacker seeks to identify an impossible differential $(\Delta\alpha, \Delta\beta)$ for the function ϕ_k. Second, using this discovered impossible differential, the attacker filters through the subkeys of the final round.

During the second phase, the attacker selects the input difference $\Delta\alpha$ and executes conventional queries across the entire encryption function to acquire a predetermined number of ciphertexts. Subsequently, for each potential key s of the final round, the attacker decrypts these ciphertexts, deriving corresponding output differences of ϕ_k. If $\Delta\beta$ is present among these output differences, the attacker eliminates s as a potential subkey.

Xie and Yang [61] uses miss-in-the-middle technique which finds extensive application in conventional impossible differential cryptanalysis, having been employed in schemes such as Skipjack, IDEA, DEAL, and others. The fundamental principle of this technique involves linking two differential paths, each with a probability of one, where the input and output differences do not correspond, resulting in the derivation of an impossible differential.

To identify Impossible differentials, they execute the Bernstein–Vazirani (BV) algorithm [5] $p(n) = O(n)$ times to uncover the linear structures. Here $p(n)$ is a polynomial. From these linear structures, they identify a differential pair that satisfies $\phi_k(\alpha \oplus \Delta\alpha) \oplus \phi_k(\alpha) \neq \Delta\beta$ for the maximum number of messages $\alpha \in P$. To ascertain an impossible differential with high likelihood, the BV algorithm is executed $O(n^2)$ times. The attack employs the same number of qubits, either $2n + 1$ or $3n$, as conventional differential cryptanalysis.

Truncated Impossible Differential Cryptanalysis.

Xie and Yang [62] introduce various techniques for attacking block ciphers. Initially, they introduce a quantum algorithm designed to identify the linear structures within a vector function, leveraging the Bernstein–Vazirani (BV) algorithm [5] as a subroutine. Subsequently, they adapt and refine this initial algorithm to create distinct versions, employing them in diverse attack strategies.

They developed quantum distinguishers customized for the three-round Feistel scheme [48]. In addition, they introduced a new quantum attack to recover a partial key of the Even-Mansour construction [18]. While these methods share similarities with those developed by Kuwakado and Morii [36], they opt for the Bernstein–Vazirani (BV) algorithm [5] over Simon's algorithm [55]. Despite this alteration leading to a slight increase in complexity, it renders the approach more versatile. For instance, by crafting functions with distinct linear structures, they can derive various distinguishers for the three-round Feistel [48] scheme. The BV algorithm [5] offers a notable advantage: it can unveil both periods of a function as well as its other linear structures.

Researchers discovered that encryption functions exhibit patterns that can be exploited for analysis. They developed three quantum cryptanalytic techniques: quantum differential analysis, quantum small probability differential cryptanalysis, and quantum impossible differential cryptanalysis. By evaluating the performance and efficiency of these methods, they assessed their potential to break encryption algorithms.

One thing to note is that the quantum algorithms used in these analyses run quite efficiently. Unlike the usual method of analyzing encryption, where it's difficult to extend the patterns and analyze many rounds, this new approach treats the encryption as a whole, making it easier to work with.

3.4 *Quantum Related Key Attack*

Robert and Martin [59] initially introduced the related-key attack model, wherein the key relationship is constrained to bit-flips. Let's say E and D represent encryption and decryption oracles, respectively. To find a secret $s \in \mathbb{F}_2^k$ the attacker interacts with these two oracles.

Encryption oracle: E. It returns the encryption $E_{s\oplus\alpha}(m)$, where $\alpha(m)$ is the bitmask of the plaintext message, $\alpha \in \mathbb{F}_2^k$ and plaintext $m \in \mathbb{F}_2^n$.

Decryption oracle: D. It computes the decryption $E_{s\oplus\alpha}^{-1}(c)$ from a given ciphertext $c \in \mathbb{F}_2^n$ and a bitmask $\alpha \in \mathbb{F}_2^k$.

Following interactions with these oracles, the attacker must generate a vector $s' \in \mathbb{F}_2^k$ as an approximation of s. Success for the attacker is achieved only when $s' = s$.

Xie and Yang's attacks, as described in [63], do not require access to the decryption oracle D. Despite this, it enables the attacker to query the encryption oracle E using superposition of keys. Basically the attacker use the quantum oracle O_E as follows:

$$\sum_{\alpha,m,y} |\alpha, m, y\rangle \rightarrow \sum_{\alpha,m,\beta} |\alpha,\ m,\ \beta \oplus E_{s\oplus\alpha}(m)\rangle$$

The adversary can smoothly incorporate the oracle O_E into their circuits without difficulty. Furthermore, they provide the attacker with the ability to query an oracle that exclusively yields a bit of the cipher using superposition of keys. Let us define, $E_{s\oplus\alpha} = (E_{s\oplus\alpha,j})_{j=1}^n$. The attacker use the quantum oracle O_{E_j}:

$$\sum_{\alpha,m,\beta} |\alpha, m, \beta\rangle \rightarrow \sum_{\alpha,m,\beta} |\alpha,\ m,\ \beta \oplus E_{s\oplus\alpha,j}(m)\rangle$$

Researchers have invested significant effort in exploring the concept of quantum attackers utilizing quantum superpositions to query cryptographic primitives. Having access to the oracle O_E implies that the attacker can query the encryption oracle using the target key s. This means, the attacker use the oracle O_{E_s}:

$$\sum_{m,\beta} |m, \beta\rangle \rightarrow \sum_{m,\beta} |m, \beta \oplus E_s(m)\rangle$$

To achieve this, they simply need to query O_E with the state $\sum_{m,\beta} |0\rangle|m\rangle|\beta\rangle$ and then remove the first register. Thus, the quantum related-key attack model also visualize as an extension of the quantum CPA model. BV algorithm [5] reveals a linear structure of this encryption scheme, it basically exposes the secret key. Now, consider an encryption map $E_s : \mathbb{F}_2^n \rightarrow \mathbb{F}_2^n$. It encrypts an n-bit plaintext m using the secret key s and give an n-bit ciphertext c, i.e., $E_s(m) = c$. Now formulate a new function $\phi_s^m : \mathbb{F}_2^k \rightarrow \mathbb{F}_2^n$ from the encryption function E_s. Here, the k-bit input x is termed a key mask. Note that, the function ϕ_s^m inherently possesses a linear structure so that $\phi_s^m(x) \oplus \phi_s^m(x \oplus s) = 0$ remains valid.

3.5 Quantum Meet in the Middle Attack

Hosoyamada and Sasaki [24] developed a quantum version of the Demiric-Selçuk Meet-in-the-Middle (DS-MiTM) technique for six-round Feistel constructions. The DS-MiTM approach consists of two phases: queried data analysis and distinguisher analysis. In the distinguisher analysis, the attacker collects sets of differential characteristics that satisfy a truncated differential. Additional rounds can be added to create the queried data analysis section, where plaintext–ciphertext pairs are gathered based on specified differential characteristics.

The quantum DS-MiTM attack can be formulated as a claw-finding problem, where the goal is to find $x \in \{0, 1\}^p$ and $y \in \{0, 1\}^q$ such that $f(x) = g(y)$. The distinguisher analysis and queried data analysis correspond to different aspects of this problem. The quantum DS-MiTM attack can be implemented in both semi-quantum and fully quantum models. In the semi-quantum model, one function is queried classically, while the other is queried quantumly. In the fully quantum model, both functions allow superposition queries.

The proposed approach for solving both semi-quantum and fully quantum claw-finding problems uses parallel Grover's search and Simon's algorithm [55]. This leads to two types of quantum DS-MiTM attacks: semi-quantum and fully quantum versions.

In the semi-quantum version, known as classical online and quantum offline, data is gathered by querying the function classically (meaning the classical oracle is online), while the amassed information is handled using quantum computers. In DS-MiTM [17], the rounds encompassed by the queried data analysis are regarded as the function g, responsible for handling classical queries, resembling the classical attack approach. On the contrary, rounds included in the distinguisher analysis are enclosed within a black-box capable of accommodating quantum superposition queries. In this framework, the claw-finding algorithm is employed to retrieve subkeys corresponding to the rounds addressed by the queried data analysis. The semi-quantum attack recovers subkey through $2^{n/2}$ classical queries and operates with a time complexity of $2^{n/q}$, where q represents the number of available qubits for parallelizing Grover's search. In the fully quantum version of the attack, since both attack phases benefit from quantum computation acceleration, Grover's search is employed to retrieve the correct key, with its accuracy confirmed through Simon's algorithm [55]. The correctness of the key signifies that the two attack phases reveal a claw (x, y) satisfying Simon's promise, namely $x \oplus s = y$. The fully quantum attack requires $2^{3n/4}$ quantum queries and offers a time complexity of $2^{3n/4}$.

3.6 *Quantum Square Attack on AES*

The Square attack, initially utilized in the Square cipher [12], constitutes a chosen-plaintext attack. Concerning AES [14], a Δ-set comprises 256 states, wherein the XOR difference between any two bytes across different states is non-zero, while it is zero for other bytes. Let U denote a Δ-set, then for any $p, q \in U$, the following conditions hold:

$$\begin{cases} p_{i,j} \neq q_{i,j} & \text{when } (i, j) \text{ is active.} \\ p_{i,j} = q_{i,j} & \text{when } (i, j) \text{ is passive.} \end{cases}$$

Here, i and j range over 0, 1, 2, 3. Let $m^l_{i,j}$ represent a byte at position (i, j), where i and j take values from 0, 1, 2, 3; of the lth state within a Δ-set following the application of $\mathcal{S}_3^{AK}$. Then,

$$\sum_{l=0}^{256} m^l_{i,j} = 0$$

This equation signifies that the outputs of the third round ($\mathcal{S}_3^{AK}$) are balanced.

Bonnetain et al. [7] proposes quantum version of square attack. Employing the square property as a three-round distinguisher, each structure imposes a condition on one byte. Despite requiring only 5 Δ-sets, they opt for 8 Δ-sets to ensure that only the correct key guess successfully passes the test, with a high probability. Consequently, the number of Grover iterations remains precisely determined. Since each partial decryption necessitates 5 S-Boxes, the classical time complexity in S-Boxes is:

$$2^{40} \times 2^{32} \times 8 \times 5 \leq 2^{78}$$

Meanwhile, the quantum time complexity is:

$$\left\lceil \frac{\pi}{4} \times 2^{20} \right\rceil \times (2^{32} \times 2 \times 8 \times 5) \leq 2^{58}$$

As the input data is purely classical, the attack algorithm adheres to Q1 (no superposition queries are necessary). By employing the attack framework [12], they successfully retrieve the key with 2^{37} chosen-plaintext queries, equivalent to a quantum time complexity of 2^{121} reversible S-Boxes. This method necessitates a small qubit count, 2^{37} classical memory, and does not use qRAM. Initially, the key search space is expanded to encompass 20 unknown bytes, necessitating more chosen-plaintext queries compared to previous scenarios. Employing 25 sets of 2^{32} plaintexts proves adequate and ensures a high probability of obtaining a single result. Grover search is then conducted over a search space of size $2^{20\times8}$ (representing the partial key bytes), with the anticipation of a single solution. Hence, the quantum time complexity is calculated as $c \cdot 2^{20\times4}(2^{32} \times 2^5 \times 10) \leq 2^{121}$ S-Boxes.

This method is designed as an attack specific to AES-256 [14], as its time complexity exceeds Grover's $2^{138.04}$ S-Boxes. Although, it exceeds the Grover search complexity for AES-192, which stands $2^{105.25}$.

3.7 Quantum Partial Sum Attack on AES

Ferguson et al. [19] introduce an improved version of the square attack known as the partial sum attack, demonstrating notable success in breaking reduced rounds of AES [14]. In partial sum attack, the attacker guess the key and compute the Δ-set sum in a form of partial calculation, rather than finding it as a whole as in square attack, thereby reducing the overall time and memory requirement.

The quantum counterpart [7] of the classical partial sum attack employs nested Amplitude Amplification procedures and utilizes random-accessible memory. Both classical and quantum memory requirements are identical. The algorithm requires 8×2^{24} 32-qubit quantum RAM registers, accessible in superposition, and 2^{35} 256-bit classical memory registers to accommodate the chosen-plaintext queries. Denoting $\pi \approx \frac{22}{7}$ as c, the time complexity, derived from its classical counterpart, can be expressed as:

$$\lceil c \cdot 2^8 \rceil \left(2 \times 2^{36} + 2\lceil c \cdot 2^4 \rceil \left(2 \times 2^{27} + 2 \times \lceil c \cdot 2^4 \rceil \left(2 \times 2^{19} + 2\lceil c \cdot 2^4 \rceil \times 2 \times 2^{11}\right)\right)\right)$$

Additional factors account for uncomputations, approximating the number of iterations for each subprocedure as $\lceil c \cdot 2^4 \rceil = 13$ and $\lceil c \cdot 2^8 \rceil = 201$. Using 8 structures guarantees only one solution at each step (with high probability), allowing us to determine the precise number of Grover iterations and facilitating efficient error corrections. Consequently, we attain a quantum time complexity equivalent to $2^{44.73}$ reversible S-Boxes.

For seven round quantum partial sum attack Bonnetain et al. [7] achieve improved time complexity by embedding the classical partial sum attack within a Grover search [21] targeting the additional key bytes. In the case of AES-256 [14], an additional 15 bytes need to be searched for in the "outer" Grover search, resulting in $c \cdot 2^{60}$ iterations. As a result, this results in a total of 2^{107} reversible S-Boxes, accounting for the need for extra structures, as well as 2^{29} 32-qubit quantum RAM registers. On the other hand, for AES-192 [14], there is one fewer key byte to guess, due to key-schedule properties. Thus, a total of $2^{103.4}$ S-Boxes are obtained, exceeding the $2^{105.25}$ S-Boxes necessary for Grover search.

3.8 Quantum Slide Attacks

Classical slide attacks [6] derive their effectiveness from the structural self-similarities inherent in ciphers. Consider the basic instance of a one-round self-similar cipher $C_k : \{0, 1\}^n \to \{0, 1\}^n$, constructed by iterating a round function $\phi(x, k) : \{0, 1\}^n \times \{0, 1\}^m \to \{0, 1\}^n$ for a total of r rounds, employing the same round key k. This arrangement embodies a straightforward equality, known as the slide property:

$$C_k(\phi(x, k)) = \phi(C_k(x), k)$$

The attacker aims to identify two pairs x, y that satisfy $\phi(x, k) = y$. According to the birthday paradox, within $O(2^{n/2})$ plaintext–ciphertext couples pt, ct, there exists a slide pair: (pt_0, ct_0) and (pt_1, ct_1) such that $\phi(pt_0, k) = pt_1$. Consequently, we also have $\phi(ct_0, k) = ct_1$. If ϕ exhibits sufficient weakness, the key k can be recovered from these two equations. Thus, the simplest attack scenario involves conducting $O(2^{n/2})$ queries, subsequently verifying for each pair (pt_0, ct_0) and (pt_1, ct_1) if it constitutes a slide pair. If so, the key k is returned.

Kaplan et al. [35] demonstrated that in certain instances, the slide property can be transformed into a promise problem, effectively solvable with Simon's algorithm [55]. In this scenario, slide pairs exhibit the property: $pt_1 = \phi(k \oplus P_0)$, or equivalently, $ct_1 = k \oplus \phi(C_0)$. Consequently, $ct_1 \oplus \phi^{-1}(pt_1) = pt_0 \oplus \phi(ct_0)$ enables the detection of a slide pair, allowing for the recovery of k in $O(2^{n/2})$ time and poly(n) memory.

Kaplan et al. [35] observed that the self-similarity can be reformulated as: $\forall x, \phi(E_k(x)) \oplus k = E_k(\phi(x \oplus k))$. Consequently, defining the two random functions $g_1(x) = \phi(E_k(x)) \oplus x$ and $g_2(x) = E_k(\phi(x)) \oplus x$, we obtain $g_1(x) = g_2(x \oplus k)$. Using Simon's algorithm [55], k can be recovered in $O(n^3)$ time with $O(n)$ Q2 queries.

4 Challenges and Future Directions

The field of quantum cryptanalysis is undergoing rapid development, with ongoing exploration and emergence of novel methods.

Cryptanalysis of Other Block/Stream Ciphers and Hash/AEAD

Lightweight ciphers are tailored for devices with limited resources, such as IoT devices, where reducing computational load and enhancing efficiency are crucial. Quantum computing has the potential to significantly boost certain cryptanalysis tasks, particularly in brute-force key searches and some specialized cryptanalysis methods. Algorithms like Grover's algorithm can search through possibilities much

Table 3 Comparing various quantum-resistant algorithms, their strengths, weaknesses, and current development status

Algorithm	Strengths	Weaknesses	Development status
Lattice-based (e.g., Kyber, Dilithium)	Strong security guarantees; Efficient key generation and encryption; Well-studied mathematical foundations	Larger key sizes and ciphertexts compared to classical algorithms; Some schemes are computationally intensive	Advanced stage in NIST PQC standardization process, with several candidates in the final round [47]
Code-based (e.g., McEliece)	Long-standing security foundation; Very large key sizes provide robustness against quantum attacks	Extremely large public keys; Inefficient in terms of storage and communication	Part of the NIST PQC competition; Practical use limited due to large key sizes [46]
Multivariate polynomial (e.g., Rainbow)	Fast signature generation and verification; Smaller key sizes than lattice-based algorithms	Vulnerable to certain algebraic attacks; Complexity in security proofs	Was a candidate in NIST PQC but has faced recent cryptanalytic challenges [58]
Hash-based (e.g., SPHINCS+)	Security based on well-understood hash functions; Stateless variants are simpler and more robust	Large signature sizes and relatively slow signing speeds; Limited to specific applications (digital signatures)	Considered mature with ongoing optimizations; Included in NIST PQC standardization [4]
Isogeny-based (e.g., SIKE)	Small key sizes; Potential for efficient implementation on constrained devices	Slow key generation and encryption; Recent vulnerabilities to quantum attacks such as SIDH	Research ongoing with recent attacks influencing its future viability [16]

Table 4 Comparing the computational complexity of classical cryptographic algorithms with quantum-resistant algorithms for various cryptographic tasks, such as encryption, decryption, and key generation

Algorithm type	Task	Classical complexity	Quantum-resistant complexity
RSA	Key generation	$O(n^3)$ using classical algorithms	Vulnerable to $O((\log n)^3)$ with Shor's algorithm on a quantum computer [53]
RSA	Encryption/decryption	$O(n^3)$ (modular exponentiation)	Quantum attacks reduce complexity to polynomial time $O((\log n)^3)$ [53]
AES	Encryption/decryption	$O(n)$ where n is the number of rounds (e.g., 10 rounds for AES-128)	$O(2^{n/2})$ with Grover's algorithm which halves the effective key length [21]
Lattice-based (e.g., Kyber)	Key generation	$O(n^2)$ for matrix operations	$O(n^2)$, but designed to be resistant to quantum attacks [47]
Lattice-based (e.g., Kyber)	Encryption/decryption	$O(n^2)$ for polynomial multiplications	$O(n^2)$; Computational efficiency comparable to classical algorithms, with quantum resistance [47]
Hash-based (e.g., SPHINCS+)	Signature generation	$O(n)$ (hash function calls)	$O(n)$ with no known quantum speedup in key generation [4]
Hash-based (e.g., SPHINCS+)	Signature verification	$O(n)$ (hash function calls)	$O(n)$ with hash-based security remaining intact against quantum computers [4]
Code-based (e.g., McEliece)	Key generation	$O(n^2)$ for encoding operations	$O(n^2)$, secure against quantum computers but with large key sizes [46]
Code-based (e.g., McEliece)	Encryption/decryption	$O(n^2)$ for encoding and decoding operations	$O(n^2)$; Resistant to quantum attacks but computationally demanding due to key sizes [46]

faster than classical methods, posing a threat to the security of lightweight block ciphers.

However, quantum cryptanalysis is still in its infancy, and we do not yet have practical quantum computers capable of breaking cryptographic systems. The impact of quantum computing on lightweight ciphers will depend on several factors, including the specific cipher algorithm, key length, and cryptographic characteristics. Researchers are actively studying how quantum computing could affect cryptography, including lightweight ciphers, by developing quantum-based cryptanalysis techniques and assessing the robustness of current cryptographic schemes against quantum threats.

As quantum technology progresses, it's essential to adjust and improve cryptographic systems to ensure they remain secure in an evolving landscape. Numerous studies in the literature analyze lightweight ciphers under the quantum threat, such as [2, 3, 26, 27, 30–32, 38, 49, 50, 64]. Additionally, research focuses on finding efficient implementations of cryptographic components, as seen in [10, 11, 15, 39, 52, 60, 65]. For further insights, see [1].

Quantum Version of New Attacks

Quantum versions of new attacks leverage the principles of quantum computing to develop novel approaches for breaching cryptographic systems. These attacks exploit the unique properties of quantum mechanics to potentially overcome classical cryptographic defenses. As quantum computing capabilities advance, researchers explore how quantum algorithms can be adapted or developed to break encryption schemes, hash functions, and other cryptographic primitives more efficiently than classical counterparts. These quantum attacks pose significant challenges to the security of current cryptographic protocols and motivate the development of quantum-resistant cryptographic techniques.

Attack on Cryptographic Hash Functions

Hosoyamada and Sasaki [25] applied collision attack on SHA-256 and SHA-512 [23]. One potential avenue for future exploration involves investigating applications to other cryptographic hash functions. The same approaches can be applied for broader applications. Additionally, exploring optimizations of the classical semi-free-start collision attack's differential characteristics in relation to the transition to the quantum collision attack presents an intriguing area for research.

Improved DS-MiTM Attack

Bonnetain et al. [7] propose future directions by employing various differentials in intermediate stages and retaining the transitions can lead to attacks with diminished data requirements, albeit with a partial increment in the previously reduced memory. They also anticipate that novel and compelling trade-offs may emerge from this combination.

Quantum key-recovery attack beyond six rounds of AES

Enhancing both classical and quantum key-recovery attacks represents an intriguing open question, particularly concerning their implications for the security analysis of AES [14].

Attacks Applying Q2 Model

Q2 attacks, operating within a quantum adversary framework, permit queries to keyed primitives in superposition, treating them as black boxes during quantum computations. This model represents a potent capability, yet to date, Q2 attacks have shown restricted effectiveness. While certain quantum linear and differential attacks, as described by Kaplan et al. [35], fall under the Q2 category, they have not yielded more than quadratic speedups.

Design of Quantum Secure Block Ciphers

Creating quantum-secure block ciphers requires developing cryptographic primitives that withstand attacks from both classical and quantum adversaries. Recent quantum cryptanalysis research has shown that some symmetric-key schemes, previously considered secure, are vulnerable in the quantum realm. For example, Liskov, Rivest, and Wagner proposed secure tweakable block ciphers [40] based on secure classical block ciphers. However, Kaplan et al. [35] discovered that their scheme is susceptible to polynomial-time quantum superposition attacks, even when the underlying block ciphers are quantum-secure.

5 Towards Quantum Resistant Algorithms

Researchers are actively developing novel public-key algorithms that are resilient to quantum computer attacks. These advancements bolster confidence in certificate authorities, digital signatures, and encrypted communications. The National Institute of Standards and Technology (NIST) [57] is conducting a rigorous evaluation of 69 potential algorithms for post-quantum cryptography. The aim is to establish a draft standard that will be integrated into web browsers and other internet systems. Table 3 presents a comparative analysis of various quantum-resistant algorithms, encompassing their strengths, weaknesses, and current development status. Table 4 provides a comparison of the computational complexity of classical cryptographic algorithms with quantum-resistant algorithms across various cryptographic tasks, such as encryption, decryption, and key generation.

6 Conclusion

The exploration of quantum cryptanalysis in symmetric ciphers offers a profound insight into the potential vulnerabilities and strengths of modern cryptographic systems. Through this chapter, we have explored the theoretical framework of quantum computing and its implications for breaking symmetric encryption algorithms. Quantum algorithms such as Grover's algorithm pose a significant threat to symmetric ciphers by drastically reducing the search space required to find a key. This amplifies the importance of utilizing longer key lengths to maintain security against quantum attacks.

Furthermore, we've to remember the limitations of quantum cryptanalysis, including the necessity for quantum computers with a sufficient number of qubits and the practical challenges associated with error correction and decoherence. As we stand at the intersection of quantum computing and cryptography, it's clear that the field is rapidly evolving. Continued research into post-quantum cryptography and the development of quantum-resistant algorithms are essential for ensuring the long-term security of our digital infrastructure. Ultimately, the study of quantum cryptanalysis serves as a reminder of the ever-present need for vigilance and innovation in the face of emerging technological threats. By staying informed and adapting our cryptographic protocols accordingly, we can strive to maintain the confidentiality and integrity of sensitive information in an increasingly quantum-powered world.

References

1. A. Baksi, K. Jang, *Implementation and Analysis of Ciphers in Quantum Computing*. Springer (Apr. 2024). https://link.springer.com/book/10.1007/978-981-97-0025-7
2. A. Baksi, K. Jang, G. Song, H. Seo, Z. Xiang, Quantum implementation and resource estimates for rectangle and knot. Quantum Inf. Process. **20**(12) (Dec. 2021). https://doi.org/10.1007/s11128-021-03307-6
3. B.N. Bathe, R. Anand, S. Dutta, Evaluation of Grover's algorithm toward quantum cryptanalysis on ChaCha. Quantum Inf. Process. **20**(12), 394 (2021). https://doi.org/10.1007/s11128-021-03322-7
4. D.J. Bernstein, D. Hopwood, A. Hülsing, T. Lange, R. Niederhagen, L. Papachristodoulou, M. Schneider, P. Schwabe, Z. Wilcox-O'Hearn, SPHINCS+: submission to the NIST post-quantum cryptography standardization project (2019). https://sphincs.org/
5. E. Bernstein, U.V. Vazirani, Quantum complexity theory. SIAM J. Comput. **26**(5), 1411–1473 (1997). https://doi.org/10.1137/S0097539796300921
6. A. Biryukov, D.A. Wagner, Slide attacks, in *Fast Software Encryption, 6th International Workshop, FSE '99, Rome, Italy, March 24–26, 1999, Proceedings*. Lecture Notes in Computer Science, vol. 1636, ed. by L.R. Knudsen (Springer, 1999), pp. 245–259. https://doi.org/10.1007/3-540-48519-8_18
7. X. Bonnetain, M. Naya-Plasencia, A. Schrottenloher, Quantum security analysis of AES. IACR Trans. Symmetric Cryptol. **2019**(2), 55–93 (2019).https://doi.org/10.13154/TOSC.V2019.I2.55-93
8. J.W. Bos, L. Ducas, E. Kiltz, T. Lepoint, V. Lyubashevsky, J.M. Schanck, P. Schwabe, D. Stehle, R. Steinfeld, Crystals—kyber: A CCA-secure module-lattice-based Kem. Submission to the NIST Post-Quantum Cryptography Standardization Project (2017). https://pq-crystals.org/kyber/
9. G. Brassard, P. Hoyer, A. Tapp, Quantum algorithm for the collision problem (1997). arXiv:quant-ph/9705002
10. J. Chen, Q. Liu, Y. Fan, L. Wu, B. Li, M. Wang, New SAT-based model for quantum circuit decision problem: searching for low-cost quantum implementation. IACR Commun. Cryptol. **1**(1) (2024). https://doi.org/10.62056/anmmp-4c2h
11. M. Chun, A. Baksi, A. Chattopadhyay, Dorcis: depth optimized quantum implementation of substitution boxes. Cryptology ePrint Archive, Paper 2023/286 (2023). https://eprint.iacr.org/2023/286

12. J. Daemen, L.R. Knudsen, V. Rijmen, The block cipher square, in *Fast Software Encryption, 4th International Workshop, FSE '97, Haifa, Israel, January 20–22, 1997, Proceedings*. Lecture Notes in Computer Science, vol. 1267, ed. by Biham, E. (1997), pp. 149–165. https://doi.org/10.1007/BFb0052343
13. J. Daemen, V. Rijmen, *The Design of Rijndael: AES—The Advanced Encryption Standard*. Inf. Secur. Cryptogr. (2002). https://doi.org/10.1007/978-3-662-04722-4
14. J. Daemen, V. Rijmen, *The Design of Rijndael: AES—The Advanced Encryption Standard (Information Security and Cryptography)*, 1st edn. Springer (2002)
15. V.A. Dasu, A. Baksi, S. Sarkar, A. Chattopadhyay, LIGHTER-R: optimized reversible circuit implementation for SBoxes. *32nd IEEE International System-on-Chip Conference, SOCC 2019, Singapore, September 3–6, 2019* (2019), pp. 260–265. https://doi.org/10.1109/SOCC46988.2019.1570548320
16. L. De Feo, D. Jao, J. Plût, Towards quantum-resistant cryptosystems from super singular elliptic curve isogenies. J. Math. Cryptol. **8**(3), 209–247 (2014)
17. H. Demirci, A.A. Selçuk, A meet-in-the-middle attack on 8-round AES, in *Fast Software Encryption, 15th International Workshop, FSE 2008, Lausanne, Switzerland, February 10–13, 2008*, Revised Selected Papers. Lecture Notes in Computer Science, vol. 5086, ed. by K. Nyberg (2008), pp. 116–126. https://doi.org/10.1007/978-3-540-71039-4_7
18. S. Even, Y. Mansour, A construction of a cipher from a single pseudorandom permutation. J. Cryptol. **10**(3), 151–162 (1997). https://doi.org/10.1007/S001459900025
19. N. Ferguson, J. Kelsey, S. Lucks, B. Schneier, M. Stay, D.A. Wagner, D. Whiting, Improved cryptanalysis of Rijndael, in *Fast Software Encryption, 7th International Workshop, FSE 2000, New York, NY, USA, April 10–12, 2000, Proceedings* (2000), pp. 213–230. https://doi.org/10.1007/3-540-44706-7_15
20. M. Grassl, B. Langenberg, M. Roetteler, R. Steinwandt, Applying Grover's algorithm to AES: quantum resource estimates (2015)
21. L.K. Grover, A fast quantum mechanical algorithm for database search, in *Proceedings of the Twenty-Eighth Annual ACM Symposium on the Theory of Computing, Philadelphia, Pennsylvania, USA, May 22–24, 1996*, ed. by G.L. Miller (ACM, 1996), pp. 212–219. https://doi.org/10.1145/237814.237866
22. E. Grumbling, M. Horowitz, *Quantum Computing: Progress and Prospects* (The National Academies Press, Washington, D.C., 2019). https://doi.org/10.17226/25196, https://www.nap.edu/catalog/25196/quantum-computing-progress-and-prospects
23. H. Handschuh, *SHA Family (Secure Hash Algorithm)* (Springer US, Boston, MA, 2005), pp. 565–567. https://doi.org/10.1007/0-387-23483-7_388
24. A. Hosoyamada, Y. Sasaki, Quantum demiric-selçuk meet-in-the-middle attacks: applications to 6-round generic feistel constructions, in *Security and Cryptography for Networks—11th International Conference, SCN 2018, Amalfi, Italy, September 5–7, 2018, Proceedings*. Lecture Notes in Computer Science, vol. 11035, ed. by D. Catalano, R.D. Prisco (Springer, 2018), pp. 386–403. https://doi.org/10.1007/978-3-319-98113-0_21
25. A. Hosoyamada, Y. Sasaki, Quantum collision attacks on reduced SHA-256 and SHA-512, in *Advances in Cryptology—CRYPTO 2021—41st Annual International Cryptology Conference, CRYPTO 2021, Virtual Event, August 16–20, 2021, Proceedings, Part I*. Lecture Notes in Computer Science, vol. 12825, ed. by T. Malkin, C. Peikert (Springer, 2021), pp. 616–646.https://doi.org/10.1007/978-3-030-84242-0_22
26. K. Jang, A. Baksi, J. Breier, H. Seo, A. Chattopadhyay, Quantum implementation and analysis of default. Cryptology ePrint Archive, Paper 2022/647 (2022). https://eprint.iacr.org/2022/647
27. K. Jang, A. Baksi, H. Kim, H. Seo, A. Chattopadhyay, Improved quantum analysis of SPECK and lowmc (full version). IACR Cryptol. ePrint Arch (2022), p. 1427. https://eprint.iacr.org/2022/1427
28. K. Jang, A. Baksi, H. Kim, G. Song, H. Seo, A. Chattopadhyay, Quantum analysis of AES. Cryptology ePrint Archive, Paper 2022/683 (2022). https://eprint.iacr.org/2022/683
29. K. Jang, S. Lim, Y. Oh, A. Baksi, S. Chakraborty, H. Seo, Quantum implementation and analysis of SHA-2 and SHA-3. Cryptology ePrint Archive (2024). https://eprint.iacr.org/2024/513

30. K. Jang, G. Song, H. Kim, H. Kwon, H. Kim, H. Seo, Efficient implementation of present and gift on quantum computers. Appl. Sci. **11**, 4776 (2021). https://doi.org/10.3390/app11114776
31. K. Jang, G. Song, H. Kim, H. Kwon, H. Kim, H. Seo, Parallel quantum addition for Korean block ciphers. Quantum Inf. Process. **21**(11), 373 (2022)
32. K. Jang, G. Song, H. Kwon, S. Uhm, H. Kim, W.K. Lee, H. Seo, Grover on PIPO. Electronics **10**(10), 1194 (2021)
33. S. Jaques, M. Naehrig, M. Roetteler, F. Virdia, Implementing Grover oracles for quantum key search on AES and LowMC, in *Advances in Cryptology—EUROCRYPT 2020—39th Annual International Conference on the Theory and Applications of Cryptographic Techniques, Zagreb, Croatia, May 10–14, 2020, Proceedings, Part II*. Lecture Notes in Computer Science, vol. 12106, ed. by A. Canteaut, Y. Ishai (Springer, 2020), pp. 280–310. https://doi.org/10.1007/978-3-030-45724-2_10, https://doi.org/10.1007/978-3-030-45724-2_10
34. M. Kaplan, G. Leurent, A. Leverrier, M. Naya-Plasencia, Quantum differential and linear cryptanalysis (2015). arXiv:1510.05836
35. M. Kaplan, G. Leurent, A. Leverrier, M. Naya-Plasencia, Quantum differential and linear cryptanalysis. IACR Trans. Symmetric Cryptol. **2016**(1), 71–94 (2016).https://doi.org/10.13154/TOSC.V2016.I1.71-94
36. H. Kuwakado, M. Morii, Quantum distinguisher between the 3-round Feistel cipher and the random permutation, in *2010 IEEE International Symposium on Information Theory* (IEEE, 2010), pp. 2682–2685
37. H. Li, L. Yang, Quantum differential cryptanalysis to the block ciphers (2015). CoRR **abs/1511.08800**. http://arxiv.org/abs/1511.08800
38. D. Lin, Z. Xiang, R. Xu, X. Zeng, S. Zhang, Quantum circuit implementations of sm4 block cipher based on different gate sets. Quantum Inf. Process. **22** (Jul. 2023). https://doi.org/10.1007/s11128-023-04002-4
39. D. Lin, C. Yang, S. Xu, S. Tian, B. Sun, On the construction of quantum circuits for s-boxes with different criteria based on the sat solver. Cryptology ePrint Archive, Paper 2024/565 (2024). https://eprint.iacr.org/2024/565
40. M.D. Liskov, R.L. Rivest, D.A. Wagner, Tweakable block ciphers. J. Cryptol. **24**(3), 588–613 (2011). https://doi.org/10.1007/S00145-010-9073-Y
41. F. Liu, F. Liu: Universal forgery and key recovery attacks: application to FKS, FKD and Keyak. Cryptology ePrint Archive (2017)
42. F. Liu, F. Liu, Universal forgery with birthday paradox: application to block cipher-based message authentication codes and authenticated encryptions. Cryptology ePrint Archive (2017)
43. Q. Liu, B. Preneel, Z. Zhao, M. Wang, Improved quantum circuits for AES: reducing the depth and the number of qubits. Cryptology ePrint Archive, Paper 2023/1417 (2023). https://eprint.iacr.org/2023/1417
44. Malviya, A.K., Tiwari, N.: Linear approximation of a Vectorial Boolean function using quantum computing. Eur. Lett. **132**(4), 40001 (2020). https://doi.org/10.1209/0295-5075/132/40001
45. M. Matsui, On correlation between the order of s-boxes and the strength of DES, in *Advances in Cryptology—EUROCRYPT '94, Workshop on the Theory and Application of Cryptographic Techniques, Perugia, Italy, May 9–12, 1994, Proceedings*. Lecture Notes in Computer Science, vol. 950, ed. by A.D. Santis (Springer, 1994). https://doi.org/10.1007/BFb0053451
46. R.J. McEliece, A public-key cryptosystem based on algebraic coding theory. DSN Progress Report (1978)
47. NIST: Post-quantum cryptography standardization (2023). https://csrc.nist.gov/projects/post-quantum-cryptography
48. K. Nyberg, Generalized Feistel networks, in *Advances in Cryptology—ASIACRYPT '96, International Conference on the Theory and Applications of Cryptology and Information Security, Kyongju, Korea, November 3–7, 1996, Proceedings*. Lecture Notes in Computer Science, vol. 1163, ed. by K. Kim, T. Matsumoto (Springer, 1996), pp. 91–104. https://doi.org/10.1007/BFB0034838
49. Y. Oh, K. Jang, A. Baksi, H. Seo, Depth-optimized quantum circuits for ascon: aead and hash. Mathematics **12**(9) (2024). https://doi.org/10.3390/math12091337

50. Y. Oh, K. Jang, H. Seo, Quantum implementation of LSH. Cryptology ePrint Archive, Paper 2024/1082 (2024). https://eprint.iacr.org/2024/1082
51. R.L. Rivest, A. Shamir, L. Adleman, A method for obtaining digital signatures and public-key cryptosystems. Commun. ACM **21**(2), 120–126 (1978)
52. S. Roy, A. Baksi, A. Chattopadhyay, Quantum implementation of ASCON linear layer. NIST LWC Workshop (2023). https://csrc.nist.gov/csrc/media/Events/2023/lightweight-cryptography-workshop-2023/documents/accepted-papers/06-quantum-implementation-ascon-linear-layer.pdf
53. P.W. Shor, Algorithms for quantum computation: discrete logarithms and factoring, in *Proceedings 35th Annual Symposium on Foundations of Computer Science*, pp. 124–134 (1994)
54. P.W. Shor, Polynomial-time algorithms for prime factorization and discrete logarithms on a quantum computer. SIAM J. Comput. **26**(5), 1484–1509 (1997). https://doi.org/10.1137/s0097539795293172
55. D.R. Simon, On the power of quantum computation. SIAM J. Comput. **26**(5), 1474–1483 (1997)
56. V. Srivastava, N. Gupta, A. Jati, A. Baksi, J. Breier, A. Chattopadhyay, S.K. Debnath, X. Hou, Ascon-sign. NIST PQC Additional Round 1 Candidates (2023). https://csrc.nist.gov/csrc/media/Projects/pqc-dig-sig/documents/round-1/spec-files/Ascon-sign-spec-web.pdf
57. N.I. of Standards and Technology, Security requirements for cryptographic modules. Tech. Rep. Federal Information Processing Standards Publications (FIPS PUBS) 140-2, Change Notice 2 December 03, 2002, U.S. Department of Commerce, Washington, D.C. (2001https://doi.org/10.6028/nist.fips.140-2
58. R. Team, Cryptanalysis of the rainbow signature scheme (2022). https://cryptanalysis.com/rainbow
59. R. Winternitz, M. Hellman, Chosen-key attacks on a block cipher. Cryptologia **11**(1), 16–20 (1987). https://doi.org/10.1080/0161-118791861749
60. Z. Xiang, X. Zeng, D. Lin, Z. Bao, S. Zhang, Optimizing implementations of linear layers. Cryptology ePrint Archive, Report 2020/903 (2020). https://eprint.iacr.org/2020/903
61. H. Xie, L. Yang, Using Bernstein-Vazirani algorithm to attack block ciphers. Des. Codes Cryptogr. **87**(5), 1161–1182 (2019). https://doi.org/10.1007/S10623-018-0510-5
62. H. Xie, L. Yang, Using Bernstein-Vazirani algorithm to attack block ciphers. Des. Codes Cryptogr. **87**(5), 1161–1182 (2019). https://doi.org/10.1007/S10623-018-0510-5
63. H. Xie, L. Yang, A quantum related-key attack based on the Bernstein-Vazirani algorithm. Quantum Inf. Process. **19**(8), 240 (2020)
64. Yang, Y., Jang, K., Baksi, A., Seo, H.: Optimized implementation and analysis of Cham in quantum computing. Appl. Sci. **13**(8) (2023). https://doi.org/10.3390/app13085156, https://www.mdpi.com/2076-3417/13/8/5156
65. Y. Yuan, W. Wu, T. Shi, L. Zhang, Y. Zhang, A framework to improve the implementations of linear layers. IACR Trans. Symmetric Cryptol. **2024**(2), 322–347 (2024). https://doi.org/10.46586/tosc.v2024.i2.322-347, https://tosc.iacr.org/index.php/ToSC/article/view/11633
66. Q. Zhou, S. Lu, Z. Zhang, J. Sun, Quantum differential cryptanalysis. Quantum Inf. Process. **14**(6), 2101–2109 (2015). https://doi.org/10.1007/S11128-015-0983-3

Cybersecurity in the Quantum Era: Advancements and Challenges in Quantum Cryptography and Post-Quantum Solutions

Pradeep Bedi, Sanjoy Das, S. B. Goyal, Anand Singh Rajawat, and Sardar M. N. Islam

1 Introduction

1.1 Quantum Computing

Quantum computers, a technology developed in the late 1970s and early 1980s, have revolutionized computing efficiency. Theorizations about its capabilities and algorithms, such as Shor's Algorithm and Wang et al.'s algorithm, have demonstrated its potential in cybersecurity. Recent advancements in quantum computing technology have led to numerous potential applications, with initial benefits expected as early as 2025 and greater benefits after 2030. Despite the difficulties in sensing and controlling quantum particles, quantum computing has demonstrated potential in cyberspaces such as machine learning and cryptography. A race to the top could result

P. Bedi
Department of Computer Science and Engineering, Graphic Era (Deemed to be) University, Dehradun, Uttarakhand, India

P. Bedi · S. Das
Department of Computer Science, Indira Gandhi National Tribal University—Regional Campus Manipur, Kangpokpi, Manipur, India

S. B. Goyal (✉)
Chitkara University Institute of Engineering and Technology, Chitkara University, Chandigarh-Patiala National Highway (NH - 64), Rajpura, Punjab 140401, India
e-mail: drsbgoyal@gmail.com

A. S. Rajawat
School of Computer Sciences and Engineering, Sandeep University, Nashik, India

S. M. N. Islam
Institute for Sustainable Industries & Livable Cities, Victoria University, Melbourne, Australia
e-mail: sardar.islam@vu.edu.au

S. B. Goyal et al. (eds.), *Quantum Computing, Cyber Security and Cryptography*,
https://doi.org/10.1007/978-981-96-4948-8_10

from a lack of understanding on how new algorithms can use quantum operations. Utilize superposition features in daily practice and master quantum programming to stay ahead of the game [1]. In the next five to ten years, quantum computing is predicted to become a reality and have an impact on network security, SSL/TLS, and e-commerce. This prognosis is validated by research advancements, which notify cyber security specialists about the consequences of quantum computation and its ability to supplant asymmetric algorithms [2]. August 2022: Belgium was promised a USD 50,000 reward by Microsoft if it could solve one of the four post-quantum cryptography (PQC) methods. The NIST's five-year effort to create a quantum-safe cryptography standard is called into question by this. The survival of the two PQC algorithms which are currently in use is uncertain, and the threat posed by quantum computing is far from over. Zero-vulnerability computation (ZVC) is an encryption-neutral technique that could make computers quantum-resistant by limiting access from outside parties and creating robust, energy-efficient a solid-state software on a chip [3]. The study described in [4] addresses the threat posed by cutting-edge hacking tools by suggesting a modified AES algorithm that uses quantum computing to encrypt/decrypt AES picture files. Quantum random walk causes an unusual shift in encryption performance, indicating how quantum computing may improve encryption security. Quantum computing presents a serious risk to cyber security and may lead to the failure of cryptographic methods. The present examination is still in its early stages, and scientists know very little about quantum technologies and their uses. It can take a while for classical cryptography to switch to quantum-resistant methods [5]. High-complexity computer-based solutions, like the simultaneous operation of numerous computers, algorithm processing times, and high-performance hardware, are predicted to undergo a revolution thanks to quantum computers.

These quantum computers shown in Fig. 1, can perform high-resolution image processing for artificial intelligence applications, improve cryptosystems, speed up internet server searches, solve optimization issues, and analyse massive amounts of data. The research in [6] assesses both harmful and non-malicious applications as it looks at the possibilities of quantum techniques and how they affect security systems. By taking advantage of quantum–mechanical concepts like entanglement and superposition, quantum computing (QC) provides a computational advantage over classical computing. This advantage is expected to aid in the resolution of difficult problems in a variety of domains, including finance, data science, and renewable energy, and medical research, production of industrial chemicals, safe interaction, and quantum chemistry among others. Still, there are issues like quantum decoherence and quit interconnection [7]. It can improve the speed and accuracy of machine learning systems. However, security systems are susceptible to the dangers that quantum computing presents, especially asymmetric key cryptography. This has led to the replacement of symmetric key cryptography with PKI-based trust mechanisms. The goal of developments in quantum key distribution is to create quantum-resistant cryptosystems that could enable secure networks for 5G and beyond [8]. Research is being done on quantum control systems to reduce security flaws in 5G networks. Quantum mechanics can be utilized to circumvent the flaws in classical cryptography, which is no longer sufficient. Using quantum key distribution (QKD) as a backbone

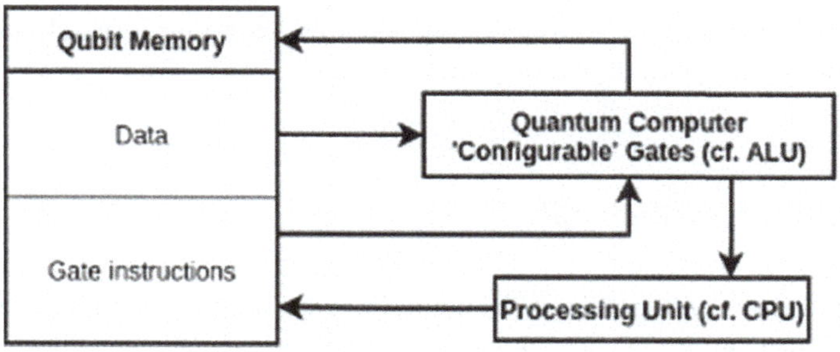

Fig. 1 Quantum computers and their capabilities [1]

of the MG distributed control system, [9] examines current quantum solutions for 5G security concerns. A measurement-device-independent QKD framework makes this framework suitable for industrial applications by protecting it against side-channel attacks. The proposed control strategy in [10] comprises a scalable QKD network, a quantum-secure control architecture based on QKD, and a deep neural network (DNN)-based method for quick parameter optimization.

Quantum computers, a technology developed in the late 1970s and early 1980s, have revolutionized computing efficiency and hold the potential to transform numerous industries. The goal of this research is to investigate the significant effects of quantum computing in the field of cybersecurity, an important field in which conventional approaches to data protection and encryption are becoming more and more insecure. The demand for strong, quantum-resistant security solutions is growing as quantum computing develops. The goal of this research is to better understand the opportunities and new problems that quantum computing presents for cybersecurity, with an emphasis on creating plans to reduce the risks involved. This chapter is structured to guide the reader through a comprehensive examination of quantum computing and its impact on cybersecurity. Section 2 begins by exploring the development of quantum computing, highlighting key algorithms such as Shor's Algorithm and recent advancements that demonstrate its potential in cybersecurity. Section 3 delves into quantum cryptography, examining how quantum mechanics can be harnessed to create secure communication channels resistant to both classical and quantum attacks. Section 4 moves to post-quantum cryptography, focusing on strategies to develop cryptographic systems that remain secure in the face of quantum computing threats, including an overview of ongoing efforts to standardize these systems. Section 5 examines quantum-secured communication networks, discussing how quantum key distribution and other quantum technologies can enhance the security of modern and future networks, such as 5G and 6G. Section 6 outlines how quantum technologies can be integrated into existing cybersecurity frameworks to protect against emerging threats, providing insights into practical applications and future directions. Section 7 continues with an exploration of quantum-safe security solutions, detailing the development of new cryptographic techniques and protocols that ensure data security in a quantum-enabled world. Finally, Sect. 8 concludes the chapter with an analysis of the potential threats and vulnerabilities introduced by quantum computing, offering recommendations for future research and policy development to safeguard digital infrastructure.

2 Quantum Cryptography

Figure 2, quantum cryptography is the umbrella term for a range of methods created to protect information and communication using the ideas of quantum physics. Notable methods include Quantum Key Distribution (QKD), such as the BB84 and E91 protocols, which allow secure key exchange by detecting any eavesdropping. Quantum Digital Signatures (QDS) guarantee the validity and non-repudiation of messages, while Quantum Secure Direct Communication (QSDC) permits the direct transfer of data without the need for key distribution. Innovative techniques such as Measurement-Device-Independent QKD (MDI-QKD) and Device-Independent QKD (DI-QKD) enhance security by mitigating device weaknesses. Additionally, Quantum Random Number Generation (QRNG) provides the true unpredictability needed for cryptographic applications, while Quantum Oblivious Transfer (QOT) and Quantum Bit Commitment (QBC) offer safe multiparty computations and fundamental cryptographic protocols. Every one of these methods is essential to the advancement of quantum-secure communications.

By utilizing the inherent properties of quantum physics, data is conveyed and secured in an unbreakable way with quantum cryptography. Figure 2 Encrypting and safeguarding data is a form of cryptography that makes it impossible for anyone

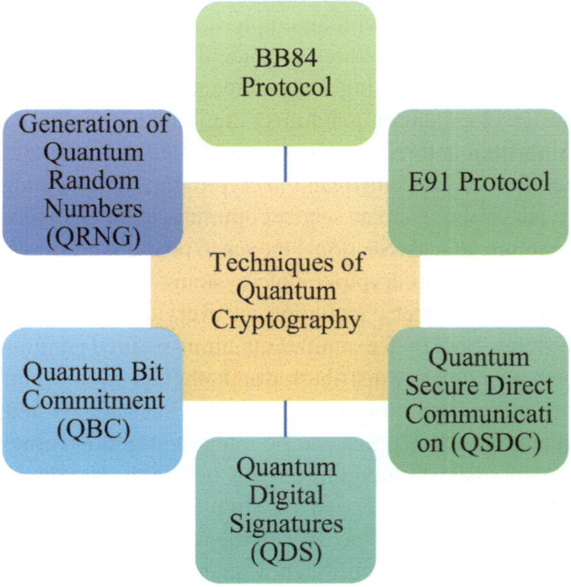

Fig. 2 Different techniques used in quantum cryptography

but the owner of the correct secret key to decrypt it. Because of quantum physics' special ideas, particles can exist simultaneously in several states at once, due to their uncertainty: Particles are capable of existing in several states at once. Furthermore, it is impossible to determine the exact quantum state of such particles. Photons are used in quantum communications to transmit data over long distances. For photonic quits, photons have persistent superposition phases due to their great resilience to shocks. They can also travel in optic fibers that have an impedance of 0.2 dB/km at 1.55 μm. Quasi-cryptography focuses on applying quantum mechanics to cryptographic operations. It is the most famous example of quantum cryptography to provide an information-theoretically safe key exchange mechanism. By utilizing quantum cryptography, authors can solve a wide range of cryptographic problems that were previously either impossible or highly unlikely to be solved with classical, or nonquantum, methods of communication. An example of this would be that it is not possible to make copies of quantum data. Reading encoded data will cause the quantum state to shift (no-cloning theorem) if the wave function collapses. Detection of ever-dropping quantum keys (QKD) could be done with this approach.

The field of quantum cryptography is expanding quickly, with novel protocols and steadily improved security proofs. This field encompasses continuous-variable infrastructure, quantum repetitions and the networks, device autonomy, satellite difficulties, and distribution of quantum keys protocols. Both theoretical and experimental improvements are being made in this area. Beyond standard distribution, examples of quantum cryptography include electronic signatures and quantum random number generators (QRNG) [11]. Quantum cryptography uses the ideas of quantum physics to process information and communicate securely. Unlike classical cryptography, which depends on computational hardness assumptions, this approach guarantees that messages stay unreadable by specified recipients. In this physical notion, secure communication and quantum key distribution are the main topics of concern [12]. Cryptosystems and online communication depend on quantum cryptography. Large quantum computers could, nevertheless, crack a lot of cryptosystems. Post-quantum cryptography seeks to maintain security by recognizing mathematical processes that benefit from the speed advantage of quantum algorithms. One of the challenges is maintaining confidence while maintaining the flexibility and utility of cryptography [13]. Theoretically, quantum cryptography leverages quantum mechanical effects for cryptographic operations such as quantum key distribution (QKD). It also looks at restrictions and difficulties posed by adversaries with quantum capabilities, like quantum bit commitment and quantum rewinding [14]. One significant technological advancement for improving network communication security is quantum cryptography. Public-key cryptography, authentication, and quantum key distribution are all included. Ensuring secure communication requires an understanding of QC from various angles, including quantum information processing, protocols, and assaults [15]. Cryptography uses encryption to disguise information and manage secret knowledge by turning plain text messages into unintelligible messages. The use of quantum mechanics in quantum cryptography is intended to address important issues, while classical cryptography is intended to ensure the security of data exchange by using quantum mechanics. Using either approach for safe communication offers advantages in certain situations [16]. In addition to shielding messages from unauthorized access, classical cryptography is also secure. In international communication, quantum cryptography enables complete data security by removing the constraints of classical encryption based on quantum physics. Both quantum and semi-quantum users are used in a semi-quantum communications paradigm. This field of study has grown and is still evolving with the development of novel protocols, security proofs, and experimental implementations, as well as the application of cryptography beyond the distribution of keys. Semi-quantum cryptography work requires a new understanding of constrained protocols, which can then be applied to quantum information science in general. It is important to examine the relationship between quantum and classical processing [17] in order to clarify important theoretical issues. Secure communications can be achieved through quantum cryptography, which uses quantum mechanics and light properties. Public key encryption and digital signatures are made more secure by addressing quantum adversary weaknesses. According to the unchangeable laws of quantum mechanics, quantum cryptography encrypts data using photons or quantum particles, which are also known as qubits [18, 19].

A thorough summary of the numerous facets of quantum cryptography, comprising case studies, difficulties, solutions, and application domains, is given in this Table 1. It highlights how quantum cryptography is evolving and being applied across different sectors to ensure secure communication and data protection.

3 Post-Quantum Cryptography

The last several decades have seen a significant increase in research interest in quantum computers because of its capacity to solve mathematical puzzles that conventional computers are unable to. If large-scale quantum computers were ever constructed, they could break many public-key encryption systems in the state they are in. Digital Communication on the World Wide Web and elsewhere could be seriously jeopardized by building an infrastructure for digital communication that is devoid of integrity and security. The development of cryptographic methods that are compatible with current networks and communication protocols and that provide security against both quantum and traditional computer attacks is the aim of post-quantum cryptography, also referred to as quantum-resistant cryptography. A field of study called post-quantum cryptography seeks to keep systems secure even when powerful quantum computers are presented in [13]. It focuses on finding mathematical processes that have the least amount of speed benefit over quantum algorithms, all the while preserving flexibility and usability without sacrificing assurance. As demonstrated in Fig. 3, Post-Quantum Cryptography (PQC) is a revolutionary cryptosystem that seeks to theoretically resist quantum computer attacks while still being capable to be implemented using traditional semiconductor technology. The National Institute of Standards and Technology, also known as NIST, has started standardizing PQC with 26 candidates from Round 2. An overview of PQC schemes, difficulties,

Table 1 Quantum cryptography overview

Aspect	Description	Case studies	Challenges	Solutions	Application areas
Data protection in quantum cryptography	Data is secured using quantum mechanics, making it impenetrable without the correct secret key	BB84 protocol, E91 protocol	Ensuring perfect implementation without loopholes; scalability over long distances	Advanced hardware implementations; continuous monitoring for eavesdropping	Government and military communications, financial transactions
Quantum state characteristics	Particles exist in multiple states simultaneously, with unpredictable quantum states	No-cloning theorem	Maintaining coherence in quantum states over long distances; error correction	Development of error correction codes and stable quantum channels	Quantum communication systems, advanced cryptographic applications
Quantum communication	Uses photons to transmit messages over long distances, with resilience to disturbances	Quantum satellite communications (e.g., Micius satellite)	High cost and complexity of maintaining photon coherence; susceptibility to atmospheric disturbances	Development of satellite-based quantum communication systems; enhancing fibre optic networks	Secure global communications, intercontinental quantum networks
Propagation in optical fibres	Photons in optical fibers have little resistance (0.2 dB/km at 1.55 μm) and transmit efficiently	Secure quantum-optical networks	Attenuation over long distances; integrating with existing infrastructure	Use of quantum repeaters; development of quantum-secured fibre optic networks	Long-distance secure communication, high-capacity data transfer
Quasi-cryptography	Applies quantum mechanics to cryptographic operations, ensuring safe key exchange	Implementation of Quasi-cryptography in secure key exchange systems	Complexity in practical implementation; need for quantum-resistant algorithms	Development of robust algorithms that utilize quantum mechanics effectively	Secure communication protocols, advanced cryptographic systems

(continued)

Table 1 (continued)

Aspect	Description	Case studies	Challenges	Solutions	Application areas
Theorem without cloning	Since quantum information cannot be duplicated without changing its original state, unapproved access can be identified	Protocols for Quantum Key Distribution (QKD)	Putting in place safe, attack-resistant QKD systems	Enhancement of QKD protocols with advanced error correction and detection mechanisms	Secure key exchange, authentication systems
Distribution of quantum keys (QKD)	Detects interference or eavesdropping to ensure safe key exchange	BB84 protocol, E91 protocol	Limited to short distances; requires specialized hardware	Development of long-range QKD using satellite systems; improving efficiency in QKD devices	Secure communications in critical infrastructure, financial institutions
Expansion of quantum cryptography	Rapid development of new protocols and security proofs, including quantum repeaters and networks	Continuous-variable QKD, device-independent QKD	Managing quantum decoherence; scalability and integration with existing systems	Development of quantum repeaters; hybrid classical-quantum systems	Quantum-secured networks, global communications, high-security systems
Digital signatures and random numbers	Extends to quantum random number creation and secure digital signatures	Generators of Quantum Random Numbers (QRNGs)	Ensuring true randomness; preventing tampering in the generation process	Integration with classical cryptographic systems; development of tamper-proof QRNGs	Cryptographic key generation, secure financial systems
Quantum-post (PQC) cryptography	Aims to develop cryptography systems that can withstand quantum computer attacks	Cryptography based on lattices, CRYSTALS-Kyber	High computational resource requirements; complexity in algorithm design	Optimization of algorithms for efficiency; development of hybrid cryptographic systems	Securing IoT devices, automotive systems, financial transactions

(continued)

Table 1 (continued)

Aspect	Description	Case studies	Challenges	Solutions	Application areas
Challenges in quantum cryptography	Includes maintaining security, flexibility, and addressing quantum adversary capabilities	Development of secure communication protocols	Ensuring robustness against quantum attacks; scalability and integration with classical systems	Adoption of hybrid systems combining quantum and classical cryptography	Secure communication, advanced cryptographic research
Public-key authentication and cryptography	Uses quantum physics to increase the safety of cryptography with public keys and authentication	Quantum Digital Signatures	Complex implementation; high resource requirements	Development of efficient and scalable quantum authentication systems	Secure authentication in legal, financial, and governmental sectors
Semi-quantum communications	Combines quantum and classical users, requiring new protocols and understanding of constrained interactions	Semi-quantum key distribution protocols	Ensuring compatibility between quantum and classical systems; preventing downgrading of security	Development of robust protocols for semi-quantum communications	Hybrid quantum–classical communication systems, secure data transfer
Quantum mechanics in cryptography	Ensures secure communication by leveraging quantum mechanics, such as encryption with qubits	Implementation of quantum cryptography in networks	Addressing quantum decoherence and error rates; high implementation complexity	Quantum correction of error code application and steady quantum channel development	Secure communications, cryptographic operations, advanced data security

novel approaches, and architectures for hardware implementation is given in the work in reference [20]. It supports the PQC standardization process and acts as a guide for future study.

A logical extension of the main logics of Easy Crypt can encompass a broad class of post-quantum cryptography proofs, as the work in [21] shows. They validate the post-quantum security of three traditional designs by putting into practice Easy PQC, a modified version of Easy Crypt for post-quantum security proofs. Because of the hyperconnectivity of the Vehicle-to-Everything (V2X) environment and the growing need for secure communication, automotive systems are becoming more significant computer platforms. Because automobile products have a lengthy lifespan, both present and potential assaults must be taken into account. Automotive electronics require effective, reliable, and long-lasting security solutions to handle

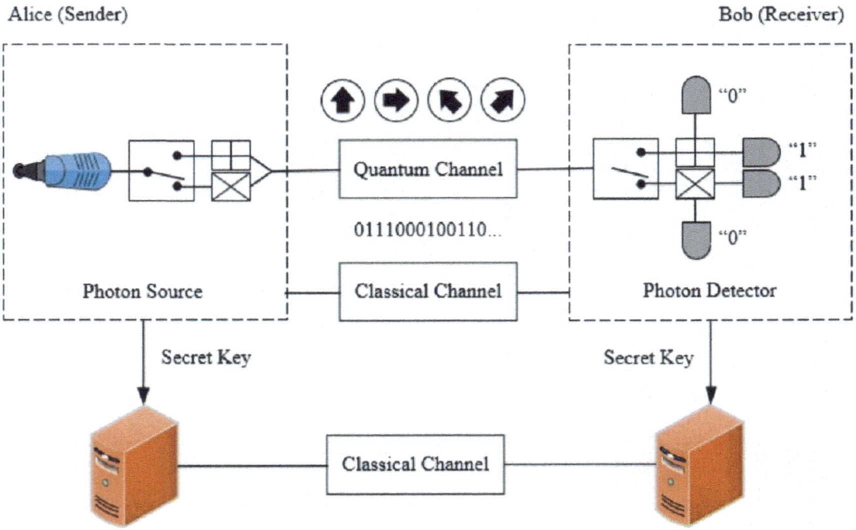

Fig. 3 Quantum key distribution process [11]

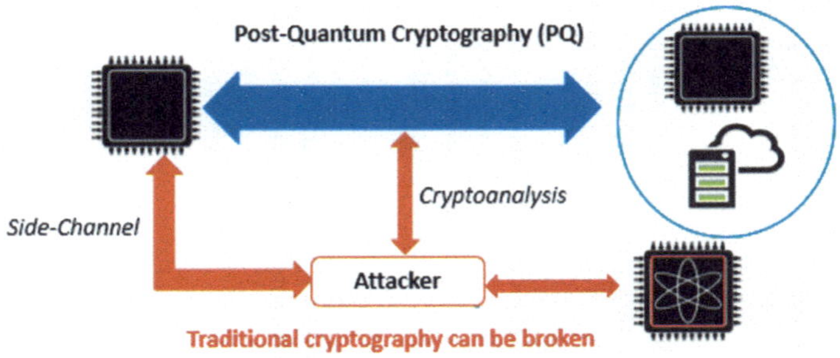

Fig. 4 Cyber security chair advances post-quantum cryptography [20]

this. Lattice-based cryptography is a promising PQC technique that has been included into general-purpose microcontrollers but has not been given much thought in automotive settings. Two contributions are shown in the work in [22]: a performance analysis of four KEM/encryption algorithms based on lattices and implemented on the automotive microcontroller AURIX, namely CRYSTALS-KYBER, NTRU, Sabre, and Three Bears. According to the investigation, the AURIX microcontroller can execute every algorithm with competitive performance. The super single isogenies Daffier-Hellman (SIDH) key exchange, which is the first isogenies-based cryptography hardware implementation, is presented in [23]. By reducing pipeline stalls, developing a powerful throughput multipliers unit, and paralleling arithmetic in F p2, the design is optimized for performance. Even on Intel Haswell processors, the results are faster than the libraries of software running linear SIDH. 1.5 times faster than existing software implementations, the approach produces shared secrets and ephemeral public keys with 1.655 million phases for Alice and 1.490 million cycles for Bob. In the coming ten years, it is anticipated that quantum computers—which can execute Short's algorithm—will become accessible, potentially upending established digital signature protocols utilized in embedded systems. In long-term applications like automobile systems or key infrastructures, this puts these systems at danger [24]. Quantum computer-based digital signature techniques must take the place of traditional ones in order to reduce this risk. The suitability of the NIST competition for a future post-quantum secure digital signature standard for secure boot and intermediate key protection in embedded systems is assessed. Techniques such as smaller, distributed SRAM blocks, efficient polynomial coefficient multiplication, and serial double-sponge functions were adopted in [25]. These optimizations resulted in significant speed improvements, with the 256-bit wide accelerator achieving up to 1.78 × faster performance for key-generation, encapsulation, and decapsulation operations compared to the baseline design. Various multiplication techniques, including Naive schoolbook, Karatsuba, and Toom-Cook, within classical and quantum contexts are explored in [26]. An 8-way Toom-Cook quantum multiplier was developed, improving space and time efficiency, and achieving the best asymptotic performance with a qubit count of approximately $O(n^{1.245})$ and Toffoli counts of $112n\log_8(15)-128n$. This design also reduces depth to approximately $n^{1.0569}$, significantly enhancing resistance to CPA-based SCA attacks. Customized SIMD (Single-Instruction-Multiple-Data) architecture for Lattice-Based Cryptography (LBC) in IoT devices, enabling efficient 256-bit polynomial operations and 320-bit Keccak acceleration was introduced in [27]. The design includes data shuffling hardware and a dual-issue memory path to optimize performance. The end result is a processor with a power consumption of as low as 2.01 mW for Kyber KEM 512, which makes it highly efficient for IoT security. It delivers over 10 × speedup in comparison to a baseline RISC-V processor and over 5 × speedup versus ARM Cortex M4 implementations. Utilizing the Karatsuba approach, Vedic multipliers, and modular reduction approaches, customized Lattice-DSP (L-DSP) was first introduced in [28] to meet the module multiplication needs of lattice-based post-quantum encryption. The L-DSP can efficiently execute integers multiplier and modular reduction for 77, 87, and 101 SLICEs at 283 MHz, 272 MHz, and 256 MHz, respectively, with different configurations. Furthermore, the design removes the N − 1 multiplier in the Inverse-NTT, reducing the overall size of the Butterfly Unit in CRYSTAL-Kyber by roughly 104 SLICEs. The proposed L-DSP significantly boosts speed and area efficacy and is the first DSP created especially for lattice-based PQC (Table 2).

4 Quantum-Secured Communication Networks

Entanglement-based quantitative security protocols that operate over and are made possible by entanglement-based quantum systems are collectively referred to as quantum safe communications or QSC. This is a physics-based approach that depends on the entanglement property. There are three types of single-point failures: wormholes, Sybil attacks, and single-point failure attacks among the security risks brought on by the expansion of optical network infrastructure and services. Security is essential given the rise in e-commerce and e-services. There are three types of single-point failures: wormholes, Sybil attacks, and single-point failure attacks. One kind of the technology of distributed ledgers that is used to shield transactions from these kinds of attacks is called blockchain technology. However, because quantum computers have developed, their security is now based on computational difficulty and is instantly vulnerable. To improve network security, researchers are now concentrating on fusing block chain technology with quantum key distribution (QKD) as shown in Fig. 5. According to [29], quantum-secured block chains are used to create secure optical networks in future research and development, along with a thorough overview of quantum-secured block chain technology and a discussion of different attacks in optical networks.

In order to facilitate the next generation of quantum communication networking, the suggested heterogeneous quantum communications network (QCN) mixes low Earth orbit satellite quantum communications networks and terrestrial QCNs [30]. Teleporting quantum states, enabling distributed quantum computing, quantum sensing networks, and cyber security systems can all be accomplished using this cluster state technique. QCNs such as those suggested by this article offer an unsurpassed level of security that can be applied to any wireless network, Internet of Things, optical network, as well as self-driving vehicles. It is also important to note that they are resistant to channel impairments as well. A quantum key distribution (QKD) system has been showcased in Burbank, Russia, as providing theoretical information security for communication. BB84 QKD protocol is used by hardware devices to generate quantum-generated keys in order to create a quantum-secured VPN tunnel that is based on the decoy-state BB84 QKD protocol. It has been shown that the approach proposed in [31] offers long-term data protection and that it can be integrated into the current architecture for information security. For securing communication in the quantum era, quantum key distribution (QKD) stands out as one of the most advanced methods of key distribution. Resilience gaps, nevertheless, make practical use difficult. The study in [32] uses software-defined networking (SDN) to increase the resilience of QKD-based microgrids.

Table 2 Post-quantum cryptography techniques

Algorithm/technique	Strengths	Weaknesses	Practical applications
Lattice-Based Cryptography (LBC)	Strong security foundation, efficient on microcontrollers, scalable	Challenging to implement efficiently in resource-constrained environments	Automotive systems, IoT devices, secure communications in critical systems
CRYSTALS-Kyber	High efficiency, competitive performance, standardized by NIST	Requires significant computational resources for optimal performance	Key encapsulation in secure communications
CRYSTALS-Dilithium	Strong security guarantees, standardized by NIST, optimized for performance	Higher power consumption, area overhead concerns in constrained environments	Digital signatures in secure boot, key protection
NTRU	Robust against quantum attacks, efficient implementation on various platforms	Complexity in hardware design, larger key sizes	Secure encryption in embedded systems
SIDH (Supersingular Isogeny Diffie-Hellman)	Efficient arithmetic, optimized for speed, hardware implementations available	Susceptible to specific quantum attacks, computationally intensive	Key exchange protocols in secure communications
Karatsuba and Vedic multiplication	Improved space and time efficiency in modular arithmetic, efficient polynomial multiplication	Implementation complexity, higher resource usage in complex operations	Post-quantum cryptographic operations, modular arithmetic in LBC
Eight-way toom-cook multiplier	Reduced qubit count, optimal asymptotic performance, and Toffoli complexity	Requires sophisticated design and optimization, complex to implement	Quantum cryptanalysis circuits, PQC acceleration
SIMD architecture for LBC	High speedup (10 × versus RISC-V, 5 × versus ARM Cortex M4), optimized for IoT with low power consumption	Requires advanced hardware design, increased complexity	IoT security, efficient cryptographic operations
Lattice-DSP (L-DSP)	High speed (up to 283 MHz), efficient modular multiplication and reduction, reduced area overhead	Limited flexibility, primarily focused on specific cryptographic tasks	Lattice-based PQC implementations, hardware accelerators

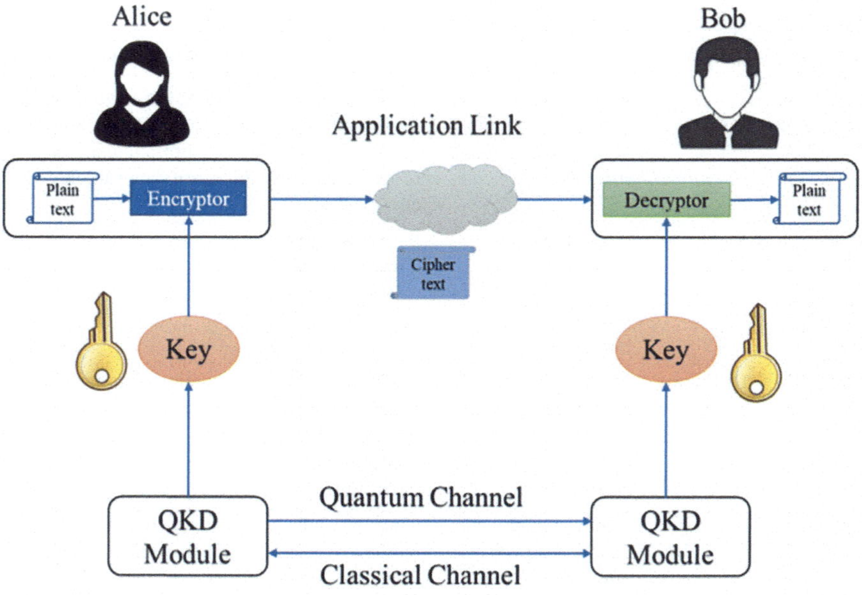

Fig. 5 Communication security through quantum key distribution [29]

The author build an improved QKD simulator, a novel communication architecture, and an SDN-based resilience augmentation approach. Numerous case studies attest to SDN's ability to enhance QKD-based microgrid resilience. Virtualization and networking that is software-defined are two examples of network services that are essential for meeting the latency, capacity, and dynamic needs of developing 5G networks. Services related to networks must be transparently deployed across several domains in order to enable end-to-end connectivity, which presents security vulnerabilities. It has been suggested that quantum key distribution (QKD) is an effective and future-proof method of preventing tampering with the hardware of the network in the future. Based on the study in [33], a quantum-secured optical network that connects virtual network functions (VNFs) across several network domains can be developed to support dynamically built network services based on virtual network functions (VNFs). In this work, a new flex-grid quantum-switched reconfiguration of a optical add drop multiplexer is presented in the form of a q-ROADM (quantum switched add drop multiplexer). It has been reported that quantum-secured optical channels in metro-scale operating settings are capable of carrying high capacities thanks to quantum key distribution technology used in [34] based on experimental work on building high-capacity quantum-secured optical channels. It describes how an 800 Gaps quantum-protected optical channel was established for the first time at a distance of 100 km and how financial transactions were secured over it. This opens the door for the application of QKD in high-capacity operating contexts. Over the past 20 years, mobile networks and spectrum technology have advanced, highlighting the need for expanding frequency bands, billions of devices being connected, reduced latency, and more artificial intelligence intervention. The need for a dependable, strong, clever, safe, and user-friendly infrastructure for data generation, storing, and transmission has increased as a result. To overcome these obstacles, the suggested 6G technological concept is built on quantum secured 6G enabled communication. Nevertheless, since the development of quantum computers, every public-key-based communication security method already in use in smart networks is vulnerable to quantum assaults [34]. Network slicing, software-defined networking, cloud computing, and other cutting-edge communication technologies are used in fifth generation (5G) networks. These networks do, however, pose difficulties with regard to privacy, security, and trust. Secure methods and mechanisms are required to address this. It has been suggested that quantum walks (QWs) are a great way to generate cryptographic keys. For 5G network applications, the study in [35] suggests two effective hash function techniques and two effective security protocols based on QWs, called QWHF-1 and QWHF-2. The results of performance assessments and simulations demonstrate how safe, effective, and resilient these methods are against different types of attacks.

Table 3 gives a summary of the characteristics of various quantum-secured network technologies. Different aspects of each technology are contrasted, such as application domains, key features, use cases, and integration with existing technologies. The table compares quantum-secured blockchain, QCN, QKD, SDN, QKD for microgrids, and quantum-secured optical networks. QKD is noted for its theoretical information security and practical difficulties when used with optical networks, for example, while quantum-secured blockchains can be applied to optical networks and address security attacks therein. There are some unique benefits associated with each network type, while others have unique challenges, demonstrating how quantum security measures can be applied across a wide range of technological sectors.

5 Incorporating Quantum into Cybersecurity Strategies

By utilizing the peculiarities of quantum mechanics, quantum cybersecurity creates encryption keys that are impenetrable by both classical and quantum computers. It enables the creation of a secret key between two parties that is totally unknown to an eavesdropper. Cybersecurity is being revolutionized by quantum computers, which offer a model for creating cryptographic algorithms influenced by quantum mechanics. A novel encryption and authentication protocol is being developed using quantum walks (QIQW) to enable safe data transfer between Internet of Things devices. This protocol replaces hash functions using conventional cryptography with quantum hash functions based on QIQW, enabling IoT nodes to quickly communicate any data while maintaining complete sovereignty over their records. To further ensure secure data transfer between IoT devices, the proposed protocol in [36] also protects against impersonation and messaging attacks. In the digital age, this quantum-inspired method of cryptographic algorithms is essential for improving cyber security and guaranteeing the security of data transfer. In the logistics sector, cyber security is essential, but as it has grown in popularity, the attack surface has

Table 3 A comparative analysis of quantum-secured networks

Aspect	Quantum-secured blockchain	Heterogeneous (QCN)	(QKD)	(SDN)	QKD for microgrids	Quantum-secured optical network
Application	Optical networks	Quantum communication networking	Communication security	Network resilience	Microgrid resilience	Optical network security
Key features	Security, distributed architecture	Mixes satellite and terrestrial QCNs, Quantum state teleportation	Theoretical information security	Resilience enhancement, improved communication architecture	Increased resilience, case studies	Dynamic network services support
Use cases	Autonomous vehicles, 5G/6G networks, IoT, and optical networks	Quantum computing, quantum sensing networks, cyber security systems	Communication security	5G networks, end-to-end connectivity	QKD-based microgrids	Dynamically created network services
Integration with current technology	Employing quantum-secured blockchain	Integrating with the information security architecture of the present	Integrating with the information security architecture of the present	Integrating with the information security architecture of the present	Integrating with the information security architecture of the present	Chaining virtual network functions
Challenges addressed	Security attacks in optical networks	Autonomous vehicles, 5G/6G networks, IoT, and optical networks	Practical use difficulties	Network infrastructure tampering	Practical use difficulties	Security risks in network infrastructure

grown as well, which could affect supply chain efficiency. There is limited quantitative study on cyber security in logistics, and existing studies hardly ever use genuine cyber security data. In an era where quantum computing techniques predominate, one-way encryption schemes ignore possible risks, most studies concentrate on preventive measures, and block chain technologies are still in their infancy [37]. Cybersecurity is crucial in the information era because of the rapid growth of the Internet as well as the IoT (Internet of Things). Large-scale data analytics, or BDA, is the practice of handling, storing, and gathering large volumes of data for a later analysis. Because of its growing value, big data is becoming an extremely valuable target for cyber security. The study in [38] examines recent big data research projects, emphasizing how data is safeguarded and employed as a cyber security tool. By outlining patterns, open research difficulties, and difficulties, it provides a comprehensive understanding of cyber security in the large-scale data era. The Quantum Internet is an interconnected system of quantum devices that symbolizes the final stage of the quantum revolution. The principles of quantum mechanics, on the other hand, place further limitations on network architecture. To fully capitalize on the idiosyncrasies of phenomena like no-cloning, quantum measurements, entanglement, and quantum teleportation, a significant paradigm shift is necessary [39]. The work in [40] investigates the post-quantum safety of this recognized block cipher, AES-256, a basic block cipher having 128 bits of post-quantum security. Both classical and quantum assaults are included in the authors' novel structured search methodology, which makes it possible to efficiently calculate their complexity. The best strategy is a quantum Demirci-Selçuk meet-in-the-middle attack, which can reduce memory usage in some AES-256 and AES-128 attacks. According to the authors, AES has a higher security margin against quantum generic assaults and is a resistant primitive in both the post-quantum and classical worlds. Open access networks jeopardize the security of industrial devices, which is why data acquisition in SCADA networks is so important [41]. Although security standards have been created to safeguard node-to-node communication, researchers have suggested strategies to get around flaws. Future SCADA network research may be necessary to meet issues like quantum attacks. In particular, the convergence of space-air-ground technologies and the growth of IoT technologies provide a variety of challenges to mobile networks, especially 6G. New challenges posed by these technologies include attacks against ubiquitous intelligence and radio technology [42]. To reduce these dangers, promising techniques such as deep network slicing, AI, quantum-safe communication, platform-agnostic safety, real-time adaptive protection, physical layer security, and special mechanisms for protecting information are needed. Smart communication technologies like the Internet of Things (IoT) have revolutionized several industries, but they also expose users to cybersecurity risks and vulnerabilities [43]. Quantum computing represents a revolutionary shift in the way we approach computation, fundamentally differing from classical computing in its basic principles and capabilities. In order to do calculations, classical computers use transistors and binary codes (0 and 1), but quantum computers use the concepts of quantum mechanics and employ qubits, which are particles that can exist in numerous states at once. This comparison highlights the distinctions between the two methods and shows how quantum computing has significant disadvantages as well, such as higher error rates and more stringent operating conditions, in addition to the possibility for exponential growth in processing power. Because quantum computing can solve complex issues beyond the scope of conventional systems, it is therefore positioned to bring in a new era in computational technology.

Figure 6 provides a comparison between classical and quantum computing across four key aspects: computing units, computing capacity, error rates & environment, and suitability. Classical computing uses transistors that represent binary states (0 and 1), leading to linear increases in computing capacity as the number of transistors grows. It operates with low error rates and can function at room temperature, making it suitable for routine processing tasks. Quantum computing, on the other hand, makes use of qubits, which may simultaneously represent 0 and 1, enabling exponential expansion in computer capability with the introduction of more qubits. However, quantum systems often experience high error rates and require extremely low temperatures to operate, making them more suitable for complex processing tasks that go beyond the capabilities of classical computing.

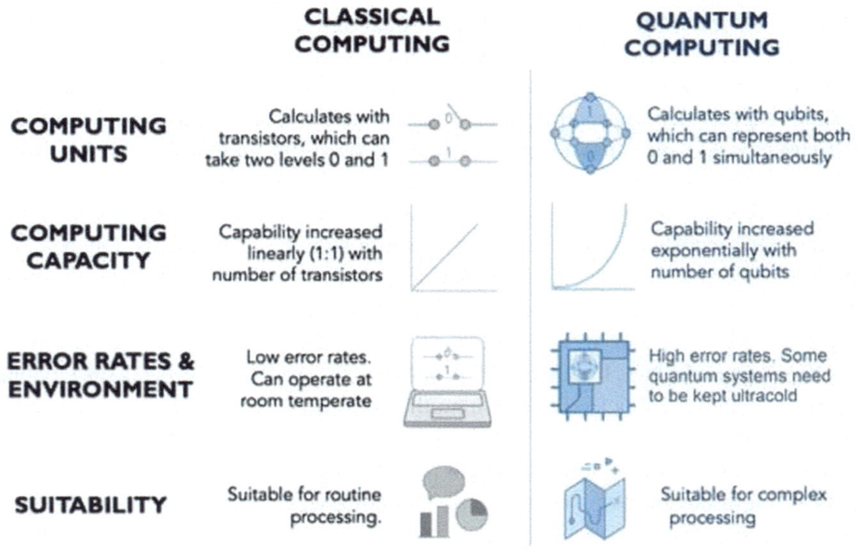

Fig. 6 Classical versus quantum computing

6 Quantum-Safe Security Solutions

It is anticipated that cloud services would gain importance in Industry 4.0, IT, digitalization, healthcare, and agriculture. In these cloud systems, data security is critical, and traffic encryption between clouds computing and sensors is standard. In the era before quantum computing, data security was guaranteed by cryptographic algorithms such as Rivets-Shamir-Adelman (RSA). But some algorithms become vulnerable to upcoming quantum computing, which is why post-quantum cryptography is a desirable option [44]. Public key cryptography, which Daffier and Hellman discovered in 1976, has supplanted traditional encryption, which was based on symmetric keys. This technique, which is now widely used for digital signatures, encrypted interpersonal interaction, and e-commerce authentication, guarantees secure information transmission over an unreliable route. The majority of public key cryptosystems rely on discrete logarithms or integer factorization; however, with the development of quantum computers, these methods have security flaws and efficiency issues [45]. Quantum computing represents a major advancement in contemporary ICT and is moving from high-tech research to commercial use. Due to its efficiency, quantum computers pose a risk to existing cryptographic methods since they could be able to crack asymmetric cryptography and address key distribution issues. Cryptography is essential to the safe operation of emerging technologies such as the Internet of Things, blockchain, 5G, autonomous cars, artificial intelligence, and robots (Fig. 7).

The development of these technologies will face major obstacles upon the realization of quantum computing. This study addresses the preliminary aspects of quantum technology, possible effects on developing technologies, and existing quantum-safe cryptosystems [46]. Promising technology for observing environmental or physical factors and improving communication between dispersed nodes include wireless sensor networks. But in the upcoming decades, security will become much

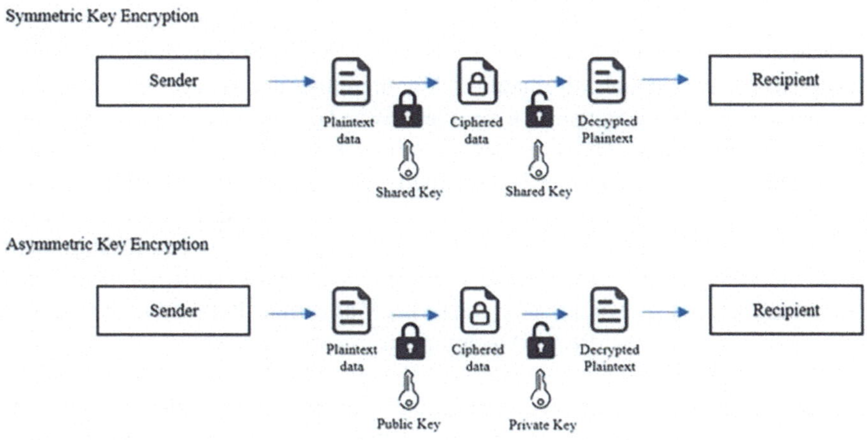

Fig. 7 Quantum safe security solution

more important as quantum computers pose a threat to sophisticated attacks. Due to resource limitations, intrusion detection systems and cryptographic techniques are not appropriate for sensors. The processor phase pervasive filtering approach is the foundation of the resource-conscious, future-proof, and lightweight security plan for sensor networks described in [47]. The Magic Number filter is assessed in order to guard against Denial-of-Service attack on CC1350 Launch Pad ARM Cortex-M3 microcontroller boards. Multiple n-qubit quantum permutations gates are combined in a quantum technique known as a quantum permutation pad (QPP). It can be implemented by quantum circuits in quantum computing systems or by pads of n-bit permutations matrices in classical computing systems. Two features set QPP apart: non-commutativity between permutations matrices and a significant Shannon information entropy. Using permutation transformations, Shannon perfect secrecy with reusability can be achieved between input and output information spaces. The study in [48] investigates AES-QPP, a quantum safe lightweight cryptography variant that combines Add Round Key and Sub Bytes using a single QPP of 16 8-bit permutation matrices. The impact of quantum computing on cryptographic protocols and network security is examined in the study in [49]. Quantum attacks can target traditional cryptographic techniques like RSA and ECC encryption. Nonetheless, immune cryptography techniques like as lattice-based algorithms and multivariate polynomial cryptography exhibit potential. Implementations of quantum key distribution (QKD) provide notable improvements in key exchange rates and error rates, indicating that QKD is a workable solution for communication channel protection up to 200 km distant. The detection accuracy of quantum-based network security systems is remarkable; Q-secured and QIDS systems demonstrate detection accuracy of 99.2% and 98.5%, respectively. The final stages of standardizing asymmetric cryptography—which is immune to attacks by quantum computers—are presently underway. In 2022–2024, a limited number of conventional schemes are anticipated, which will replace the present asymmetric ways for securing civilian and military networks [50]. Multivariate Polynomial Public Key (MPPK/DS), a quantum-safe digital signature technique, is created in [51]. The approach creates a key construction using one base multivariate polynomial and two univariate polynomials; the multivariate polynomials obscure confidential information while the univariate polynomials represent plain messages. To thwart attempts to recover private keys over the ring, the technique employs a secure prime. The authors want to make the private key recovery attack more challenging by adding more private parts and optimizing MPPK/DS to lower the size of signatures. This method is referred to as the Modular Diophantine Equation Problem (MDEP) because it allows for multivariate key recovery. In addition to the 128 bit field encryption, MPPK/DS provides extra security with 384 bit entropy. With two multivariate polynomials, P and Q, the Multivariate Polynomial Public Key with Digital Signature (MPPK/DS) technique uses modular arithmetic to guarantee that the power of P modulo p-1 in a prime Galois field GF (p) equals g to a power of P modulo p-1. In order to prevent quantum computers from solving the discrete logarithm problem, the algorithm has been concealed. Known-message attacks, key-only attacks, and chosen-message attacks can all be thwarted by it. Using the generalized safe prime p and optimal multivariate polynomial selections, it achieves all three of the NIST security standards [52]. It is anticipated that quantum computing would transform computing by facilitating quicker problem-solving and improving applications such as artificial intelligence, machine learning, and financial modelling. But it also threatens public-key encryptions, which is why Post-Quantum Cryptography (PQC) was created. With regard to Postgre SQL, an open-source client-server database management system, this paper explores quantum-safe TLS. The paper tests client-server connections and incorporates quantum-safe TLS into the PostgreSQL source code. The study assesses hybrid schemes as well, with a particular emphasis on key exchange and authentication lattice-based schemes. Positive outcomes are seen, however there is more communication overhead [53].

7 Quantum Threats and Vulnerabilities

Businesses like IBM, Rigetti, and D-Wave provide quantum computing, a promising method for resolving computationally unmanageable issues, via a cloud-based platform. These computers are unique in that they have numerous pieces of hardware with identical coupling maps, different qubits and coupling maps in quantum hardware, varied hardware quality, and the inability for the user to validate results [54]. The scheduler of the cloud provider distributes the needed hardware, saving money and promoting scientific research. Potential dangers do, however, include unreliable quantum computers from outside parties assigning subpar hardware, malicious code segments in workload scheduling algorithms, dishonest workers damaging the reputations of cloud vendors, and information theft via programs being diverted to subpar hardware. Quantum computing is a fast-developing technology that poses a danger to popular public-key encryption algorithms like Rivest-Shamir-Adleman (RSA), that are critical for high security in insecure communications channels. Both conventional computers and Internet of Things (IoT) devices employ these systems, which are a component of important Internet protocols like TLS. Nevertheless, there are several obstacles to overcome when integrating these cryptosystems into battery-operated, resource-constrained Internet of Things devices. In order to lessen this hazard, researchers are working on post-quantum Internet of Things platforms [55]. The impact of quantum computing on 5G mobile telecommunication security is thoroughly examined in [56]. This entails taking into account the usage of encryption in 5G and how the introduction of quantum computing might impact system security. This naturally leads to the specification of a set of straightforward, incremental, recommended modifications meant to prevent significant harm to the security of 5G (as well as 3G and 4G) in the event that large-scale quantum computing proves to be a viable technology. The authors were able to provide a multi-phase security upgrading method that makes the switch to a post-quantum safe system simple and smooth by utilizing backwards-compatibility aspects in the design of the 5G security system. Figure 5, a risk timeline related to quantum computing threats is shown for various industries. Telecommunications and consumer electronics, which have short data shelf lives, are at risk before 2025, while insurance, public sector, and banking,

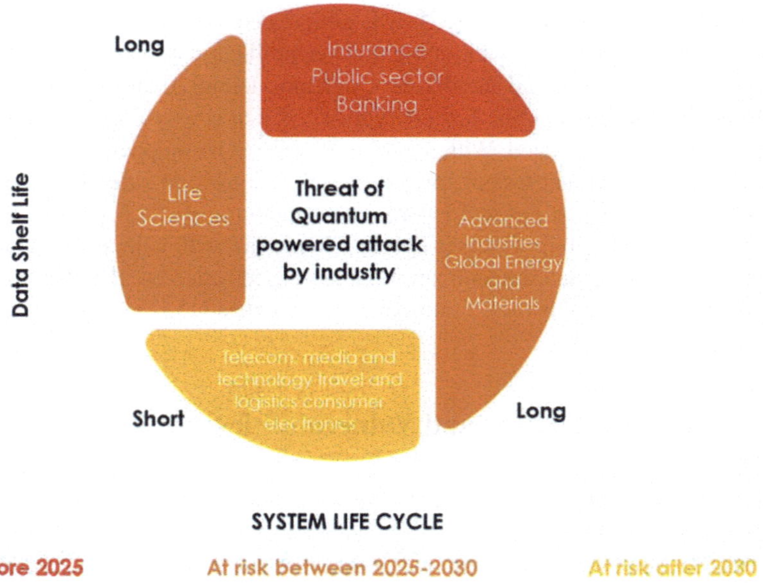

Fig. 8 Quantum computing impacts cyber security

which have long data shelf lives, face risks in-between 2025–2030. A threat will mount after 2030 to industries that have long-lasting data, including life sciences and advanced materials. In light of the longevity of the sensitivity of the data in each sector, it is imperative that each sector prepares for quantum threats (Fig. 8).

The work in [57] investigates the privacy and security consequences of 6G wireless networks, emphasizing the effect of security on systems that are imagined, possible difficulties with various technologies, and possible solutions. It provides a vision for 6G protection and key performance metrics, explains the threat scenario according to the network design, and discusses security issues in relation to requirements and potential uses. In addition, the paper sheds light on research-level initiatives and standardization attempts pertaining to 6G security, with a particular emphasis on technologies that enable the network, there are many ways in which this can be accomplished, including blockchain technology, physical security measures, distributed artificial intelligence and machine learning, visible light communication, THz, and quantum computing.. Since identity theft is a common cybercrime, authentication is essential. In this paper, a key agreement and authentication technique for cloud server architecture based on quantum identities is proposed in [58]. To safeguard privacy and confidentiality, network security necessitates the application of quantum cryptography, a technology which is founded on the principles of quantum physics. The resilience of the approach against all security assaults was proven by a rigorous security study conducted with the AVISPA tool. An entanglement-based simulation was done over 100 km to test the applicability of quantum key distribution in cloud computing with key rates of 4.11 bits/s and error rates of 9.21%. As IoT systems interact with the real world, Industry 4.0 standards may cause growing vulnerabilities. In particular, multilevel security techniques and connection encryption are necessary for receiving data streams from several intermediates. Serious security risks are presented by the systems' heterogeneity and non-institutionalized hardware and software compatibility. It is critical to pinpoint the main weaknesses, related dangers, and threats, as well as to recommend suitable defenses.

In [59], attacks against IIoT systems are described, and suggested countermeasures are examined. Traditional data analytics techniques are no longer enough due to the exponential growth in data generated by the rapid improvements in smart devices. This has made it possible for attackers to launch attacks using data analytics flaws. In order to detect and thwart both known and unknown assaults, [60] investigates machine learning (ML) and deep learning (DL) models and strategies. An architecture for Secure Data Analytics (SDA) based on DL and ML is proposed to classify attack- and normal-type input data. A model of threat has been abstracted into a system that considers factors including efficacy, latency, accuracy, dependability, and adversarial attacks in order to address research challenges. To help end users make the best decision possible, a comparative evaluation of existing SDA concepts is offered. An appealing method for utilizing quantum key distribution (QKD) to future-proof current fiber networks is the Cambridge quantum network. This method provides smooth user interaction, effective delivery, and broadband transfer of quantum keys. The Cambridge network can accommodate tens of thousands of users with key rates more than one kilobit per second per user, despite the requirement for dark fibers and bit rates. This novel technique demonstrates the application of quantum security to metropolitan fiber networks, ensuring data security and the inability to be compromised by data theft in [61]. Table 4 is a structured compilation of critical research in quantum computing's intersection with cyber security. The quantum computing-based research reports on quantum computing efforts surrounding digital security and networking technologies are divided into several categories based on the extent of the challenge they present and the level of sophistication with which they address it. The tables are composed of the findings obtained as well as the practicability of the methodologies to different industries based on the findings in the above tables. These tables are critical given the ever-nearing realization of quantum computing to improve our understanding of subsequent cyber security measures and network architecture advancement.

8 Conclusion

Quantum computing presents tremendous opportunities for enhancing cybersecurity by using quantum-resistant solutions and advanced cryptographic systems. However, it is also a threat in its own right, as it explodes the reliability and effectiveness of all existing cryptographic schemes. There is an urgent need for researchers, experts, and policymakers to address the issue ahead of a quantum leap. Quantum-resistant algorithms need to be used to create cryptographic systems and quantum cryptography

Table 4 Quantum computing, cybersecurity, and networking: a comparative analysis

Study	Focus	Key findings	Methodology	Applications
[54]	Cloud-based quantum computing	Utilization of cloud-based quantum computing platforms by IBM, Rigetti, D-Wave; challenges including hardware variations, validation issues, and potential risks	Cloud-based platform analysis, risk assessment	Scientific research, computationally intensive problem-solving
[55]	Post-Quantum Internet of Things platforms	Examining how to incorporate post-quantum cryptography into Internet of Things devices; overcoming obstacles in contexts with limited resources	Research development, experimentation	Internet of Things security and post-quantum cryptography
[56]	Quantum computing's effects on 5G security	Analysing how 5G security is affected by quantum computing and suggesting security enhancement plans to reduce concerns	Analysis of 5G encryption, quantum computing impact assessment	5G mobile telecommunication security
[57]	Security and privacy in 6G wireless systems	Analysing privacy and security issues in 6G networks; talking about possible fixes and standardization initiatives	Threat analysis, technology assessment	6G wireless networks, network security
[58]	Quantum identity-based authentication for cloud servers	Proposal of authentication technique based on quantum identities for cloud server architecture; security study demonstrating resilience against attacks	Security protocol development, AVISPA security analysis	Cloud server authentication, network security
[59]	Security issues with the industrial Internet of Things (IIoT)	Description of vulnerabilities in IIoT ecosystem; examination of attacks and recommended defenses	Vulnerability assessment, attack analysis	Industrial Internet of Things security, multilevel security techniques
[60]	Using ML and DL together to provide secure data analytics	Investigating deep learning and machine learning methods for safe data analytic; proposing a safe data analytics (SDA) architecture	Model development, threat analysis	Data analytics security, machine learning, deep learning
[61]	Quantum Key Distribution (QKD) for fiber networks	Utilization of QKD for future-proofing fiber networks; description of the Cambridge quantum network method	Network architecture analysis, experimental validation	Fiber network security, quantum key distribution

needs to be developed to protect the cyber infrastructure from quantum threats. Multidisciplinary cybersecurity techniques and innovations should be developed in order to maximize the advantages of quantum computing technology while minimizing the hazards that go along with them. To maintain the speed of technological developments in quantum computing and cyber security, ongoing research and development is vital. The creation of more reliable and expandable quantum-resistant cryptographic codes is a major area that requires focus, especially for environments with limited resources like Internet of Things devices. Another important area is the combination of quantum cryptography methods with current cybersecurity infrastructure through hybrid systems. Additionally, research should focus on enhancing quantum key distribution (QKD) networks to be more practical and widely deployable, and on studying quantum cryptanalysis to develop mitigation strategies against quantum attacks. Interdisciplinary research is crucial to develop comprehensive cybersecurity strategies, policy frameworks, and standards that anticipate the challenges posed by quantum technologies, ensuring that our cyber infrastructure remains secure in the quantum era.

References

1. K. Keplinger, Is quantum computing becoming relevant to cyber-security? Netw. Secur. **2018**(9), 16–19 (2018)
2. C. Easttom, Quantum computing and cryptography. in *Modern Cryptography: Applied Mathematics for Encryption and Information Security*, pp. 397–407. Cham: Springer International Publishing (2022)
3. F. Raheman, The future of cybersecurity in the age of quantum computers. Futur. Internet **14**(11), 335 (2022)
4. K.-K. Ko, E.-S. Jung, Development of cybersecurity technology and algorithm based on quantum computing. Appl. Sci. **11**(19), 9085 (2021)
5. H.J. Hadi, Y. Cao, M.A. Alshara, N. Ahmad, M.S. Riaz, J. Li, Quantum computing challenges and impact on cyber security. in *International Conference on Digital Forensics and Cyber Crime*, pp. 333–343. Cham: Springer Nature Switzerland (2023)
6. B. Arslan, M. Ulker, S. Akleylek, S. Sagiroglu, A study on the use of quantum computers, risk assessment and security problems. in *2018 6th International Symposium on Digital Forensic and Security (ISDFS)*, pp. 1–6. IEEE, 2018
7. S.S. Gill, A. Kumar, H. Singh, M. Singh, K. Kaur, M. Usman, R. Buyya, Quantum computing: A taxonomy, systematic review and future directions. Softw. Pract. Exp. **52**(1), 66–114 (2022)
8. V. Chamola, A. Jolfaei, V. Chanana, P. Parashari, V. Hassija, Information security in the post quantum era for 5G and beyond networks: Threats to existing cryptography, and post-quantum cryptography. Comput. Commun. **176**, 99–118 (2021)
9. C. Mangla, S. Rani, N.M.F. Qureshi, A. Singh, Mitigating 5G security challenges for next-gen industry using quantum computing. J. King Saud Univ. Comput. Inf. Sci. **35**(6), 101334 (2023)
10. R. Yan, Y. Wang, J. Dai, Y. Xu, A.Q. Liu, Quantum-key-distribution-based microgrid control for cybersecurity enhancement. IEEE Trans. Ind. Appl. **58**(3), 3076–3086 (2022)
11. S. Pirandola, U.L. Andersen, L. Banchi, M. Berta, D. Bunandar, R. Colbeck, D. Englund et al., Advances in quantum cryptography. Adv. Opt. Photonics **12**(4), 1012–1236 (2020)
12. C. Portmann, R. Renner, Security in quantum cryptography. Rev. Mod. Phys. **94**(2), 025008 (2022)
13. D.J. Bernstein, T. Lange, Post-quantum cryptography. Nature **549**(7671), 188–194 (2017)

14. A. Broadbent, C. Schaffner, Quantum cryptography beyond quantum key distribution. Des. Codes Crypt. **78**, 351–382 (2016)
15. J. Li, Y.N. Li, S.W. Zhang, D. Wei, W. Chen, W. Ma, A survey on quantum cryptography. Chin. J. Electron. **27**(2), 223–228 (2018)
16. S. Subramani, S.V.N. Santhosh Kumar, Review of security methods based on classical cryptography and quantum cryptography. Cybern. Syst. 1–19 (2023)
17. H. Iqbal, W.O. Krawec, Semi-quantum cryptography. Quantum Inf. Process. **19**(3), 97 (2020)
18. S. Mitra, B. Jana, S. Bhattacharya, P. Pal, J. Poray, Quantum cryptography: Overview, security issues and future challenges. in *2017 4th International Conference on Opto-Electronics and Applied Optics (Optronix)*, (IEEE, 2017), pp. 1–7
19. V. Padamvathi, B.V. Vardhan, A.V.N. Krishna, Quantum cryptography and quantum key distribution protocols: A survey. in *2016 IEEE 6th International Conference on Advanced Computing (IACC)*, (IEEE, 2016), pp. 556–562
20. J. Xie, K. Basu, K. Gaj, U. Guin, Special session: The recent advance in hardware implementation of post-quantum cryptography. in *2020 IEEE 38th VLSI Test Symposium (VTS)*, (IEEE, 2020), pp. 1–10
21. M. Barbosa, G. Barthe, X. Fan, B. Grégoire, S.-H. Hung, J. Katz, P.-Y. Strub, X. Wu, L. Zhou, EasyPQC: Verifying post-quantum cryptography. in *Proceedings of the 2021 ACM SIGSAC Conference on Computer and Communications Security*, pp. 2564–2586 (2021)
22. T. Fritzmann, J. Vith, D. Flórez, J. Sepúlveda, Post-quantum cryptography for automotive systems. Microprocess. Microsyst. **87**, 104379 (2021)
23. B. Koziel, R. Azarderakhsh, M.M. Kermani, D. Jao, Post-quantum cryptography on FPGA based on isogenies on elliptic curves. IEEE Trans. Circuits Syst. I Regul. Pap. **64**(1), 86–99 (2016)
24. S. Marzougui, J. Krämer, Post-quantum cryptography in embedded systems. in *Proceedings of the 14th International Conference on Availability, Reliability and Security*, pp. 1–7 (2019)
25. M. Imran, A. Aikata, S.S. Roy, S. Pagliarini, High-speed design of post quantum cryptography with optimized hashing and multiplication. IEEE Trans. Circuits Syst. II Express Briefs **71**(2), 847–851 (2023)
26. D.S.C. Putranto, R.W. Wardhani, H.T. Larasati, H. Kim, Space and time-efficient quantum multiplier in post quantum cryptography era. IEEE Access **11**, 21848–21862 (2023)
27. Z. Ye, R. Song, H. Zhang, D. Chen, R.-C. Cheung, K. Huang, A Highly-efficient lattice-based post-quantum cryptography processor for IoT applications. IACR Trans. Cryptogr. Hardw. Embed. Syst. **2024**(2), 130–153 (2024)
28. T.-H. Nguyen, C.-K. Pham, T.-T. Hoang, A high-efficiency modular multiplication digital signal processing for lattice-based post-quantum cryptography. Cryptography **7**(4), 46 (2023)
29. P. Sharma, K. Choi, O. Krejcar, P. Blazek, V. Bhatia, S. Prakash, Securing optical networks using quantum-secured blockchain: An overview. Sensors **23**(3), 1228 (2023)
30. I.B. Djordjevic, On global quantum communication networking. Entropy **22**(8), 831 (2020)
31. A.V. Duplinskiy, E.O. Kiktenko, N.O. Pozhar, M.N. Anufriev, R.P. Ermakov, A.I. Kotov, A.V. Brodskiy et al., Quantum-secured data transmission in urban fiber-optics communication lines. J. Russ. Laser Res. **39**, 113–119 (2018)
32. Z. Tang, P. Zhang, W.O. Krawec, Enabling resilient quantum-secured microgrids through software-defined networking. IEEE Trans. Quantum Eng. **3**, 1–11 (2022)
33. R. Wang, R.S. Tessinari, E. Hugues-Salas, A. Bravalheri, N. Uniyal, A.S. Muqaddas, R.S. Guimaraes et al., End-to-end quantum secured inter-domain 5G service orchestration over dynamically switched flex-grid optical networks enabled by a q-ROADM. J. Light. Technol. **38**(1), 139–149 (2019)
34. M. Pistoia, O. Amer, M.R. Behera, J.A. Dolphin, J.F. Dynes, B. John, P.A. Haigh et al., Paving the way toward 800 Gbps quantum-secured optical channel deployment in mission-critical environments. Quantum Sci. Technol. **8**(3), 035015 (2023)
35. K. Prateek, N.K. Ojha, F. Altaf, S. Maity, Quantum secured 6G technology-based applications in internet of everything. Telecommun. Syst. **82**(2), 315–344 (2023)

36. A.A. Abd EL-Latif, B. Abd-El-Atty, S.E. Venegas-Andraca, W. Mazurczyk, Efficient quantum-based security protocols for information sharing and data protection in 5G networks. Futur. Gener. Comput. Syst. **100**, 893–906 (2019)
37. K.-F. Cheung, M. GH Bell, J. Bhattacharjya, Cybersecurity in logistics and supply chain management: An overview and future research directions. Transp. Res. Part E Logist. Transp. Rev. **146**, 102217 (2021)
38. D.B. Rawat, R. Doku, M. Garuba, Cybersecurity in big data era: From securing big data to data-driven security. IEEE Trans. Serv. Comput. **14**(6), 2055–2072 (2019)
39. A.S. Cacciapuoti, A. Caleffi, F. Tafuri, F.S. Cataliotti, S. Gherardini, G. Bianchi, Quantum internet: Networking challenges in distributed quantum computing. IEEE Network **34**(1), 137–143 (2019)
40. X. Bonnetain, M. Naya-Plasencia, A. Schrottenloher, Quantum security analysis of AES. IACR Trans. Symmetric Cryptol. **2019**(2), 55–93 (2019)
41. S. Ghosh, S. Sampalli, A survey of security in SCADA networks: Current issues and future challenges. IEEE Access **7**, 135812–135831 (2019)
42. V.-L. Nguyen, P.-C. Lin, B.-C. Cheng, R.-H. Hwang, Y.-D. Lin, Security and privacy for 6G: A survey on prospective technologies and challenges. IEEE Commun. Surv. Tutor. **23**(4), 2384–2428 (2021)
43. M. Gupta, M. Abdelsalam, S. Khorsandroo, S. Mittal, Security and privacy in smart farming: Challenges and opportunities. IEEE Access **8**, 34564–34584 (2020)
44. O. Grote, A. Ahrens, C. Benavente-Peces, Small quantum-safe design approach for long-term safety in cloud environments. in *2021 International Conference on Engineering and Emerging Technologies (ICEET)*, (IEEE, 2021), pp. 1–5
45. A.-D. Stefan, I.-P. Anghel, E. Simion, Quantum-safe protocols and application in data security of medical records (2023), Preprint at Cryptology ePrint Archive
46. A. Abuarqoub, S. Abuarqoub, A. Alzu'bi, A. Muthanna, The impact of quantum computing on security in emerging technologies. in *The 5th International Conference on Future Networks & Distributed Systems*, pp. 171–176 (2021)
47. M. Heigl, M. Schramm, D. Fiala, A lightweight quantum-safe security concept for wireless sensor network communication. in *2019 IEEE International Conference on Pervasive Computing and Communications Workshops (PerCom Workshops)*, (IEEE, 2019), pp. 906–911
48. R. Kuang, D. Lou, A. He, A. Conlon, Quantum safe lightweight cryptography with quantum permutation pad. in *2021 IEEE 6th international conference on computer and communication systems (ICCCS)*, (IEEE, 2021), pp. 790–795
49. S. Ambika, V. R.T.R. Balaji, P.N. Periyasamy, N. Kamal, Explore the impact of quantum computing to enhance cryptographic protocols and network security measures. in *2024 IEEE International Conference on Computing, Power and Communication Technologies (IC2PCT)*, vol. 5, (IEEE, 2024), pp. 1603–1607
50. M. Strand, A status update on quantum safe cryptography. in *2021 International conference on military communication and information systems (ICMCIS)*, (IEEE, 2021), pp. 1–7
51. R. Kuang, M. Perepechaenko, Optimization of the multivariate polynomial public key for quantum safe digital signature. Sci. Rep. **13**(1), 6363 (2023)
52. R. Kuang, M. Perepechaenko, M. Barbeau, A new quantum-safe multivariate polynomial public key digital signature algorithm. Sci. Rep. **12**(1), 13168 (2022)
53. A. Tueno, D. Boehm, S.H. Choe, Integrating and evaluating quantum-safe TLS in database applications. in *IFIP Annual Conference on Data and Applications Security and Privacy*, (Cham: Springer International Publishing, 2022), pp. 259–278
54. K. Phalak, A. Ash-Saki, M. Alam, R.O. Topaloglu, S. Ghosh, Quantum puf for security and trust in quantum computing. IEEE J. Emerg. Sel. Top. Circuits Syst. **11**(2), 333–342 (2021)
55. T.M. Fernández-Caramés, From pre-quantum to post-quantum IoT security: A survey on quantum-resistant cryptosystems for the Internet of Things. IEEE Internet Things J. **7**(7), 6457–6480 (2019)
56. C.J. Mitchell, The impact of quantum computing on real-world security: A 5G case study. Comput. Secur. **93**, 101825 (2020)

57. P. Porambage, G. Gür, D.P.M. Osorio, M. Liyanage, A. Gurtov, M. Ylianttila, The roadmap to 6G security and privacy. IEEE Open J. Commun. Soc. **2**, 1094–1122 (2021)
58. G. Sharma, S. Kalra, Identity based secure authentication scheme based on quantum key distribution for cloud computing. Peer-To-Peer Netw. Appl. **11**, 220–234 (2018)
59. K. Tsiknas, D. Taketzis, K. Demertzis, C. Skianis, Cyber threats to industrial IoT: A survey on attacks and countermeasures. IoT **2**(1), 163–186 (2021)
60. R. Gupta, S. Tanwar, S. Tyagi, N. Kumar, Machine learning models for secure data analytics: A taxonomy and threat model. Comput. Commun. **153**, 406–440 (2020)
61. J.F. Dynes, A. Wonfor, WW-S. Tam, A.W. Sharpe, R. Takahashi, M. Lucamarini, A. Plews et al., Cambridge quantum network. NPJ Quantum Inf. **5**(1), 101 (2019)

Quantum Cryptographic Key Distribution Models and Protocols and Its Progress

Mohamed Zaid and Ramani Selvanambi

1 Introduction

A vulnerability in computer security is a weakness that a threat actor or an attacker could use to gain access to a computer system and carry out unauthorized activities. In order for that, an attacker needs to have at least one useful tool or technique that they can use to match to the system's weakness in order to exploit that vulnerability. One such useful tool is the development of quantum computers, which can be used to launch effective attacks on well established, conventional techniques such as the popular public key cryptography [1].

Quantum cryptography, or quantum key distribution, is a protocol that is unconditionally secure against all future improvements in algorithms and computational technology, particularly the development of quantum computers. Quantum key distribution is a subset of quantum cryptography, along with quantum random number generation, closed group digital signatures, long-term secure data storage, and multi-party secure computation [2].

Quantum cryptography will sustain security because, in its very core, it has the capability to distribute an encryption key in a secure way. Here the secret key will be shared in encrypted form: encryption of a single photon is done at the per-photon level to irreversibly alter information coded on that photon in case any possible eavesdropper intercepted the photon-most probably for probing or observing [3].

Encryption should be like a trapdoor, with forwarding operations being simple to implement and backward operations being challenging. One of the most well-known trapdoors is RSA, it encrypts the data using a key that is created by multiplying

M. Zaid · R. Selvanambi (✉)
School of Computer Science and Engineering, Vellore Institute of Technology, Vellore, Tamil Nadu, India
e-mail: ramani.s@vit.ac.in

M. Zaid
e-mail: mohamed.zaid2020@vitalum.ac.in

S. B. Goyal et al. (eds.), *Quantum Computing, Cyber Security and Cryptography*,
https://doi.org/10.1007/978-981-96-4948-8_11

numerous prime numbers. However, Shor's algorithm created a new issue for the field of cryptography. It has been applied to break RSA, which is a quantum algorithm. Shor's algorithm puts at risk the systems that depend on the product of prime numbers as their way of getting the keys [4].

Some new points of attack have identified that are Trojan-horse attack, multi-photon emission, imperfect encoding, Bright-light attack, Back-flash attack, Efficiency mismatch and time-shift attack. These are to be seen as would-be threats, which heighten a call for increased safety precautions [5, 6]. On the part of Quantum cryptography technology the Heisenberg uncertainty principle has been widely used so as to make the cryptography un-noticeable. It is by implementing the principles of quantum physics that sees to it that data transmission would indeed be totally confidential.

1.1 Quantum Cryptography

QKD, in most usages, stands for quantum key distribution: it's a secure communication protocol that is based on the quantum mechanics properties and uses a cryptographic protocol. This could be used for finally generating a key between two users that are later in time, truly secret, random, and known only to the key creators and could be used for encryption and decryption purposes. In fact, although this is the most well-known protocol for quantum cryptography, it is to date, inappropriately so called [7]. Quantum key distribution encryption bases its security on principles of quantum mechanics, whereas classical public-key cryptography is based on the computational difficulty of certain mathematical functions, although there is no mathematical proof on the complexity in practice of the process reverse of the used one-way function. QKD offers provable security based on information-theoretic principles and forward secrecy.

Quantum Cryptography, more correctly called as Quantum key distribution (QKD), involves a series of photons or light particles that carry information over fiber-optic cable between two locations. The key and the safety of its use can be determined by comparing the measurements of fraction of those photons' properties at the two endpoints.

Let's simplify this a bit more to understand the process:

1. That means, the sender sends photons through a filter or polarizer that would randomly assign them one of four-bit designations and polarizations. These are: 45 degrees left-zero bit, 45 degrees right-one bit, or vertical- one bit, and horizontal- zero bit.
2. The photons arrive at a receiver; at the receiver, the polarization of each photon is "measured" by utilising two beam splitters-one for horizontal/vertical, one for diagonal. The receiver has to make a guess which beam splitter to use, since it doesn't know.

3. The sender of the photons now compares the sequence of polarizers he had to use in order to transmit the key with the information from the receiver regarding which beam splitter was used for every single photon within the sequence it was sent. That sequence of bits, finally, is the key, and only they do so where the reading is done with the correct beam splitter.

The no cloning theorem basically states that if the eavesdropper in any form reads or copies this photon, then the state of the photon will change and it will be observable at the end point. Its translation at this point into laymen terms is that nobody can read the photon, transfer it, and copy it without being detected.

1.2 The Necessity of Quantum Cryptography

Peter Shor in 1994 developed an algorithm to factor prime numbers which placed current public key infrastructure (PKI) in danger as PKI is considered safe because large prime numbers factoring is difficult or not possible. For instance, an RSA key of 1024 bit that can take millions of years to break on classical computers but can break in seconds using a quantum computer. Due to advancements in quantum computing (QC), many security agencies like NSA announced changes from classical algorithms to protect data to quantum resistant algorithms. So, quantum cryptography is required to encrypt messages in a way which cannot be decrypted using quantum computers unlike today's encryption algorithms.

Quantum cryptography also helps to complete different tasks of cryptography which are impossible by classical cryptography. It uses principles of quantum mechanics to encrypt messages which cannot be decoded or decrypted by anyone not allowed. Such a level of security is the need of the hour which is in threat due to use of classical cryptography in every domain. Unlike classical cryptography, quantum cryptography can encrypt data for larger periods of time like for 100 years minimum. In the current scenario there is a need for unbreakable encryption to protect data for a longer period which can only be fulfilled by quantum cryptography. With advancements in quantum computing, current encrypted data is at risk. These all facts relate to the need for quantum cryptography.

1.3 Classical Versus Quantum Cryptography: A Comparison on Various Cryptographic Algorithms

Analysis of four cryptographic algorithms which are DES, TDES, AES and Blowfish in classical and quantum cryptography shows that algorithms perform far better in the quantum regime.

DES—DES stands for Data Encryption standard. In this, algorithm and cryptographic key is applied to one bit at a time as well as to data block together. It groups plain text messages into 64-bit blocks to encrypt them.

TDES—It takes three 64-bit keys for all 192-bit key lengths. It uses the exact same encryption technique as DES, only it repeats the process three times, hence the name "triple DES."

AES—It is based on some principles of design called permutation and substitution of networks, using combinations of these software is past.

Blowfish—It is a cipher block used for encryption of data to safeguard it. Although it suffers from weak keys, no successful attacks are possible against it (Table 1).

One of the most general quantum threats of symmetric-key algorithms is Grover's algorithm. This algorithm reduces the effective security level by 50%. To eliminate this risk, longer key lengths like 256 bits or more are recommended as such the limitations and challenges of each algorithm are mentioned below:

DES—This algorithm is inherently considered insecure in both classical and quantum perspective due to its short key length.

TDES—In this algorithm they have attempted to solve the limitation present in DES by tripling the encryption in DES but this brings a new problem to light in terms of scalability as key generation is slow.

AES—Since Quantum computing is still in its infant stage, long term impact on AES would require a lot of computing on AES-256. It is because of this that AES is considered as the gold standard of symmetric encryption.

Blowfish—This algorithm has not been tested thoroughly in Quantum context, but Grover's algorithm theoretically can reduce its security by half.

In order to determine the encryption and decryption times, throughput, and avalanche effect of algorithms, these algorithms are run on various sizes of data files. Analysis and testing reveal that TDES (Quantum) integrates well with the

Table 1 A brief overview of classical and quantum cryptography

Algorithm	Classical cryptography	Quantum cryptography
DES	Vulnerable to brute force attacks with classical computers	Vulnerable to Grover's algorithm, which can reduce the search space from 2^n to √(2^n) operations. Still secure with long key lengths (e.g., 128 bits or more)
TDES (3DES)	More secure than DES but slower	Vulnerable to Grover's algorithm, as it's based on DES. Triple-encrypting may provide some level of quantum resistance
AES	Secure and widely used with various key lengths (128, 192, and 256 bits)	No known quantum algorithms to efficiently break AES when using long key lengths
Blowfish	Relatively fast and secure when used with an adequate key length (e.g., 128 bits or more)	No known quantum algorithms to efficiently break Blowfish when using long key lengths

BB84 protocol, requiring less time for encryption and decryption. In comparison to other algorithms, it also has a higher average throughput rate of 0.31. With a key change of one bit, it has less of an avalanche effect.

2 Quantum Key Distribution

Quantum key distribution, or QKD, recently has been receiving a lot of attention, especially from security experts, and for a good reason. We need a more secure encryption method because quantum computers are currently posing a threat to conventional encryption techniques [1]. It is a safe process; the symmetric encryption key is transmitted between different entities based on very well settled and validated laws of quantum physics. Quantum Key Distribution, QKD, was one of the technologies proposed by the Quantum-Safe Security Working Group to protect and future-proof data against developments in computer power, new attack strategies, weak random number generators, and the advent of quantum computers.

Among the major concerns of this digital society is the security of data during communications and storage. This can be manifested in e-banking, e-business, e-health, and e-government channels. Considering the fact that quantum computers are developing with leaps and bounds, the threat to one and all is increasing faster every day and can break the security. At present, QKD handles the issue of security in a long run; that in health terms might imply records that have to be kept secure for one's life. Quantum technologies come into play again in the mitigation of this risk within the critical infrastructure that provides something as basic as energy [8, 9].

Although these systems have been on the market for a while, advancements kept making them more affordable, smaller, and able to function over greater distances. These are all essential features considered by governments and businesses to adopt these technologies. The current hurdle is implementing these QKD systems into the current network infrastructure. In order to do so diverse groups consisting of scientists, network operators, critical infrastructure providers, telecommunication equipment manufacturers, and QKD equipment providers are working on this.

With QKD, secret keys can be established and shared for executing cryptographic protocols. The key ingredient is that the secret key must remain secret between communicating parties. It is for this very reason that we use a property that was originally regarded as a challenge for quantum systems: their sensitivity—this means that if you "look" or in any way "disturb" them, you destroy their quantum properties. As seen in Fig. 1, the information is usually encoded on an individual photon-by-photon basis. Whereby Alice sender sends the data in a "bit sequence" and can opt for the way that bit sequence is encoded in one of two states, that is either horizontal H or vertical V, or Alice can also encode in two other different states, that are the other two different combinations of the states labeled as $+45^\circ$ and -45°. Bob now selects one of two "bases" in which to measure the bit sequence: Bob measures either H, V or $+45^\circ$, -45°. When sender and receiver both have opted for the same base, they will end up having a perfectly correlated result; Bob finds H, and Alice had sent

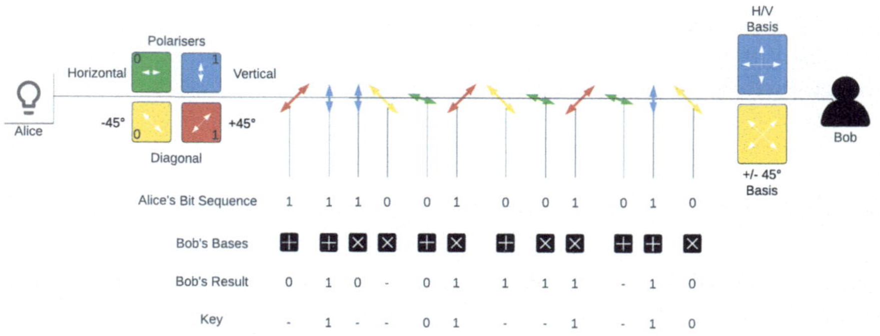

Fig. 1 Quantum key distribution

H, and these are kept. If Bob measured in a different base than Alice had prepared in, his result is random, and he discards it. To proceed, Alice and Bob must publicly discuss which base was used, but they do not publish the outcome, which is now the secret key. There are several variants at present, but this is one way it might be done [2].

As seen in Fig. 1, this will, however, produce only a secret key that needs to be integrated with cryptographic protocols for security to be ensured in various applications that they are put to. The beauty of this solution lies in the fact that any attempt at spying or intercepting the process of key generation by a hacker causes errors and exposes them before any data is communicated or encoded.

2.1 QKD Working and Ensuring Unbreakable Security

QKD over fiber optic cables allows for the fact that millions of polarized light particles, or photons, flow from one entity to another. Each photon will have a random quantum state summing up to a bit stream of ones and zeros. Unlike other communication protocols, QKD is the only one that stands out as the only provably secure communication method, as it relies on the principles of physics to encrypt data rather than mathematical formulas.

The unbreakable nature of QKD is attributed to its unique security features. What characterizes this system's security is the capability of the system to identify if any intruder has entered on to the QKD transmission. This is due to the fact that the photons, because of their unique and fragile characteristics, change state whenever a third party tries to read or make copies of them. This alteration is readily detected by the endpoints, alerting them to potential tampering with the key. In response, the compromised key is discarded, and a new key is transmitted. Additionally, the randomness inherent in the generated keys provides protection against future hacking attempts, making sure the robustness and stability of the quantum communication system.

2.2 *Quantum Key Distribution Attacks*

A. **Faked-State Attack**

The ability of the system to transmit and recessive single photons is the foundation of BB84. However, it is difficult to convey single photons, therefore current devices only send a faint light signal [10]. A single photon may or may not be present in these signals. This dependence on flimsy signals results in security lapses. Bob's weakness is exploited by Eve in a simulated state attack, where she forces Bob to measure using the same criteria as herself [10]. The QKD apparatus component known as the diode has a flaw that can be exploited to enable the faked-state attack [11].

A single photon is what an avalanche photodiode (APD) detector ideally detects. However, between the detection of two successive photons, there needs to be a recharge period of roughly 1 μs. In ideal circumstances, there wouldn't be any issues since the rare occurrence of weak laser signals would allow the detector adequate time to recharge. Makarov says that "when an APD receives light of a fixed level, the detector doesn't have time to recharge and then it behaves as a classical photodiode". Hence Eve can get his beam to pre-determine when the detector clicks or when it is blinded from seeing the valid data.

We now have four diodes, which stand for four different polarizations. Eve will first deceive Bob by intercepting Alice's photon and sending it to him in a false condition. This counterfeit state indicates that Bob is not sent a quantum state by Eve; rather, it is a bluff that fools its detector into thinking such. Eve detects that the base and bit are inverted, and this prepares the fake state.

For example, Eve modifies 0 by setting it to 1 with D base if 0 is identified in R base. By blinding Bob's 1-bit detector, Eve will make use of Bob's detector. Bob has an identical chance of finding a 0 or nothing if he uses the same base, R, as Eve does. It becomes proven to detect nothing at all if Bob switches bases D. Table 2 provides an illustration of the method. By doing this, Eve will be able to operate Bob's equipment and ensure that, just as she did, the identical bit is identified using the same base. Which intern would cause a concern as it would increase the error bits.

B. **Photon Number Splitting (PNS) Attack**

As previously stated, perfect single photon sources are hard to produce. Thus, cryptographic devices use Weak Coherent Pulses (WCP). These gadgets fall short of expectations and fail to produce ideal outcomes. The Poisson Distribution describes the distribution of photon counts in a pulse.

A noisier weak coherent pulse has a lower mean photon number due to its fewer photons in total. The source weakness, however, may be balanced by PNS, Photon number splitting. According to [6], Rahul Agarwal says that one of the strong attacks, which focuses on realistic photon source, is PNS. As for the strategy of attack, after the interception of a portion of the pulse, Eve forwards to Bob a residual multiphoton pulse. She then bides her time until the communicators' base sequences are made

Table 2 Fakes state attack demonstration

Alice	Eve's base	Eve's measure
R0	R	0
R0	R	0
R0	D	0
R0	D	0
R0	D	1
R0	D	1
Eve sends	Bob's base	Bob's measures
D1	R	0/Nothing
D1	D	Nothing
R1	R	Nothing
R1	D	0/Nothing
R0	R	Nothing
R0	D	1/Nothing

public. This helps Eve measure the photon that was intercepted using the appropriate basis (Fig. 2).

Even though it is a very destructive attack, it is very challenging to execute. Few pulses in the real-world application of WCP are multiple-photon beams; most pulses in the implementation contain zero or no photons. To verify whether the pulse contains multiple photons, Eve must maintain complex hardware and an algorithm. But after taking care of these prerequisites, Bob finds it extremely difficult to detect Eve's presence.

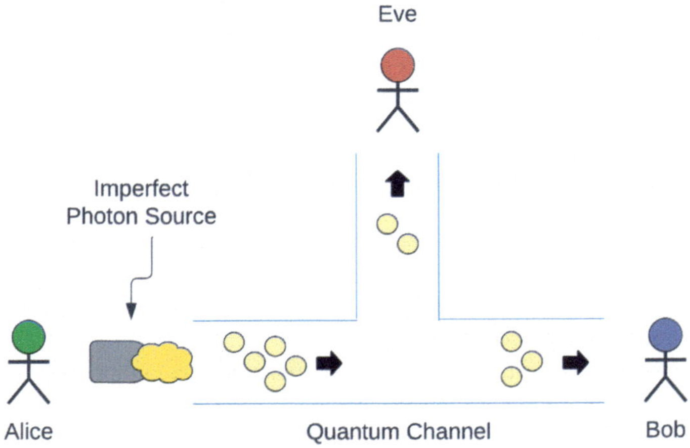

Fig. 2 PNS attack demonstration

C. **Man-In-The-Middle Attack**

QKD doesn't escape any technique of classical communication in that it falls prey to a man-in-the-middle attack if it is not authenticated. Authentication gives us the information to discern an ally from an enemy. Similar to current security procedures, Alice and Bob would need a way to confirm one another's identities in order to guarantee authentication. A pre-shared secret key may be a part of this. In addition to quantum key distribution, we can use this key to create unconditionally secure authentication schemes.

2.3 *Vendor Product for QKD*

The United States firm that operates in the field of quantum information processing is MagiQ Technologies [12]. It has then further used this work to implement the first security system based on the quantum cryptography. This company develops integrated devices in collaboration with another company Cavium Network located in California [12]. Quantum private networks are mixed components that hold a communication channel which is encrypted and has various quantum cryptography capabilities [11]. It is a system which is embedded and contains both software and hardware developed by this company MagiQ. In Europe, id Deckpoint and Quantique jointly developed and implemented the first network which is data archiving which is secured by quantum cryptography [12]. It is actually the very first company to have ever commercialized quantum cryptography. One such device designed by another French startup, SmartQuantum, is a 2 Gb/s device that combined the functions of encrypted data transfer and key distribution into one fiber.

3 Quantum Key Distribution Protocols

This is from the fact that Quantum Modulation might be able to come up with a provably secure way to generate and distribute with which a sequence of genuinely random numbers in order to create a shared trusted key has been developed. Now that quantum mechanics can solve the classical problem of the key distribution process, the information that a quantum is carrying is destroyed by a quantum measurement [13]. It was the so-called BB84 scheme, put forward by Charles Bennet and Gilles Brassard, that asserted that one could encode a quantum system in its quantum state information with at most only one user who was able to receive that information reliably [13]. They came to realize that the coherent superposition of quantum is capable of mapping two fragments of information on one quantum system using a single channel. Availing the information to the receiver using Heisenberg's indeterminacy principle [13]. Following this theory, it was found that pairs of quantum non-commuting observables are readily accessible and adhere to the Heisenberg principle. Additionally, a more complex system than a single quantum can transmit quantum information that can be prepared as a coherent superposition [13]. The telecommunication language was used to translate the ideas mentioned earlier. In essence, bright, laser-like beams function as a complex system.

3.1 BB84 QKD Protocol

Well-known protocols enabling QKD include, in detail, the BB84 [14], B92, Six-State, SARG04 [5], and Ekert91. Among them, BB84 is the most general and common practical key distribution system. The protocol BB84 was independently proposed by Bennett and Brassard in the year 1984.

The protocol is conducted based on the two big channels through which data are transmitted:

- Quantum channel: One-Way communication.
- Classical channel: Two-way communication.

Under ideal conditions, BB84 enables two communicating parties at either end of the protocol to exchange polarized photons through a shared key string. The figure below shows how keys would be exchanged and then sifted:

Using Quantum Channel (Key Exchange)

- In this process, as shown in Fig. 3, the sender will encode the data with random bits of 0 and 1 by sending the bits using randomly chosen bases-rectilinear or diagonal. Repeating this whole process, he sends all the photons to the receiver who each time selects a random base for every photon.
- Since he does not know on which bases the sender had encoded the photons, the receiver at the other end randomly chooses either the rectilinear or diagonal basis to perform the measurement on the received photons. The receiver now uses the public channel to communicate with the sender in order to do key sifting after receiving all the photons.

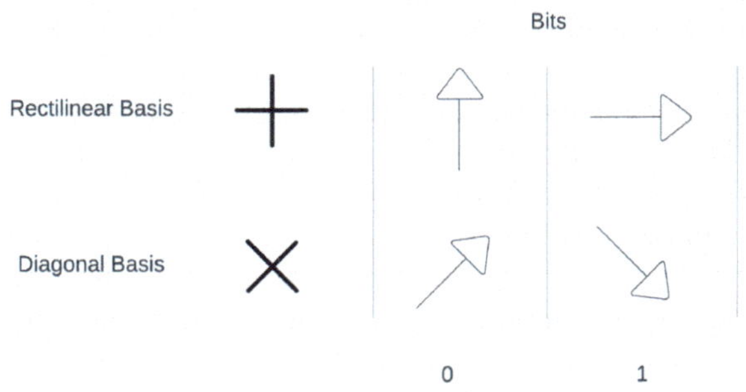

Fig. 3 Photon polarization using base

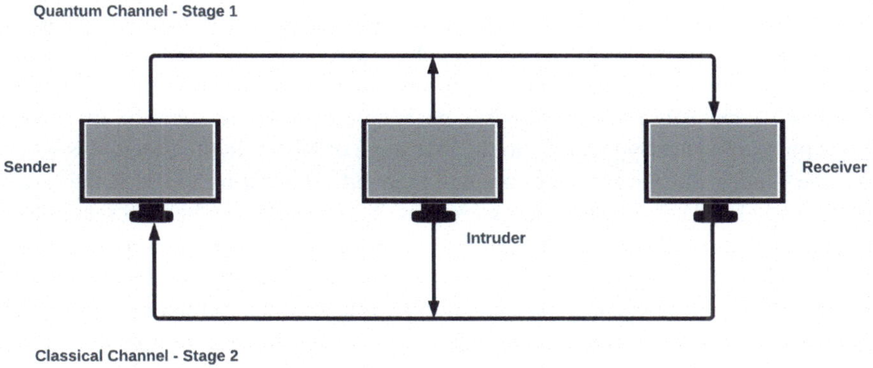

Fig. 4 Various channels involved in QKD

Using Classical Channel (Key Sifting)

- If the receiver now tells the sender which bases were used in the measurement of the photons, the sender asks whether these were the same.
- Without disclosing the information's true value, they both agree that the bases used were correctly matched. After eliminating all the non-matching data on the polarizer bases, two key strings consisting of shorter sequences are left, which are referred to as the raw keys (Fig. 4).

The secret raw key is made up of the bits that match, which is not the complete key, after which as part of the ongoing communication between the two the following steps take place:

(1) **Error estimation**: A comparison is made between the raw keys to verify that no eavesdropping has taken place. If there is an intrusion, fault will be inserted to one of the raw keys. Then when these two are compared, they will become different. Therefore, error estimation should urgently be done. If the quantity of errors surpasses the QBER threshold, the key would be abandoned and the data try to resend the key.

(2) **Error correction (Reconciliation)**: This step is performed for removing errors from the raw key in order to get the common key, using one of the various available protocols. Amongst the most popular are the Winnow protocol, relying on parity exchange and using a Hamming hash function to help with the correction of individual errors, and Cascade, which uses optimal linear codes, releases fewer data, and performs parity-based error correction.

(3) **Privacy Amplification**: Finally, both parties will be left with strings of bits that are equal but not wholly private, for instance, some partial knowledge of the string can indeed be acquired by the eavesdropping third party. It is this latter information that the process of privacy amplification aims to reduce enough for the third party to then recreate a valid secured secret key.

In 2004 Scarani, Acin, Ribordy and Gisin published the SARG04 as an enhancement to the BB84 protocol. SARG04 is much more robust than BB84 is with regards to attacks involving photon-number splitting. In fact the information carriers in SARG04 like in BB84 are laser pulses not single-photon sources. The first stage of this protocol is nearly identical to the first stage of BB84. In this stage, it is when both the sender and the receiver compare their bit information to see if the bases are the same for them to extract the secret key. This process is a bit different in this protocol as compared to the BB84 since the sender gives out two bases to be considered, and they will just use one of those bases instead of giving out the actual basis information used for encoding. This means the right state measurement requires the Receiver to apply the correct bases used by the sender. In case he uses the wrong bases, he wouldn't have matched any of the states of the sender and therefore not be able to have access to the data.

A. **Discrete Variable Quantum Cryptography**

At first when Quantum key distribution protocols were being researched on Discrete variable quantum cryptography was one of the first to be discussed as a reliable cryptography technique. The single photon states, entangled photon pairs and photo-detection were used, which showed security on tests on asymptomatic notational limits [13]. DV-QKD remains fully secure in concord with the no cloning theorem and the indistinguishability theorem of four out of arbitrary quantum states. The no cloning theorem simply states that it is impossible to clone exact arbitrary states or even with the help of quantum computers [13]. That effectively means that none of them would ever be able to decrypt the encrypted message if one does not have the decryption key before transferring the message. Indistinguishability theorem defines that states that are non-orthogonal are not differentiable among each other. Single-photon detectors set one of the major disadvantages of this technique, its dead time, which for different producers is approximately in the range between 10 microseconds to 10 ns. The dead time, if more, increases the dark time in the sensors which reduces the key generation rate which is very crucial as we want new random keys to be generated as frequently as possible to prevent any kind of intervention from any attacker [13] (Fig. 5).

B. **Continuous Variable Quantum Cryptography**

As stated above, quantum physics firmly secures the public key for all QKD protocols, and the first DV coding protocol was proposed. In 1999-that is fifteen years after the BB8 paper-Ralph knew that quantum information could be encoded onto bright beams as the phase and amplitude modulation of two entangled carrier beams. More generally, the field quadrature in the CV coding is understood to denote a pair of non-commuting observables representing, respectively, phase and amplitude of an optical field. CV does not provide new protocols but only shares some of them with DV, such as the necessity of classical communication at the end to distill the key from a bigger data set, exploitation of several measurements, and random variables introduced both at the transmitter and the receiver. The main point to focus on is the modulation itself. Methods and techniques that are already incorporated into the

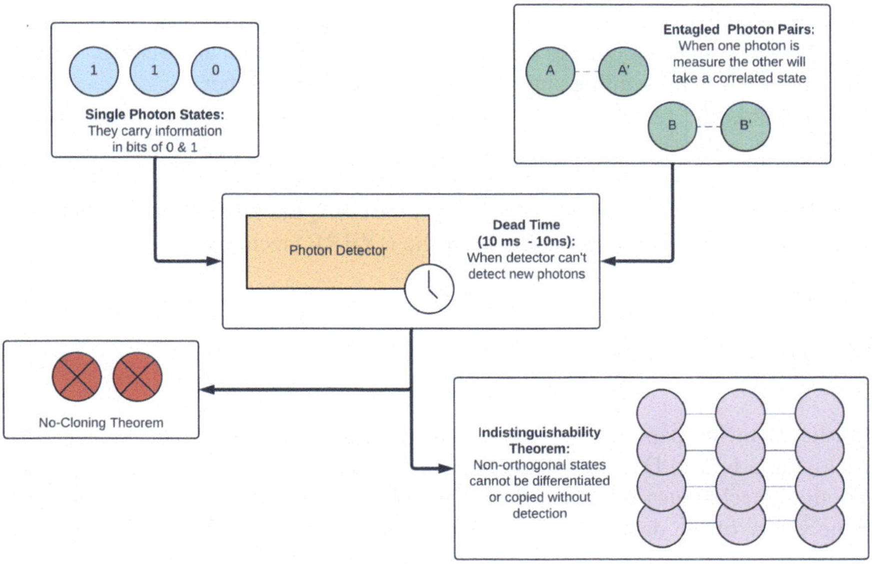

Fig. 5 Key concepts of DV-QKD

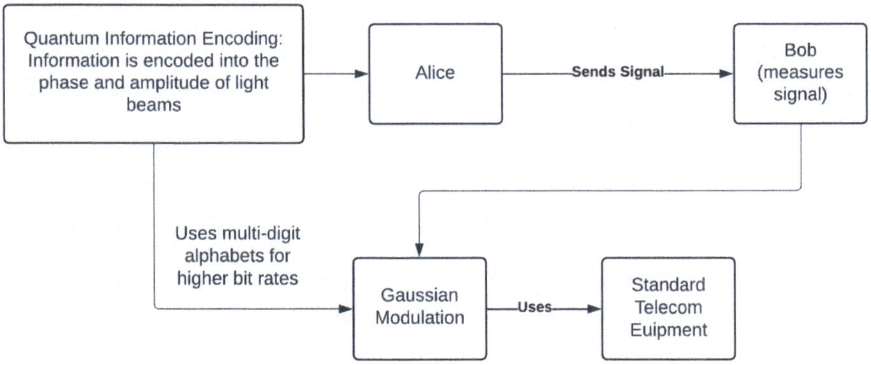

Fig. 6 Continuous variable quantum cryptography (CV-QKD)

telecom network, such as homodyne/heterodyne detection and modulation coding/decoding. A bit later, some advanced modulation techniques appeared—for example, Gaussian modulation, which was found very advantageous and largely improved the performance of a system. However, all the proposals coming out up to the advent of Gaussian modulation have used binary alphabets. But multi-digit alphabets use highly bit rates.

In other words, in this CV QC scheme, Alice modulates the input to a quantum system with some random variable and transmits it to Bob. Later, by processing the raw datum each party holds, Alice establishes a secret key, and Bob measures the signal to obtain b that corresponds to the value of a held by Alice. There do, however exist specific systems needed in the protocols of Single photon and discrete variable which are capable of sending single particles through optic fiber.

This yields the factor bit rate, which reduces the key generation rate. Where decoherence reduces only the distance a CV link can travel, for DV the problems show up only in its key rate. The trade-off between distance and bit rate is what keeps us in doubt about which among the two, DV or CV QKD methodologies, is better; hence, we continue our search to look for the HYBRID technology (Fig. 6).

3.2 *Hybrid QKD Protocol*

Djordjevic in his paper [15] discussed a new technique in which Ey suggested employing a hybrid DV-CV QKD protocol to counter the disadvantages we faced in both the techniques we saw earlier. The DV QKD protocol though gave a surety of the safety of the data but faced a limitation in the transmission distance of the photons and significantly low SKR (secret key rate). However, the CV provided a faster SKR even though it was unable to provide security proof. The introduction of a DV QKD subsystem allows the DV QKD subsystem's unconditional security as well, which is the main concept of the approach suggested in this paper. It is crucial for the CV subsystem's SKR to be in the range of the DV subsystem for them to function together [7] because the DV subsystem's SKR is in orders of values lower than the CV subsystems. As well the DV subsystem doesn't need to be as spectrally efficient but just match the CV subsystem.

What is being done is that Alice sends message by performing both DV (Time phase based) and CV (DM based) QKD to achieve the most efficient and accurate encoding. Bob will randomly select one of the quantum key distribution protocols based on the optimized probability received, part of which becomes the single-photon delay time of the SPD of a dual-variable quantum key distribution protocol.

The prepare and measure (PM) hybrid DV-CV QKD protocol is then discussed in the paper. The results of the SKR analysis indicate that this protocol performs better than both DV and CV methods, providing a higher SKR as well as a longer transmission distance.

4 Novel Approaches in Quantum Key Distribution

Developing a multi-party secret sharing scheme via QKD

Noor UI has attempted to present a protocol that is dependent on quantum cryptography in this paper [9]. This protocol makes sure that nobody can listen in on the conversation or obtain any portion of the confidential message. This protocol ensures that eavesdropping or extraction of any part of secret message is not possible.

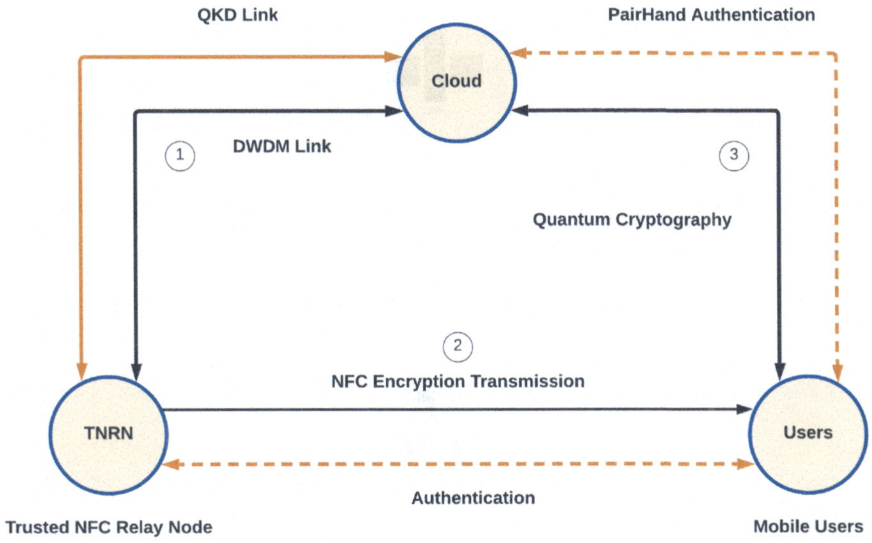

Fig. 7 A QKD model in mobile cloud computing

Because it uses quantum teleportation and quantum entanglement swapping to provide inherent security. Entanglement swapping creates an entanglement between two non-interacting quantum systems. On the other hand, a qubit is sent across a channel in quantum teleportation. To ensure Security, distribute EPR pairs are generated by the sender himself, to be used in the scheme. The result is the development of the QSS scheme, which is safe from both internal and external threats. Since quantum cryptography is a relatively new field, many questions remain to be resolved. The protocols defined in these papers are complete, but there are some ways in which quantum secret sharing can be achieved [9]. Super dense coding can be used to encode the classical information to be sent. Another thing which can be explored further is "Quantum Digital Signature" [9] (Fig. 7).

Securing IoT devices through Blockchain and QKD

The authors Shalini Dhar, Ashish Khare, Ashutosh Dhar Dwivedi, and Rajani Singh propose fusing blockchain technology and quantum cryptography to build a robust security framework for Internet of Things (IoT) [16].

Blockchain technology is integrated into the framework to produce a decentralized, impenetrable system. The integrity and transparency of transactions involving IoT devices are guaranteed by this blockchain layer. The authors hope to stop unauthorized access and data manipulation by using blockchain's built-in cryptographic mechanisms and decentralizing control [16].

The framework is also impregnated with quantum cryptography. Quantum cryptography, though based on a couple of the basic principles lying behind postulates of quantum mechanics, has gained some form of reputation in recognizing the possible

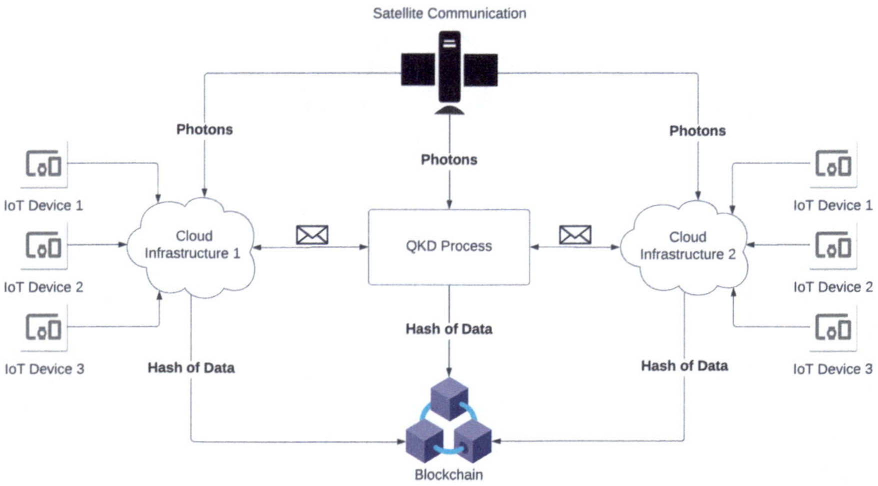

Fig. 8 Execution of the blockchain model

presence of eavesdroppers and establishing secure channels of communication. Apart from conventional cryptographic methods, the embedment of quantum cryptography adds a heightened layer of security in the communication aspects of IoT devices (Fig. 8).

In summary, the suggested framework in Fig. 6 incorporates quantum cryptography for secure communication and uses blockchain for decentralized control and data integrity. Together, these strategies seek to address the particular security issues that IoT devices present and provide a complete solution for protecting private information and communications in IoT ecosystems.

Cryptographic Analysis of Algorithms in Local and Cloud Environment using QKD

One of the greatest options available today for security models is cloud computing. However, these security models make a number of assumptions. For cloud computing, quantum key distribution (QKD) offers security through the QKD protocol. It offers data dynamics, secure cloud storage, and the advantages of quantum mechanisms. In this work, Murali and Sivaram compare the effectiveness of several cryptographic algorithms in a local and cloud environment utilizing both classical and quantum cryptography [17]. The throughput of various algorithms in relation to varying file sizes is shown in Fig. 7.

This paper aims to analyze and compare different quantum cryptography techniques in order to determine which local or cloud environment is optimal [17]. It has been identified that throughput in cloud environments is higher than in local environments (Fig. 9).

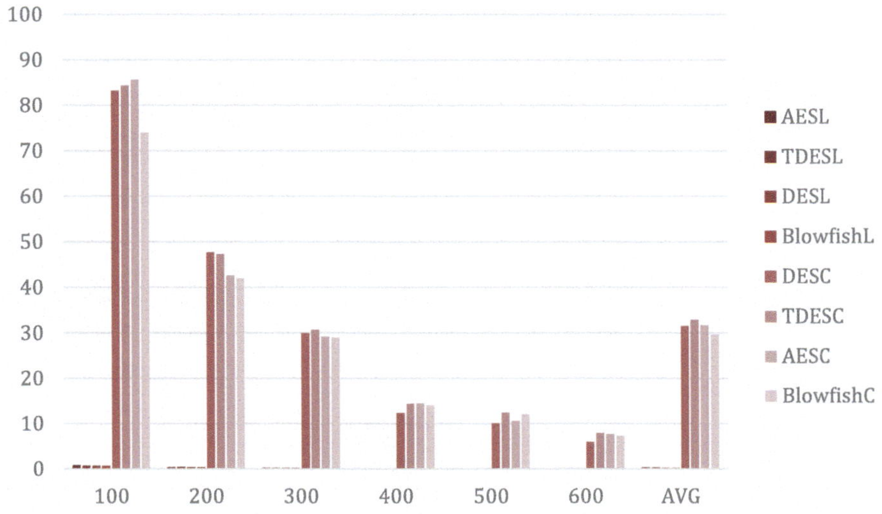

Fig. 9 Throughput with different file sizes [17]

Using Syndrome Measurement for Asymmetric Quantum Error Correction

The paper outlines some inadequacies of the usual symmetric quantum error correction procedures and brings up a new theme in this area, that is, asymmetric quantum error correction, AQEC [18]. Phase flip, Z, errors have been found experimentally more frequent in comparison to the bit flip errors, X, hence justifying the rationale behind AQEC. Paper [18] discusses a decohering, syndrome measurement and encoding apparatus for an effective AQEC process. While it would reduce the circuit depth by 48% in contrast, it would improve the fidelity of the circuit at most 85.89%. Since phase flip errors very frequently occur, asymmetric error correction is optimized to correct bit flip and phase flip errors in a far more efficient way. A bounding function facilitates syndrome measurement by identifying the precise qubit that has experienced an error.

Despite the challenges faced by the decoherence and noise in quantum systems, the proposed method as seen in Fig. 8 performs effectively. The Qiskit tool's implementation on the IBM quantum system yields promising results in terms of reduced error rates, increased fidelity, and circuit depth. Conclusions were drawn on the potential use of AQEC in safe data transmission and other quantum technologies: quantum key distribution, quantum teleportation, and quantum random number generation [18]. To further enhance outcomes, the authors propose integrating entanglement in future research (Figs. 9 and 10).

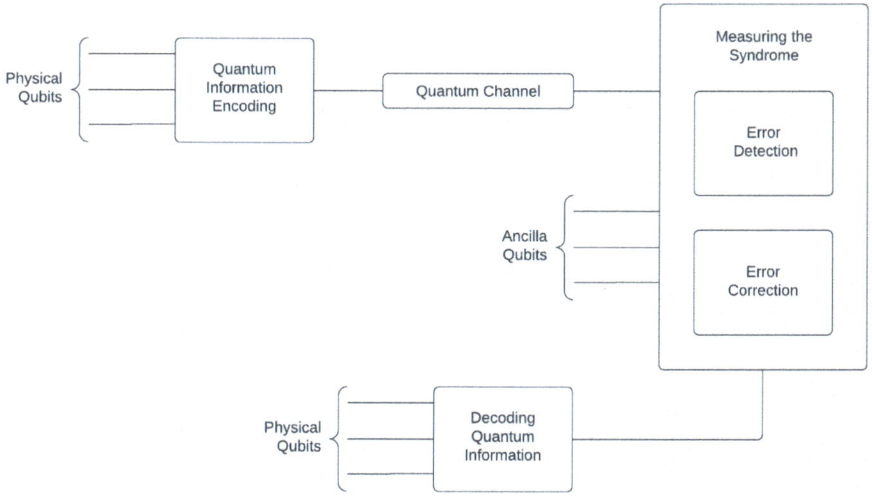

Fig. 10 Abstract architecture of the proposed system [18]

5 Cryptosystems Vulnerable to Quantum Algorithms

This section addresses the implications of quantum algorithms on modern-day cryptography. Further, it describes Grover's Algorithm and Shor's algorithm. Especially the following subsection on an explanation of Shor's algorithm provides significantly more general overview how this specific algorithm endangers security in systems based on difficulty of factorization or computation of discrete logarithms.

For this reason, modern electronic communication systems are still based at their core on cryptography. Be the intention to secure e-mails, passwords, bank transactions, and even electronic voting systems, the two main goals in security are confidentiality and integrity [19]. The fact that messages that are encrypted can only be decrypted by parties that have exchanged keys-through authentication-shows just how central cryptography is in maintaining secure electronic communications.

Being more powerful computationally, quantum computers are far more threatening to the basic objective of secure and authenticated communication compared to classical, traditional computers. It is hard for traditional computers to rapidly crack cryptographic keys with any challenging computation or search that a quantum computer could make. Because of this, the security of communication channels between legitimate parties (sender and receiver) is threatened, making it possible for eavesdroppers to take advantage of weaknesses. On the other hand, a traditional computer cannot complete the task computationally [1] (Table 3).

A. **Shor's Algorithm**:

It was within this setting that the mathematician Peter Shor, in his 1994 paper titled "Algorithms for Quantum Computation: Discrete Logarithms and Factoring" 26,

Table 3 Impact of quantum computing on current cryptographic schemes

Cryptographic algorithm	Type	Purpose	Impact from quantum computer
NTRUEncrypt	Symmetric key	Encryption	Resistant to quantum attacks (lattice-based)
SHA-3 (quantum resistant)	Hash functions	Hashing	Designed to be resistant to quantum attacks
NTRUEncrypt (signatures, key establishment)	Public key	Signatures, key establishment	Resistant to quantum attacks (lattice-based)
Lattice-based ECC variant	Public key	Signatures, key exchange	Quantum-resistant variant using lattice-based cryptography
SPHINCS +	Public key	Signatures, key exchange	Considered quantum-resistant (hash-based)
DHE, RSA (in TLS/SSL)	Public key	Key exchange in secure Comm	No longer secure
HMAC (Hash-based Message Auth. Code)	Symmetric key	Authentication, integrity	Secure
ChaCha20-Poly1305 (in TLS/SSL)	Symmetric key	Encryption, authentication	Secure
Post-quantum cryptographic schemes	Various	Preparing for quantum threats	Developing

demonstrated that the appearance of quantum computers has strikingly revised the feasibility of factoring large integers. In this day and age, asymmetric cryptosystems take either big prime integer factorization as the prime factors of 15. This will illustrate how Shor's algorithm factors large primes unmanageable task for current computers or else that of the discrete logarithm problem. We will now need to find, and to do this requires a 4-qubit register. Since 15 in binary is 1111, for the computation of prime factorization a 4-qubit register will suffice. Computation on a register can be regarded, according to Bone and Castro, as computation carried out simultaneously on all possible values a register can take [20]. The only thing a quantum computer needs to do is this one step.

Mentioned below is an example of how the algorithm works:

1. **Initialization**:
   - The number to factorize is $n = 15$.
   - Choose a random number such that $1 < x < n - 1$.
   - In this example, $x = 2$.

2. **Quantum Superposition**:
   - Raise x to the powers of the 4-qubit register and divide by n = 15.
   - Store the remainders in a second 4-qubit register.
3. **Observation of Results**:
   - Observe the results in the second register after performing the quantum operations.
   - Note the repeating sequence of 4 numbers: 1, 2, 4, 8.
4. **Determination of f**:
   - Identify the period *f* of the quantum function, which is the length of the repeating sequence.
   - In this case, f = 4 since the sequence repeats every 4 steps.
5. **Calculation of Possible Factor**:
   - Use the formula P = x(f/2)−1 to calculate a possible factor.
   - For x = 2 and f = 4, the possible factor P = 2(4/2)−1 = 22−1 = 3.
6. **Result**:
   - The possible factor obtained is P = 3, which could be a non-trivial factor of n 15.

If the first computation does not give a prime answer, then other values of f are tried. The application of Shor's algorithm extends to the resolution of discrete logarithm problems. For instance, Vazirani [21] carefully described, in the case of a procedural explanatory development of the method by Shor, how to derive a new superposition from a random superposition of two integers by successive applications of a set of Fourier transformations. What this new superposition achieves with high likelihood is providing two integers such that the equation yields, by formula, the value of r, the "exponent" which is the unknown in DLP.

B. **Grover's Algorithm**:

In fact, a quantum computing uses a search algorithm for unsorted databases developed by Lov Grover, which finds an entry in the unsorted database of N entries in up to $\sqrt{N}$ searches [22]. The corresponding solution on the conventional computer would take as many as N/2 searches. It may affect breaking the DES, based on a 56-bit key, since Grover's algorithm, as shown by Bone and Castro needed only 185 searches to determine the key [20]. For now, the strategy to render such techniques resistant to breaking consists of embedding more key bits into the password. In this manner, one is nothing but expanding the key space. As such the exponential growth of the amount of searches necessary to brute force a password acts as the deterrent. Grover's algorithm works in symmetric cryptosystems according to Buchmann et al., though it is not as quick relative to Shor's algorithm [23].

C. **Symmetric Encryption Schemes**:

Other than Grover's algorithm, no other quantum computing threat has been identified; this shows a speed increase proportional to the square root over traditional brute-force algorithms. Thus, symmetric cryptography is threatened. Even more illustratively, a quantum computer uses −2n = 2n/2 for n-bit ciphers. That is, an n = 128-bit key symmetric cipher, like AES-128, would provide approximately 64 bits of security. Remember, 80 bits are considered the norm for a secure level of security. In the context of quantum computations, AES can be regarded as a robust cryptographic primitive, in particular when it uses key sizes of at least 192 or 256 bits. Another guarantee given to AES for the security of classified information at security levels SECRET and TOP SECRET is that the NSA only accepts the use of AES cipher with key sizes of 192 and 256 bits in all cases, respectively [24].

D. **Asymmetric Encryption Schemes**:

All public-key algorithms in use today, when actually deployed, are based on two problems: factorization of large numbers, which underlies RSA, and the computation of discrete logarithms used in DSA signatures and ElGamal encryption. Shor's algorithm decrypts both in polynomial time, and also their mathematical structure is quite similar. More modern algorithms, like the elliptic curve-based variants of ECDSA, depend on variants of the discrete logarithm problem, hence are in no better relative position regarding quantum computers than their ancestors. According to Kirsch and Chow [25], data encrypted with ECC can be decrypted using one of the variants of Shor's algorithm. The authors highlight the fact that much weaker key space is used in ECC compared to RSA, which practically makes it much more vulnerable due to quantum attacks. Proos and Zalka [26] also showed that a 1000-qubit quantum computer will be able to break 160-bit elliptic curve while breaking 1024-bit RSA requires 2000 qubit quantum computers. The qubits required to break a cryptosystem depend on the algorithm. They also provide quite substantial details of how the Shor's algorithm will break the ECC over GF(p).

E. **Hash Functions**:

As it happens, the family of hash functions shares with symmetric ciphers the unfortunate property that its security is tied to a fixed length output. Using Grover's algorithm may be insecure since it was able to find a collision in steps that were the square root of its original length. That is no harder than searching an unsorted database by an exhaustive search. It should not be a great surprise that Grover's algorithm and the birthday paradox were combined somehow. This is made clear by Brassard et al. [27], who showed a quantum birthday attack that works because an attacker can succeed using the preparation of a table with O complexity and applying Grover's algorithm to find collisions. As such, a hash function should produce at least 3b bits to have a level of security of b bits against Grover's quantum algorithm. This implies that most of the hash algorithms currently in use shall be non-useful during the quantum era. However, it is resistant to the quantum attack when set up with longer outputs for both SHA-2 and SHA-3.

6 Various Implementation of Quantum Cryptography

A. Drone-based Quantum Key Distribution

This paper explores the use of QKD between Unmanned Aerial Vehicles to address the need for secure wireless communication during flight [28]. By using quantum communication protocols, security is enhanced as compared to classical approaches by reducing the possibility of attacks such as spoofing and eavesdropping. Complying with the drone's small Size, Weight, and Power (SWaP) requirements is one of many difficulties.

This will be realized by a QKD source using resonant-cavity LEDs in combination with a fiber-coupled polarization modulator and several other important subsystems. In addition, the PAT system is equipped with single-photon detectors, an FPGA-based time-tagger, a new postprocessing synchronization algorithm, and fast-steering mirrors and IR beacons or cameras for alignment [28].

A QKD Transmitter Drone (TX—Alice) and a QKD Receiver Drone (RX—Bob) are part of the suggested architecture. Compact Decoy-State QKD sources are used in the TX drone, while 3D-printed receive optics and reciprocal PAT systems are used in the RX drone. An FPGA-based time tagger records detector clicks while the QKD states are transferred over free-space optical paths. In order to enable quantum applications like entanglement distribution and Quantum Positional Verification (QPV) in the context of UAVs, the paper describes the progress made thus far and presents the proposed architecture for drone-to-drone QKD [28].

B. QKD in telecommunication Network

The ancient way of data and control communication in telecommunication networks is highly risky, mainly because of the advances in computing. Because it uses public encryption algorithms. Therefore, there was a need for technology that was immune to this. QKD uses random bits as security key, which are based on quantum mechanics and therefore cannot be predicted with any computational algorithm. In the paper "Quantum cryptography networks in support of path verification in service function chains" an overview on Madrid Quantum Network has been provided, MQK is a real QKD network, on which experiments have been done by Telefonica Spain's Production. Additionally, it was recently installed in a network encompassing Madrid's downtown metropolitan area. The key was generated by this QKD network in accordance with SDN principles. The methods currently in use for applying PoT into practice have been discussed, along with the scheme's vulnerability. An IETF working group document is the most recent PoT method (WGD). An improvement to the vulnerability has been suggested, it makes use of conventional symmetric encryption algorithms and offers flexibility in terms of providing varying degrees of assurance (LoA). The previous nested encryption proposal has been replaced with these improvements, which are intended to bring about order to the IETF draft. The standard MQN has also been implemented to show how the enhanced solution uses less techniques.

C. **Security of Big Data: Neural versus Quantum Cryptography**

Big data is the data, which is so massive in size, that local Database management systems have troubles while handling it. When the data increases, the risk of losing the data increases. And hence comes the question of security. This paper "Big Data Protection Via Neural and Quantum Cryptography" initiated itself with the overview classical cryptography followed by Quantum Cryptography. Quantum cryptography depends upon the principle of polarization of photons and Heisenberg's Uncertainty Principle. According to Heisenberg's Uncertainty Principle, more than one pairs of physical properties cannot be measured out simultaneously. The Photon polarization principle states that an un polarized light is polarized when it enters vertically aligned filters. In 1984, the quantum key distribution protocol known as BB84 was created. It was simple to identify the attacker's presence thanks to the BB84 protocol. Under Neural Cryptography, it is mentioned under the paper that it is a model taken from biology. In the human body the brain and its neurons are connected in neural networks. In neural networks, many highly organized processing elements are arranged in a way to implement a solution. Neuron is the basic computational part. Each neuron is initially given a random weight, and it changes according to the input and output while training. The process of neural exchange depends on the synchronization of the 2 Tree parity Machine.

D. **Post-Quantum Cryptography and 5G Security**

There is a shift in the implementation of cryptography for fifth generation mobile broadband (5G). The security of previous generations of mobile broadband is based on standard key cryptography. There is a shared key which is used for the authentication of the user equipment to the network in SIM [29]. On the other hand, 5G is utilizing cloud architecture and is moving toward a PKI (Public Key Infrastructure) based model by utilizing protocols like OAuth 2.0 and TLS for authentication and authorization [29]. This change indicates the development of quantum computers creating a security vulnerability in established cryptography algorithms like RSA and others. There is a need to advance within 5G standards, considering the standards of NIST post-quantum cryptography. There is a requirement for post-quantum cryptography algorithms that can replace the traditional RSA algorithms, which thus propose a threat to security with growth in quantum computers. However, several challenges exist while developing such algorithms—signature size, public key size, auxiliary functions which can serve as alternatives, etc. Since post-quantum cryptography does not depend on the same mathematics as RSA and other conventional algorithms, it is more resistant to rapid advancement in quantum computers. There are different types of PQC techniques such as code based, lattice based, and hash-based cryptography [29].

E. **Robust Protocols for Secure Quantum Cryptography**

A new protocol for secure quantum cryptography is being discussed in [30]. By creating an unbreakable secret key, quantum key distribution enables secure communication between two parties. Further, in case of attack on this system leads to a noise signature which can be detected by a proper protocol. Thus, required protocols should be developed to observe attacks on communication through quantum cryptography. In this work [30], a QKD protocol is considered as the list of decoders and encoders. Encoders choose a quantum state to be sent depending on the key bit choice by the first party. These values (probability of execution, state choice) can be improved by genetic algorithms. Decoders being controlled by the second party get a quantum state by the first party. It performs measurement on the received state. Finally, decoders and encoders share a message which is classical to each other. If the required message matches at both ends, it shows that no attack has happened. If it does not match, then everything sent or received is discarded. This is the proposed protocol in [30] trying to overcome limitations of attack on secure quantum cryptography. Many problems and limitations persist like practical devices and effects of finite key and requires more advanced simulations of protocols to check goodness.

F. **Implementing Quantum Cryptography**

High-speed quantum key distribution is used by Harvard, Boston University, and BBN to provide network security on the first quantum network in history, DARPA [12]. This network is tested against a variety of extremely complex eavesdropping attacks. As suggested, this network can be installed using telecom fibers in metro areas. Users at the universities can access a fiber optic loop or system through this network, which is protected by modern quantum cryptography [12]. The model of this network DARPA is a cryptography based virtual private network. Conventional virtual private networks use symmetric cryptography and public keys for achieving authentication and security. Public key mechanism and symmetric mechanism together provide authentication and confidentiality. In the proposed DARPA network, the virtual private network key agreement strategy is completely updated by quantum cryptography keys [12].

G. **Quantum cryptography approach CNN models used in smart cities**

The above paper introduces a new method in the case of smart cities by taking advantage of the benefits of quantum cryptography together with CNN. In smart cities, most applications require secure communications; therefore, this meets that demand in the form of applications with increased security. The proposed method encrypts data using CNN after the key has been obtained by using QKD for secure key exchange. Data transmission in smart cities is made even more secure by this CNN-based encryption. Simulations show that when compared to conventional techniques, CNN-based encryption greatly improves communication's security. For large-scale smart city applications, the method is thought to be effective and scalable, providing a viable option for secure communication.

A hybrid deep learning model for traffic management in smart cities was presented using CNN combined with BLSTME [31]. The proposed model efficiently predicts the traffic congestion, minimizes holdups, and monitors passenger transportation. The suggested model has the potential to provide effective traffic management in smart cities, as it performs better than current models in terms of accuracy, precision, and recall when it comes to collision prediction. More advancements and the hybrid incorporation of predictors are planned for the future in order to improve network performance during collision anticipation.

H. **Reducing Information Leakage for Quantum Key Distribution**

The study addresses the issues of maximizing transmission distance and enhancing the secure key rate (SKR) in QKD to provide unconditionally secure communication. The goal is to minimize information leakage in QKD, specifically by looking at the areas where the quantum and post-processing phases overlap. The authors suggest a novel method that does away with the necessity of repeatedly considering data from multi-photon pulses during the quantum and post-processing phases [32]. This is based on the quantum phase of photon-number-splitting attack, Eve already possesses complete knowledge of multi-photon pulse data. The study offers formulas for estimating information leakage in two QKD protocols: sending-or-not-sending twin-field and decoy-BB84. It also shows, via simulations, that their method greatly increases maximal transmission distance and SKR in real-world experimental settings.

In conclusion, the authors summarize their approach, emphasizing its potential to enhance SKR by addressing the overlap in information leakage between the quantum and post-processing stages [32]. The efficiency of the suggested method is confirmed by simulation results for particular QKD protocols, and the paper implies that the approach might have consequences for continuous-variable QKD protocols.

7 Recent Progress in Quantum Cryptography

Indian Space Research Organization (ISRO) showcased a successful space quantum key distribution for more than 300 m in March 2021. To achieve this major feat many key technologies were developed like the NAVIC receiver to synchronize the time between the receiver and transmitter modules, and gimbal mechanism systems for optical alignment.

The demonstration included a live video call which was done using quantum encrypted signals. Communication was unconditionally secure and a major achievement in this field. The demonstration was done at the Space Application Centre (SAC) Ahmedabad, on the campus between two buildings which were in line of sight. The experiment was done in the nighttime so that direct sunlight doesn't interfere.

The University of Science and Technology of China showcased an experiment in June 2020 [33] where an open-air demonstration of a new quantum cryptography system was done for the first time which was called MDI-QKD. The communication took place successfully between detectors placed at over 19kms of distance with a measurement station in between them. The experiment was not one hundred percent secure because hacking could still be done by an attacker which could compromise the quantum key generating lasers. But in theory communication was entirely secure.

The qubit is the quantum equivalent of the standard mechanical bit. In ordinary mechanical and classical computing, information donates a string of bits zero or

1-two values in all these. A quantum bit can be in any of the four states: 1, 0, and a linear superposition of both states, since it can be in both o state as well as in 1 state at the same time—in common language, this phenomenon is known by the term SUPERPOSITION.

When we run a computer, we want it to hold its state while we are away from it so that when we come back to it, we can resume our work. The biggest problem with qubits is that they lose their state because of the environment they are kept in. If a computer changes its state over time, it's nothing but useless. Superconducting transmon qubits, which are made of superconducting materials such as niobium and aluminum over a silicon base, are the physical qubits used in IBM Quantum's quantum computers. For this qubit material to exactly work as a qubit as we want them to in our computers, we must have them at super low temperatures in order of 15 millikelvins. This is done to minimize the ambient noise that might excite the superconducting material to change the state of the qubit.

Now there is an ever-increasing competition between leading giants for quantum computing like google and IBM where both the parties want to dominate the field before the other and are constantly coming up with newer technologies to outperform each other. In an article in September 2020, IBM announced their road map in which they aim to have a supercomputer containing about 1000 qubits by 2023. IBM's largest quantum computer contained 65 qubits. This also worked as a reply to google who proclaimed quantum supremacy by making their own 53 qubit computer a year ago [34]. Quantum supremacy is a state where a computer is developed that defeats the fastest supercomputer in terms of computations. IBM then challenged this claim of google as the experiment when again conducted by IBM on supercomputers using a different method, did it faster than the Google's quantum computer 'Sycamore' and was much more efficient with it [35]. But with the 53 qubit Sycamore google could perform tasks in fast polynomial tasks, tasks such as random circuit sampling which takes long polynomial time by the best supercomputers out there owned by IBM. But with the new road map of IBM, we can surely say that this competition between computing giants is far from over. And there is a huge scope of development in quantum computing as well as quantum cryptography.

In addition, specific quantum innovation hubs or accelerators can be set up that integrate academia, industry experts, and startups, where extensive prototyping and commercialization of quantum cryptographic technologies can be done. Such innovation centers could provide more internships and fellowships for students and researchers who wish to get hands-on experience while working on real projects. This would also create collaborative platforming in workshops, hackathons, and conferences on quantum cryptography that would allow the flow of knowledge and ideas across sectors. Finally, well-defined pathways for technology transfer and commercialization, including easy licensing agreements and public–private partnerships, would serve to ensure that breakthroughs in academic research can be rapidly and efficiently brought to market to drive mass market adoption of quantum cryptographic technologies. This would help fully exploit the potential that industry-academia collaboration could have in accelerating the adoption of quantum cryptographic technologies.

8 Merits, Challenges and Limitations of Quantum Cryptography

Merits:

- Since data encoded in a quantum state cannot be copied, QKD offers an efficient way of detecting eavesdropping: any attempt at determining the data encoded in the quantum state alters it, hence signaling a potential breach.
- The quick adoption of quantum cryptography systems for protecting trade and governmental details is a result of their ongoing performance improvements.
- By utilizing the fundamental principles of computing technologies and algorithms, quantum cryptography transforms secure communication and offers a strong basis for security.

Limitations:

- The reason quantum communication is so expensive is that it needs a dedicated channel between the source and the destination. Quantum channels do not allow for multiplexing, a common technique in classical communication; thus, different channels for each destination are necessary.
- Through a guided medium, the maximum distance that QKD supports is roughly 250 km at 16 bps.
- A variety of external factors may cause polarization changes in photons during transmission through the channel, such as air or optical fiber.
- The absence of fundamental functions such as certified mail and digital signatures limits the use of quantum cryptography in specific situations.

Challenges:

- Trust and Maturity of Algorithm: Quantum cryptography algorithms are still quite new and untested; hence, people doubt their security and practical viability [36].
- Higher Costs and Infrastructure Requirements: Most importantly, quantum cryptography requires expensive new infrastructure because the existing networks cannot support the data rate and low transmission loss which this cryptography requires [36].
- Performance and Transmission Problems: Long-distance quantum state transfer is quite challenging to achieve with low-intensity light sources due to high computational requirements, which boil down to slow speed [36].
- Key sizes and Interoperability: Larger keys require more storage and bandwidth, making their integration painful within current systems and increasing interoperability and upgrade costs [36].
- Adoption and Development Uncertainty: Uncertainty over the exact timing of the quantum threat makes the question of 'when' to transition very problematic, while the constant need for innovation within secure protocols further adds to complexity [36].

8.1 Inconsistency in Policy Concern and Technicality

Quantum computing does threaten, for instance, systems like RSA and smaller-key AES; however, it does not make encryption obsolete. The 256-bit AES is still secure, and enhancements will make it even stronger. In fact, post-quantum cryptography is working on quantum-safe schemes, and it is expected that NIST will standardize these in 2024. The NSA's CNSA Encryption Suite also provides strong security against quantum and classical threats [37].

Quantum Key Distribution is said to offer advantages over classical cryptography by detecting eavesdropping, but this is overstated since eavesdropping isn't a fundamental problem it solves. Both QKD and classical methods prevent an eavesdropper from extracting useful plaintext without the decryption key and thus are equally secure.

Quantum computing will not be able to break secure communications after post-quantum encryption schemes are standardized. Secure communications will continue to protect against advanced computers. Adversaries will not gain increased security by using quantum cryptography. The "race for quantum supremacy" has more to do with economic dominance than with national security and is, therefore, less urgent [37].

9 Future Research Directions

Future quantum technology research directions listed below are expected to drive critical development across many industries such as secure telecommunications, consumer electronics, autonomous vehicles, and commercialization of quantum solutions and many more.

Telecommunication: Quantum key distribution is going to make data security in telecommunications much more distinct. QKD will, over time, become the standard for secure communication, with trials by BT, Equinix, Verizon, and in partnership with organizations such as SK Telecom and ID Quantique. Later, this technology will find broad applications in telecommunication networks for protecting private information against dynamically changing cyber threats [38].

Consumer Electronics: The inculcation of Quantum Random Number Generation (QRNG) into consumer electronics, like Samsung's Galaxy Quantum series, introduces a huge leap into personal security. The most trivial digital activities would be much safer once quantum chips become more available, especially in mobile banking applications. It is very likely that consumer electronics, including smartphones, will adopt quantum technology-based features of security in the future [39].

Automotive Industry: The collaboration between IonQ and Hyundai Motor Company shows well the role of quantum machine learning in the future of mobility.

Quantum technologies would result in autonomy in the cars, ensuring higher accuracy of image classification and 3D object detection in real-time, hence more safe and reliable autonomous automobiles.

Commercialization of Quantum Technologies: In this respect, agreements between KIST and SDT marked the move into commercialization to realize quantum technologies. We will get to see QRNG, and QKD quantum-based security solutions, which shortly will become very common in highly secure areas of sectors about finance, government, or critical infrastructure. This kind of commercialization encourages further innovation and integration of quantum technologies into various industries [40].

Some of the research questions that could be addressed in Quantum Computing.

Q1: How can post-quantum cryptographic techniques be integrated into QKD systems to enhance their resilience against future quantum computing threats?

Q2: How can existing QKD protocols be optimized for better key distribution efficiency and security?

Q3: What are the challenges and solutions for the seamless integration of QKD into existing communication infrastructures?

Q4: What are the strategies for implementing QKD over long-distance communication or diverse network links, such as undersea cables or Mesh networks and Star networks?

10 Conclusion

One unique and important characteristic of quantum cryptography is its ability to identify whether a third party is tapping in on a conversation between two users. Network security could be completely transformed by quantum cryptography's security foundation, which is based in quantum mechanics. Through the combination of traditional internet technology and Quantum Key Distribution (QKD) methods, a highly secure communication framework can be established for real-world applications. Even though quantum cryptography has advanced significantly in the last ten years, there are still significant obstacles to be solved before key distribution in business, government, and academic settings can widely use it. These difficulties include the requirement for more sophisticated hardware to enable quantum key exchange with higher quality and longer transmission distances. The work of research and development in quantum cryptography will continue to be driven by ever-enhanced computer processing power and the continued threat to the already existing cryptographic systems. As a matter of fact, technology has immense potential for tremendous enhancement in security in the public, commercial, and academic domains.

References

1. W. Buchanan, A. Woodward, Will quantum computers be the end of public key encryption? J. Cyber Secur. Technol. **1**(1), 1–22 (2016)
2. P. Lakshmi, G. Murali, Comparison of classical and quantum cryptography using QKD simulator. in *International Conference on Energy, Communication, Data Analytics and Soft Computing (ICECDS-2017)*, (2017), 978-1-5386-1887-5/17
3. A. Porzio, Quantum cryptography: Approaching communication security from a quantum perspective. in *2014 Fotonica AEIT Italian Conference on Photonics Technologies*, Naples, Italy, 2014, pp. 1–4. https://doi.org/10.1109/Fotonica.2014.6843831
4. P.W. Shor, Algorithms for quantum computation: discrete logarithms and factoring. in *Proceedings of the 35th Annual Symposium on Foundations of Computer Science, ser. SFCS'94*. Washington, DC, USA: IEEE Computer Society, (1994), pp. 124–134
5. V. Scarani, A. Acin, G. Ribordy, N. Gisin, Quantum cryptography protocols robust against photon number splitting attacks for weak laser pulse implementations. Phys. Rew. Lett. **92**, 057901 (2004)
6. R.K. Sinha, M. Mishra, S.S. Sahu, Quantum key distribution: Simulation of BB84 protocol in C. Int. J. Electron. Electr. Comput. Syst. (IJEECS) **6**(1), 57–60 (2017). ISSN 2348-117X
7. A. Young, The future of cryptography: Practice and theory. IEEE IT Prof. J. 62–64 (2003)
8. A. Sharma, D. Sharma, Big data protection via neural and quantum cryptography. in *2016 3rd International Conference on Computing for Sustainable Global Development (INDIACom)*, (New Delhi, India, 2016), pp. 3701–3704
9. N. Ul Ain, A novel approach for secure multi-party secret sharing scheme via quantum cryptography. in *2017 International Conference on Communication, Computing and Digital Systems (C-CODE)*, (2017), pp. 112–116. https://doi.org/10.1109/C-CODE.2017.7918912
10. C. Pacher, A. Abidin, T. Lorünser, M. Peev, R. Ursin, A. Zeilinger, J.Å. Larsson, Attacks on quantum key distribution protocols that employ non-ITS authentication. Quantum Inf. Process. **15**, 327–362 (2016)
11. S. Liu, J. Sullivan, J. Ormaner, A practical approach to enterprise IT security. IEEE IT Prof. J. **9**(3), 35–42 (2001)
12. M.S. Sharbaf, Quantum cryptography: A new generation of information technology security system. in *2009 Sixth International Conference on Information Technology: New Generations*, (2009), pp. 1644–1648. https://doi.org/10.1109/ITNG.2009.173
13. S. Choe, Quantum computing overview: discrete vs. continuous variable models (2022), Preprint at arXiv:2206.07246
14. N.A. Muhamad, Z.A. Zukarnain, Implemetation of BB84 quantum key distribution protocol's with attacks. Eur. J. Sci. Res. **32**(4), 460–466 (2009)
15. I.B. Djordjevic, Hybrid QKD protocol outperforming both DV- and CV-QKD protocols. IEEE Photonics J. **12**(1), 1–8, Art no. 7600108 (2020). https://doi.org/10.1109/JPHOT.2019.2946910
16. S. Dhar, A. Khare, A.D. Dwivedi, R. Singh, Securing IoT devices: A novel approach using blockchain and quantum cryptography. Internet Things 101019 (2023)
17. G. Murali, R.S. Prasad, Comparison of cryptographic algorithms in cloud and local environment using quantum cryptography. in *2017 International Conference on Energy, Communication, Data Analytics and Soft Computing (ICECDS)*, (2017), pp. 3749–3752, https://doi.org/10.1109/ICECDS.2017.8390165
18. M. Swathi, B. Rudra, A novel approach for asymmetric quantum error correction with syndrome measurement. IEEE Access **10**, 44669–44676 (2022)
19. M. Campagna, C. Xing, Quantum safe cryptography and security: An introduction, benefits, enablers and challenges. ETSI Tech. Rep. **8** (2015)
20. S. Bone, M. Castro, A brief history of quantum computing. Imperial College in London (1997). http://www.doc.ic.ac.uk/~nd/surprise_97/journal/vol4/spb3
21. U. Vazirani, On the power of quantum computation. Philos. Trans. R. Soc. Lond. Math. Phys. Eng. Sci. **356**(1743), 1759–1768 (1998)

22. L. Grover, A fast quantum mechanical algorithm for database search. Tech. Rep. Bell Labs, New Jersey (1996)
23. D.J. Bernstein, Introduction to post-quantum cryptography. in *Post-quantum cryptography*, (Berlin, Heidelberg: Springer Berlin Heidelberg, 2009), pp. 1–14
24. C.N.S.S. Policy No. 15, Fact Sheet No. 1, National Policy on the Use of the Advanced Encryption Standard (AES) to Protect National Security Systems and National Security Information (2003)
25. Z. Kirsch, M. Chow, Quantum computing: The risk to existing encryption methods (2015). http://www.cs.tufts.edu/comp/116/archive/fall2015/zkirsch.pdf
26. J. Proos, C. Zalka, Shor's discrete logarithm quantum algorithm for elliptic curves (2003), Preprint at quant-ph/0301141
27. G. Brassard, P. Høyer, A. Tapp, Quantum cryptanalysis of hash and claw-free functions. in *LATIN'98: Theoretical Informatics: Third Latin American Symposium Campinas, Brazil, April 20–24 1998*. Proceedings 3, pp. 163–169. Springer Berlin Heidelberg (1998)
28. A. Conrad, S. Isaac, R. Cochran, D. Sanchez-Rosales, B. Wilens, A. Gutha, P. Kwiat, Drone-based quantum key distribution (QKD). in *Free-space laser communications XXXIII*, vol. 11678, pp. 177–184. SPIE (2021)
29. T.C. Clancy, R.W. McGwier, L. Chen, Post-quantum cryptography and 5G security: tutorial. in *Proceedings of the 12th Conference on Security and Privacy in Wireless and Mobile Networks (WiSec'19)*. Association for Computing Machinery, New York, NY, USA, (2019), p. 285. https://doi.org/10.1145/3317549.3324882
30. W.O. Krawec, S.A. Markelon, Discovery of robust protocols for secure quantum cryptography. in *Proceedings of the Genetic and Evolutionary Computation Conference Companion (GECCO'19)*. Association for Computing Machinery, New York, NY, USA, (2019), pp. 379–380. https://doi.org/10.1145/3319619.3321945
31. N.J. Mohammed, Quantum cryptography in convolution neural network approach in smart cities. J. Surv. Fish. Sci. **10**(2S), 2043–2056 (2023)
32. H.K. Mao, Q. Zhao, Y.C. Qiao, B.Z. Yan, B.J. Xu, A.A.A. EL-Latif, Q. Li, A novel approach to reducing information leakage for quantum key distribution (2021). Preprint at arXiv:2104.13580
33. Y. Cao et al., Long-distance free-space measurement-device-independent quantum key distribution, securing a wireless link with quantum physics **125**(26) (2020). https://doi.org/10.1103/PhysRevLett.125.260503
34. F. Arute, K. Arya, R. Babbush et al., Quantum supremacy using a programmable superconducting processor. Nature **574**, 505–510 (2019). https://doi.org/10.1038/s41586-019-1666-5
35. E. Pednault, J.A. Gunnels, G. Nannicini, L. Horesh, R. Wisnieff, Leveraging secondary storage to simulate deep 54-qubit sycamore circuits (2019), Preprint at Quantum Physics
36. A. Horpenyuk, I. Opirskyy, P. Vorobets, Analysis of problems and prospects of implementation of post-quantum cryptographic algorithms. in *CQPC* (2023), pp. 39–49
37. J. Tibbetts, *Quantum computing and cryptography: Analysis, risks, and recommendations for decisionmakers* (No. LLNL-TR-790870). Lawrence Livermore National Lab (LLNL), Livermore, CA (United States, 2019).
38. H. Vella, The race for quantum-resistant cryptography [quantum-cyber security]. Eng. Technol. **17**(1), 56–59 (2022)
39. G.N. Pham, B.A. Nguyen, V.Q. Tran, K.D. Ta, P.H. Nguyen, D.M. Nguyen, Simulation of resources for quantum algorithms and quantum communication protocols based on a novel framework (2020)
40. M.H. Ullah, R. Eskandarpour, H. Zheng, A. Khodaei, Quantum computing for smart grid applications. IET Gener. Transm. Distrib. **16**(21), 4239–4257 (2022)

Enhancing UAV Security Through Quantum Cryptography: Current Strategies and Future Pathways

Hayat Ullah, Mohammad Faisal, and Ikram Ali

1 Introduction

The creation of secure communication methods has become crucial in the modern era, particularly with the growing usage of Unmanned Aerial Vehicles (UAVs) for everything from military operations to citizen surveillance. The foundation of secure communication has long been cryptography [1, 4], an ancient information encoding method. Cryptography [4] is a technique in which unwanted data is hidden from those you don't want to share information or hide something from them. Now in networking, it is a method used to secure information or communication by encoding. It has been used from the almost the beginning of time to secure info from unwanted people in one way or another. The technique of cryptography [4] has changed from one day to another for centuries. Some common are RSA, AES, and DES [9–13]. Although established cryptographic techniques like RSA, AES, and DES have offered a strong foundation for data transmission security, the efficiency of these techniques has been progressively called into question by evolving cyber threats [1–3]. Everyone wants to keep their life as private and secure as possible; for that, the UAVs must obey the ethical side if it has to do surveillance because no one want to sacrifice their privacy [4, 5]. For these purposes, a UAV must be protected from intruders, a UAV must be protected from intruders to hack it to steal data and must be designed to be very secure. For that security, one of the promising sides is quantum, which has the highest guaranty of success because of its quantum mechanics, which cannot be hacked or copied as whole but the intruder or change of data will cautions the people. In the future, UAVs can better serve with AI as security

H. Ullah · M. Faisal (✉)
Department of Computer Science, University of Malakand, Malakand, Pakistan
e-mail: mfaisal@uom.edu.pk

I. Ali
National Engineering Laboratory for Big Data System Computing Technology, Malakand, Pakistan

S. B. Goyal et al. (eds.), *Quantum Computing, Cyber Security and Cryptography*,
https://doi.org/10.1007/978-981-96-4948-8_12

identifying threats, but it will need some time and new research for the outcome that we truly desire [6–8].

UAVs have many applications in todays, from military to simple surveillance. It can help with simple security tasks like identifying threats and patrolling. It can serve as security guards and many more. But it will need huge data to be processed, and it is impossible without the quantum computing, which can guarantee exponential power. Once we get to that exponential power, it will be easier to process data and we will get a lot of new potential and help humanity. UAVs can help a lot with quantum's potential speed.

Nevertheless, there are significant obstacles to integrating quantum cryptography [4] into UAV communication systems. Sophisticated hardware that can handle quantum states and overcome environmental defects and noise is necessary for practical implementation. Furthermore, UAV designs need to consider the special needs of quantum communication, which include using lightweight components with low power consumption. Notwithstanding these difficulties, there is no denying that quantum cryptography [4] has the capacity to bolster the UAV communication systems' dependability and security. You can hide things from someone or hide information from someone you don't want to share with them by using cryptography [4]. These days, networking uses this approach to encrypt data and safeguard communication. Since almost the beginning of time, people have utilized it in one form or another to keep information safe from prying eyes. For millennia, the method of cryptography [4] has undergone constant modifications. RSA, AES, and DES are a few popular ones [9–13]. While well-known cryptography methods such as RSA, AES, and DES have provided a solid basis for data transmission security, new cyber threats have gradually questioned the efficacy of these methods [1–3].

These days, UAVs are used for a wide range of purposes, from basic observation to military use. Simple security duties like patrolling and threat identification can benefit from its assistance. It may perform numerous roles, including security guards. However, processing massive amounts of data will be necessary, and this cannot be done without quantum computing, which can provide exponential power. Reaching that exponential capacity will facilitate data processing, unlock a plethora of new possibilities, and benefit all of humanity. With regard to quantum's potential speed, UAVs can be quite helpful.

2 Literature Review

As unmanned aerial vehicles (UAVs) [1] transform a number of industries; it is critical to make sure that communication routes are safe. Conventional cryptography techniques are insufficient for the changing threat landscape because they are vulnerable to brute-force attacks and advances in computing power. Researchers have focused on quantum secure communication as a solution to these weaknesses since it uses the ideas of quantum physics to attain previously unheard-of degrees of security. The notion of quantum key distribution (QKD) was first established in a fundamental paper by Bennett and Brassard in 1984, which laid the groundwork for quantum secure communication. Based on the ideas of quantum physics, QKD uses the quantum characteristics of photons to create a secure key between communication parties, ensuring total security.

Among the noteworthy developments are free-space quantum communication lines customized for unmanned aerial vehicles. By implementing QKD over a free-space optical channel, Xu et al. (2018) successfully overcame atmospheric turbulence and other obstacles that are common in UAV communication situations. The implementation took place between a base station and a UAV. They accomplished dependable and secure communication by utilizing adaptive optics and error-correction strategies, opening the door for quantum-secured UAV networks [4].

Furthermore, to strengthen data transmission and reception, the integration of quantum encryption techniques with unmanned aerial vehicles has been investigated. Specifically for UAV communication systems, Li et al. (2020) presented a unique quantum encryption technique based on entangled photon pairs. While quantum secure communication has great potential for unmanned aerial vehicles, there are still a number of real-world obstacles and restrictions. The incorporation of quantum hardware into UAV systems that are both lightweight and energy-efficient represents one such problem. Current quantum technologies are bulky, which limits UAV payload capacity [5] and power consumption. Therefore, more research into shrinking and optimization approaches is needed. A growing interest in improving the security of communication systems used in aerial applications is reflected in the significant progress made recently in the integration of quantum cryptography with Unmanned Aerial Vehicles (UAVs). This section summarizes important research and their contributions to the field while reviewing recent developments in this topic [8]. A quantum communication system prototype that uses the BB84 protocol in UAVs was recently demonstrated. The goal of this study was to address the unique difficulties posed by using BB84 in an airborne environment.

The study addressed the problem of air interference, which has the potential to impede quantum signal transmission. Through the utilization of sophisticated error-correction methods and the optimization of the quantum signal's encoding, the researchers were able to exhibit enhanced signal dependability and quality, even when there were atmospheric disturbances [12].

Integration with Conventional Data Lines: Another significant issue addressed was the integration of quantum communication with currently in use data lines. The researchers suggested a hybrid strategy that takes use of the increased security offered by the BB84 protocol to guarantee smooth data transfer by combining quantum key distribution with conventional communication routes. The E91 protocol in military UAV networks, emphasizing its capacity to strengthen cyber threat resistance and safe key distribution [4].

Secure Key Distribution: The research showed how E91, which depends on quantum entanglement, improves key distribution security by almost completely preventing eavesdroppers from intercepting the key without being noticed. This was especially important for military uses where secure communication is essential.

Resistance to Cyber Threats: The researchers also looked at how resilient the E91 protocol was to different types of cyber threats. They demonstrated that E91 offers a higher level of security than traditional approaches by evaluating the protocol's effectiveness in simulated military scenarios. This lowers the possibility of cyberattacks on sensitive data transmitted by UAV.

Furthermore, the significance of continuing research in quantum-resistant cryptography and quantum error-correction codes is highlighted by the susceptibility of quantum systems to specific attacks, such as side-channel attacks and quantum hacking. Resolving these issues is essential to achieving quantum secure communication for UAVs' full potential in practical applications [5].

3 Methodology

For Unmanned Aerial Vehicles (UAVs), quantum secure communication offers a promising path to improve data transmission security in contemporary aerial networks. This work presents a thorough implementation approach for quantum secure communication protocols designed with unmanned aerial vehicles in mind. By utilizing the special qualities of quantum mechanics, like quantum entanglement and quantum key distribution, this approach seeks to create extremely secure communication channels that are impervious to eavesdropping and interception. The suggested method combines theoretical ideas with real-world concerns to handle the particular difficulties presented by the UAV environment, such as constrained computational capacity and fluctuating network conditions. This technique offers a road map for developing, implementing, and assessing quantum secure communication systems for unmanned aerial vehicle (UAV) applications through a methodical analysis of crucial elements and operational protocols.

The remainder of this chapter discusses the following issues of the topic of this chapter:

1. Importance of Secure Communication in UAV
2. Quantum Cryptography [4] as a Possible Solution
3. Real-World Applications
4. Application to Industry
5. E91 Protocol
6. Introduction to Quantum Cryptography [BB84 protocol [4]
7. UAV Communication Basics and Design
8. Challenges in UAV Communications
9. Classical Solutions
10. Modern Solutions
11. Integration of QKD in UAV Communication Systems
12. Problems and Solutions in UAV cryptography [4]
13. Attacks on Quantum-Based UAVs
14. Cutting-Edge Technologies

15. Classical vs Quantum Cryptography [4] Analysis
16. Ethical Considerations and Privacy
17. International Privacy Standards
18. Conclusion and Future Directions

4 Importance of Secure Communication in UAV

Why UAVs as we all can see how human life is important to save it? The UAVs were introduced to use them in those place places or circumstances where humane life can be saved. UAVs are using mostly in critical sides in military and for that the security and integrity of UAVs are more important than most of the things [14–19]. As sometime, they can be life-sensitive critical. If the communication is not secure, then it can be compromised and can caused a lot of damage to a lot of people. Keeping the integrity in UAVs is essential as example: what if you send a drone to enemy territory and they hacked it and manipulate it and it come back as a spy what you think will happen? Yes, simple you are doomed.

From above example, it is clear that its integrity and security is as important as a baby life for future [20–23].

5 Quantum Cryptography [4] as a Possible Solution

Quantum cryptography [1, 4] is the next big revolution in the world for securing communication and is known for fast processing through principles of physics. Quantum cryptography [4] utilizes the inherent properties of quantum particles [8] to achieve unbreakable encryption and secure key distribution. Quantum cryptography [4] is the encryption with quantum mechanics properties to secure communication or transfer of data. Now it can be an advancement and a future to see because quantum cryptography [2, 4] cannot be hacked. So, it is a possible solution for secure communication in UAVs [24, 25].

6 Real-World Applications

6.1 Military Surveillance UAVs

The BB84 protocol was deployed in military surveillance UAVs as part of a pilot project carried out to improve secure communication channels. As part of the project, ground-based quantum communication stations and a quantum communication module installed on UAVs were established. Results showed improved mission security and a notable decrease in the chance of data interception. Among the lessons learnt are the significance of tackling signal attenuation issues and optimizing quantum communication gear for UAV situations [3].

6.2 *Disaster Response UAVs*

A case study explored the use of quantum cryptography in disaster response UAVs equipped with E91 protocol. The project aimed to secure communications between UAVs and ground stations during emergency operations. The implementation highlighted the effectiveness of quantum entanglement in maintaining secure communication channels despite environmental challenges. Key takeaways include the need for robust entanglement generation and detection methods suitable for the operational conditions of disaster response scenarios. As human is prone to errors and mistakes, these UAVs can work much better than human [4].

7 Application to Industry

7.1 *Military Operations*

For military UAVs, quantum cryptography offers a high degree of security, shielding critical communications and intelligence from possible cyberattacks. The incorporation of QKD protocols, such as BB84 and E91, guarantees that sophisticated attackers are unable to compromise the encrypted communication channels [5].

7.2 *Logistics and Supply Chain*

UAVs are employed in logistics to move expensive cargo. Quantum cryptography can be used to encrypt communication between logistics centers and unmanned aerial vehicles (UAVs), preventing data breaches and guaranteeing the accuracy of cargo information [8].

7.3 *Surveillance*

For surveillance UAVs used in border security, quantum cryptography enhances the protection of surveillance data. By securing communication channels, quantum cryptography ensures that sensitive information remains confidential and tamper-proof. One of the most important technologies that can help in surveillance is the micro or mini drones that are just the size of bugs it can really help in this field [6].

7.4 BB84 Protocol

Charles Bennett and Gilles Brassard invented BB84, the first and most used QKD protocol, in 1984. It enables the safe exchange of a cryptographic key between two parties (sometimes referred to as Alice and Bob), even when Eve is present as an eavesdropper [6].

7.5 States of Quantum Optics

BB84 makes use of the no-cloning theorem and the Heisenberg uncertainty principle, two fundamental ideas of quantum physics [6].

Alice uses two bases to prepare qubits in one of the four possible states:

Z-basis, or rectilinear: $|0\rangle$ and $|1\rangle$

Diagonal, also known as the X-basis: (equal superposition with a phase shift) and $|+\rangle$ (equal superposition of $|0\rangle$ and $|1\rangle$).

For every bit in the key, Alice selects a basis (Z or X) at random and sends Bob the matching quantum state [7].

7.6 Measurement

Bob selects a basis (Z or X) at random for measuring each qubit that is received.

Bob will correctly calculate the bit value if his measurement basis matches Alice's preparation basis because of the uncertainty principle. Bob's outcome will be random if the bases are not the same [6].

7.7 Key Shifting

Alice and Bob openly compare the bases they used without disclosing the actual bit values once Bob receives and measures the qubits.

They remove the bits in which their bases diverge, leaving a shorter sequence of bits in which they are in agreement. The "sifted key" is the name given to this string [5].

Challenges

1. **Photon Loss and Detection**: Key bits can be significantly lost in real-world implementations due to detector inefficiencies and photon loss during transmission. One of the biggest challenges is dealing with these losses while maintaining high key rates [5].
2. **Quantum Bit Error Rate (QBER)**: Transmission errors may result from defective sources, detectors, or outside interference. Although noise may be the cause of a high QBER and result in false alarms, it can also be an indication of the existence of an eavesdropper [5].

7.8 E91 Protocol

Quantum entanglement is the foundation of the E91 protocol, which was put forth by Artur Ekert in 1991 and provides secure key distribution. It is an application of Bell's theorem and is strongly related to the Einstein–Podolsky–Rosen (EPR) paradox [5].

7.9 Entangled Pairs

In E91, a source produces entangled pairs of qubits, and Alice and Bob receive one qubit each pair [4].

Generally, the entangled state is a Bell state, like $|\Phi+\rangle = (|00\rangle + |11\rangle)/\sqrt{2}$.

7.9.1 Measurement

Bob and Alice each measure their qubits independently using one of three randomly selected bases. In order to guarantee that the correlation between the measurement outcomes depends on the choice of bases, the measurement basis must be selected, and this decision is done after the qubits are in transit [3].

7.9.2 Bell's Inequality

Alice and Bob are able to compare their measurement settings and outcomes in public after taking numerous measurements. After that, they can calculate a number associated with Bell's inequality. If this inequality is broken, it verifies that the qubits were entangled in the first place, and any attempt to listen in would have caused measurable disruptions to the system [2].

7.9.3 Importance

Cutting-edge QKD systems like BB84 and E91 use the special abilities of quantum physics to provide secure communication. Both, nevertheless, have serious technical issues that must be resolved before they can be widely used. Although E91 relies on entanglement and Bell's theorem, it may be able to provide even greater security assurances than BB84, which is more straightforward and commonly used. The advancement of quantum technologies, including robust entanglement distribution techniques and quantum repeaters, is crucial to surmount the existing constraints and render these protocols feasible for high-security, long-range applications [1].

7.9.4 Practical Challenges in Implementation

Quantum Noise and Signal Attenuation: Deal with the issues that arise in aerial contexts related to quantum noise and signal attenuation, as these could potentially affect the dependability of QKD algorithms. Integration with Conventional Communication Systems: Talk about the challenges and potential solutions for integrating quantum communication systems with the current conventional systems utilized in unmanned aerial vehicles. In today's world, it is difficult, but tomorrow, it will be one the most important technology in the world as world proceed forward, we can need new innovation and, for this, it is one the most important promising technology of the future world.

8 Introduction to Quantum Cryptography [4]

It is the process, technique, or art in which data or information is encoded so that only intended people can use it. Quantum cryptography [4] uses the quantum mechanics laws [9] instead of mathematics. It is based on basically two factors.

Heisenberg uncertainty principle photon polarization: Heisenberg uncertainty principle says that some physical properties are related in a way that measuring one property may prevent knowing the other property concurrently.

When an unpolarized light enters a vertically aligned filter, it absorbs some of the light and polarizes the rest in the vertical direction. Photon polarization is the process by which a subsequent filter slanted at an angle q absorbs part of the polarized light and transmits the remaining portion, giving it a new polarization [17].

Quantum entanglement is a phenomenon where two particles become interconnected such that the state of one particle instantaneously affects the state of the other, no matter how far apart they are. This unique property is used in quantum cryptography to create secure communication channels.

8.1 Qubit

It can be known as quantum bit. It is the basic unit of information [1] in quantum computing. This works as a binary bit of memory for storing data or information but its acts on quantum mechanics.one main point in qubit is that it can be 0 or 1 concurrently [22].

9 Quantum Superposition

In quantum mechanics, a system can exist in multiple states simultaneously until it is measured.

Example: Unlike classical bits, a quantum bit (qubit) can be both 0 and 1 concurrently until observed [19].

9.1 Entanglement

It can be known as the property "No matter how far away two quantum particles are, their states instantly affect one another due to their coupling" [4]. For example, if two entangled electrons have opposite spins, you can measure one to find the other's spin.

In some QKD processes, entangled particles allow for a significant correlation between the particle states. This characteristic aids in identifying any attempt to tampering the quantum states [13].

9.2 Quantum Uncertainty

Quantum mechanics principles [2] introduce uncertainty, making it challenging for an eavesdropper to gain information about transmitted quantum states without detection [13, 14].

9.3 No-Cloning Theorem

This theorem asserts that an unpredictability unknown state of quantum cannot be copied perfectly, providing a means to detect any attempts to intercept and clone quantum states [18, 19]. A significant benefit of quantum cryptography [4] is that it's security on all the above factors, which are basics of quantum physics. The main advantage of quantum is that it can detect an eavesdropper easily as a piece of cake [20, 21].

9.4 UAV Communication Basics and Design

The using of UAVs has been increasing from day to day in almost a lot of industries, from military to simple delivery package systems and what's make them ideal is that it can cause no harm to human life if failed. Thus, efficiency, integrity, and security are key factors for the UAVs. UAVs are aerial vehicles ranging from small ones used in homes as toys to huge aircraft used for surveillance. These platforms typically consist of several key components, including the airframe, propulsion system, avionics, and communication subsystems [1, 3, 4].

The communication system of UAV has three main components:

- UAV
- Ground control station (GCS)
- Communication link

9.5 UAV

It is also known as the unmanned vehicle. An aircraft that operates without a human pilot is known as a UAV. Sometimes, it also called drone.

9.6 GCS

It is the base station from which the UAV is operated. From GCS, the UAVs are monitored and sometimes even piloted.

9.7 Communication Link

It serves as a conduit between the GCS and UAV through which data is communicated from GCS to UAVs

Example: radio signal, satellite communication, etc.

Transmitter is responsible for sending and receiving data it has the following components:

- Antennas
- RF modules

- Signal processors
- Amplifiers

9.8 Communication Protocols

UAVs have some set of protocols which is obeyed for the transmission of data. Example: MAVLINK, also known as micro air vehicle communication protocol.

9.9 Signal Processor

UAVs have a processor for data. A data processor to handle its security and encryption to safeguard the data and keep its integrity and confidentiality.

9.10 Fail Safe

UAVs also have a failsafe design for the control if any communication lost or in any critical situation.

9.11 Challenges in UAVs Communications

As UAVs are still developing day to day, it has a lot of vulnerabilities and challenges. Big two are processing speed jamming which can be hacked. These are basic problems and require unique but highly tentative solutions and the one we have in hand is quantum which can solve a lot of problems in UAVs communication.

9.12 Classical Solutions

The security of UAVs mostly depends on the classical techniques of encryption, e.g., protocols hashing, but with time, new threads arise and the conventional techniques are no longer safe but are easy to crack for hackers. They are not sufficient with the current time and are vulnerable to hackers, so new solutions are required.

9.13 Quantum Solutions

One of the best solutions for UAVs communication security employs quantum cryptography [4] as it is based on quantum physics instead of math it can't be changed or copied and also gives the leverage of no eavesdropping, which is crucial for security as, in classical, there is certain chance of an intruder or data alternation or even hijacking the whole UAV.

One of the possible solutions is to use QKD (Quantum Key Distribution). It offers much more security than classical key distribution and speed than classical key distribution as it is based on quantum principles, [7] and it can be done with the following steps:

- The sender sends polarized photos to the receiver with rectilinear or diagonal bases.
- The receiver then checks the direction of photons with random bases.
- Through public channels the receiver then informs the sender about his bases.
- The sender then sends the correct bits or the same bases.

Now, with the above process, a unique and secure key is shared for communication and if it's hacked, the sender will know, thus giving the reliance of security what is needed for our solution.

Imagine if we achieved this level of security what we can do with this imagine the speed the support. Not only were we able to communicate with the drones but also can communicate with one another and make fast decision-making processing. It can lead to a substantial number of applications for UAVs, so we just need to give it a little time.

Quantum mechanics: Quantum mechanics is the basis for quantum cryptography [4]. First, we need to understand how quantum mechanics works. Basically, "Quantum mechanics is the study of particles in the least or smallest way possible [4]."

Quantum states: They are simply referred to as the spin, momentum, and location of particles. Moreover, they can exist in numerous states in quantum mechanics, a phenomenon called superposition. The best idea in quantum cryptography is quantum superposition [4], which links two bodies simultaneously. Imagine if you are not here and know everything here it seems too impossible, but some believe the whole universe depends on it [4, 8, 13].

9.14 Quantum Computing

Quantum computation is known as the computation using quantum mechanics principles [5, 10] or simple by qubits or quantum bits. They are faster than classical computing and thus can solve many problems very easily with exponential power as it can crack the RSA encryption in few seconds which can be cracked by classical in millions to billions of years. This is nearly impossible to wait so quantum gives much more power security and reliance which is needed for the upcoming future as with time new threads arrive and need new solutions for them, and this is a big step forward [7, 10, 21].

9.15 Quantum Key Distribution (QKD) [2]

One of the primary security techniques in quantum cryptography [4] is quantum key distribution, which involves sending a secure key from sender to recipient. In this principal, quantum mechanics is used instead of classical math. It uses the properties of uncertainty and no-cloning theorem to safely share the key if there is an intruder. It will be caught and detected because of these properties [7, 10, 21].

9.16 Protocols: There Are a Lot of Protocols Presented Such as BB84 and E91

They have their own advantages and disadvantages but the most common is BB84, presented by Bennett and Brassard in 1984. It is done through horizontal and vertical bases. It is one of the most used protocols of quantum cryptography [4]. It is useful in preventing the intruder from getting the key or eavesdropping [14, 15].

9.17 Integrating QKD in UAVs Communication Systems

Integrating QKD in UAVs is a big step forward as it can enhance security with a lot of processing speed, which is useful for decision-making. But it comes with a lot of challenges as implementation is not that easy and not a piece of cake as it will need the essential hardware capable of handling such adaptations and synchronization with everything.

It will need experts in the following skill:

- Quantum physics
- Aerospace engineering
- Cybersecurity

Along these, there will also be the need of many more expertise to simply work with this.

Its look hard but once it achieved it can be very useful imagine having a UAV which cannot be hacked and it just do the work for you and it's

- Smarter

- Faster
- Better

Thus, with these it can be a dream for many and just the required thing.

9.18 Problems and Solutions in UAV Cryptography [4]

Implementation of quantum in UAVs is not that easy and can have a lot of challenges from practical side. Some of them are given below:

9.19 Practical QKD Implementation

The application of quantum [2] in real life is not an easy task with latest inventions the technology able to some extent to implement some part of it but it will need more time. Some of them are given below:

- **Hardware capability**
- **Noise reduction and error correction from environment**
- **Advance photonic devices**

With advancement in technology, many can be achieved up to some extent but it will still need a lot of future research to be able to solve the current problems.

9.20 Hardware Capability

As today's hardware are not as robust as required and the main problem is the limitation of the computation power, we have which can't handle much of the speed of communication and thus can be a problem in decision-making. Example:

Imagine we are facing a situation in which you have to act in slit seconds and it goes to processing and decides what to do and a life is lost, then what is benefit of this security if it's not fast so we need security but with fast hardware which can be able to process all the processes it can be given [18, 19].

9.21 Noise Reduction and Error Correction from the Environment

In quantum communication, noise in the environment can lead to the loss of photons, decoherence and, thus the communication can malfunction so this noise could lead to errors, and this error can lead to disasters. So, the atmospheric factors should be kept in mind.

9.22 Advanced Photonic Devices

Recently, some inventions can smit photons more efficiently, but it is not enough and more research is needed for the problem.

In the above problems, we see that it hard to develop in today's technology, but with some time, it will be achieved. One concept is a quantum repeater for long-distance communication will play crucial role in solving the photon loss and decoherence problem.

In developing such UAVs, the following factors should be preamble:

- Resource allocation
- Network management
- Fast processing hardware

Without these and proper management, it will be hard to implement such UAVs with this much potential, which can lead to a new growth of technology.

9.23 Integration of QKD in Quantum Cryptography [4]

Integration in QKD is the overcoming of various problems in UAVs communication systems, but it is not that simple and would require several studies, capabilities and skills with sophisticated knowledge.

Hardware for QKD: In QKD, special hardware is required to detect, generate, and change quantum states.

The following are the main requirements for QKD hardware:

- Photon source
- Photon detector
- Optical fibers managing hardware
- Quantum number generator

With these hardware, it is essential for the UAV to be lightweight, power-sufficient, and fast enough to synchronize with all of the above. Without one, all other will be useless and will not work; thus, it is essential to make a model and keep all the aspects in mind.

9.24 Integration Challenges in UAV Design

In designing the UAV, every aspect must be in mind to keep the UAV in balance. Some of these aspects are as follows:

- Size
- Weight
- Power and aerial constraints
- Compact and robust design

With the above challenges as there are requirements for quantum, these aspects should be kept in mind to achieve the outcome we want. These are essential aspects for atmospheric and aerial flights in UAVs. Mostly, its design is made to address these challenges. Some drones can fly in hilly and challenging areas, while some can go in extreme weather condition there with quantum requirements; the design should be kept in mind and not disturbed by it.

Example: Imagine you have one of the best security drones that cannot be hacked but cannot fly. Then what is the reason why it is even called a UAV? Therefore, these aspects are for proper UAVs.

9.25 Testing of Quantum UAV Systems

If these are even implemented, then they should be tested and evaluated by experts to see if they can do what they are made for. They are critical states to these new innovations. Testing is the basic step, and new progress cannot be achieved without it.

The following aspects should be checked:

- Reliability
- Security
- Performance

These are the basics on which the whole quantum idea is based. These are the preamble to be achieved. Different aspects can be achieved with different methods through different trials and tests. The more we test it, the better it becomes. It should be exposed to be checked to see the security of the UAV communication system. It should be checked in field trials and in extreme environment for which it is created for. Also, it design should be tested to see if the performance of the flight and decision is affected or not.

It all depends on testing. More testing can provide a better and refined product.

9.26 Threats in UAV Communication

UAV communication has substantial quantity of vulnerabilities. This may result in a lot of risks like failure in work, hijacking, spying, and many more, which can lead to failure of the mission or the task for which it is called.

Some of them are given below:

- Eavesdropping
- Jamming
- Spoofing
- Cyberattacks

An attacker may come from any aspect or vulnerable side, such as

- Communication protocol
- Hardware components
- Network Infrastructure

With these, an attacker can come from any aspect, but the damage it can cause is a lot to take or simply the damage is huge. This may result into the following:

- Integrity of the UAV
- Data confidentiality
- Availability

These are the main basics for many things in most organizations. At any cost, these should not be compromised.

With these problems, classical is not safe anymore, but quantum is a game changer and can lead to success up to a big extent. However, there are still some attacks possible on quantum-based UAVs communication systems which should be solved.

9.27 Attacks on Quantum-Based UAVs

As quantum-based UAVs are extremely safe against classical attacks but are still vulnerable to quantum-based attacks, which can still lead to the system's failure. Some of the main are given below:

- Photon number splitting attack
- Trojan horse attacks
- Side channel attack

The quantum UAVs must have all the counter measures for the quantum-based attacks to fully deliver their functions, for it is created once it's achieved, the new UAVs will be impenetrable from any side like imagine you can't do anything with UAV it is just above you that's the target we should be achieving, these attacks should be neutralized at all cost no matter what because without it the classical problem will arise some of the best preventions are including the following in the design of the UAVs [2, 15, 24].

- Quantum resistant algorithm
- Intrusion detection
- Intrusion prevention

These should be implemented in the design and monitored in testing phase to see how much it is open for invasion and how much it can withstand attacks.

10 Cutting-Edge Technologies

10.1 Quantum Repeaters

By overcoming the drawbacks of signal deterioration over extended distances, quantum repeaters are a new technology that will dramatically increase the range of quantum communication. Secure communication over broader operational regions will be made possible by their use in UAV networks.

10.2 Advanced Quantum Sensing

New developments in quantum sensing technology, like quantum-enhanced navigation and imaging, will expand UAV capabilities and improve situational awareness and operational precision.

10.3 Importance

Enhancing UAV capabilities can be achieved through two promising areas of research and development: quantum repeaters and advanced quantum sensing. The problem of secure long-distance communication is tackled by quantum repeaters, which enable the maintenance of strong quantum networks over large operational areas. However, cutting-edge quantum sensing technologies provide better imaging and navigation, giving UAVs more accuracy and situational awareness. When taken as a whole, these developments will increase the efficacy and security of UAV operations, expanding their possible uses and operational range.

10.4 Classical Versus Quantum Cryptography [4] Analysis

Classical cryptography [4] UAVs are less secure and more exposed to threats because of the mathematical solutions in cryptography [4] but quantum is that easy to implement in UAVs, so both have their own strengths and weaknesses. However, quantum guarantees more security and computational power than that of classical, and the best possible option is quantum.

10.5 Ethical Considerations and Privacy

While developing the quantum UAVs, all the ethical frameworks must be kept and it should not violate the privacy concerns and international standards. It should comply with all the legal frameworks governing in an organization or international standard because if it violates the ethical side, it can lead to many problems. It should be deployed within an lawful organization, and a license should also be used.

10.6 Data Protection and Privacy

Data protection and privacy are one of life's most important things. For some it can be more important than life because when someone loses its privacy it can be unimaginable for some and can lead to many problems. So, creating a UAV with a quantum communication system provides strongly guaranteed data security and protection. It can provide the following.

- Confidentiality
- Integrity of data
- Prevent unauthorized access

These are very important factors and should provide a core of principle relied on this, and it shall have privacy implemented. Advanced UAVs raise the ethical concerns of privacy, civil rights and integrity and should not violate human dignity and protection of human rights. It shall obey all the laws of ethics and data should be available to authorized users only.

10.7 International Privacy Standards

These are the basic rules and principles that all over the world is accepted and are considered as standard ethical rules. This quantum design UAV must obey the laws of ethics which on global scale. Everyone has privacy from people to y, and it should not be avoided, which can lead to conflicts and do a lot of damage. Some of the main principles are [7, 8]:

- Human rights and data protection
- Data integrity
- Violating border laws

These can lead to huge problems and can be dangerous.

10.8 International Privacy Standards

The use of quantum cryptography in UAVs presents a number of ethical and privacy issues, such as the potential abuse of secure communication channels and the need to strike a balance between civil liberties and security. In order to establish ethical deployment criteria, it is necessary to safeguard individual rights, maintain openness, and stop unauthorized usage of security technologies. Creating global guidelines for moral quantum cryptography practices and encouraging responsible innovation in UAV applications are two recommendations [1].

11 Conclusion and Future Aspects

Quantum-based UAVs are one the big things in the future; as it is based on quantum, it can be life-changing in security surveillance and help protect people and lives. It has several applications and once it reaches its full potential it can lead to a new future of UAVs. But it needs time to do the job and as quantum is its first step, just like a computer was in the 1950s.

But once it achieves the hardware required technology, it will a step toward the future.

Imagine having a UAV making decisions and saves life. Yes, it can be very beautiful and a blessing for some. It can be helpful in dangerous terrain and wars surveillance and general security of people. It can also play an important role in hostile situations, which are very dangerous if you don't have the right data and can lead to much damage.

So, integration of quantum-based communications in UAVs is a big step to the future and can bring many new opportunities which can facilitate life decisions through secure communication. We can say that it can be a big step in automation, which can be useful.

Combining quantum with AI will be a more crucial and more advanced step.

With AI, these UAVs can be used in security and detect threats, criminal activities, and many more. Once we overcome the challenges, it can play crucial roles.

References

1. W. Abdallah, A physical layer security scheme for 6G wireless networks using post- quantum cryptography [4]. Computer Communications (2024)
2. M. Adil et al., $ R3ACWU $: A lightweight, trustworthy authentication scheme for UAV-assisted IoT applications. IEEE Trans. Intell. Transp. Syst. (2024)
3. S. Akhai, V. Kumar, Quantum Resilience and Distributed Trust: The Promise of Blockchain and Quantum Computing in Defense, in *Sustainable Security Practices Using Blockchain, Quantum and Post-Quantum Technologies for Real Time Applications*. (Springer, 2024), pp.125–153
4. H. Al-Hraishawi et al., Characterizing and utilizing the interplay between Quantum technologies and non-terrestrial networks. IEEE Open J. Commun. Soc. (2024)
5. L. Chen et al., PUF-based dynamic secret-key strategy with hierarchical blockchain for UAV swarm authentication. Comput. Commun. (2024)
6. J.O. del Moral et al., Cybersecurity in Critical Infrastructures: a Post-quantum cryptography Perspective. arXiv preprint arXiv:2401.03780 (2024)
7. U. Dubey et al., A review on practical challenges of aerial Quantum communication. Physics Open, 100210 (2024)
8. A. Farouk, N.A. AbuAli, S. Mumtaz, Quantum-computing-based channel and signal modeling for 6g wireless systems. IEEE Commun. Mag. **62**(2), 64–70 (2024)
9. A.A.M. Gnanaraj et al., Hyperelliptic curve based authentication for the internet of drones. Int. J. Reconfigurable Embedded Syst. **13**(1), 133–142 (2024)
10. A. Gupta, D. Dhawan, N. Gupta, Review on UAV-based FSO links: recent advances, challenges, and performance metrics. Opt. Eng. **63**(4), 041204–041204 (2024)
11. P.-Y. Kong, UAV-assisted quantum key distribution for secure communications with resource limited devices. IEEE Trans. Veh. Technol. (2024)
12. M. Korrapati et al., Design and modeling of a pat system for freespace optical links in quantumkey distribution, in *2024 16th International Conference on COMmunication Systems & NETworkS (COMSNETS)*. IEEE (2024)
13. J. Li et al. Encryption and return method of inspection image of UAV converter station based on chaotic sequence, in *International Conference on Algorithm, Imaging Processing, and Machine Vision (AIPMV 2023)*. SPIE (2024)
14. C. Maraveas et al., Harnessing Quantum computing for smart agriculture: empowering sustainable crop management and yield optimization. Comput. Electron. Agric. **218**, 108680 (2024)
15. A. Mehmood et al., Advances and vulnerabilities in modern cryptographic techniques: a comprehensive survey on cybersecurity in the domain of machine/deep learning and quantum techniques. IEEE Access **12**, 27530–27555 (2024)
16. R. Moskowitz et al., Trustworthy device identifiers for unmanned aircraft and other mobile things,in *2024 2nd International Conference on Unmanned Vehicle Systems-Oman (UVS)*. IEEE (2024)
17. T. Pradhan, P. Patil, Quantum cryptography [4] for secure autonomous vehicle networks: a review. in *2024 IEEE International Students' Conference on Electrical, Electronics and Computer Science (SCEECS)*. IEEE (2024)
18. L. Shi et al., Practical performance analysis of mdi quantum key distribution with orbital angular momentum on UAV relay platforms. Xing yu and Deng, Zhifeng and Tang, Jie and Cao, Yuexiang and Hu, Hao Ran and Wang, Ya and Liu, Ying and Yu, Huicun and Wei, Jiahua and Lun, Hua Zhi, Practical Performance Analysis Of Mdi Quantum Key Distribution With Orbital Angular Momentum on Uav Relay Platforms
19. H.A. Signature, X. Pan, F. Li, Check for heterogeneous aggregate signature for unmanned aerial vehicles Xiangyu Pan and Fagen Li () School of computer science and engineering, university of electronic science and, in *Frontiers in Cyber Security: 6th International Conference*, FCS 2023, Chengdu, China, August 21–23, 2023, Revised Selected Papers. Springer Nature (2024)
20. A. Tuncal, E. Ufuk, Integrating unmanned aerial vehicles in airspace: a systematic review. J. Aviation Res. **6**(1), 89–115 (2024)

21. D. Wang et al., Authentication and key agreement based on three factors and PUF for UAVs-assisted post-disaster emergency communication. IEEE Internet Things J. (2024)
22. Z. Wei, W. Guo, Control layer security: exploiting unobservable cooperative states of autonomous systems for secret key generation. IEEE Trans. Mobile Comput. (2024)
23. H. Wen et al., Exploring the future application of UAVs: face image privacy protection scheme based on chaos and DNA Cryptography [4]. J. King Saud Univ. Comput. Inform. Sci. **36**(1), 101871 (2024)
24. Q. Xiao et al., Securing NextG networks with physical-layer key generation: a survey. Security and Safety **3**, 2023021 (2024)
25. M. Xu et al., Generative AI-enabled quantum computing networks and intelligent resource allocation. arXiv preprint arXiv:2401.07120 (2024)

Hayat Ullah is an undergraduate student in Computer Science at the University of Malakand. He would like to express his gratitude to Dr. Mohammad Faisal for their guidance and support throughout the research process.

Mohammad Faisal currently working at the Department of Computer Science and IT, University of Malakand Pakistan, received his M.S. degree in Information Security Management from SZABIST, Pakistan, in 2012, and the Ph.D. degree in Network Security from the Department of Computer Science and Information Technology, University of Malakand, in 2018. His research interests include quantum cryptography, ML, and security of wireless ad hoc networks such as MANETs, VANETs, IoT, cloud, fog, edge, blockchain, and digital forensics.

Enhancing Internet of Things (IOT) Security Through Quantum Cryptography

V. Sakthivel, Vibha Rao, P. Prakash, and Jae-Woo Lee

Keywords Glossary

- **Internet of Things**: A network of devices and systems which are interconnected with each other, that communicate and exchange data with each other over the internet, enabling automation and enhanced connectivity.
- **IoT Security**: Measures and practices meant to protect IoT devices, networks, and data from cyber-attacks, illegal access, and other malicious security threats.
- **Quantum Cryptography**: A secure communication method that uses the fundamental principles of quantum mechanics to create encryption systems that are protected against eavesdropping and tampering.
- **Post-Quantum Cryptographic Algorithms**: Cryptographic algorithms that provide security against the potential threats posed by quantum computers, which can break the classical traditional encryption methods.
- **Quantum Key Distribution (QKD)**: A quantum cryptographic technique that enables secure communication by sharing encryption keys through quantum channels, making sure that any interception is detectable.
- **Quantum Threats**: Risks and vulnerabilities associated with quantum computing technologies, which can potentially break existing cryptographic systems and compromise data security and integrity.

V. Sakthivel (✉) · V. Rao · P. Prakash
Vellore Institute of Technology, Chennai, Tamilnadu, India
e-mail: sakthivel.v@vit.ac.in

V. Rao
e-mail: vibha.rao2022@vitstudent.ac.in

P. Prakash
e-mail: prakash.p@vit.ac.in

J.-W. Lee
Konkuk University, Seoul, South Korea
e-mail: jwlee@konkuk.ac.kr

S. B. Goyal et al. (eds.), *Quantum Computing, Cyber Security and Cryptography*,
https://doi.org/10.1007/978-981-96-4948-8_13

1 Introduction: Increasing IOT Use and Its Future Threats

The Internet of Things (IoT) has changed how we connect ourselves with technology and the world, but this increased connectivity has led to IoT systems being incredibly vulnerable to cyber-attacks. Quantum attacks pose a huge threat to the security of IoT devices. The increasing use of IoT devices across various sectors such as healthcare, transportation, smart homes, and industrial systems has led to a vast network of interconnected devices generating and transmitting a large amount of data. While this interconnectedness has brought many advancements in efficiency, productivity, and convenience, it has also opened up new possibilities for potential security breaches and cyber-attacks which pose a threat to data security [1]. To further illustrate the issue, the average number of weekly attacks on IoT devices per organization rose by 41% within just the first two months of 2023 compared to 2022, which is quite concerning given the brief period. Additionally, 54% of organizations on average are experiencing attempted cyber-attacks each week, underscoring the pressing need for improved security measures.

Quantum attacks are a formidable threat to the security of IoT devices. These attacks exploit the power of quantum computers to compromise traditional cryptographic methods which were previously considered secure. With further advancement in quantum computing, the risk posed by quantum attacks to the integrity of IoT systems becomes increasingly significant.

In response to the urgent need for enhanced security measures, one of the promising solutions that has emerged to fortify and strengthen the resilience of IoT systems against malicious agents is the integration of quantum cryptography. Quantum cryptography leverages the fundamental principles of quantum mechanics and generates secure communication channels which are resistant to interception and tampering. By utilizing the quantum key distribution mechanism and algorithms that are quantum resistant, IoT networks can establish a robust security framework that safeguards sensitive data and communications from potentially dangerous quantum threats.

Furthermore, the implementation of quantum-safe cryptographic protocols can introduce a new layer of protection for an IoT device, making it extremely secure and reliable in the world of continuously evolving cyber threats. As IoT constantly continues to expand its reach and influence the adoption of quantum cryptography, it becomes increasingly imperative to maintain the trust and integrity of IoT systems in the digital age. By staying ahead of potential vulnerabilities and embracing quantum-safe security measures, we can encourage a more secure and resilient IoT ecosystem that empowers innovation while safeguarding privacy and data integrity.

To navigate the complex landscape of IoT security in the quantum era, this chapter is structured to provide a comprehensive exploration of key topics. We begin by discussing the IoT Revolution, which has significantly transformed industries through enhanced connectivity. We then examine the IoT Attack Surface to identify vulnerabilities and potential cyber threats. Following this, we delve into Quantum

Hacking, analyzing how quantum computers can compromise traditional cryptographic methods. The chapter continues with a comparison of Pre-Quantum and Post-Quantum IoT Security methods, assessing the effectiveness of both current and emerging techniques. We also cover Standardization Attempts, exploring efforts to establish robust quantum-safe cryptographic protocols. The discussion includes Potential Post-Quantum Cryptographic Solutions, highlighting promising algorithms for securing IoT devices. We further explore the Crux of the Quantum Key Distribution Technique and its role in enhancing IoT Security. The chapter addresses the Need for Enhancement in IoT security measures and evaluates Proposed Cryptosystems versus Physical Attacks. It then considers the Challenges for the IoT Future, focusing on obstacles to implementing quantum cryptography. We conclude with insights into Future Trends and Practical Implementations of Quantum Cryptography in IoT Scenarios, before summarizing the key findings in the Conclusion. This structured approach tries to provide a clear conception of how quantum cryptography can address the evolving security needs of IoT systems.

2 IOT Revolution

Reference [2] It predicts that in the future, the number of devices interconnected to the internet will exceed the number of people connected with Internet services. As for now, there are more than 200 billion IOT devices interconnected worldwide. This would bring with it a huge influx of data since IoT-based applications generate a huge amount of data per second. Therefore, the need to device methods to safeguard and protect this data is greater than ever.

3 IOT Attack Surface

Reference [3] To develop a complete understanding of the intricacies of IoT hacking, it is crucial to have a clear understanding of its attack surface. Different IoT devices have different kinds of vulnerabilities due to significant differences in their own embedded structures (Table 1). For example, Smart Home Security Systems include vulnerabilities such as communication protocols (Wi-Fi, Zigbee), mobile app security, cloud infrastructure security, and physical components like cameras and sensors. Industrial Control Systems, on the other hand, have distinct vulnerabilities which include programmable logic controllers, human-machine interfaces, communication protocols like Modbus, DNP3, and network infrastructure security.

Despite these differences, a commonality among all IoT devices is their Printed Circuit Boards (PCBs). PCBs are used to interconnect components and wire them down, but they also pose significant vulnerabilities. Attackers with physical access to IoT devices can easily modify and tamper with exposed components on the PCB, potentially leading to signal interception attempts, firmware reverse engineering, and

Table 1 Various attack entry points in an IOT device including hardware attacks, network interface attacks, firmware vulnerabilities and cloud communication vulnerabilities

IOT attack surface	Attack types	About the attack
IOT hardware attacks	Debug pin tampering	Unauthorized access through interfaces like JTAG or SWD to extract data or inject malicious code
	Memory scraping	Directly accessing and extracting sensitive data from memory chips, potentially bypassing encryption
	Bus snooping	Monitoring data transfers over system buses to capture critica information such as encryption keys
IOT network interface attacks	Man-in-the-Middle (MITM)	Intercepting and possibly altering communications between the IoT device and the network
	Wi-Fi decryption	Using advanced techniques, including quantum computing, to break Wi- Fi encryption and access network traffic
	BLE spoofing	Impersonating a legitimate Bluetooth device to steal data or control the IoT device
IOT firmware vulnerabilities	OTA update interception	Capturing and modifying firmware updates to introduce malicious software
	Firmware signature forgery	Exploiting weaknesses in validation mechanisms to pass malicious updates as legitimate
	API key exploitation	Misusing shared API keys across devices to extend attacks to multiple IoT devices
Cloud communication vulnerabilities	TLS certificate spoofing	Altering or injecting data during communication between IoT devices and cloud services
	MQTT authorization bypass	Exploiting weaknesses in MQTT authorization to push unauthorized commands or malicious software
	Data tampering	Altering or injecting data during communication between IoT devices and cloud services

side-channel attacks. Additionally, deliberate and targeted supply chain attacks could be initiated to tamper with the PCB layout and design, introducing abnormalities that could compromise routing and essential security features, leaving the IoT device completely vulnerable.

In each of these areas, quantum attacks prove to be a major threat to IoT device security. Quantum computers have a tremendous potential. They can exploit vulnerabilities easily and break widely used encryption algorithms efficiently. If IoT devices rely on these algorithms for securing data stored in memory or transmitted over networks, quantum attackers could exploit this weakness to intercept sensitive information or inject malicious code into the firmware during memory manipulation attacks. Therefore, the security of IoT devices must evolve to withstand potential quantum threats, demanding the adoption of quantum-resistant cryptographic algorithms and robust security measures across all layers of modern-day IoT systems.

3.1 IOT Hardware Attacks

Reference [3] Hacking a PCB involves gaining an understanding of each of the wired-down components and testing their security. This involves tampering with the debug pins (JTAG, SWD), unauthorized data fetching from the memory, and monitoring sensitive data over system buses. J-Link is an on-chip debugger that acts as a link between the microcontroller and the development environment. It allows developers to interact with the microcontrollers during the testing and development processes. However, with the constant development of quantum computing, cryptographic algorithms commonly used for securing IoT devices may become vulnerable to attacks. For instance, commonly used encryption algorithms like RSA and ECC can be attacked and broken easily if not for quantum cryptographic algorithms. If IoT devices rely on these algorithms for securing data stored in memory or transmitted over networks, quantum attackers could exploit this weakness to intercept sensitive information or inject malicious code into the firmware during memory manipulation attacks.

3.2 IOT Network Interface Attacks

Reference [3] IOT devices are usually connected by network interfaces such as Wi-Fi, BLE, and ZigBee. Wi-Fi serves as the primary medium of communication between the internet and an IOT device. With numerous networks available at a particular time, a device needs a way to be able to connect to the right Wi-Fi network [3]. In order to do so, a device either exposes its own specific access point (AP) or it communicates over Bluetooth Low Energy (BLE). It is essential to encrypt data at the transport or application layer when a device hosts its own access point.

Failure to do so can lead to numerous security risks and vulnerabilities. This could potentially lead to man-in-the-middle attacks, data interception, tampering, spoofing and eavesdropping. However, with the accelerated progress of quantum computing, encryption schemes used in Wi-Fi (e.g., WPA2) and BLE could become vulnerable to attacks. Malicious agents with access to quantum computing capabilities could decrypt intercepted traffic, perform man-in-the-middle attacks, or forge authentication messages, undermining the security of IoT devices connected over these protocols.

3.3 IOT Firmware Vulnerabilities

Reference [3] The running firmware of devices poses the highest number of security vulnerabilities. Over-the-air updates (OTA) are often initiated remotely on an IOT device. During OTA updates, the communication channel should be encrypted, and the firmware signatures should be validated. Failure to do this can lead to man-in-the-middle attacks that modify the OTA update package or perform malicious software updates. API keys are present in the device firmware and can be shared among various other devices. Due to the sharing of API keys, one maliciously affected device can affect all other devices connected to it. An OTA update shared key could be extracted from one IOT device and used to device malicious firmware for other devices. However, with the advancements in quantum computing, vulnerabilities in firmware signing mechanisms used during OTA updates could be exploited. If IoT devices rely on traditional cryptographic algorithms vulnerable to quantum attacks for verifying firmware authenticity and integrity, attackers could manipulate OTA update packages, inject malicious firmware, or compromise device functionality without detection.

3.4 Cloud Communication Vulnerabilities

Reference [3] IOT comprises much more than just the IOT end device. The IOT device constantly communicates with gateway devices, mobile applications, and backend cloud services. Devices communicate with cloud services either through HTTPS or MQTT. Often, problems arise in how an IOT device screens a server TLS certificate, which ultimately gives an opportunity to strategic man-in-the-middle attack scenarios. In case of any discrepancy in MQTT authorization, a single device can potentially update all other devices with malicious software within a split second. However, quantum computers could undermine the security of TLS certificates used for establishing secure communication between IoT devices and cloud services. Adversaries with quantum capabilities could exploit vulnerabilities in cryptographic algorithms underpinning TLS, allowing them to impersonate legitimate servers, intercept encrypted traffic, or tamper with data exchanged between IoT devices and cloud environments. Furthermore, quantum attacks could target the authorization mechanisms employed by MQTT-based communication frameworks. Weaknesses in MQTT authentication processes, compounded by vulnerabilities in cryptographic algorithms susceptible to quantum cryptanalysis, might enable attackers to compromise the integrity of MQTT messages, gain unauthorized access to IoT resources, or propagate malicious software across interconnected devices within the MQTT network.

4 Quantum Hacking

"Quantum hacking" refers to the theoretical possibility of using extremely powerful quantum computers to target vulnerabilities in modern cryptography systems. This could be used to target vulnerable IoT-based devices which are much easier to maliciously access compared to other complex advanced devices. Quantum hacking includes various techniques that prove to be a giant threat to today's cryptographic systems in terms of their safety and security, some of which are as follows.

4.1 Quantum Cryptanalysis

Reference [4] Quantum cryptanalysis is a technique that poses a threat to widely used encryption methods like RSA and ECC. It relies on advanced quantum algorithms such as Shor's algorithm to undermine the security of encrypted data. One particular concern is its impact on IoT devices, which are especially vulnerable to these attacks.

4.2 Side-Channel Attacks

Reference [4] Side-channel attacks take advantage of unintentional signals emitted by cryptographic systems. Their objective is to gain insight into secret encryption keys, compromising the security of encrypted data. IoT devices are often targeted by these attacks, making them more susceptible to privacy breaches.

4.3 Quantum Brute Force

Reference [4] Quantum brute force leverages the immense processing power of quantum computers to quickly test every possible encryption key. This method allows cybercriminals to gain unauthorized access to confidential information. IoT devices are specifically targeted by quantum brute force attacks, increasing the risk of data exposure.

IOT devices, being interconnected and often lacking robust security measures, are prime targets for various attacks. To counter these threats, continuous efforts are underway to create and approve post-quantum cryptographic algorithms that can resist quantum attacks. Such algorithms should provide secure encryption methods resistant to quantum adversaries. It is critical to develop vigorous security measures for IoT devices to prevent the loss of sensitive information. Ensuring user privacy and improving IoT device security should be a top priority.

5 Pre-quantum and Post-Quantum IOT Security

In the world of encryption, there are two approaches: pre-quantum cryptography and post-quantum cryptography. These approaches deal with the security challenges posed by quantum computers. Let's explore the key differences between them. Comparison of Classical vs. Quantum-Resistant Cryptographic Algorithms is highlighted in the Table 2.

Pre-quantum cryptographic encryption methods are currently in use and were designed to be a safety mechanism against classical computers. These methods employ specific algorithms to transform human-readable data into secret code, making it difficult to reverse-engineer. The primary challenge of pre-quantum cryptography lies in creating encryption ciphers that are both easy to understand and hard to break. The security of pre-quantum cryptography depends on the computational complexities of certain mathematics-based problems. These methods are, however, vulnerable to attacks from quantum-based computers, which have the capability to solve such problems exponentially faster than regular computers.

Post-quantum cryptography on the other hand is quantum resistant and in turn provides safety against attacks. It is a type of encryption devised to provide protection against attacks from both regular and quantum computers. Its objective is to develop new cryptography-based algorithms that provide safety against attacks by quantum computers. Such algorithms depend on complex mathematics-based problems that are believed to be of very high complexity, proving to be difficult to solve even for quantum-based computers. Some of the popular post-quantum cryptographic algorithms include lattice-based, code-based, and multivariate-based cryptographic techniques. The ultimate goal of post-quantum cryptography is to ensure that encrypted data remains secure even if quantum computers become powerful enough to invade the barriers of current encryption methods. While post-quantum cryptography slowly keeps evolving, efforts are being made to establish standards in this domain. Organizations like NIST play a crucial role in advancing and regulating post-quantum cryptography-based algorithms [5]. The aim is to create robust and widely accepted encryption standards that can endure attacks from both classical and quantum computers. While this standardization process is ongoing, it is important to stay up to date about the latest advances in post-quantum cryptography

Table 2 Comparison of classical versus quantum-resistant cryptographic algorithms

Cryptographic task	Classical algorithm	Computational complexity (Classical)	Quantum-resistant algorithm	Computational complexity (Quantum-Resistant)	Performance metrics
Public key encryption	RSA	$O(n^3)$	Lattice-Based Cryptography (e.g., NTRU)	$O(n^2)$ or $O(n^3)$ based on the algorithm	Lattice-based algorithms are quantum resistant but may require significant computational resources, particularly with larger key sizes, which can impact encryption and decryption speed
Digital signatures	ECDSA	O(n)	Hash-Based Signatures (e.g., SPHINCS+)	O(n)	Hash-based signatures offer strong quantum resistance but typically have larger signature sizes, leading to increased storage requirements and potential bandwidth constraints
Key exchange	Diffie-Hellman	O(n)	Lattice-Based Key Exchange (e.g., Kyber)	O(n)	Lattice-based key exchange protocols are secure against quantum attacks but often require larger key sizes, which can slow down the key exchange process, especially in constrained environments
Symmetric encryption	AES	O(n)	AES with Larger Key Sizes (e.g., AES-256)	O(n)	AES is considered quantum resistant with larger key sizes (e.g., AES-256), though using larger keys may slightly reduce encryption/decryption speed. Grover's algorithm effectively reduces the key length by half, so using AES-256 is recommended
Error correction (for data integrity)	Reed-Solomon	O(n log n)	Code-Based Cryptography (e.g., McEliece)	O(n log n)	Code-based cryptosystems like McEliece offer robust quantum resistance but often involve very large key sizes, which can increase memory usage and slow down the overall system performance

Table 3 NIST post-quantum cryptography standardization timeline

NIST post-quantum cryptography standardization	
Round 1 (2017)	Call for proposals issued in December 2016
	69 submissions received
	Categories: public-key encryption, digital signatures, key-establishment protocols
Round 2 (2019)	26 algorithms selected from Round 1
	Advanced to second round in January 2019
	Categories: encryption, key exchange, digital signatures
Round 3 (2020–2021)	15 algorithms selected from Round 2
	Proceeded to third round in July 2020
	Further analysis: performance testing, security considerations
Round 4 (2022–2023)	Submission deadline for Round 4: November 30, 2022
	Evaluation of Round 4 submissions: December 2022–June 2023
	Selection and announcement of Round 4 finalists: After completion of evaluation process in mid-2023
	Work on finalization of standardization process for selected algorithms: Mid-2023 onwards
Finalization (Expected 2024)	Submission of candidate algorithms: Ongoing process of evaluating and selecting potential post-quantum cryptographic algorithms
	Round 3 candidates: Evaluation and selection of finalists from the round 3 submissions
	Public review and analysis: Community feedback on the selected finalists to ensure robustness and security
	Final standardization process: Formalization of the selected post-quantum cryptographic algorithms into NIST standards by 2024

to ensure data protection in the midst of evolving technologies. Some of the most significant standardization attempts have been elaborated in the Table 3.

6 Standardization Attempts

The Standardization attempts over the years-ETSI, NIST, IETF & CRFG and other standardization methods are listed in the Table 4.

Table 4 Standardization attempts over the years-ETSI, NIST, IETF & CRFG and other standardization methods

Quantum Cryptography Standardization Initiatives	
ETSI	White papers on quantum security
	Collaboration with IQC for quantum-safe cryptography workshops
	Technical committee cyber working group on quantum-safe cryptography deliverables
NIST	Reports on quantum threat
	Call for proposals for post-quantum public-key cryptosystems
	Ongoing standardization process until 2024
IETF & CFRG	Internet-drafts related to quantum cryptography
	Covering topics like post-quantum cryptosystems for TLS and IKEv2 extension
Other standardization initiatives	ISO/I EC JTC 1/SC27 and IEEE P1363 project standards
	Addressing security risks of quantum computing in finance industry

6.1 ETSI

Reference [5] The European Telecommunications Standards Institute (ETSI) has taken steps toward quantum security by publishing documents related to quantum cryptography-based security. It has also partnered with the Institute for Quantum Computing since 2013, to organize the Cryptography Workshops based on quantum safety. ETSI has a Technical Committee Cyber Working Group on Cryptography which is quantum protected. It has released reports addressing topics such as the effects of quantum computing attacks across various sectors and the limitations of quantum computing in the realm of symmetric cryptography.

6.2 IETF

Reference [5] The Internet Engineering Task Force is collaborating with the Crypto Forum Research Group on several Internet drafts focused on quantum cryptography. These drafts cover topics such as post-quantum cryptosystems for TLS and expanding IKEv2 to be post-quantum. Some of their other drafts focus on the shift from classical modern-day quantum cryptography to post-quantum cryptography and applications of specific schemes like XMSS and Leighton-Micali Hash-Based Signatures.

6.3 NIST

Reference [5] The National Institute of Standards and Technology has organized workshops on the standardization of post-quantum cryptography and has released reports on the imminent quantum threat. NIST declared an open call for proposing candidates for post-quantum public-key cryptosystems and received numerous candidates. The standardization process is ongoing, with multiple rounds expected to take place until 2024.

6.4 Other Standardization Initiatives

Reference [5] Other significant standardization initiatives include those led through the ISO/IEC JTC 1/SC 27 by the International Organization for Standardization and the Institute of Electrical and Electronics Engineers with the P1363 project. These efforts have led to the publication of standards for lattice cryptography, which is based on public-key principles. Additionally, they have tackled the security risks posed by quantum computing for particular sectors, such as finance, through a document released by the American National Standards Institute's Accredited Standards Committee (ASC) X9.

7 Potential Post-Quantum Cryptographic Solutions

Numerous post-quantum cryptographic algorithms have been proposed till date. Most of them can be categorized into pre-existing cryptosystem categories. The most common IoT-based cryptosystems are elaborated with the help of the Tables 5 and 6.

7.1 Code-Based Cryptosystems

Reference [5] Code-based cryptosystems, like McEliece's scheme, are based on error-correction codes such as binary Goppa codes. These cryptosystems rely on the syndrome decoding problem, which is NP-complete. However, one challenge with McEliece's scheme is the large size of the public and private keys, making it difficult to implement on resource-constrained IoT devices. To address this issue, decompression and compression techniques can be investigated, and alternative adaptations of the McEliece scheme can be put forth. This can involve using different codes like LDPC, MDPC, or quasi-cyclic codes. Additionally, coding techniques like puncturing can be employed to optimize the scheme for IoT devices.

Table 5 Post quantum cryptographic solutions-code-based, lattice-based, and isogeny-based approaches designed to resist quantum attacks

Potential post-quantum cryptographic solutions for loT devices	
Code-based cryptosystems	
McEliece's scheme	Compression and decompression techniques
	Alternative versions using different codes (LDPC, MDPC, quasi-cyclic codes)
Code-based signing algorithms	Variants of Niederreiter and CFS cryptosystems
	loT signature schemes derived from the Fiat-Shamir transformation
Lattice-based cryptosystems	
Lattice-based cryptosystems	Use of lattices in N-dimensional spaces with a periodic structure
	Efficient storage and operation with large keys and ciphertext overheads
	Exploration of loT-optimized lattice-based cryptosystems based on polynomial algebra and the Learning With Errors (LWE) problem
Isogeny based cryptosystems	
Supersingular elliptic curve isogeny cryptosystems	Use non-commutative supersingular curves
	Estimated key sizes in the order of a few thousand bits

There are also code-based signing algorithms, such as variants of Niederreiter and CFS cryptosystems. These variants are similar to McEliece's scheme but have their own considerations for IoT developments. For example, CFS variants generate short signatures that can be verified quickly, but they require large key sizes and inefficient signature generation. Another approach is the advancement of Fiat-Shamir transformation-based IoT signature schemes, which have shown promising results in surpassing CFS.

7.2 *Lattice-Based Cryptosystems*

Reference [5] Lattice-based cryptosystems utilize clusters of points in n-dimensional spaces, characterized by a unique periodic structure known as lattices. These cryptosystems rely on the difficulty of solving lattice-based problems, such as the Shortest Vector Problem and the Closest Vector Problem. Unlike quantum algorithms, current methods are unable to efficiently address these challenges. Generally, lattice-based cryptosystems provide robust security and are relatively straightforward, quick, and efficient to implement.

Table 6 Comparing various quantum-resistant algorithms strengths, weaknesses, and current development status

Cryptosystem type	Algorithm example	Strengths	Weaknesses	Current development status
Code-based cryptosystems	McEliece	Proven security over decades Based on the Nondeterministic Polynomial time-complete syndrome decoding problem	Large public and private key sizes Implementation challenges in IoT devices	Active research on compression techniques (e.g., puncturing, LDPC) Exploration of alternative codes like MDPC and quasi-cyclic codes
	Niederreiter, CFS	Short signatures with fast verification (CFS)	Large key sizes Inefficient signature generation	Investigations into Fiat-Shamir transformation-based IoT signature schemes
Lattice-based cryptosystems	NTRU, NewHope	Strong security based on lattice problems (e.g., SVP, CVP) Smaller key sizes compared to code-based systems Efficient and fast	Larger key sizes than traditional cryptosystems Requires storage and computational optimization for IoT	Optimization of polynomial algebra-based cryptosystems Research on Learning with Errors (LWE) and variants for IoT adaptation Focus on energy-efficient protocols for IoT devices
	Short Integer Solution (SIS)	Manageable key sizes Efficient signing operations	Requires further optimization for energy efficiency in low-power IoT devices	Demonstrated efficiency on embedded devices Continued development for faster operations
Isogeny-based cryptosystems	Super singular isogeny Diffie-Hellman (SIDH)	High potential for post-quantum cryptography Small key sizes (thousands of bits)	Computationally intensive Difficult to optimize for resource-constrained IoT	Research on compression techniques Efforts to reduce computational load for IoT integration Focus on optimizing performance for energy-efficient IoT applications

(continued)

Table 6 (continued)

Cryptosystem type	Algorithm example	Strengths	Weaknesses	Current development status
Multivariate quadratic cryptosystems	Unbalanced oil and vinegar (UOV), Rainbow	Fast encryption and decryption Small signature sizes	Complex key generation Potential vulnerabilities in specific attack models	Active research in improving security against differential attacks Implementation trials for IoT device compatibility
Hash-based cryptosystems	SPHINCS+, XMSS	Strong security based on hash functions Stateless operation (SPHINCS+)	Larger signature sizes Slower signing speed compared to traditional systems	NIST standardization efforts Optimization for faster signing and smaller signatures in IoT contexts

However, a significant requirement for future post-quantum lattice-based cryptosystems in Internet of Things (IoT) devices is the need for efficient storage and compatibility with large keys and ciphertexts. One potential solution is to investigate IoT-optimized polynomial algebra-inspired lattice-based cryptosystems, particularly focusing on the Learning With Errors problem and its various forms.

To effectively utilize lattice-based cryptosystems in IoT applications, it is essential to employ compression techniques and optimizations, given the larger key sizes compared to pre-quantum systems. Nonetheless, lattice-based schemes are an attractive option since their key sizes are generally smaller than those of code-based or multivariate public-key cryptosystems. Noteworthy lattice-based cryptosystems suitable for IoT include NTRU and NewHope, which typically feature key sizes ranging from around one thousand bits to a few thousand bits.

In regard to lattice-based signature schemes, the ones based on the Short Integer Solution show promise for signing operations. These schemes allow for manageable key sizes and have demonstrated efficiency on specific embedded devices. However, further optimization is needed for fast and operations associated with energy-efficient signing in IoT devices that are low-power. Similarly, in order to reduce energy consumption and conserve computational resources, lattice-based key exchange protocols need to be adapted.

7.3 *Isogeny Based Cryptosystems*

Reference [5] Supersingular isogeny-based cryptosystems are derived from the ordinary elliptic curve's isogeny protocol. These cryptosystems use non-commutative supersingular curves. This property makes them apt post-quantum systems.

Supersingular elliptic curve isogeny cryptosystems are estimated to have key sizes ranging from around one thousand bits to several thousand bits. It is crucial to investigate compression techniques and modifications to minimize key sizes in the development of IoT applications. These cryptosystems can also be used to help design post-quantum digital signature schemes, which need further optimizations in resource-constrained devices.

Applying Supersingular Isogeny Diffie-Hellman in resource-constrained IoT devices and employing supersingular elliptic curve isogeny cryptosystems that are energy-efficient pose numerous challenges. Essential compression schemes for isogeny-based cryptosystems require steps that are very computationally intensive; therefore, optimizing performance becomes crucial. Addressing these challenges is essential for the successful integration of supersingular elliptic curve isogeny cryptosystems in IoT applications.

Therefore, cryptosystems that are code-based, lattice-based, and super singular elliptic curve isogeny cryptosystems are all potential post-quantum cryptographic solutions for IoT devices. Further research and development are needed to ensure the successful implementation of these cryptosystems in IoT applications, as each approach has its own considerations and challenges, such as key size, compression techniques, and optimization for resource-constrained devices (Table 6).

8 The Crux of Quantum Key Distribution Technique

Reference [6] Quantum key distribution technique is a fundamental technique in quantum-based cryptography. It uses photons with different spin properties, such as horizontal, vertical, 45° diagonal, and -45° diagonal, to transmit data. The spin of a photon is measured using filters, with horizontal and 45° filters representing binary 1, and vertical and -45° filters representing binary 0.

However, the Heisenberg Uncertainty principle states that measuring the spin of a photon disturbs its state. This means that if we try to measure the spin, it will change and potentially alter the value of the photon. This can indicate the presence of an unauthorized entity interrupting the communication.

Reference [7] In the quantum key distribution process, Alice forwards a stream of polarized photons to Bob (Fig. 1), who randomly chooses between + and x bases to measure the photons. Bob then communicated the bases he used back to Alice. Alice informs Bob which bases were similar to the ones she used, discarding the measurements where Bob used a different basis. By guessing the correct basis with a 50% possibility, Bob can obtain a similar polarization to Alice's. The key is then derived from the remaining photons by interpreting them as 0 s and 1 s. In device-independent quantum cryptography, the aim is to ensure security even when using untrusted third-party devices. This is important because intermediate measurement devices in communication networks may be compromised. To address this, self-testing devices and cross-checking of polarizations and their probability distributions are used as solutions.

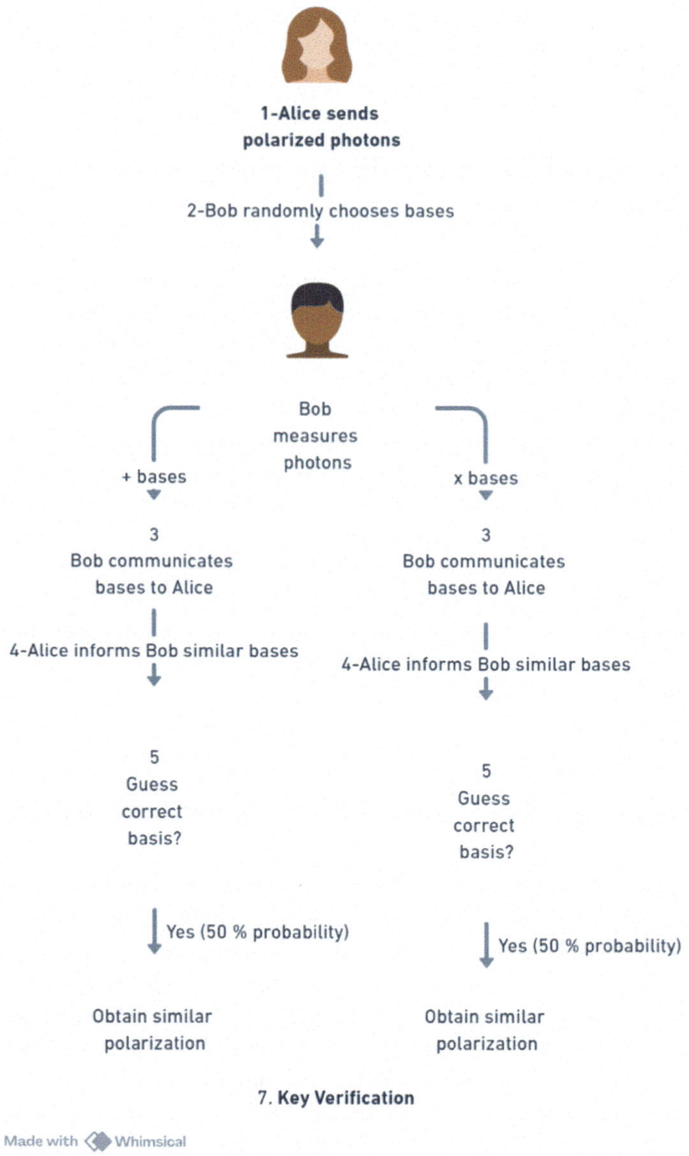

Fig. 1 Quantum Key Distribution (QKD): A Step-by-Step Guide, 1- Photon Generation: Alice, the sender, generates a series of polarized photons. The polarization of each photon can be either horizontal (+) or vertical (x). 2- Random Basis Selection: Bob, the receiver, randomly chooses a measurement basis for each photon. He can either measure in the horizontal (+) or vertical (x) basis. 3- Photon Measurement: Bob measures each photon using the basis he selected. 4- Basis Communication: Bob communicates the basis he used for each photon to Alice. 5- Basis Comparison: Alice and Bob compare the bases they used for each photon. 6- Key Generation: If the bases used by Alice and Bob for a particular photon match, they have a 50% chance of obtaining the same polarization. These matching photons form the basis for the shared secret key. 7- Key Verification: Alice and Bob can perform additional steps to verify the security of the shared key, such as error correction and privacy amplification

Therefore, the quantum key distribution technique provides a secure approach for distributing cryptographic keys.

9 Enhancing IoT Security with Quantum Key Distribution Technique

The Quantum key distribution technique, a fundamental aspect of quantum-based cryptography, offers a notable advantage over classical cryptography by enabling the detection of eavesdroppers in the system architecture. Nevertheless, implementing quantum cryptographic protocols faces challenges related to the maximum distance that photons, the carriers of quantum information, can travel without distortion. Moreover, quantum devices are large, extremely bulky, and costly, making their widespread adoption challenging. The current quantum key distribution protocols are devised for two end devices, which are not viable in the current scenario where IoT systems connect thousands of devices.

Advancements in quantum key distribution (QKD) have shown promise in enhancing the security of IoT systems. However, there are limitations that need to be addressed for successful QKD implementation in the IoT infrastructure. To overcome them and ensure the effective integration of QKD with IoT systems, several key areas need to be focused on:

9.1 Combining Classical and Modern Quantum Techniques

A hybrid approach that combines classical and modern quantum techniques can be employed to address the challenges. One proposed solution involves utilizing currently existing semiconductor chips while leveraging quantum-based techniques to generate lengthy and distinctive cryptographic keys for every single device. Quantum random number generation can be employed to create highly random noise sources efficiently and quickly, enhancing the security of communications.

By merging the plus points of both QKD and traditional cryptography, it is possible to achieve a robust security framework for IoT. QKD can be used to establish secure cryptographic keys, while classical cryptography can handle the bulk encryption and decryption operations. This integration ensures a balance between security and efficiency.

9.2 Creating a WAN with Intermediary Nodes

In order to overcome the limited transmission distance of QKD, it is necessary to establish a network of intermediate trusted nodes. These nodes can act as relays to extend the reach of secure key distribution, enabling secure communication over longer distances in IoT systems.

9.3 Simultaneous Key Distribution and Data Transmission

To improve data transmission rates, it is essential to develop methods for executing quantum key distribution simultaneously with data movement over the optical channels. This approach would enable efficient and secure communication in IoT systems with large volumes of data.

9.4 Broadcasting in a QKD Network

In IoT systems where multiple devices need to communicate securely, the ability to broadcast secure keys becomes crucial. Developing protocols and mechanisms for broadcasting within a QKD network would enhance its applicability in IoT scenarios.

9.5 Overall Cost Reduction

The high cost associated with dedicated QKD hardware poses a challenge to widespread adoption. Efforts should be made to explore cost-effective solutions, such as integrating QKD functionality into existing hardware or optimizing the manufacturing process to reduce production costs.

9.6 Device-Independent Quantum Cryptography

The reliability of devices can be proven by employing device-independent quantum cryptography. This approach ensures that the security of the system does not rely on the specific characteristics of the devices, providing an additional layer of security.

Despite these challenges, considering the immense amount of data involved in IoT systems and the need for robust security measures, QKD remains a viable solution. By addressing the limitations and focusing on the specific requirements of IoT infrastructure, QKD can contribute significantly to guaranteeing secure communication and protecting sensitive IoT data.

10 Need for Enhancement

Reference [5] Post-quantum cryptography presents unique challenges and quantum cryptanalysis when applied to the Internet of Things (IoT). These challenges include:

10.1 Rapid Evolution in Quantum Computing

Quantum computing is speedily progressing, and it's uncertain whether the post-quantum cryptosystems being developed for IoT will withstand future advancements in quantum algorithms and attacks. IoT developers need to closely monitor the progress in quantum computing and adapt their systems accordingly.

10.2 Large Key Size

Post-quantum algorithms often demand key sizes which are much larger in comparison to existing public-key cryptosystems. This can be problematic for resource-constrained IoT devices. To address this challenge, a trade-off needs to be made between key size, security level, and performance. Additionally, post-quantum lattice-based cryptosystems that are energy-efficient should be designed and employed on IOT devices to efficiently control the operation and storage dynamics with large keys.

10.3 Key Generation Issues

Post-quantum cryptosystems often set a limit to the quantity of messages that can be authorized with the same key to evade attacks. This necessitates generating a new set of keys for every group of messages which are signed, which might not have been effectively handled by conventional IoT devices. Methods to optimize post-quantum key production mechanisms and reduce energy intake need to be devised.

10.4 Consumption of Unwarranted Energy, Time or Computational Resources

Post-quantum algorithms may consume unwarranted energy, time, or computational resources for decrypting, encrypting, signature authentication, and signing. In order to ensure practical usability, accurate measurements need to be made, and ineffective cryptosystems or their implementations should be rejected.

10.5 Standardization Attempts that Are Currently Ongoing

IoT developers may have their attention misaligned on such post-quantum cryptosystems that might not even be eventually chosen by industry and academic circles for standardization. In order to align with the interests of the industry and academic circles, it is important to closely monitor the output generated by entities involved in standardization initiatives, such as NIST and others mentioned earlier.

10.6 Urgent Need for Standard Security Level Benchmarks

The modern-day benchmarks do not take into consideration the safety of the algorithms against quantum cryptanalytic techniques. Experts working on post-quantum cryptography methods must come to a consensus on how to assess security against quantum attacks and establish acceptable key lengths. Striking a balance between security and the cost of implementing quantum-resistant algorithms is crucial.

10.7 Rapid IoT Hardware Evolution

IoT developers face the challenge of selecting IoT platforms for evaluation. The devices currently considered low-end may not be representative of the devices that will come up in the future to come. To address this, post-quantum cryptographic systems need to be developed to cater to the different types of IoT devices, be it near-term, middle-term, or long-term, considering performance, computational ability, and energy utilization.

10.8 Requirement of Further Optimizations Which are IoT Specific

Post-quantum algorithms need to be optimized for IoT devices. For example, lattice-based cryptosystems require accelerating recurrent lattice operations and minimizing energy consumption. IoT microcontroller assembly code must also be optimized according to the architecture of the chosen embedded devices.

10.9 Optimizing Cryptographic Software

Cryptographic software running on other IoT network devices also requires optimizing. These optimizations should aim to increase speed by reducing the CPU cycles required by cryptographic systems and ideally reduce battery-dependent device energy consumption.

10.10 Unsuitability of IOT Node Hardware

Certain post-quantum cryptosystems that are theoretically designed may not be suitable for some IoT devices due to the computational resources they demand or the energy they consume. To address this issue, it is important to establish rigorous computational and power utilization requirements during the theoretical proposal phase. By doing so, at least one group of IoT devices will be able to suitably run the suggested algorithms.

10.11 Improvement in IoT Device Physical Security

Beyond analyzing the mathematical strength of post-quantum cryptosystems, it is essential to examine their physical security as well. Although the proposed post-quantum systems can offer moderate resistance to mathematical attacks, their implementation may be susceptible to physical threats. Attackers who gain physical access to IoT devices running these algorithms could take advantage of possible weaknesses. Therefore, it is crucial to evaluate and enhance the physical security of IoT devices to ensure the overall security of the system.

Addressing these challenges is extremely crucial to ensure the security and privacy of IoT devices and systems in the face of advancements in quantum computing.

11 Proposed Cryptosystems Versus Physical Attacks

Reference [5] When designing and evaluating proposed cryptosystems, it is crucial to consider and address the main physical attacks that can compromise their security. These attacks include.

11.1 Timing Attacks

Timing attacks take advantage of the correlation between the time it takes to execute certain operations and the secret data being processed or the operations performed on that data. By observing variations in the time taken to perform cryptographic operations, it might be possible to determine the entire secret key. Statistical analysis of timing measurements has demonstrated the feasibility of such attacks across networks.

11.2 Power Analysis Attacks

Power analysis attacks involve observing the power consumption of a hardware device, such as a CPU or cryptographic circuit, to obtain detailed information about the cryptographic operations being executed. These attacks are categorized into simple power analysis (SPA) and differential power analysis (DPA), and they can provide insights into the nature of the underlying computations.

11.3 Fault Attacks

Fault attacks can create errors in the circuitry of an IoT device by manipulating factors such as the power supply voltage, applying strong electric or magnetic fields, overclocking the device's clock, or exposing the device to high temperatures. These induced faults can be exploited to extract information about the private key.

To counter these attacks, several strategies can be implemented, including:

- Introducing redundant noise to balance power consumption.
- Splitting the private key into shares that can only be reconstructed by combining all shares or a minimum subset.
- Creating constant-time implementations to prevent timing variations.
- Measuring the power consumption of these countermeasures to find the optimal balance between security and energy efficiency.

12 Challenges for the IOT Future

Reference [2] The current highly vulnerable, resource-constrained IOT devices will soon face the following problems if immediate action is not taken:

- The evolution of quantum computing methods presents various obstacles to traditional cryptography-enabled IOT technologies. There is no solid assurance that the currently devised cryptography-based IOT applications will be able to withstand malicious quantum attacks. Therefore, it is very important that we address this issue.
- Current IoT networks typically use small keys, with a maximum size of 4096 bits, whereas post-quantum algorithms require significantly larger keys. Consequently, integrating IoT networks with post-quantum cryptographic algorithms necessitates a thorough analysis of key sizes.
- Resource-constrained IOT devices can cause significant delay when integrating cryptography protocols. This creates difficulty in synchronizing communication between IOT devices and fast computing servers. Due to this, there might be a limited number of signatures in hash-based cryptosystems. This results in the requirement to generate new keys for subsequent groups of messages. This proves to be a highly challenging task as it consumes excessive energy and yields poor results. Therefore, it is important to reinvent post-quantum approaches for resource-constrained IoT devices to efficiently handle network performance and quality of Service.
- Most post-quantum cryptosystems focus on only a few parameters, the primary parameter being security. Other important parameters like power consumption, latency, and resource utilization are often ignored. A real-time approach must include multiple parameters rather than just a few.
- There is a pressing need to identify new factors and their priority levels for IoT applications. It is crucial to develop methods for assessing the security of post-quantum cryptosystems against quantum attacks. Achieving this requires the establishment of frameworks by standards organizations like NIST or collaboration among researchers in the field to create a comprehensive framework.
- Although in the near future, resource-constrained IOT devices are anticipated to remain low-computational, their computation ability is expected to gradually increase after a decade or two. Post-quantum approaches which are energy efficient among all scenarios without compromising on security are anticipated to be the most beneficial for future purposes.
- Currently, the lattice-based cryptosystem in post-quantum cryptography is the major contender for IOT devices. Although this cryptosystem is currently the best suited for IOT devices, it needs further improvement in various sectors. Its polynomial-based calculation needs to be sped up, and the energy consumption and execution time need to be reduced. Likewise, other cryptography mechanisms need further optimization, too.

13 Future Trends

13.1 Novel Methodologies for Quantum Computing Impact

Reference [2] in the context of IoT, novel methodologies are needed to address the evaluation and quantification of quantum computing impact on the theoretical aspects of an IoT network. For example, these methodologies could involve studying the potential vulnerabilities of IoT devices to quantum attacks and developing strategies to secure IoT networks against quantum threats.

13.2 New Theoretical Insights on Post-Quantum Devices

Further research is required on the testing, strategy, and performance of post-quantum devices in IoT. For instance, this could involve exploring the development of post-quantum cryptography-based algorithms suitable for resource-constrained IOT devices and investigating their practical implementation in IoT environments.

13.3 Methodologies for Analyzing and Validating Algorithms

Testing and validating quantum-resistant algorithms for IoT devices, typically those that are energy efficient, is crucial. For example, researchers may need to develop testing methodologies to evaluate the functionality and safety of post-quantum cryptographic algorithms in resource-constrained IoT environments, ensuring their suitability for real-world deployment.

13.4 Design and Implementation of Energy-Efficient Architectures

Designing and implementing energy-efficient post-quantum IoT architectures for small, medium, and large-scale scenarios is essential. For instance, this could involve creating IoT architectures that can hold up against quantum attacks and guarantee the energy efficiency of cryptographic operations in resource-constrained IoT devices across various scales of deployment.

13.5 Development of Post-Quantum Internet Protocols

Designing and implementing post-quantum internet protocols targeted to devices that have constraints in resources is crucial. For example, this could involve developing communication protocols that are resilient to quantum attacks and optimizing their energy consumption for resource-constrained IoT devices.

13.6 Optimization of Coding Schemes

Designing and optimizing new coding schemes and techniques for developing post-quantum IoT code-based cryptosystems is necessary. For instance, researchers may need to explore the development of efficient error-correcting codes suitable for post-quantum cryptographic algorithms in IoT applications, ensuring data integrity and security.

13.7 Mathematical Modeling of Performance and Power Consumption

Mathematically modeling IoT device performance and power consumption is necessary. For example, this could involve developing mathematical models to analyze the performance and energy consumption of post-quantum cryptographic-based operations in primarily IoT devices, which are resource constrained, aiding in the optimization of IoT system designs.

13.8 Exploration of Quantum Key Distribution (QKD)

Studying physics-based methods for quantum-resistant key establishment, such as Quantum Key Distribution (QKD), is crucial. For instance, researchers may need to explore the integration of QKD protocols into IoT security architectures to establish secure communication channels resilient to quantum attacks in resource-constrained IoT environments.

13.9 Collaborative Efforts

Reference [8] The Post-Quantum Cryptography Alliance (PQCA) was launched in February 2024 by the Linux Foundation. This group includes major companies and researchers who are working together to create open-source software for post-quantum cryptography. Founding members are AWS, Cisco, Google, IBM, and IntellectEU. This initiative is crucial for developing and standardizing encryption methods that can withstand quantum computer attacks.

13.10 New Technologies and Standards

Reference [9] On August 13, 2024, the US National Institute of Standards and Technology (NIST) introduced three new cryptographic standards to protect against quantum computer threats: ML-KEM, ML-DSA, and SLH-DSA. These standards, known as FIPS 203, 204, and 205, will help secure IoT networks from upcoming quantum risks.

13.11 Funding Opportunities

There will likely be more funding available as the need for quantum-resistant cryptography grows. Both government and private sectors are expected to invest in research and development in this area, fostering innovation and the development of strong post-quantum cryptographic solutions.

13.12 Industry Partnerships

Reference [10] Collaborations, like the one between Quantum Xchange and Juniper Networks, are becoming more important. This partnership aims to provide secure data and communication networks that can resist quantum threats for Managed Service Providers, government agencies, and businesses.

13.13 Industry Initiatives

Reference [11] The Emerging Payments Association Asia (EPAA) has created a group focused on post-quantum cryptography. This group brings together major financial service providers to promote awareness and develop solutions for secure transactions in a world with quantum computing. Founding members include EPAA, IBM, HSBC, AP+, and PayPal.

14 Practical Implementations of Quantum Cryptography in IoT Scenarios

Quantum cryptography is gradually on the verge of becoming a cornerstone in securing IoT frameworks. The convergence of quantum cryptography and IoT is not just theoretical; several research projects and pilot implementations are actively exploring its practical applications.

14.1 Quantum-Resistant Schemes in 5G-Enabled IoT

Reference [12] Recent studies have emphasized the integration of quantum-resistant cryptographic schemes within 5G-enabled IoT environments. These schemes are designed to counter the potential threats posed by quantum computing to traditional cryptographic methods. Lightweight ciphers such as ARX and ASCON are being advocated for resource-constrained IoT devices to enhance security without compromising performance. These ciphers, coupled with Quantum Key Distribution (QKD), provide a robust framework to secure communications against eavesdropping, a significant concern in IoT systems.

14.2 Pilot Implementations in Smart Cities

Smart city initiatives are at the forefront of applying quantum cryptography in real-world IoT scenarios [13]. For example, in a pilot project in Europe, QKD is being integrated into the smart grid systems of a city to secure communication between sensors and control units. This integration ensures that any attempt at eavesdropping or tampering is instantly detected, making the energy distribution network more resilient to cyber-attacks.

14.3 IoT and Quantum Sensors in Smart Farming

Quantum sensors, capable of unprecedented precision, are being utilized in IoT frameworks for applications like intelligent farming and innovative medical infrastructure. These sensors, when paired with quantum cryptography, ensure that the data transmitted is not only accurate but also secure from any unauthorized access [14]. In smart farming in rural areas, for instance, quantum-enhanced sensors monitor environmental conditions, and the data is encrypted using quantum keys to maintain the confidentiality and integrity of the information.

14.4 IoT-Enabled Healthcare Ecosystem in Quantum Computing

Reference [15] A notable development in healthcare IoT security is the introduction of a lightweight two-factor authentication protocol tailored for IoT-enabled healthcare ecosystems. This protocol is built to withstand both current and future quantum threats. It offers several key features, including ease of use without requiring memory, ensuring user anonymity, providing mutual authentication, and protecting against various types of attacks such as tampering, theft of biometric data, and insider threats.

15 Conclusion

The quick expansion of the Internet of Things exhibits both opportunities and significant security challenges. As the number of interconnected devices continues to rise, the need to safeguard the vast amount of data generated by IoT applications has become more critical than ever.

Organizations across various industries are actively investing in research and development efforts to bolster IoT security through quantum cryptography. Collaborative initiatives between academic institutions, research laboratories, and industry stakeholders are driving advancements in quantum-resistant algorithms and cryptographic protocols tailored for IoT environments. Moreover, partnerships between quantum technology firms and IoT solution providers are paving the way for the integration of quantum-secure communication methods into existing and future IoT infrastructures.

As IoT devices become more integrated into our daily lives, addressing security concerns and implementing quantum-safe security measures will be paramount in building a resilient IoT ecosystem. By recognizing the impact of quantum attacks on IoT and embracing the potential of quantum cryptography, we can make way for a safer and more reliable IoT landscape. Ultimately, the adoption of quantum-safe security measures holds the potential to empower innovation while guaranteeing the reliability and privacy of IoT data in the midst of evolving cybersecurity challenges.

References

1. The Tipping Point: Exploring the Surge in IoT Cyberattacks Globally (2023, April 11). https://blog.checkpoint.com/security/the-tipping-point-exploring-the-surge-in-iot-cyberattacks-plaguing-the-education-sector/
2. Securing the Future Internet of Things with Post-Quantum Cryptography (2021, December 9). https://doi.org/10.48550/arXiv.2206.10473
3. Methodology for Hacking IoT: From Chip to Cloud (2021, June 22). https://www.praetorian.com/blog/how-to-hack-iot/

4. Quantum Cryptography for Internet of Things Security (2019, September 3). https://medium.com/@cybertec/quantum-hacking-unravelling-the-threat-landscape-9c8f913e2f6a
5. From Pre-Quantum to Post-Quantum IoT Security: A Survey on Quantum-Resistant Cryptosystems for the Internet of Things (2020, July). https://ieeexplore.ieee.org/document/8932459
6. Securing IOT Network through Quantum Key Distribution (2019, April). https://www.ijitee.org/wp-content/uploads/papers/v8i6s4/F11410486S419.pdf
7. Quantum Cryptography for Internet of Things Security (2019, July 1). https://www.sciencedirect.com/science/article/pii/S1674862X19300345
8. Post-Quantum Cryptography Alliance Launches to Advance Post-Quantum Cryptography (2024, February 06). https://www.linuxfoundation.org/press/announcing-the-post-quantum-cryptography-alliance-pqca
9. NIST Releases Three Post-Quantum Cryptography Standards (2024, August 20). https://www.hklaw.com/en/insights/publications/2024/08/nist-releases-three-post-quantum-cryptography-standards
10. Quantum Cryptography Market Size, Share, Growth & Industry Analysis, By Offerings (Solutions, Services), By Application (Network Security, Application Security, Database Security), By End-Use Industry (Government & Defense, BFSI, Healthcare, Telecommunication, Other End Use Industry) and Regional Analysis, 2024–2031 (2024, May). https://www.kingsresearch.com/quantum-cryptography-market-747#:~:text=March%202024%20(Partnership):%20Quantum,data%20security%20and%20consumer%20privacy.
11. Charting a path to quantum-safe transformation through industry initiatives (2024, April 25). https://www.ibm.com/quantum/blog/quantum-safe-consortia
12. A roadmap from classical cryptography to post-quantum resistant cryptography for 5G-enabled IoT: Challenges, opportunities and solutions (2023, December). https://www.sciencedirect.com/science/article/abs/pii/S2542660523002731
13. Linking QKD Testbeds across Europe (2024, January 31). https://www.mdpi.com/1099-4300/26/2/123
14. A Quantum Leap Towards More Effective and Efficient Smart Farming in Rural Areas Using IOT (2023, March 7–9). https://www.researchgate.net/profile/Nasser-Almuraqab/publication/371723981_A_Quantum_Leap_Towards_More_Effective_and_Efficient_Smart_Farming_in_Rural_Areas_Using_IOT/links/6491e8ba95bbbe0c6edd355d/A-Quantum-Leap-Towards-More-Effective-and-Efficient-Smart-Farming-in-Rural-Areas-Using-IOT.pdf
15. Lightweight Two-Factor-Based User Authentication Protocol for IoT-Enabled Healthcare Ecosystem in Quantum Computing (2022, September 22). https://doi.org/10.1007/s13369-022-07235-0

Quantum-Powered Intelligence: Unleashing the Potential of IoT in the Quantum Computing Era

Pawan Whig, Balaram Yadav Kasula, Nikhitha Yathiraju, Anupriya Jain, Seema Sharma, and Sandeep Kautish

1 Introduction

The convergence of quantum computing and the Internet of Things (IoT) marks a groundbreaking paradigm shift in the realm of technology [1]. As we stand at the threshold of this transformative era, it becomes imperative to explore the intricate interplay between these two cutting-edge fields [2]. This introduction provides a comprehensive overview of the synergy between quantum computing and IoT, unraveling the potential impact on various aspects of our digital landscape [3].

Quantum computing, with its foundation in the principles of quantum mechanics, stands poised to revolutionize traditional computing paradigms [4]. Unlike classical computers that rely on bits for processing information, quantum computers leverage qubits, which can exist in multiple states simultaneously [5]. This inherent capability allows quantum computers to perform complex calculations at unprecedented speeds, offering a tantalizing glimpse into the future of computational power [6]. As quantum computing progresses from theoretical concepts to practical applications, its potential to reshape industries becomes increasingly evident [7]. From optimization problems to cryptography and drug discovery, quantum computing holds promise in solving challenges that have long stumped classical computers [8]. The introduction of quantum supremacy—a milestone where quantum computers outperform classical counterparts in specific tasks—has ignited a new era of exploration and innovation.

P. Whig (✉)
Vivekananda Institute of Professional Studies-TC New Delhi, New Delhi, India
e-mail: pawanwhig@gmail.com

B. Y. Kasula · N. Yathiraju
University of The Cumberlands, Williamsburg, KY, USA

A. Jain · S. Sharma
School of Computer Applications MRIIRS Faridabad, Faridabad, India

S. Kautish
LBEF Campus, Kathmandu, Nepal

S. B. Goyal et al. (eds.), *Quantum Computing, Cyber Security and Cryptography*,
https://doi.org/10.1007/978-981-96-4948-8_14

The Rise of the Internet of Things (IoT)

In parallel, the rise of the Internet of Things has ushered in an era of unprecedented connectivity. IoT is characterized by a vast network of interconnected devices, from smart home appliances and wearables to industrial sensors and autonomous vehicles [9]. These devices generate an immense volume of data, forming the backbone of a data-driven society. The seamless integration of physical devices with digital systems has paved the way for enhanced automation, efficiency, and insights across various domains [10].

As IoT proliferates, so do the challenges associated with its massive data processing requirements, security concerns, and the need for efficient algorithms [11]. The traditional computational infrastructure faces limitations in addressing these challenges, prompting the exploration of innovative solutions. This is where the synergy with quantum computing becomes particularly intriguing [12].

Unveiling the Synergy: Quantum Computing Meets IoT

The intersection of quantum computing and IoT holds the promise of overcoming some of the most pressing challenges faced by both fields [13]. One of the primary areas of convergence lies in data security. As IoT devices continue to permeate various aspects of our lives, securing the sensitive information they generate becomes paramount. Quantum computing's prowess in cryptography offers a potential solution by leveraging quantum key distribution and other quantum-safe encryption methods [14].

Moreover, the exponential growth in data generated by IoT devices demands efficient processing capabilities. Quantum computing's parallelism and ability to handle massive datasets make it an ideal candidate to meet the computational demands of IoT applications [15]. From optimizing supply chains to analyzing vast datasets for actionable insights, the collaboration between quantum computing and IoT opens avenues for unprecedented efficiency gains [16].

Navigating the Landscape of Quantum-Powered IoT Applications

As we delve deeper into this synergy, it becomes crucial to explore the practical applications that arise from the integration of quantum computing and IoT. Quantum-powered intelligence has the potential to redefine various industries, including healthcare, finance, logistics, and more [17].

In healthcare, for instance, the collaboration between quantum computing and IoT can lead to advancements in personalized medicine, drug discovery, and medical diagnostics. Quantum algorithms can sift through vast genomic datasets generated by IoT-enabled devices, providing insights into individualized treatment plans and disease prevention [18].

In the financial sector, where data security and transaction speed are paramount, quantum computing can enhance encryption methods and optimize complex financial models [19, 20]. This not only fortifies the integrity of financial transactions but also opens avenues for innovative financial services. As we embark on this exploration of quantum computing and IoT synergy, the chapters that follow will delve into specific aspects, challenges, and opportunities presented by this convergence. From addressing the security concerns of a hyperconnected world to unlocking the full potential of quantum-powered IoT applications, this journey promises to unravel the mysteries of a future where quantum computing and IoT intertwine seamlessly. Join us as we navigate the quantum landscape, unlocking the doors to a future where intelligence is not just artificial but quantum-powered.

This Table 1 literature review highlights the main focus, key findings, and research gaps identified in each reference. The research gaps point to areas where further investigation and exploration are needed to enhance the understanding of the impact of quantum computing on IoT and related technologies.

2 Foundations of Quantum Computing

Quantum computing is built upon the principles of quantum mechanics, a branch of physics that explores the behavior of matter and energy at the smallest scales—atomic and subatomic levels. At its core, quantum computing diverges from classical computing by utilizing quantum bits, or qubits, as the fundamental unit of information as shown in Fig. 1. Unlike classical bits, which can exist in a state of 0 or 1, qubits can exist in multiple states simultaneously, a phenomenon known as superposition. This unique property allows quantum computers to process a vast number of possibilities simultaneously, enabling them to perform certain calculations exponentially faster than their classical counterparts. Quantum gates, analogous to classical logic gates, manipulate qubits to perform operations, and entanglement, another quantum phenomenon, enables qubits to be correlated in such a way that the state of one qubit instantaneously influences the state of another, regardless of the distance between them [19–24]. These foundational principles, superposition, entanglement, and the use of qubits, distinguish quantum computing and set the stage for its potential to revolutionize computational capabilities across various domains.

2.1 Quantum Mechanics Primer

Quantum Mechanics serves as the theoretical foundation for understanding the behavior of matter and energy at the quantum level. At this microscopic scale, classical mechanics no longer holds, and instead, quantum mechanics describes the probabilistic nature of particles. Key principles include superposition, where particles can exist in multiple states simultaneously, and wave-particle duality, highlighting the dual nature of matter and energy. The Uncertainty Principle, formulated by Werner Heisenberg, underscores the inherent limits in simultaneously measuring certain pairs of properties, such as position and momentum. Understanding these quantum principles is crucial for grasping the fundamentals that govern the behavior of particles, forming the basis for the design and functionality of quantum computing systems.

Table 1 Literature review with research gap

References	Main focus	Key findings	Research gap
[1]	Quantum computing integration with other ICCT technologies	Advances in quantum computing technology	Limited exploration of specific ICCT technologies integrated with quantum computing
[3]	ML-based RIS-enhanced IoT systems	Comprehensive review on ML-based RIS-enhanced IoT systems	Lack of exploration on the impact of quantum computing on ML-based RIS-enhanced IoT systems
[5]	Secure IoT in the era of quantum computers	Identification of bottlenecks in securing IoT against quantum threats	Limited discussion on practical solutions and countermeasures to mitigate identified bottlenecks
[8]	Pre-quantum to post-quantum IoT security	Survey on quantum-resistant cryptosystems for IoT	Limited discussion on the practical implementation and challenges of deploying post-quantum cryptographic solutions in IoT
[10]	Quantum computing applications for IoT	Exploration of quantum computing applications for IoT	Limited coverage of challenges and limitations in implementing quantum computing applications in IoT
[12]	Hardware Security for IoT in the Quantum Era	Survey on hardware security challenges in IoT with the advent of quantum computing	Limited exploration of practical solutions and strategies for securing IoT hardware in the quantum era
[14]	Use of quantum communications for securing IoT devices	Investigation of quantum communications for IoT security	Lack of discussion on scalability and practical challenges in implementing quantum communications for widespread IoT deployment
[16]	Quantum computing in Machine Learning	Challenges and opportunities of quantum computing in machine learning	Limited exploration of practical challenges in integrating quantum computing with machine learning for IoT applications
[18]	Impact of Quantum Computing on Security in Emerging Technologies	Examination of the impact of quantum computing on security in emerging technologies	Limited discussion on the specific security challenges posed by quantum computing in various emerging technologies
[19]	Analysis of Quantum Computing Spanning IoT and Image Processing	Analysis of quantum computing in IoT and image processing	Lack of exploration on the combined impact of quantum computing on both IoT and image processing

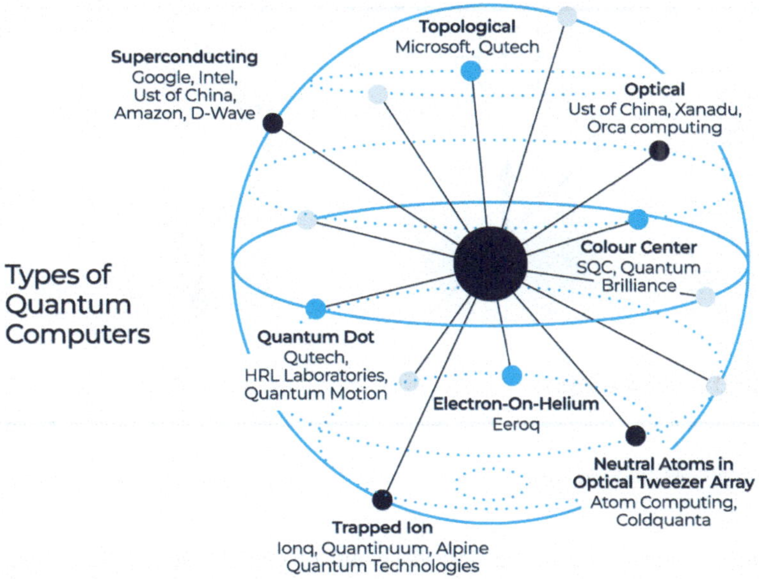

Fig. 1 Foundations of quantum computing

2.2 *Qubits and Quantum Gates*

In the realm of quantum computing, quantum bits, or qubits, serve as the fundamental building blocks as shown in Fig. 2. Unlike classical bits, qubits can exist in a superposition of states, representing both 0 and 1 simultaneously. This property exponentially expands the computational possibilities. Quantum gates, the analogs of classical logic gates, manipulate qubits to perform quantum operations. Operations on qubits leverage principles such as superposition and entanglement, enabling the creation of quantum circuits. The unique computational power of quantum computers arises from their ability to process information in parallel through these qubits and gates. Understanding qubits and quantum gates is essential for designing algorithms that harness the advantages of quantum parallelism and entanglement.

2.3 *Quantum Supremacy and Its Implications*

Quantum supremacy represents a milestone where a quantum computer outperforms the most powerful classical computers in solving specific problems. This achievement, first demonstrated by Google in 2019, signifies the practical viability of quantum computing. Quantum supremacy doesn't imply superiority in all types of computations, but it showcases that, for certain tasks, quantum computers can provide exponential speedup over classical counterparts. The implications of

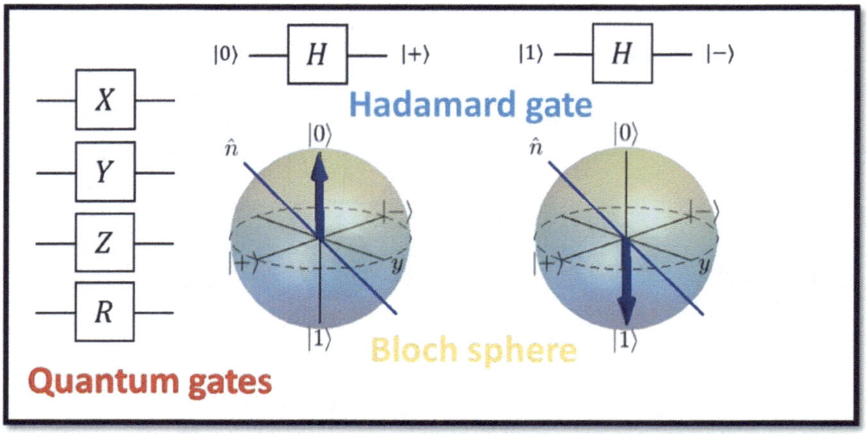

Fig. 2 Qubits and quantum gates

quantum supremacy extend beyond the realm of computation, influencing fields such as cryptography and optimization. As quantum computers advance, they have the potential to revolutionize industries by solving complex problems that were previously intractable for classical computers. However, this progress also raises security concerns, particularly regarding encryption methods that may become vulnerable to quantum attacks. Quantum supremacy, therefore, marks a significant step toward realizing the transformative potential of quantum computing and prompts a reassessment of the security landscape in the digital age.

3 Key Concepts in the Internet of Things (IoT)

The Internet of Things (IoT) is founded on a set of key concepts that define its transformative nature in the digital landscape. At its core, IoT represents a network of interconnected devices that communicate and share data through the internet. These devices, ranging from everyday objects like home appliances and wearable devices to industrial sensors and machinery, are embedded with sensors, actuators, and communication technologies. The key concept of connectivity enables these devices to seamlessly exchange information, creating a web of data that can be harnessed for various purposes. Another crucial aspect is the notion of data generation, where IoT devices continuously produce vast amounts of data, providing insights into real-time processes and user behaviors. The architecture of IoT involves the integration of hardware, software, and communication protocols to facilitate this interconnected ecosystem. Additionally, the concept of autonomy allows IoT devices to operate independently, making decisions based on the data they collect. Security and privacy considerations are integral key concepts, given the sensitive nature of the data transmitted between devices. The convergence of these key concepts shapes the foundation of IoT, ushering in a new era of connectivity, automation, and data-driven decision-making across diverse sectors.

3.1 Evolution of IoT

The Evolution of the Internet of Things (IoT) traces a remarkable journey from its inception to the present landscape of interconnected devices. The concept emerged as a vision of a seamlessly connected world, where everyday objects could communicate and share data to enhance efficiency and convenience. The early stages saw the integration of RFID tags and sensors, paving the way for the evolution of smart devices. The proliferation of internet connectivity, advancements in wireless communication technologies, and the development of low-power, high-performance sensors have been instrumental in propelling the evolution of IoT. From smart homes and wearables to industrial applications, the evolution of IoT reflects a dynamic progression, continually expanding its scope and impact on various facets of our lives.

3.2 IoT Architecture and Components

The architecture of the Internet of Things (IoT) encompasses a sophisticated framework designed to facilitate seamless communication and data exchange among connected devices. The core components of IoT architecture include the perception layer, which involves sensors and actuators that gather data from the physical environment. The data is then transmitted to the network layer, where communication protocols ensure efficient and secure transfer. The third layer, the middleware, manages data processing, storage, and device management. The application layer utilizes processed data for specific use cases, offering insights and enabling real-world applications. Additionally, edge computing has emerged as a critical component, allowing data processing to occur closer to the source, reducing latency and improving overall system efficiency. The intricate interplay of these components forms the foundation of IoT architecture, driving its functionality and adaptability.

3.3 Challenges in Traditional IoT Computing

While the Internet of Things (IoT) has witnessed substantial growth, it also faces a set of challenges in traditional computing paradigms. One significant challenge lies in the massive volume of data generated by IoT devices, leading to issues related to data storage, transmission, and processing. Traditional computing architectures often struggle to handle the sheer scale and diversity of IoT data. Security concerns pose another challenge, as the interconnected nature of IoT devices creates potential vulnerabilities, making them susceptible to cyber-attacks. Furthermore, the need for efficient energy utilization in IoT devices is crucial, especially in applications with limited power resources. Addressing these challenges requires innovative solutions, such as the integration of edge computing, advancements in security protocols, and the development of more energy-efficient hardware. As IoT continues to evolve, overcoming these challenges becomes imperative to unlock its full potential and ensure a robust and secure connected ecosystem.

4 Security in the Quantum-IoT Landscape

The intersection of quantum computing and the Internet of Things (IoT) introduces both opportunities and challenges, particularly in the realm of security. As traditional cryptographic methods face the risk of being compromised by quantum algorithms, ensuring the security of IoT devices and data becomes a paramount concern.

Quantum computing poses a unique threat to conventional encryption techniques commonly employed in securing IoT communications. Classical encryption algorithms, such as RSA and ECC, rely on the difficulty of certain mathematical problems, which can be efficiently solved by quantum algorithms like Shor's algorithm. This implies that the encryption methods widely used in current IoT systems could become vulnerable to rapid decryption once large-scale, fault-tolerant quantum computers become a reality.

To counteract this potential threat, the Quantum-IoT landscape explores quantum-safe cryptographic algorithms. These algorithms leverage the principles of quantum mechanics to create encryption methods that remain secure even in the face of quantum attacks. Quantum key distribution (QKD) is one such quantum-safe solution that utilizes the quantum properties of particles to establish secure communication channels, detecting any eavesdropping attempts.

Moreover, the integration of quantum-resistant algorithms is not the sole consideration. Quantum entanglement, a phenomenon allowing correlated states between particles, opens new possibilities for secure communication in the Quantum-IoT landscape. The entanglement-based communication protocols offer a level of security that is theoretically unattainable with classical communication systems.

However, with these advancements in quantum security, challenges persist. The practical implementation of quantum-resistant algorithms and entanglement-based protocols in resource-constrained IoT devices remains a hurdle. The computational demands and energy requirements of quantum-safe encryption can strain the limited resources of many IoT devices, necessitating the development of efficient solutions for secure communication.

The security landscape in the Quantum-IoT fusion is evolving to confront the unique challenges posed by the advent of quantum computing. As researchers strive to develop robust, efficient, and quantum-resistant security protocols, the Quantum-IoT paradigm holds the promise of ushering in a new era where the security of interconnected devices is fortified against the potential threats posed by quantum advancements.

4.1 Quantum Cryptography

Quantum cryptography represents a paradigm shift in securing communication channels by leveraging the principles of quantum mechanics. Unlike classical cryptography, which relies on the complexity of mathematical problems, quantum cryptography uses the inherent properties of quantum particles for secure communication. One key concept is the use of quantum key distribution (QKD) protocols, which involve the transmission of quantum states between communicating parties. The uncertainty principle ensures that any attempt to intercept the quantum states would disturb them, alerting the communicating parties to the presence of an eavesdropper. Quantum cryptography thus provides a fundamentally secure method for key exchange, laying the groundwork for robust encryption in the Quantum-IoT landscape.

4.2 Quantum Key Distribution

Quantum Key Distribution (QKD) is a pioneering technique within quantum cryptography designed to secure communication channels by distributing cryptographic keys in a quantum-safe manner. QKD leverages the principles of quantum mechanics, such as superposition and entanglement, to enable the secure exchange of encryption keys between two parties. The quantum nature of the key transmission ensures that any attempt to intercept the key introduces detectable disturbances. If interference is detected, the parties can abort the key exchange, preventing the creation of a compromised key. QKD holds significant promise in enhancing the security of IoT communication, providing a robust foundation for key management in the Quantum-IoT landscape.

4.3 Securing IoT Devices with Quantum Technologies

Securing Internet of Things (IoT) devices in the Quantum-IoT landscape involves integrating quantum technologies to fortify the overall security posture. Quantum-safe cryptographic algorithms play a pivotal role in ensuring the confidentiality and integrity of data transmitted between IoT devices. As quantum computing poses a potential threat to traditional encryption methods, the adoption of quantum-resistant algorithms becomes imperative. Additionally, the principles of quantum entanglement offer innovative possibilities for securing communication channels between IoT devices. Entanglement-based communication protocols can enhance the resilience of IoT networks against eavesdropping attempts, providing an extra layer of security. Implementing quantum technologies in securing IoT devices is a dynamic area of research, aiming to address the unique challenges posed by the intersection of quantum computing and the Internet of Things. The goal is to establish a quantum-safe foundation for the increasingly interconnected and data-driven IoT ecosystem, ensuring that the security measures evolve in tandem with the advancements in quantum technologies.

5 Quantum Computing Algorithms for IoT

The integration of quantum computing algorithms into the Internet of Things (IoT) holds the promise of unlocking unprecedented computational power and addressing complex problems that are challenging for classical computers. Several quantum algorithms show particular relevance in enhancing the capabilities of IoT applications:

1. **Quantum Machine Learning (QML)**: Quantum machine learning algorithms can significantly impact IoT by providing efficient solutions for data analysis and pattern recognition. These algorithms leverage quantum parallelism to process vast datasets simultaneously, offering advantages in tasks such as anomaly detection, predictive maintenance, and optimization within IoT environments.
2. **Quantum Optimization Algorithms**: Quantum optimization algorithms, like the Quantum Approximate Optimization Algorithm (QAOA), are well-suited for solving combinatorial optimization problems often encountered in IoT scenarios. This includes optimizing resource allocation, route planning for connected devices, and maximizing efficiency in complex networks.
3. **Quantum Search Algorithms**: Grover's algorithm, a quantum search algorithm, can be applied in IoT to accelerate the search process within large datasets. This is particularly valuable for tasks like database querying, enabling IoT systems to retrieve information more efficiently than classical search algorithms.
4. **Quantum Random Number Generation**: Quantum computers can generate true random numbers using quantum properties such as superposition and entanglement. This capability enhances the security of cryptographic protocols and ensures robust key generation for securing communications within the IoT ecosystem.
5. **Quantum Fourier Transform**: The Quantum Fourier Transform (QFT) is crucial for applications involving signal processing and frequency analysis within IoT sensor networks. QFT can efficiently extract meaningful information from sensor data, enabling more accurate and rapid analysis in applications such as environmental monitoring and industrial sensors.

While these quantum algorithms showcase the potential benefits for IoT, it's important to note that the practical implementation of quantum computing in IoT systems is still in its early stages. Overcoming challenges such as error correction, coherence time, and hardware scalability is essential for realizing the full potential of quantum algorithms in the IoT landscape. Nevertheless, as quantum technologies continue to advance, the synergy between quantum computing and IoT holds the promise of transforming the way we process, analyze, and derive insights from the vast amounts of data generated by interconnected devices.

5.1 *Optimization Algorithms*

Optimization algorithms play a pivotal role in enhancing the efficiency and performance of Internet of Things (IoT) systems. In the Quantum-IoT landscape, quantum optimization algorithms offer a quantum advantage in solving complex optimization problems. One notable algorithm is the Quantum Approximate Optimization Algorithm (QAOA), which leverages quantum parallelism to explore multiple solutions simultaneously. In the context of IoT, optimization algorithms can be applied to tasks such as resource allocation, energy management, and routing optimization. By harnessing the power of quantum parallelism, these algorithms hold the potential to revolutionize the decision-making processes within interconnected IoT environments, leading to improved resource utilization and overall system efficiency.

5.2 *Machine Learning in Quantum Computing*

The fusion of machine learning with quantum computing presents a transformative synergy for IoT applications. Quantum machine learning (QML) algorithms, designed to run on quantum computers, bring advantages in processing large datasets and solving complex problems. In the context of IoT, where massive amounts of data are generated continuously, QML algorithms can provide faster and more efficient solutions for tasks such as anomaly detection, predictive maintenance, and data classification. The quantum parallelism inherent in these algorithms allows for the exploration of multiple solutions simultaneously, offering a quantum speedup compared to classical machine learning approaches. Integrating quantum machine learning into IoT ecosystems has the potential to unlock new insights and improve the decision-making capabilities of interconnected devices.

5.3 Quantum Search Algorithms for IoT Big Data

Quantum search algorithms, with Grover's algorithm being a prominent example, address the challenge of efficiently searching through vast amounts of data—an issue particularly relevant in the context of IoT Big Data. Grover's algorithm provides a quadratic speedup over classical search algorithms, making it a powerful tool for IoT applications where quick retrieval of relevant information is crucial. In scenarios where IoT devices generate enormous datasets, such as in smart cities or industrial IoT deployments, quantum search algorithms can significantly reduce the computational time required to find specific information. This can lead to more rapid decision-making processes, improved response times, and enhanced overall efficiency within IoT systems dealing with substantial volumes of data.

As quantum computing continues to advance, the exploration and refinement of optimization, machine learning, and search algorithms in the Quantum-IoT landscape hold the key to unlocking new possibilities and addressing the evolving challenges of interconnected and data-intensive environments.

6 Practical Applications of Quantum-IoT Integration

The integration of quantum computing into the Internet of Things (IoT), often referred to as Quantum-IoT integration, opens the door to a wide array of practical applications across various industries. Here are several examples of how this convergence can be applied in real-world scenarios:

1. **Enhanced Security with Quantum Key Distribution (QKD)**: Quantum-IoT integration enhances security by employing Quantum Key Distribution (QKD) protocols. These protocols use the principles of quantum mechanics to secure communication channels between IoT devices, providing a quantum-safe foundation for protecting sensitive data in applications like healthcare, finance, and critical infrastructure.
2. **Optimized Supply Chain Management**: Quantum computing's optimization algorithms can revolutionize supply chain management within the IoT ecosystem. Tasks such as route optimization for delivery vehicles, inventory management, and demand forecasting can benefit from quantum optimization algorithms, leading to more efficient and cost-effective supply chain operations.
3. **Drug Discovery and Healthcare Optimization**: Quantum computing's ability to process complex datasets is particularly valuable in drug discovery and healthcare applications within IoT. Quantum machine learning algorithms can analyze genomic data from IoT-connected medical devices, facilitating personalized medicine, identifying potential drug candidates, and optimizing treatment plans for individuals.

4. **Financial Services and Portfolio Optimization**: In the financial sector, Quantum-IoT integration can be leveraged for optimizing investment portfolios. Quantum optimization algorithms can efficiently solve complex financial modeling problems, leading to improved risk management, asset allocation, and the development of innovative financial products.
5. **Smart Cities and Infrastructure Management**: Quantum computing's capabilities can be harnessed in managing smart cities and infrastructure. Quantum search algorithms can quickly process and analyze data from various IoT sensors, facilitating real-time decision-making for traffic management, energy consumption optimization, and environmental monitoring.
6. **Climate Modeling and Environmental Monitoring**: Quantum-IoT integration is instrumental in addressing environmental challenges. Quantum computing algorithms can process and analyze vast datasets from IoT devices deployed for climate modeling, weather prediction, and environmental monitoring. This contributes to more accurate predictions and proactive measures for climate change mitigation.
7. **Enhanced Machine Learning for IoT Analytics**: Quantum machine learning algorithms can improve the analytics capabilities of IoT systems. Tasks such as anomaly detection, pattern recognition, and predictive maintenance can benefit from the quantum parallelism and computational speedup offered by quantum computing, leading to more accurate and timely insights.
8. **Efficient Energy Management in Smart Grids**: Quantum-IoT integration can optimize energy management in smart grids. Quantum optimization algorithms can address complex problems related to energy distribution, load balancing, and grid stability, contributing to more efficient and sustainable energy systems.

As quantum technologies continue to mature and become more accessible, the practical applications of Quantum-IoT integration are expected to expand, offering innovative solutions to challenges in diverse industries and contributing to the advancement of interconnected, intelligent systems.

6.1 Healthcare

In the healthcare sector, the integration of quantum computing into the Internet of Things (Quantum-IoT) holds significant promise for transformative applications. Quantum machine learning algorithms can analyze massive datasets generated by IoT-connected medical devices, facilitating personalized medicine, drug discovery, and treatment optimization. Quantum cryptography ensures secure communication of sensitive patient data, enhancing privacy and confidentiality. Additionally, the use of quantum algorithms for optimization can improve healthcare logistics, such as efficient routing of medical supplies and equipment within hospital networks.

6.2 Finance

Quantum-IoT integration has profound implications for the financial industry. Quantum computing's optimization algorithms can revolutionize portfolio management, risk assessment, and algorithmic trading. In IoT-connected financial systems, quantum-safe cryptography ensures the secure transmission of financial data. Quantum machine learning can enhance fraud detection, credit scoring, and customer relationship management. The intersection of quantum computing and IoT in finance contributes to more efficient and secure financial operations.

6.3 Logistics and Supply Chain

The logistics and supply chain industry benefits significantly from Quantum-IoT integration. Quantum optimization algorithms can address complex logistics challenges, such as route optimization for delivery vehicles, warehouse management, and demand forecasting. In an IoT-connected supply chain, quantum cryptography secures the communication and exchange of critical supply chain data. This integration enhances the overall efficiency, reliability, and transparency of logistics and supply chain operations.

6.4 Smart Cities and Infrastructure

Quantum-IoT integration plays a crucial role in the development and management of smart cities and infrastructure. Quantum search algorithms can efficiently process data from various IoT sensors, enabling real-time decision-making for traffic management, energy distribution, and environmental monitoring. Quantum machine learning contributes to predictive maintenance of infrastructure assets, optimizing resource utilization. Quantum-safe cryptography ensures the secure communication of data in interconnected smart city systems, safeguarding against potential cyber threats.

The application of Quantum-IoT technologies in healthcare, finance, logistics, and smart cities showcases the versatility and transformative potential of this convergence. As quantum technologies continue to advance, the integration with IoT is poised to address complex challenges and unlock innovative solutions in these critical sectors.

7 Case Study

Comparative Analysis of Classical and Quantum Computers in RSA Decryption for IoT Security

In the rapidly evolving landscape of the Internet of Things (IoT), security is a paramount concern. One of the widely used cryptographic algorithms for securing communications in IoT devices is RSA (Rivest–Shamir–Adleman). However, with the advent of quantum computing, traditional RSA encryption faces new challenges. This case study delves into the comparative analysis of classical and quantum computers in the context of RSA decryption without private keys for IoT security.

RSA encryption relies on the difficulty of factoring the product of two large prime numbers. Classical computers use algorithms like the General Number Field Sieve (GNFS) for factoring large numbers. On the other hand, quantum computers leverage algorithms like Shor's algorithm, which can potentially solve the factorization problem exponentially faster than classical algorithms.

Methodology:

To assess the impact of classical and quantum computing on RSA decryption in the IoT context, a series of experiments were conducted. The experiments involved the use of classical computers with state-of-the-art algorithms and quantum computers using simulated quantum processors.

Results:

1. **Classical Computing**:

   With classical computers, the decryption of RSA-encrypted messages without the private key is a time-consuming process.

   The efficiency of classical algorithms like GNFS depends on the size of the RSA key, making it computationally expensive for larger key sizes commonly used in IoT devices.
2. **Quantum Computing**:

   Shor's algorithm, when implemented on a quantum computer, exhibits a significant speedup in the factorization process.

   Quantum parallelism allows for the simultaneous evaluation of multiple possibilities, drastically reducing the time required to decrypt RSA-encrypted messages.
3. **Security Implications**:

   The advent of quantum computers poses a potential threat to the security of RSA-encrypted communications in IoT devices.

   As quantum computers advance, larger RSA key sizes may be necessary to maintain security, increasing computational requirements for both classical and quantum decryption.

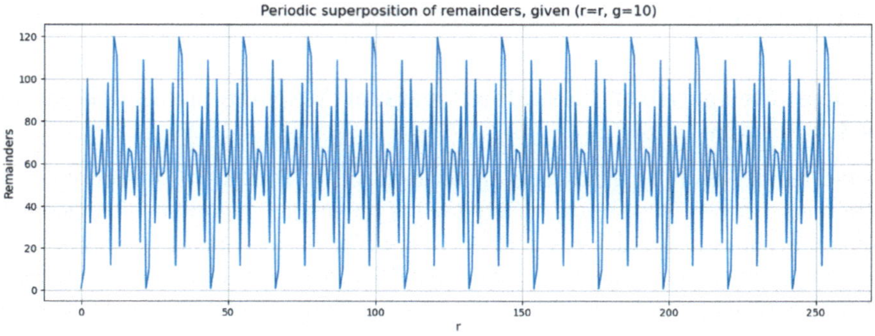

Fig. 3 Periodic superposition of remainders

4. **Practical Considerations**:

   The practical implementation of quantum computers for RSA decryption is currently in the experimental stage, with challenges such as error correction and decoherence mitigation.

The provided Python code utilizes the Matplotlib library to generate a graphical representation of periodic superposition of remainders. The Fig. 3 is created with a specified figure size of 15 by 5 inches. The x-axis corresponds to the variable "r," and the y-axis represents the "Remainders." The data for the plot is assumed to be stored in the variables "superposition_A" and "superposition_B." The title of the plot is dynamically generated based on the values of "r" and "g," with the caption indicating that it represents the periodic superposition of remainders given a specific relationship (r=r, g=g). The x-axis represents the variable "r," and the y-axis depicts the corresponding remainders. The grid lines aid in visual clarity. This visualization is valuable for understanding the periodic patterns and relationships between remainders under the specified conditions, providing a visual insight into the mathematical relationships encoded in the data.

In the evolving landscape of IoT security, the comparative analysis highlights the potential vulnerabilities of RSA encryption in the face of quantum computing. While classical computers struggle with the computational intensity of RSA decryption, quantum computers pose a significant threat by leveraging Shor's algorithm for rapid factorization. As quantum computing technology progresses, the need for quantum-resistant cryptographic algorithms becomes imperative to ensure the long-term security of IoT communications. This case study underscores the importance of staying ahead in cryptographic advancements to address emerging threats in the realm of IoT security.

8 Conclusion

The exploration of Quantum-IoT integration in this chapter has illuminated a path toward a future where the amalgamation of quantum computing and the Internet of Things (IoT) stands as a cornerstone of technological progress. Through the lens of healthcare, finance, logistics, supply chain, and smart cities, the symbiotic relationship between quantum computing and IoT has emerged as a catalyst for innovation. Quantum algorithms, cryptography, and optimization techniques showcased their transformative potential, enhancing security, computational efficiency, and offering groundbreaking applications. As we draw conclusions from this journey, it is evident that Quantum-IoT integration transcends mere collaboration; it signifies a paradigm shift, a quantum leap into a realm where interconnected devices operate with unprecedented intelligence and security. This chapter sets the stage for a future where the fusion of quantum computing and IoT continues to redefine the boundaries of what is possible in the digital era.

9 Future Scope

Looking ahead, the future scope of Quantum-IoT integration encompasses several key areas. Advancements in quantum algorithms tailored for IoT applications are paramount, ensuring the continual evolution of optimization, machine learning, and search algorithms. Simultaneously, improvements in quantum hardware, including qubit coherence and error correction, are vital for scalability and reliability. Innovations in quantum-safe cryptographic protocols will fortify the security foundations of IoT ecosystems, paving the way for a more secure interconnected world. The integration of Quantum-IoT with edge computing holds promise for more efficient, real-time solutions, addressing latency concerns. Industry-specific applications, interdisciplinary collaboration, and a holistic approach to research will collectively drive Quantum-IoT toward a future where its transformative potential is fully realized. In this dynamic landscape, Quantum-IoT is poised to reshape industries, redefine computational paradigms, and usher in a future where quantum-enhanced intelligence becomes an integral part of our technological fabric.

References

1. P.S. Aithal, Advances and new research opportunities in quantum computing technology by integrating it with other ICCT underlying technologies. Int. J. Case Studies Bus. IT Educ. (IJCSBE) **7**(3), 314–358 (2023)
2. A. Daniel, M. Arvindhan, K. Bellam, N. Krishnaraj (Eds.).*Principles and Applications of Quantum Computing Using Essential Math*. IGI Global (2023)
3. S.K. Das, F. Benkhelifa, Y. Sun, H. Abumarshoud, Q.H. Abbasi, M.A. Imran, L. Mohjazi, Comprehensive review on ML-based RIS-enhanced IoT systems: basics, research progress and future challenges. Comput. Netw. **224**, 109581 (2023)

4. A. Kaushik, R. Narwal, Integration of quantum computing with IoT. Int. J. Eng. Adv. Technol. **9**(4), 1307–1311 (2020)
5. M. Schöffel, F. Lauer, C.C. Rheinländer, N. Wehn, Secure IoT in the era of quantum computers—where are the bottlenecks? Sensors **22**(7), 2484 (2022)
6. H. Alyami, M. Nadeem, W. Alosaimi, A. Alharbi, R. Kumar, B.K. Gupta, R.A. Khan, Analyzing the data of software security life-span: quantum computing era. Intell. Autom. Soft Comput. **31**(2), 707–716 (2022)
7. Z.S. Ageed, S.R. Zeebaree, R.H. Saeed, Influence of quantum computing on IoT using modern algorithms, in *2022 4th International Conference on Advanced Science and Engineering (ICOASE)*, pp. 194–199. IEEE (2022
8. T.M. Fernández-Caramés, From pre-quantum to post-quantum IoT security: a survey on quantum-resistant cryptosystems for the Internet of Things. IEEE Internet Things J. **7**(7), 6457–6480 (2019)
9. M. Younan, M. Elhoseny, A.A. Ali, E.H. Houssein, Quantum chain of things (qcot): a new paradigm for integrating quantum computing, blockchain, and internet of things, in *2021 17th International Computer Engineering Conference (ICENCO),* pp. 101–106. IEEE (2021)
10. M.S. Peelam, A.A. Rout, V. Chamola, Quantum computing applications for internet of things. IET Quant. Commun. (2023)
11. O.S. Althobaiti, M. Dohler, Cybersecurity challenges associated with the Internet of things in a post-quantum world. IEEE Access **8**, 157356–157381 (2020)
12. D. Dione, B. Seck, I. Diop, P.L. Cayrel, D. Faye, I. Gueye, Hardware security for IoT in the quantum era: survey and challenges. J. Inf. Secur. **14**(4), 227–249 (2023)
13. P. Agarwal, M. Alam, Exploring quantum computing to revolutionize big data analytics for various industrial sectors, in *Big Data Analytics*, pp. 113–130. Auerbach Publications (2021)
14. H.A. Al-Mohammed, E. Yaacoub, On the use of quantum communications for securing IoT devices in the 6G era, in *2021 IEEE International Conference on Communications Workshops (ICC Workshops),* pp. 1–6. IEEE (2021)
15. L. Malina, P. Dzurenda, S. Ricci, J. Hajny, G. Srivastava, R. Matulevičius, ...Q. Tang, Post-quantum era privacy protection for intelligent infrastructures. IEEE Access **9**, 36038–36077 (2021)
16. P. Sajwan, N. Jayapandian, Challenges and opportunities: quantum computing in machine learning, in *2019 Third International conference on I-SMAC (IoT in Social, Mobile, Analytics and Cloud)(I-SMAC),* pp. 598–602. IEEE (2019)
17. A.K. Sharma, M.S. Peelam, B.K. Chauasia, V. Chamola, QIoTChain: Quantum IoT-blockchain fusion for advanced data protection in Industry 4.0. *IET Blockchain* (2023)
18. A. Abuarqoub, S. Abuarqoub, A. Alzu'bi, A. Muthanna, The impact of quantum computing on security in emerging technologies, in *The 5th International Conference on Future Networks & Distributed Systems*, pp. 171–176 (2021)
19. P. Kamaleswari, A. Daniel, An analysis of quantum computing spanning IoT and image processing, in *Principles and Applications of Quantum Computing Using Essential Math*, pp. 107–124. IGI Global (2023)
20. R. Deepa, P. Nandakumar, Artificial intelligence in IoT and QC: future computing. Journal of Xidian University **5**(4), 338 (2021)
21. S. Kautish, A. Reyana, A. Vidyarthi, SDMTA: attack detection and mitigation mechanism for DDoS vulnerabilities in hybrid cloud environment. IEEE Trans. Industr. Inf. **18**(9), 6455–6463 (2022)
22. R. Kaur, S. Kautish, Multimodal sentiment analysis: a survey and comparison. Int. J. Serv. Sci. Manag. Eng. Technol. (IJSSMET) **10**(2), 38–58 (2019)
23. A. Naim, S.K. Kautish, (Eds.), *Building a Brand Image Through Electronic Customer Relationship Management*. IGI Global (2022)
24. S. Kautish, Online banking: a paradigm shift. E-Bus. ICFAI Publ. Hyderabad **9**(10), 54–59 (2008)

Quantum Computing and Generative Adversarial Networks (GANs): Navigating Ethical and Security Challenges

Wasswa Shafik

1 Introduction

Quantum computing[1] and Generative Adversarial Networks (GANs)[2] herald a new era in technology, promising unprecedented capabilities and opportunities while posing profound ethical, privacy, and security challenges [1]. With its roots in the mysterious field of quantum physics, quantum computing presents the alluring possibility of computational capability exponentially beyond that of conventional computers. In the meantime, GANs, a byproduct of the quickly developing science of artificial intelligence, have revolutionized our ability to produce fake data and material that is startlingly realistic [2]. The unexplored domain of quantum computing presents an enticing prospect of resolving unsolvable challenges for conventional computers. As we explore the complexities of quantum bits, superposition, and entanglement, we unveil the possible applications of these phenomena in several fields, for example, cryptography, optimization, drug discovery, and other domains [3]. Nevertheless, the recent advancements in computational abilities raise ethical considerations about safeguarding cryptographic systems and the conscientious use of these capacities.

The notion of privacy, which has historically been fundamental to safeguarding personal and data security, assumes a novel perspective within this framework. The implications of quantum computing on cryptography necessitate reassessing our current encryption standards and pursuing solutions that are immune to quantum attacks [4]. GANs pose a simultaneous challenge to conventional conceptions of

[1] https://www.ibm.com/topics/quantum-computing.

[2] https://en.wikipedia.org/wiki/generative_adversarial_network.

W. Shafik (✉)
School of Digital Science, Universiti Brunei Darussalam, Brunei, Brunei Darussalam
e-mail: wasswashafik@ieee.org

Dig Connectivity Research Laboratory, P. O. Box. 600040, Kampala, Uganda

S. B. Goyal et al. (eds.), *Quantum Computing, Cyber Security and Cryptography*,
https://doi.org/10.1007/978-981-96-4948-8_15

digital privacy by producing synthetic data that can be employed for manipulation or deception. Furthermore, the continuous progression of innovation also requires a thorough analysis of security implications [5]. The advent of quantum computing has brought up new vulnerabilities and the possibility of cyber-threats that were previously unimaginable. The expeditious advancement of GANs necessitates the development of resilient countermeasures against their potential malevolent applications, given the increasing persuasiveness and availability of deepfakes and synthetic content [6, 7].

This work aims to enhance the comprehension of the ethical, privacy, and security implications inherent in the domains of quantum computing and GANs, given the intricate problems and opportunities they present. As we traverse this unfamiliar domain, we aim to not only elucidate the complexities of these technologies but also offer perspectives and suggestions that can facilitate their conscientious incorporation into our swiftly progressing digital realm [8]. Simultaneously, GANs, a machine learning (ML) application, have enabled us to generate highly realistic images, text, and audio from digital platforms. They have stimulated innovation across various disciplines, encompassing art, fashion, medical imaging, and data augmentation. It is essential to acknowledge that GANs come with ethical implications [9]. These implications arise from their ability to produce deepfakes, synthetic misinformation, and intrusions on privacy, necessitating careful monitoring and appropriate solutions to address these concerns.

The interconnection of new technologies is exemplified by the smooth synergy observed between quantum computing and GANs. Quantum computing potentially enhances the training and optimization procedures of GANs, hence facilitating the generation of synthetic data with heightened realism and increased accessibility [10]. In contrast, GANs can be utilized to support quantum researchers in generating intricate quantum states and datasets, augmenting quantum computers' functionalities. This mutually beneficial association also requires a comprehensive approach to tackling ethical, privacy, and security concerns [11]. The progress of quantum computing and GANs will result in concurrent advancements in our capacity to generate and manipulate data, engage in research activities, and interact with technological systems. Therefore, it is crucial to establish inclusive frameworks incorporating both technologies, guaranteeing their responsible and ethical progression [12].

The area of quantum computing has significant promise in addressing previously unsolvable issues, offering possible breakthroughs in various domains such as healthcare, climate modeling, and logistics [13]. GANs are expanding the limits of artistic expression and data synthesis, presenting an array of boundless prospects in art, entertainment, and scientific investigation. Nevertheless, the utilization of this revolutionary capability entails significant obligations [14]. The ethical considerations presented by modern technologies are not simply theoretical concepts but rather urgent matters requiring prompt and focused examination. As the technological superiority of quantum computing is harnessed, it becomes imperative to address the issues of justice, accountability, and transparency in decision-making procedures that heavily rely on quantum algorithms. The emergence of GANs has presented a significant problem in detecting and combating the spread of deepfakes and synthetic misinformation [15]. GANs can generate compelling material, hence necessitating the development of robust systems to identify and counteract the dissemination of such deceptive media.

1.1 The Chapter Contribution

The study contribution is listed below:

- The study presents quantum computing, explains the fundamentals of quantum computing, discusses the potential benefits of quantum computing, highlights ethical considerations related to quantum computing, and explores security and privacy concerns specific to quantum computing.
- Illustrates the GANs, explains the basics of GANs and their applications, discusses the impact of GANs on various industries, presents the ethical concerns related to the use of GANs, and examines privacy and security issues associated with GANs.
- Details the ethical considerations discussing the ethical frameworks relevant to quantum computing and GANs, analyzing the ethical implications of quantum computing and GANs, including their potential for misuse, and providing examples of ethical dilemmas in these fields.
- Explains privacy concerns; explores the privacy issues arising from quantum computing and GANs; discusses data privacy, surveillance, and the risks to personal information; and considers privacy-enhancing technologies and techniques. The security challenges detailing security challenges in quantum computing and GANs, discussing the vulnerabilities and threats associated with these technologies, and extant strategies for securing quantum computing and GANs are presented.
- Presents the regulatory and legal frameworks describing existing and proposed regulations governing quantum computing and GANs, analyzing the effectiveness of current legal frameworks in addressing ethical, privacy, and security concerns.
- Real-world examples and case studies illustrate ethical, privacy, and security challenges in quantum computing and GANs.
- It further illustrates the mitigation strategies, offering some recommendations and strategies to address the identified ethical, privacy, and security challenges, and explores encryption, authentication, and other security measures.
- Finally, the study presents the future directions, recommendations of quantum computing, GANs, and conclusions.

1.2 *The Chapter Organization*

Section 2 presents quantum computing, explaining the fundamentals of quantum computing, discussing the potential benefits of quantum computing, highlighting ethical considerations related to quantum computing, and exploring security and privacy concerns specific to quantum computing. Section 3 illustrates the GANs, explores the basics of GANs and their applications, discusses the impact of GANs on various industries, presents the ethical concerns related to the use of GANs, and examines privacy and security issues associated with GANs. Section 4 details the ethical considerations, discusses the ethical frameworks relevant to quantum computing and GANs, analyzes the ethical implications of quantum computing and GANs, including their potential for misuse, and provides examples of ethical dilemmas in these fields. Section 5 presents some case studies in quantum computing and GANs, discussing the vulnerabilities and threats associated with these technologies and extant strategies for securing quantum computing and GANs. Section 6 illustrates the mitigation strategies, offering some recommendations and strategies to address the identified ethical, privacy, and security challenges and exploring encryption, authentication, and other security measures. Finally, Sect. 7 presents the future directions, recommendations, and conclusions.

2 Quantum Computing

This section highlights potential benefits, security, and privacy concerns specific to quantum computing, and ethical considerations related to quantum computing.

2.1 *Potential Benefits of Quantum Computing*

Although quantum computing is still in its infancy and faces technological obstacles, the potential advantages are enormous. These advantages are anticipated to become more widely available and transformative across various sectors and scientific fields as researchers and scientists continue to build quantum hardware and algorithms [9].

2.1.1 Drug Discovery and Material Science

The advantages of quantum computing are immense, even though it is still in its infancy and confronts technological difficulties. These advantages should grow more widely available and transformational across a vast number of companies and scientific fields as researchers and scientists continue to build quantum hardware and algorithms [11]. The impact of quantum computing on the field of material science is significant. This computational tool facilitates the simulation of material behavior at the quantum level, hence facilitating the identification of novel materials possessing extraordinary features previously undiscovered [16]. The materials above encompass improved superconductors designed to enhance energy transmission efficiency; unique battery materials engineered to prolong their lifespan; and lightweight, resilient materials tailored explicitly for aircraft use. The capacity to engineer materials with precise characteristics and capabilities holds significant ramifications across various sectors, encompassing renewable energy and electronics, fundamentally transforming the approach to material development and utilization [17].

2.1.2 Optimization

Quantum computing demonstrates exceptional proficiency in addressing intricate optimization problems pervasive across diverse industries. In logistics and transportation, using AI technology can enhance several aspects, such as route planning, vehicle scheduling, and supply chain management. This can result in cost reduction and a decrease in environmental effects [17]. Quantum computing can potentially increase various aspects of finance, including portfolio optimization, risk assessment, and trading methods. This is achieved by rapidly processing extensive datasets and intricate financial models. Likewise, within the realm of manufacturing, the utilization of this technology has the potential to enhance the allocation of resources and streamline production processes, hence resulting in heightened operational efficiency and diminished wastage [18]. These applications can effectively conserve significant resources for firms and enhance decision-making processes across several sectors.

2.1.3 Climate Modeling

Technology holds significant potential in climate modeling and simulation, as it can enhance the precision and comprehensiveness of climate pattern forecasts and their associated consequences. Climate models are characterized by intricate mathematical equations and vast datasets, rendering them computationally demanding (Smith, 2020). The computational speedup offered by quantum computing can significantly enhance the efficiency of climate simulations, facilitating a more comprehensive understanding of the intricacies of climate change dynamics, extreme weather occurrences, and long-term environmental patterns. An imperative aspect in formulating well-informed policy decisions and devising efficacious methods to alleviate the consequences of climate change is the acquisition of an expanded comprehension of climate science [1].

2.1.4 Supply Chain Management

It possesses the capability to enhance intricate supply chain networks through the optimization of goods flow, reduction of transportation expenses, and mitigation of delays. The utilization of this technology may effectively tackle various challenges, including demand forecasting, inventory management, and route optimization, leading to enhanced operational efficiency and cost reduction [1]. In addition, quantum computing has the potential to augment the resilience of supply chains through its ability to respond to disruptions and efficiently redirect resources swiftly. The capacity to efficiently negotiate complexity and uncertainties is paramount in the current era characterized by a global supply chain [19]. This capability equips organizations with the necessary tools to ensure the smooth delivery of goods and services to consumers.

2.2 Ethical Considerations Related to Quantum Computing

As this revolutionary technology develops, ethical issues relating to quantum computing are becoming more crucial. The creation, implementation, and application of quantum computing raise several significant ethical issues.

2.2.1 Cryptography, Security, and Arms Race

The ethical questions about the implications of quantum computing on security and cryptography are of utmost significance. The possibility of quantum computers compromising the cryptographic underpinnings that protect sensitive data, such as personal, financial, and governmental information, is a significant concern [20]. The focal point of the ethical predicament lies in the concept of responsible disclosure within the quantum computing community concerning the appropriate timing and way breakthroughs that have the potential to compromise encryption standards should be disseminated [21]. Achieving a harmonious equilibrium between scientific advancement and data safeguarding is imperative. There is a current endeavor to build encryption algorithms resistant to quantum attacks [22]. However, this initiative raises ethical concerns regarding the appropriate speed and level of transparency with which they should be implemented to safeguard data privacy. The expeditious advancement of quantum computing technology has elicited apprehensions over a potential escalation in the competition to achieve quantum dominance. The allocation of significant resources by nations and corporations toward quantum research and development has prompted ethical deliberations regarding the potential military applications of quantum capabilities [23]. The ethical obligation resides in preventing the advancement of quantum-based weaponry or surveillance systems that can potentially encroach upon individual privacy and human rights. Establishing standards and rules to manage these hazards poses challenges for the international community.

2.2.2 Inequality, Access, and Data Privacy

The ethical aspect of the equal availability of quantum computing resources and knowledge is essential. As the progression of quantum technology continues, there exists a potential for the amplification of disparities, wherein specific groups or nations may acquire preferential access to the advantages offered by quantum advancements [24]. In contrast, others may experience a lag in their adoption. Ethical considerations encompass the imperative to promote inclusivity and accessibility of quantum computing across many communities and nations [51]. It is imperative to prioritize efforts to narrow the digital divide while also considering the potential for worsening pre-existing inequities in technological accessibility. The ethical implications surrounding data privacy arise due to the ability of quantum computing to decipher sensitive information. Safeguarding the privacy of individual and organizational data in a quantum-powered era necessitates implementing resilient encryption techniques and technologies that enhance privacy [2]. Ethical considerations pertain to protecting personal information, financial records, and confidential data from illegal access. Organizations and governments are confronted with the ethical obligation to adopt rigorous security protocols to address the potential hazards posed by quantum technology to data privacy.

2.2.3 Ethics in Research and Environmental Impact

In quantum research, ethics are crucial. Transparency, accountability, and ethical research are values that scientists and organizations studying the potential and ramifications of quantum computing must uphold. Disclosing vulnerabilities and potential implications of quantum discoveries are ethically contentious issues [25]. The more enormous ethical consequences of their work must be considered by researchers, who must balance scientific advancement and reduce potential adverse effects, such as those relating to security and privacy. Quantum computers, on a large scale, often function under conditions of exceedingly low temperatures and exhibit substantial energy consumption. The ethical issue is the potential environmental consequences associated with quantum computing infrastructure. Incorporating sustainable practices in creating and operating quantum computing infrastructure is a crucial aspect of ethical issues [26]. It is imperative to consider the careful equilibrium between the prospective advantages of quantum computing and its energy consumption and environmental impact to ensure responsible technological progress [19].

2.2.4 Dual-Use Dilemma Global Collaboration

Like several nascent advancements, quantum technologies demonstrate the capacity for dual-use applications. Ethical issues encompass the delicate task of striking a balance between promoting the progress of useful applications, such as medication discovery and climate modeling, while also exercising caution to prevent the emergence of detrimental uses, such as creating quantum-enabled cyber-weapons or surveillance technologies [27]. Achieving an optimal equilibrium between promoting scientific advancement and mitigating the risks of potential misapplication necessitate careful ethical contemplation and the implementation of ethical frameworks and norms to regulate the evolution and use of quantum technologies. Large-scale quantum computers frequently function at very low temperatures and use much energy. The ethical issue here is how quantum computing facilities will affect the environment. Adopting sustainable methods in constructing and operating quantum computing infrastructure is a matter of ethics. Quantum computing's possible advantages must be weighed against its energy requirements and environmental impact if responsible technological development is to take place [28].

2.3 Data Privacy and Security Concerns Aspect to Quantum Computing

Privacy and security concerns concerning quantum computing are emerging as this transformative technology advances within this subsection. There are detailed explanations of these concerns.

2.3.1 Cryptographic Vulnerabilities

The emergence of quantum computing poses a significant challenge to the fundamental principles of contemporary cryptography. One example of an algorithm with the capability to factor huge numbers efficiently is Shor's algorithm. This technique is fundamental to several encryption schemes, such as RSA. With the advancement of quantum computers, there is a growing concern regarding their potential ability to compromise commonly employed encryption techniques, jeopardizing the security of critical information and communications [18]. The issue, as mentioned earlier, gives rise to significant apprehensions regarding privacy and security, as traditional encryption techniques may be rendered susceptible to quantum attacks, compromising the confidentiality of sensitive information.

2.3.2 Data Privacy in a Post-Quantum World and Quantum-Safe Cryptography

Given the potential of quantum computers to compromise current encryption systems, companies must prepare for a future characterized by a "post-quantum" landscape. The use of quantum-resistant encryption technologies and the enhancement of security protocols are crucial in safeguarding data privacy [29]. Nevertheless, the shift to quantum computing poses various difficulties, such as the potential vulnerability to data breaches during the migration phase and the requirement for a well-coordinated and prompt implementation of quantum-resistant encryption across different industries. The emergence of quantum computing poses a significant challenge to data privacy and security, necessitating the urgent development of encryption technologies and cryptographic protocols that can withstand quantum attacks. Ethical concerns emerge concerning the responsible dissemination of these developments [30]. The complex ethical challenge lies in the need to compromise between the progression of quantum-safe encryption and the prevention of premature knowledge sharing, potentially empowering bad actors to exploit quantum weaknesses.

2.3.3 Espionage and Surveillance and Quantum Key Distribution

The utilization of quantum computing may have substantial ramifications in the domains of surveillance and espionage. Quantum technologies can compromise encryption systems, enabling their utilization for unlawful data retrieval and intelligence collection [26]. The issue raises significant ethical considerations about the infringement upon individuals' privacy, the potential misuse of quantum capabilities for surveillance objectives, and the necessity of implementing protective measures to deter unwanted quantum surveillance [22]. Still, quantum computing poses a significant obstacle to conventional encryption methods. However, it also presents the possibility of employing quantum key distribution (QKD) as a viable means to provide secure communication channels. QKD utilizes the inherent characteristics of quantum mechanics to guarantee the security of encryption keys [31]. Nevertheless, the implementation of QKD encounters pragmatic obstacles and ethical deliberations over the presence and reachability of QKD infrastructure and the possibility of quantum assaults on QKD systems.

2.3.4 Quantum Network Security, Impact on Cryptocurrency and Blockchain

Robust security measures are crucial for quantum networks since they facilitate safe communication by utilizing quantum key distribution. Ethical considerations arise concerning safeguarding the quantum network infrastructure from both physical and cyber-threats [16]. Preserving privacy and security is of utmost importance in guaranteeing the integrity and confidentiality of quantum information transmitted across these networks. The security of blockchain and cryptocurrency systems is contingent upon using cryptographic approaches. The potential of quantum computing to compromise cryptographic hashes and signatures presents a significant threat to the security and confidentiality of blockchain transactions [17]. Ethical issues encompass the imperative to adopt proactive steps aimed at fortifying blockchain systems against quantum threats while safeguarding cryptocurrency users' privacy and security. To effectively tackle the privacy and security concerns associated with quantum computing, it is imperative to adopt a comprehensive strategy encompassing various aspects [32].

3 Generative Adversarial Networks

This section details the discussion on the impact of GANs on various industries, presents ethical concerns related to the use of GANs, and discusses some notable privacy and security issues associated with GANs.

3.1 Potential Benefits of Generative Adversarial Networks

GANs offer various potential benefits across various domains, primarily due to their ability to generate realistic and high-quality data and detailed explanations of some key potential benefits of GANs.

3.1.1 Enhancement Image and Synthesis

GANs have demonstrated notable advancements in the domains of image creation and improvement. Convolutional neural networks can produce visuals of exceptional realism, rendering them indispensable in several domains, such as computer graphics, entertainment, and design [33]. GANs have emerged as a valuable tool for artists and designers to create visually captivating artwork and immersive visual effects for movies and video games. GANs facilitate the improvement of image quality by enhancing resolution, reducing noise, and enhancing visual aesthetics [34]. The utilization of GANs extends to the medical imaging field, whereby they can improve the visual quality and precision of diagnostic images. This enhancement contributes to the facilitation of accurate diagnoses by healthcare professionals [35].

3.1.2 Data Augmentation

In ML and DL, the availability of extensive and varied datasets is crucial for developing resilient models through training. GANs play a pivotal part in data augmentation by effectively creating synthetic data that exhibits a high degree of resemblance to real-world instances. The utilization of synthetic data has the potential to address class imbalance, mitigate overfitting, and enhance the generalization capabilities of artificial intelligence models [29]. The applications of GANs span across various domains, including computer vision and natural language processing. In computer vision, GANs are utilized to produce supplementary training images. Similarly, in natural language processing, GANs are employed to generate diverse textual variations that might enhance the performance of language models [36].

3.1.3 Artistic Expression, Style Transfer, Diagnosis, and Medical Imaging

GANs have facilitated the exploration of novel routes in artistic expression by employing style transfer techniques. These methodologies enable artists and designers to amalgamate the aesthetic components of diverse artworks, yielding visually captivating pieces [37]. GANs can creatively reinterpret renowned artworks by emulating the distinct styles of various artists. GANs can extend their artistic influence beyond paintings since they can also apply artistic styles to images and videos. The expansive creative capacity has not only facilitated the emergence of innovative modes of artistic manifestation but has also been used in advertising, marketing, and the entertainment sector [38]. GANs have demonstrated considerable potential in medical imaging by their ability to generate synthetic images that accurately imitate a range of medical diseases and anomalies. The utilization of synthetic images has the potential to enhance the training and validation processes of diagnostic models, leading to enhanced accuracy in illness diagnosis. GANs play a crucial role in promptly identifying medical illnesses such as cancer, hence facilitating timely therapies and potentially leading to life-saving outcomes [39]. Moreover, GANs can produce anatomically accurate models that can be utilized in medical education and surgical preparation, depending on the abilities, as demonstrated in Fig. 1.

3.1.4 Molecular Design, Drug Discovery, Fraud Prevention, and Anomaly Detection

The utilization of GANs in drug development and molecular design has proven advantageous for the pharmaceutical sector. GANs have demonstrated a noteworthy ability to create molecular structures and make accurate predictions regarding chemical characteristics [18]. Scientists employ these tools to investigate the extensive chemical landscape, ascertain promising pharmaceutical candidates, and enhance molecular architectures to enhance effectiveness and safety. This phenomenon expedites the drug development procedure, diminishes expenses, and potentially expedites the introduction of life-saving pharmaceuticals to the market [40]. GANs are being utilized with growing frequency in implementing anomaly detection systems. By undergoing training using genuine data, GANs can acquire the ability to discern patterns and expected behaviors. Therefore, GANs prove proficient in detecting anomalies or deviations from the established norm [41]. Within financial services, GANs assume a pivotal role in detecting fraudulent activities by identifying atypical transactions or behaviors that could signify fraudulent conduct [42]. This measure

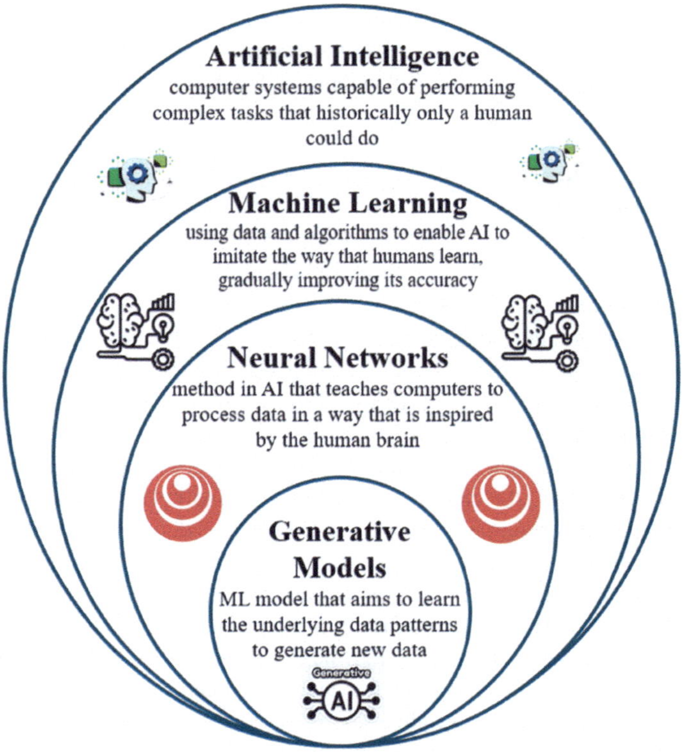

Fig. 1 Artificial intelligence capacities

improves security and safeguards individuals and businesses from potential financial losses.

3.1.5 Text-To-Image Generation, Animation, and Video Production

GANs possess the capacity to produce visual representations based on textual descriptions, hence exhibiting a wide range of potential applications. E-commerce platforms employ GANs to generate visual representations of products based on textual descriptions, enabling shoppers to envision merchandise that has not been physically photographed [43]. Content developers can transform written concepts into visual representations, hence facilitating the execution of marketing campaigns and the art of storytelling [44]. The integration of textual and visual elements can revolutionize the landscape of digital content generation and online retail interactions. The influence of GANs encompasses the domains of video and animation production. The ability to produce lifelike animated sequences, create intricate special effects, and develop three-dimensional character models is within their capabilities. The implementation of this technology optimizes the workflow of content generation, resulting in a decrease in the requirement for labor-intensive manual tasks and time-consuming rendering procedures [30]. Within film and gaming, GANs play a pivotal role in augmenting the level of immersion and visual aesthetics, enriching the narrative and entertainment value.

3.1.6 Augmented Reality and Virtual Reality, Creative Writing, and Content Generation

GANs play a crucial role in developing immersive virtual reality (VR) and augmented reality (AR) environments. The capability to produce lifelike three-dimensional (3D) models, textures, and surroundings enables developers to create immersive virtual worlds that are visually engaging [26]. Training simulations, gaming experiences, architectural visualization, and educational applications benefit from the realism and interaction that GANs provide in VR and AR. This integration enhances user engagement and improves learning results. In natural language processing, GANs have significantly advanced in generating text that closely resembles human language. The capacity significantly impacts content generation, chatbots, and creative writing. GANs can autonomously generate many forms of content, including news articles, marketing copy, and personalized content recommendations [35]. While automated content generation can enhance the process of creating material and enhance user experiences, it also raises ethical concerns about its potential exploitation for disinformation or manipulation. Achieving a harmonious equilibrium between automation and responsible content development poses a significant ethical dilemma within this field [25].

3.2 Ethical Considerations Related to Generative Adversarial Networks

These multifaceted ethical considerations require collaboration among researchers, policymakers, industry stakeholders, and ethicists. Developing and adhering to ethical frameworks and guidelines can help ensure that GANs are harnessed for the greater good while minimizing their potential for misuse and harm.

3.2.1 Deepfake Generation, Privacy, and Consent

The utilization of GANs to produce deepfakes raises a substantial ethical quandary due to their capacity to deceive and manipulate individuals and the broader public. Ethical considerations encompass a range of issues, such as the dissemination of inaccurate information, the act of defaming individuals or entities, and the possibility of instigating violent behavior [16]. Identifying and mitigating deepfakes play a crucial role in preserving trust in digital media and minimizing their possible negative consequences. Using GANs to produce synthetic images or films depicting humans without explicit authorization gives rise to significant privacy implications. These technologies can generate information that convincingly portrays individuals in compromising or invasive scenarios [43]. The safeguarding of individuals' entitlement to privacy necessitates the establishment of parameters for the utilization of GAN-generated images, as well as the assurance of obtaining consent when deemed appropriate.

3.2.2 Intellectual Property and Copyright, Bias and Discrimination

GANs have the potential to introduce complexities in the realm of intellectual property and copyright law, particularly in the context of generating artistic creations or derivative works. Ethical considerations encompass several aspects, such as assessing the authenticity of content generated by GANs, resolving issues related to credit and royalties, and preserving the artistic and intellectual rights of individuals involved in content creation [40]. Establishing ethical principles and legal frameworks holds paramount importance within this setting. GANs have the potential to perpetuate biases and reinforce societal disparities by inheriting the biases contained in their training data, resulting in biased and stereotypical outputs. As an illustration, these systems can provide visual representations that perpetuate and strengthen existing racial or gender prejudices [24]. Ethical considerations revolve around the possibility of reinforcing detrimental biases and the imperative to meticulously curate and scrutinize algorithms and data to guarantee equity and inclusiveness.

3.2.3 Deception and Misinformation, Emotional and Psychological Impact

The ethical implications arising from the ability of GANs to produce deceptive content effortlessly raise concerns regarding the propagation of misinformation and the act of deceiving individuals. The implications of this phenomenon can extend significantly to domains such as journalism, political communication, and societal interpretation [26]. Ethical considerations encompass the obligation of both platforms and users to engage in information verification and fact-checking, as well as the advancement of tools for detecting content generated by GANs. The utilization of GAN-generated content, particularly deepfake films, has the potential to elicit profound emotional reactions among individuals who encounter such material. Ethical considerations revolve around the possible psychological damage resulting from false or emotionally manipulative content [45]. When developing content and utilizing platforms, content creators and platforms must consider the potential consequences on individuals, especially those who may be more susceptible to experiencing emotional discomfort. GANs can enable illicit activities such as identity theft and fraud by creating fabricated identities, forged papers, and even manipulated voice recordings. Ethical considerations encompass the implementation of rigorous identity verification systems and the establishment of legislative frameworks aimed at addressing fraudulent actions [46]. Safeguarding persons against potential financial and reputational damage is a crucial ethical consideration.

3.2.4 Algorithmic Accountability and Environmental Impact

The ethical concept of accountability is of significant importance to GAN developers, companies, and platforms. The individuals accountable for developing and implementing GANs must conscientiously assess the societal ramifications of these technologies and adopt proactive strategies to guarantee their ethical utilization [47]. This encompasses openly acknowledging the artificial origin of content, taking responsibility for any potential misuse, and implementing effective reporting systems to identify and resolve ethical breaches. Likewise, the environmental consequences of training and operating GANs, especially when dealing with large-scale models, are substantial due to the computational resources involved [43]. The ethical dimensions of GAN research and development encompass the assessment of carbon footprint and energy use. In this context, the adoption of sustainable behaviors, such as the utilization of energy-efficient technology and the responsible allocation of resources, is seen to be morally obligatory [48], as demonstrated in Fig. 2.

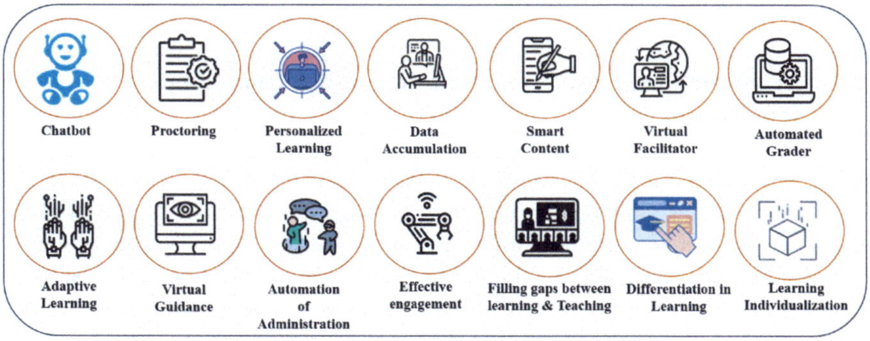

Fig. 2 Artificial intelligence benefits

3.3 Security and Data Privacy Concerns Specific to Generative Adversarial Networks

Adopting a comprehensive approach that encompasses technical remedies, ethical deliberations, legal structures, and awareness initiatives is imperative to effectively tackle the privacy and security risks associated with GANs. The following sections outline these necessary components in detail.

3.3.1 Synthetic Identity Generation and Deepfake Creation

GANs can generate remarkably authentic synthetic images depicting non-existent humans. The utilization of synthetic identities gives rise to privacy apprehensions, as unscrupulous individuals can exploit them for diverse objectives, including but not limited to impersonation, identity theft, and the fabrication of counterfeit profiles on social media platforms [25]. Identifying and preventing fraudulent activities with synthetic identities pose significant challenges, necessitating a constant state of alertness in digital identity verification. GANs have emerged as a pivotal technology in creating deepfake content, encompassing the production of manipulated or impersonated films and audio recordings involving persons. The utilization of this technology has significant privacy concerns since it has the potential to generate fabricated videos depicting individuals participating in compromising or inappropriate actions [33]. Such misuse might result in reputational harm and emotional suffering for the affected individuals. Implementing deepfake detection mechanisms and promoting awareness campaigns play a vital role in addressing and minimizing the privacy risks associated with deepfake technology.

3.3.2 Biometric Data Privacy and Data Generation from Limited Information

GANs can generate artificial biometric data, including but not limited to fingerprints, facial photos, and voice recordings, which exhibit a high degree of similarity to authentic biometric identifiers. The issue poses a significant security risk, given the prevalent utilization of biometric data for authentication and identification purposes [30]. In the event of a hack, there is a potential for unauthorized access to sensitive information, ranging from personal devices to protected facilities. To safeguard biometric data against GAN-based attacks, it is imperative to implement resilient security protocols, such as multi-factor authentication and biometric encryption [36]. GANs can produce intricate visual representations by utilizing restricted data or source images with poor resolution. This technique presents potential privacy concerns, particularly in scenarios where it is employed to improve surveillance footage or reconstruct recognizable images using incomplete data [49].

The privacy of individuals may be violated when GANs are employed to retrieve sensitive information from apparently harmless sources.

3.3.3 Content Manipulation and Misinformation, Ethical Data Usage, and Data Recovery Attacks

GANs have the potential to be utilized to manipulate content in a manner that can lead to deception or misinformation. This might encompass manipulating visual media, such as photos or films, to construct misleading narratives or forge substantiating proof [45]. The dissemination of modified content has the potential to affect individuals' reputations adversely, manipulate public sentiment, and erode trust in information sources, becoming a noteworthy security and privacy risk. The utilization of GANs to produce synthetic data in diverse applications, including data augmentation and privacy-preserving approaches, gives rise to ethical concerns about the utilization and consent of data. Privacy risks arise when GANs are trained on datasets containing sensitive information without obtaining explicit agreement or when synthetic data is utilized in manners that may unintentionally lead to the identification of persons [14]. The data generated by GANs may not consistently provide the desired level of anonymization or privacy. Sophisticated methodologies, such as data recovery assaults, employ patterns and correlations to reverse-engineer the original data from fake data [9].

3.3.4 Data Leakage, Inference Attacks, and Regulatory Compliance

The synthetic data generated by GANs, although intended to safeguard privacy, can unintentionally disclose information from the original data sources. Privacy breaches can manifest as inference attacks, wherein malicious actors employ statistical analytic techniques to infer confidential information from synthetic data [6]. Continual investigation into sophisticated privacy-preserving methodologies is needed to guarantee the intense privacy of data generated by GANs. As the utilization of GANs becomes more prevalent in handling sensitive data, enterprises face the challenge of effectively managing intricate privacy and security requirements, for example, the General Data Protection Regulation (GDPR)[3] implemented in Europe [5]. Maintaining adherence to these standards while effectively utilizing GANs for lawful objectives is a notable obstacle, given that non-compliance may lead to considerable financial penalties and legal ramifications, application of generative artificial intelligence demonstrated in Fig. 3.

[3] https://gdpr.eu/what-is-gdpr/.

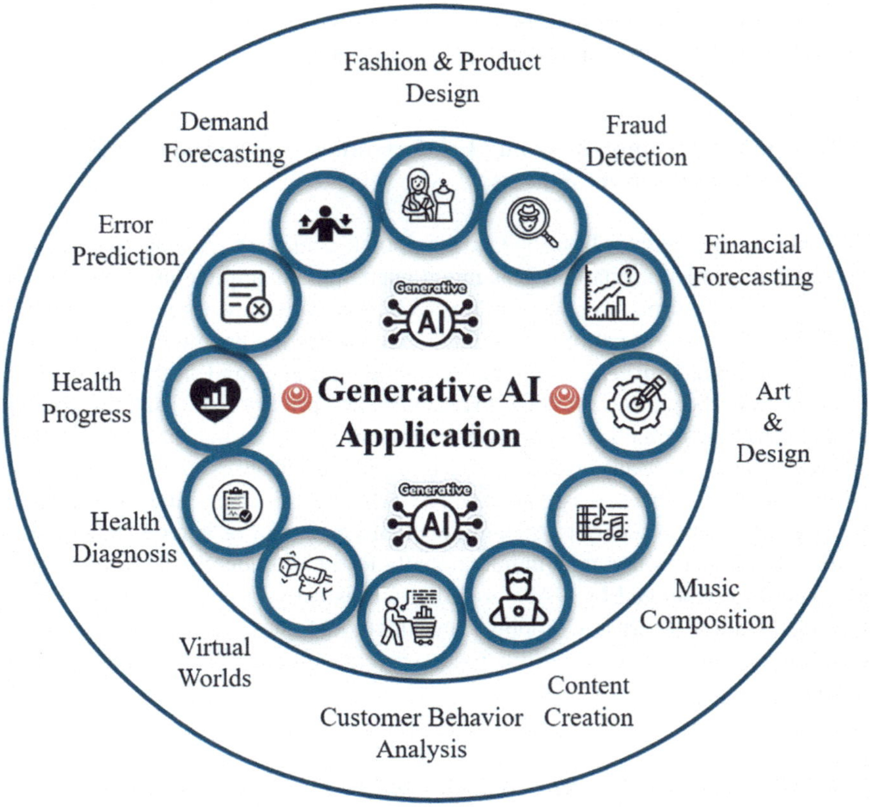

Fig. 3 Generative artificial intelligence applications

4 Regulatory and Legal Frameworks

This section describes existing and proposed regulations governing quantum computing and GANs and analyzes the effectiveness of current legal frameworks in addressing ethical, privacy, and security concerns.

4.1 Existing Regulations Governing Quantum Computing and Generative Adversarial Networks

It is imperative to acknowledge that the legislation about quantum computing is currently under development and can exhibit substantial variations across different jurisdictions. The dynamic characteristics of quantum technology and its possible ramifications across diverse domains, such as cybersecurity, encryption, and data privacy, render it a multifaceted realm of governance [7]. Policymakers and regulatory agencies are actively collaborating with experts and stakeholders to create the regulatory framework effectively, ensuring a harmonious equilibrium between innovation, security, and ethical issues.

4.1.1 Export Controls

Export controls refer to governmental rules that are implemented to limit the exportation of sensitive technologies, products, or information. The primary objective of these controls is to prevent the unauthorized acquisition of such items, particularly in situations where their potential misuse could threaten national security [3]. Export control legislation in numerous nations may impose restrictions on quantum computing technologies, specifically on quantum hardware. An illustration of this can be seen in the export control regulations of the United States, which the department of commerce oversees. These regulations stipulate that specific quantum technologies are subject to procedures for obtaining export licenses [2]. These regulations prevent the transfer of powerful quantum computing hardware and associated technology to entities or governments that may exploit them for nefarious intentions.

4.1.2 Intellectual Property Laws

Intellectual property legal framework comprises various forms of protection, including patents, copyrights, trademarks, and trade secrets. Researchers and companies engaged in quantum computing can pursue patent protection for their innovative contributions. This encompasses quantum algorithms, ideas for quantum hardware, and implementations of quantum software [3]. The grant of patent protection catalyzes fostering innovation by affording inventors and entities the privilege of exclusive rights over their quantum-related creations for a predetermined duration. The safeguarding of intellectual property rights plays a crucial role in the preservation of investments made in research and development, as well as in providing incentives for continued progress and innovation within the field of quantum technology [7].

4.1.3 Privacy Laws and Data Protection

Data protection and privacy regulations govern the various aspects of personal data, including its acquisition, storage, processing, and dissemination. An illustration of a comprehensive data protection framework worldwide is the GDPR of the European Union [4]. Quantum computing applications that include personal data processing must conform to these restrictions. To uphold the private rights of individuals, organizations that employ quantum technology for data storage, analysis, or encryption must adhere to data protection and privacy legislation [5]. Encryption techniques, especially those immune to quantum attacks, hold significance in endeavors aimed at safeguarding data.

4.1.4 Telecommunication Regulations

QKD is a quantum-based technological approach that aims to provide secure communication channels by exploiting the fundamental laws of quantum physics. The current telecommunication legislations could influence the adoption and use of QKD systems. In certain nations, regulatory authorities oversee the adoption of secure communication technologies, such as QKD, to ensure adherence to telecommunications standards and uphold the security and integrity of communication networks [5]. The significance of QKD in guaranteeing secure communication is an essential part of telecommunications regulation.

4.1.5 Certification and Standards

The involvement of established standards and certification organizations is crucial in establishing rules and benchmarks for quantum-resistant encryption and security. The National Institute of Standards and Technology (NIST)[4] in the United States is actively developing standards for post-quantum cryptography. These standards aim to mitigate the potential susceptibilities of existing encryption techniques to quantum attacks and establish a structure for ensuring secure encryption practices in a future where quantum computing is prevalent [11]. Ensuring adherence to these standards is paramount for enterprises seeking to safeguard their data from potential quantum-related risks.

4.1.6 Trade Agreements

The regulation of quantum technology can indirectly influence international trade agreements and treaties. The potential effects of these developments on cross-border research collaborations, technological transfers, and commerce in quantum-related products and services should be considered. Nations frequently harmonize their trade rules with global accords, potentially affecting the trade of quantum technology and associated intellectual property rights in terms of exports and imports [12]. The dynamic nature of global trade dynamics has the potential to influence further the regulatory frameworks governing quantum technology. The current regulations establish a comprehensive framework governing several quantum computing facets, encompassing research and development, data protection, and international trade.

[4] https://www.nist.gov/.

4.2 Proposed Regulations Governing Quantum Computing and GANs

These proposed regulations reflect the growing recognition of the need to balance innovation in quantum computing and GANs with ethical, security, and privacy considerations. Policymakers, regulatory bodies, industry experts, and stakeholders are actively shaping these regulatory frameworks to address the unique challenges these transformative technologies pose [36]. The specifics of these regulations may vary by jurisdiction, but the overarching goal is to ensure that quantum computing and GANs contribute positively to society while minimizing potential risks and harms.

4.2.1 Cybersecurity Standards and Quantum-Safe Encryption

The potential restrictions in the field of quantum computing may emphasize the necessity of implementing practical cybersecurity standards. The quantum computing standards would encompass quantum hardware, networks, and algorithms. In addition, their attention would be directed toward mitigating the security vulnerabilities that quantum computing poses to existing encryption techniques [13]. This entails the establishment of protocols enabling the advancement and acceptance of encryption methods that are resistant to quantum computing, such as lattice-based cryptography or post-quantum cryptographic algorithms.

4.2.2 Data Privacy and Quantum Encryption, Export Controls and Dual-Use Technologies

Considering the escalating risk posed by quantum attacks on conventional encryption, it may be deemed necessary for regulatory bodies to mandate companies' use of quantum-resistant encryption techniques. As mentioned earlier, the legislation will emphasize the significance of safeguarding sensitive data in an era dominated by quantum computing while advocating for adopting QKD and other encryption technologies resistant to quantum attacks [16]. Embracing quantum-safe encryption standards may be mandated for specific sectors and industries. The prospective implementation of regulatory measures for quantum computing may encompass more stringent export controls, particularly for technology with dual-use capabilities. To mitigate potential misuse, governments may consider broadening their regulatory monitoring of exports of quantum-related hardware and software. To uphold national security interests and mitigate the proliferation of sensitive quantum capabilities, it may be necessary to extend the scope of export permits to encompass a wider array of quantum technology.

4.2.3 Ethical Considerations and Responsible Research Practice Standards and Certification Processes

Formulating ethical guidelines is one potential course of action to foster responsible and ethical conduct within quantum research and development. The rules may address various factors, such as research ethics, the responsible disclosure of flaws in quantum systems, and the prevention of detrimental applications of quantum technology, particularly in domains such as encryption and quantum computing for cybersecurity [27]. Proposed regulations may encourage the development of industry standards and certification processes for quantum hardware and software. Certification would ensure that quantum technologies meet specific security and performance criteria, fostering trust among users and organizations. These standards could encompass hardware reliability, quantum-resistant encryption, and quantum key distribution protocols [28].

4.2.4 Deepfake Detection and Mitigation, Ownership Guidelines and Content Attribution

Potential legislation within the realm of GANs may necessitate implementing sophisticated deepfake detection and mitigation methods by online platforms, content publishers, and technology vendors. The primary objective of this legislation is to address the proliferation of deceptive deepfake content and safeguard individuals' reputations, privacy, and emotional well-being [49]. It is possible to construct guidelines that provide certain levels for the accuracy of deepfake detection and the effectiveness of mitigating strategies. Regulatory measures may require the implementation of explicit labeling for content created by GANs to differentiate it from genuine content [19]. Labeling is crucial in enabling users to discern synthetic content and comprehend its possible consequences, especially when disinformation or deceit may arise. Regulations may address complex content attribution and ownership issues regarding GAN-generated content. Clear guidelines could be proposed to determine ownership, royalties, and fair use of content created using GANs [30]. This includes establishing mechanisms for content creators to assert their rights and claim ownership over GAN-generated creations.

4.2.5 Data Privacy and Informed Consent, Responsible AI and Bias Mitigation

In the context of GANs, proposed regulations may require explicit informed consent for generating and using synthetic data, mainly if it involves individuals' likeness or personal information. These regulations would emphasize the importance of respecting individuals' privacy rights and obtaining consent to create synthetic content that may impact their privacy or identity [22]. Potential rules and guidelines could be implemented to manage the ethical concerns surrounding the development and utilization of GANs effectively, explicitly focusing on mitigating issues related to bias, fairness, transparency, and accountability. Organizations employing GANs may be obligated to incorporate fairness assessments, bias detection methods, and transparency procedures into their operations [26]. This is to ensure that their AI applications do not propagate detrimental biases or discriminatory outcomes.

4.3 Effectiveness of Current Legal Frameworks

The effectiveness of current legal frameworks in addressing ethical, privacy, and security concerns related to quantum computing and GANs is mixed. While some existing laws can be applied to mitigate specific issues, these frameworks often lag at the rapid pace of technological advancements [33]. Specific regulations and guidelines tailored to these emerging technologies are necessary to address their unique challenges effectively.

4.3.1 Ethical Concerns

The current legal frameworks generally do not have explicit laws pertaining to the ethical considerations associated with quantum computing. Ethical considerations in scientific research encompass responsible research techniques and prevent detrimental applications [35]. These considerations are typically upheld through self-regulation within the scientific community and adherence to ethical principles established by institutions and professional organizations. There is a necessity for developing more extensive ethical norms and regulatory frameworks to successfully tackle the ethical concerns associated with quantum computing. The ethical implications associated with GANs, including the production of deepfakes, dissemination of misinformation, and potential privacy infringement, have garnered much scholarly and public scrutiny. Although there is a lack of dedicated legislation on GANs, specific ethical concerns can be addressed by leveraging existing regulations, such as defamation and privacy laws [33]. Nevertheless, the efficacy of these frameworks exhibits variability depending on the jurisdiction, and the enforcement process can prove to be arduous, particularly in cross-border complexities.

4.3.2 Privacy Concerns

The focus of privacy concerns in the field of quantum computing pertains to safeguarding data and the possibility of unauthorized access to confidential information via quantum assaults. Current data protection and privacy regulations, such as the GDPR, need enterprises to establish security protocols to safeguard personal data, including encryption techniques [40]. The efficacy of these regulations is contingent upon entities' implementation of quantum-resistant encryption techniques and adherence. One of the primary issues with privacy in the context of GANs pertains to the unpermitted utilization of individuals' data to generate synthetic content. The current legislation on data protection establishes a legal structure aimed at safeguarding the privacy rights of individuals [39]. Nevertheless, implementing enforcement measures might present significant difficulties, especially in cases involving the fabrication of synthetic data and the potential for re-identifying persons. There is a potential need to implement new legislation that mainly targets synthetic data and content generated by GANs to strengthen privacy safeguards.

4.3.3 Security Concerns

The primary focus of security concerns in the field of quantum computing mainly pertains to the possible disruption of conventional encryption techniques through the utilization of quantum attacks. The current emphasis within legal frameworks is predominantly on export controls and intellectual property protection for quantum technologies since efforts are underway to build encryption standards resistant to quantum attacks [29]. The efficacy of these frameworks is of utmost importance in mitigating the spread of sensitive quantum capabilities and protecting national security interests. Also, one of the core security considerations associated with GANs pertains to the production of deepfake material, which has the potential to inflict damage upon individuals' reputations and manipulate public opinion. Particular security concerns can be effectively addressed by utilizing established legal frameworks, such as those about defamation and fraud [35]. Nevertheless, identifying and mitigating deepfake content frequently depends on technological interventions and cooperation between various platforms and law enforcement entities.

5 Case Studies

This section illustrates real-world examples and case studies illustrating ethical, privacy, and security challenges in quantum computing and GANs.

5.1 Case Studies Illustrating Ethical, Data Privacy, and Security Challenges in Quantum Computing

These case studies demonstrate the ethical, privacy, and security challenges associated with quantum computing in various sectors, including healthcare, finance, cybersecurity, and government applications. Addressing these challenges requires a multidisciplinary approach involving technologists, policymakers, ethicists, and legal experts to develop ethical guidelines, privacy protections, and security measures that align with the unique capabilities and risks of quantum computing.

5.1.1 Security Challenge—Post-Quantum Cryptography Adoption

They initiated a competition to select post-quantum cryptographic algorithms to replace current encryption methods vulnerable to quantum attacks. This competition reflects the security challenge of transitioning to new encryption standards to protect sensitive data in a quantum computing era [34]. Organizations and governments need to adapt their cryptographic infrastructure to ensure data security.

5.1.2 Ethical Challenge—Quantum Computing for Drug Discovery

Case Study: Pharmaceutical corporations are currently investigating the potential application of quantum computing to expedite the drug discovery process. Although there are hopeful advancements in healthcare, using patients' genetic information and medical records for quantum-based research raises ethical problems related to data privacy and permission [33]. The ethical dilemma of reconciling medical advancements with protecting patient privacy poses a substantial problem.

5.1.3 Privacy Challenge—Quantum-Safe Encryption for Financial Transactions

Case Study: Financial institutions, such as banks and payment processors, are currently engaged in extensive research endeavors about quantum-safe encryption techniques, with the primary objective of safeguarding financial transactions [39]. Privacy problems emerge when contemplating the imperative to safeguard client data and financial records against quantum attacks while guaranteeing uninterrupted and confidential transactions for customers. The preservation of confidentiality for financial information is of utmost importance.

5.1.4 Security Challenge—Quantum Computing for Cybersecurity

The emergence of quantum computing has brought forward the potential for enhanced cyberattacks. The potential utilization of quantum algorithms by nation-states and malevolent actors poses a significant threat to the security of present encryption systems and the integrity of vital infrastructure [22]. The primary concern pertains to proactively safeguarding against these potential risks and designing resilient cybersecurity strategies against quantum-based attacks.

5.1.5 Ethical Challenge—Quantum Computing for AI and Surveillance

Using quantum computing to process extensive information at unparalleled velocities gives rise to ethical considerations when employed in surveillance and artificial intelligence. Governments and organizations can employ quantum-powered AI systems for surveillance, encompassing facial recognition, and tracking activities [2]. Ethical considerations include protecting individual privacy, obtaining informed consent, and the potential implications of widespread surveillance.

5.1.6 Privacy Challenge—Quantum-Safe Elections and Voting Systems

Case Study: Preserving privacy and security in electoral processes constitutes a key tenet of democratic governance. Implementing quantum-safe encryption could be important in safeguarding electronic voting systems against the vulnerabilities of quantum attacks [3]. Nevertheless, the task of ensuring both transparency and voter privacy in electronic voting systems while also protecting against potential quantum threats presents a multifaceted privacy dilemma.

5.1.7 Security Challenge—Quantum-Safe Supply Chain Management

The effectiveness of supply chain management is contingent upon establishing robust mechanisms for secure communication and preserving data integrity. The potential of quantum computing to compromise existing encryption techniques presents a significant security concern in safeguarding the integrity and confidentiality of supply chain data [35]. Organizations must establish and implement quantum-safe safeguards to ensure the security of their supply chains.

5.2 Case Studies Illustrate Ethical, Data Privacy, and Security Challenges in GANs

These real-world examples and case studies highlight the multifaceted ethical, privacy, and security challenges associated with GANs. Addressing these challenges requires technical solutions, ethical guidelines, legal regulations, and public awareness efforts to ensure responsible and secure use of GAN technology while safeguarding privacy and combating malicious use cases.

5.2.1 Deepfake Manipulation and GAN-Generated Malware

In 2019, a mobile application known as "DeepNude[5]" garnered significant attention due to its utilization of GANs to produce authentic-looking nude depictions of women by digitally eliminating their attire in photographs. The use of GAN technology in an immoral manner has given rise to significant apprehensions regarding privacy, as it has showcased the capacity for hostile entities to manipulate GANs to produce non-consensual, sexual material, infringing upon individuals' privacy rights [37]. Studies conducted by researchers have demonstrated the capability of GANs to generate polymorphic malware, a type of malicious software that can dynamically alter its code to elude conventional antivirus systems. The detection of such malware presents a significant problem, and its ability to adapt represents a security concern for cybersecurity professionals [22]. This scenario underscores the necessity of implementing heightened security protocols to mitigate the risks posed by threats generated using GANs effectively.

5.2.2 GANs for Misinformation and Deepfake and Impersonation

GANs can produce fabricated news stories, blog posts, and social media content that exhibit a remarkably high realism. The potential for malevolent entities to leverage this technology to spread false information, manipulate public sentiment, and instill skepticism toward media outlets is evident [42]. This ethical dilemma pertains to the dual objective of countering the dissemination of inaccurate information while safeguarding the principle of freedom of expression. Deepfake[6] technology has been employed to fabricate videos that simulate the actions and statements of prominent individuals and renowned personalities, creating a deceptive impression of their engagement in activities or utterances that they have not actually performed [41]. The act of impersonation presents a significant privacy concern for persons whose likenesses are utilized without their explicit agreement, hence potentially causing detrimental effects on their reputation and privacy.

5.2.3 GANs in Art and Copyright, GAN-Enhanced Cyberattacks

The utilization of GANs in creating artistic and musical content has prompted ethical inquiries about authorship and copyright matters. The utilization of GANs in producing artistic creations or musical compositions poses a significant challenge in establishing the appropriate allocation of rights and royalties for the respective producers [40]. The auction sale of "Portrait of Edmond de Belamy,[7]" an artwork generated by artificial intelligence, has ignited discussions surrounding the worth

[5] https://www.deep-nude.ai/.

[6] https://mitsloan.mit.edu/ideas-made-to-matter/deepfakes-explained.

[7] https://en.wikipedia.org/wiki/edmond_de_belamy.

and proprietorship of AI-made creations. Case Study: GANs can potentially augment the efficacy of cyberattacks by generating persuasive phishing emails, replicating the communication patterns exhibited by trusted individuals, and fabricating counterfeit documents [38]. Using sophisticated attack techniques presents security obstacles for businesses and individuals that depend on conventional security measures.

5.2.4 Synthetic Identity Generation and Deepfake in Politics

GANs can produce artificial identities that exhibit a high degree of realism, encompassing visually authentic photographs and accompanying credentials. Synthetic identities possess the potential to be employed in a multitude of fraudulent endeavors, encompassing identity theft, online fraudulent schemes, and social engineering assaults [37]. The identification and mitigation of the misuse of synthetic identities provide significant privacy and security obstacles. The potential impact of deepfake films featuring political leaders disseminating false information on electoral outcomes and public opinion cannot be underestimated. During the pre-election period, the utilization of deepfake content can potentially undermine the credibility of political personalities and disseminate false information, hence presenting ethical dilemmas for democratic processes and the integrity of information [35].

6 Mitigation Strategies

These detailed mitigation strategies encompass a holistic approach to address the ethical, privacy, and security challenges quantum computing and GANs pose [34]. By implementing these strategies, organizations can navigate the complexities of emerging technologies while upholding ethical principles, ensuring privacy protection, and enhancing security measures.

6.1 Responsible AI Governance

To effectively tackle ethical considerations and promote responsible utilization of quantum computing and GANs, enterprises must build comprehensive frameworks for AI governance. It is imperative that these frameworks incorporate ethical principles, means for ensuring compliance, and committees tasked with overseeing and assessing AI initiatives [38]. Integrating ethical issues, including but not limited to bias reduction, transparency, and accountability, is crucial in the development processes of AI.

6.2 Privacy-Preserving Technologies and Quantum-Safe Encryption

The successful management of privacy concerns necessitates using and integrating privacy-preserving technologies. Quantum computing and GANs can benefit from incorporating techniques like differential privacy, federated learning, and homomorphic encryption. These methodologies enable enterprises to effectively handle and analyze data while safeguarding the confidentiality of people's sensitive information, hence mitigating the potential for data breaches and privacy infringements [35]. To effectively mitigate the security concerns arising from the advent of quantum computing, companies must undertake a transition toward the adoption of quantum-safe encryption techniques. It is imperative to be well informed on post-quantum cryptography advancements and revise encryption standards accordingly [33]. Preserving sensitive information necessitates implementing measures to protect data and communication security from potential quantum attacks.

6.3 Responsible for Data Handling, Regulatory Compliance, and Advocacy

To prevent privacy and security issues, it is imperative to adopt appropriate data handling procedures that effectively reduce data acquisition, retention, and sharing. It is advisable to employ anonymization or pseudonymization techniques whenever feasible to safeguard the identity of persons and mitigate the potential consequences of data breaches. Ensure adherence to the dynamic regulatory landscape of quantum computing and GANs within a given specific jurisdiction. Engage in collaborative efforts with legislators to establish ethical, privacy, and security frameworks that follow the progress of technology [33].

6.4 Public Awareness, Education, Collaboration, and Industry Standards

This proposal advocates for promoting public awareness and media literacy as effective measures to enable individuals to discern synthetic content produced by GANs and comprehend the ramifications of quantum computing. Educational endeavors should focus on students, professionals, and the wider public, enabling them to evaluate and constructively respond to developing technologies critically. Promote collaborative efforts among various stakeholders, encompassing scholars, organizations, and government agencies, to tackle emergent difficulties cooperatively [30]. Exchanging knowledge, implementing best practices, and gaining insights can significantly enhance the ability to identify and address hazards more efficiently [42].

6.5 Responsible Use, Governance, Continuous Monitoring, and Vulnerability Assessment

Develop corporate rules and governance frameworks that emphasize quantum computing and GAN technologies' responsibility and ethical utilization. This statement emphasizes the need to establish unambiguous parameters for appropriate use scenarios, critically assess the ethical ramifications, and ensure adherence to established guidelines [38]. The establishment of ethics boards or committees to oversee AI initiatives, particularly those containing sensitive data or applications with substantial social implications, should be considered. It is imperative to consistently assess the security and privacy issues associated with quantum computing and applications of GANs. Perform comprehensive vulnerability assessments and penetration testing to detect and address potential vulnerabilities [34].

7 Future Directions, Recommendations, and Conclusions

This section presents some notable future directions, lessons learned, and the conclusion.

7.1 Future Directions

These future directions encompass a range of technical, ethical, and regulatory considerations, reflecting the evolving landscape of quantum computing and GANs as they become increasingly integrated into various aspects of society and technology.

7.1.1 Quantum-Safe Cryptography and Quantum-Resistant Protocols

Quantum machine learning will mature as researchers develop quantum algorithms that outperform classical counterparts. Hybrid quantum-classical algorithms will be designed for various applications, including optimization, drug discovery, and financial modeling [50]. Future advancements will focus on the research and deployment of quantum-safe cryptographic methods, including lattice-based, code-based, and multivariate cryptography. Organizations and governments will transition to quantum-resistant encryption standards to protect sensitive data, financial transactions, and critical infrastructure from potential quantum attacks [35]. Quantum-resistant digital signatures and authentication methods will also become integral components of cybersecurity. Quantum neural networks and quantum-enhanced reinforcement learning will emerge as key areas of exploration, offering transformative solutions for AI-driven industries. The development of a quantum internet will advance, enabling ultra-secure communication through QKD. Quantum repeaters and quantum teleportation protocols will be refined to extend the range of quantum-secured communication [44]. Quantum networks will find applications in secure government communications, financial transactions, and sensitive data transfers, fostering trust in digital interactions.

7.1.2 Ethical AI and Deepfake Detection Advancements, Quantum Computing, and GANs for Scientific Discovery

To combat deepfake threats, ethical AI guidelines will incorporate stringent content verification standards and responsible AI practices. Advanced deepfake detection methods will leverage ML, natural language processing, and computer vision to identify manipulated content accurately [20]. The ethical use of AI in content generation and dissemination will be reinforced through legislation and industry standards. Quantum computing's potential for simulating quantum systems will drive breakthroughs in scientific discovery. Quantum chemistry simulations will accelerate the discovery of novel materials, catalysts, and drug compounds with applications in renewable energy, healthcare, and materials science [35]. Quantum-inspired algorithms will optimize supply chains, leading to more efficient logistics and resource management.

7.1.3 Privacy-Preserving AI and Data Sovereignty, Global Collaboration, and Regulatory Frameworks

Privacy-preserving AI techniques will continue to evolve, offering granular control over data access and utilization. Federated learning will enable collaborative model training across distributed datasets without exposing raw data [32]. Secure multi-party computation will be applied to protect individual privacy rights while allowing insights to be derived from sensitive data sources. Data sovereignty frameworks will grant individuals greater control over their data and its usage. International cooperation will be paramount in establishing cohesive regulatory frameworks for quantum computing and GANs. Cross-border partnerships among governments, industry consortia, and research institutions will facilitate the development of consistent ethical, privacy, and security standards. Harmonized regulations will promote responsible innovation, data protection, and secure technology deployment on a global scale [50].

7.2 Lessons Learned from the Chapter

The lessons learned from the discussion on ethical, data privacy, and security considerations in quantum computing and GANs emphasize the need for a holistic and proactive approach. By prioritizing ethics, privacy, and security; fostering collaboration; and staying adaptable in the face of evolving technology, we can navigate the complexities of these emerging fields responsibly and effectively.

- Ethical considerations should be at the forefront of technological development. As quantum computing and GANs continue to advance, it is crucial to prioritize ethical principles, transparency, fairness, and accountability to ensure responsible innovation.
- The importance of privacy cannot be overstated. Both quantum computing and GANs can potentially infringe on individuals' privacy rights. The lesson learned is that privacy protection mechanisms, such as data anonymization and encryption, must be incorporated into technology design.
- Security in the quantum computing and GANs landscape should be proactive, not reactive. With the potential for quantum attacks and advanced cyber-threats, organizations must stay ahead by adopting quantum-safe encryption and continually improving security measures.
- Addressing the complex challenges of these technologies requires interdisciplinary collaboration. Ethicists, policymakers, technologists, and legal experts must collaborate to develop comprehensive solutions that balance innovation and safeguard societal interests.
- Regulations governing quantum computing and GANs need to evolve in parallel with technological advancements. Policymakers should engage with industry experts to develop adaptive regulatory frameworks that address emerging ethical, privacy, and security concerns.
- Raising public awareness about the capabilities and risks associated with quantum computing and GANs is essential. Educating individuals about deepfake threats, quantum attacks, and responsible AI use can empower them to make informed decisions and contribute to a safer digital environment.
- The chapter underscores the importance of responsible innovation. Organizations and researchers should commit to responsible AI development, ethical content creation, and secure technology deployment to mitigate potential harm.
- The global nature of the challenges and opportunities in quantum computing and GANs requires international collaboration, highlighting the need for cooperation among nations, industry leaders, and research communities to establish consistent standards and regulations.

7.3 Conclusion

The integration of quantum computing and GANs offers unprecedented opportunities for innovation across various domains, from scientific research to artificial intelligence. However, it also brings forth multifaceted ethical, privacy, and security considerations that demand careful and proactive attention. As we delve into the quantum era and explore the vast potential of GANs, we must prioritize ethical principles, such as transparency, fairness, and accountability, to ensure the responsible development and deployment of these technologies. Privacy protection mechanisms, robust encryption, and privacy-preserving AI techniques will be indispensable in safeguarding individuals' rights in an increasingly data-driven world. Moreover, the lessons learned underscore the necessity of interdisciplinary collaboration, adaptive regulations, and global cooperation to effectively navigate the challenges and harness the benefits of quantum computing and GANs. By embracing ethical guidelines, staying vigilant about privacy and security, and fostering responsible innovation, we can unlock the full potential of these technologies while upholding fundamental values, promoting digital trust, and advancing toward a more secure and ethically grounded technological landscape.

References

1. Y. Xu, L. Wang, C. Wang, H. Zhu, Effective agent quantum private data query against malicious joint attack with blind quantum computing. Int. J. Theoret. Phys. **61**(4) (2022). https://doi.org/10.1007/s10773-022-05104-y
2. P. Bautista, P.S. Inventado, Protecting Student Privacy with Synthetic Data from Generative Adversarial Networks. Lecture Notes in Computer Science (Including Subseries Lecture Notes in Artificial Intelligence and Lecture Notes in Bioinformatics), 12749 LNAI (2021). https://doi.org/10.1007/978-3-030-78270-2_11
3. A. Zhao, N. Jiang, C. Wang, S. Liu, K. Qiu, synchronization optimization of chaotic laser based on generative adversarial network. Guangxue Xuebao/Acta Optica Sinica **43**(1) (2023). https://doi.org/10.3788/AOS220994
4. S. Kumari, M. Singh, R. Singh, H. Tewari, Post-quantum cryptography techniques for secure communication in resource-constrained internet of things devices: a comprehensive survey. Software—Practice and Experience (2022). https://doi.org/10.1002/spe.3121
5. L. Arulmozhiselvan, E. Uma, QKD in cloud-fog computing for personal health record. Comput. Syst. Sci. Eng. **43**(1) (2022). https://doi.org/10.32604/csse.2022.022024
6. H. Chi, U. Maduakor, R. Alo, E. Williams, Integrating Deepfake detection into cybersecurity curriculum. Adv. Intell. Syst. Comput., 1288 (2021).https://doi.org/10.1007/978-3-030-63128-4_45
7. P. Tinsley, A. Czajka, P.J. Flynn, Haven't i seen you before? assessing identity leakage in synthetic irises. 2022 IEEE International Joint Conference on Biometrics, IJCB 2022 (2022). https://doi.org/10.1109/IJCB54206.2022.10007948
8. W. Bouzeraib, A. Ghenai, N. Zeghib, A Blockchain data balance using a generative adversarial network approach: application to smart house IDS. ICAASE 2020—Proceedings, 4th International Conference on Advanced Aspects of Software Engineering (2020). https://doi.org/10.1109/ICAASE51408.2020.9380110
9. R. Raveendran, E.D. Raj, Deep generative models under GAN: variants, applications, and privacy issues. Lecture Notes Network. Syst., 494 (2023). https://doi.org/10.1007/978-981-19-4863-3_9
10. X. Chen, S. Xu, Y. Cao, Y. He, K. Xiao, AQRS: anti-quantum ring signature scheme for secure epidemic control with blockchain. Comput. Networks, 224 (2023).https://doi.org/10.1016/j.comnet.2023.109595
11. J. Chen, W. Gan, M. Hu, C.M. Chen, On the construction of a post-quantum blockchain for smart city. J. Inform. Secur. Appl., 58 (2021). https://doi.org/10.1016/j.jisa.2021.102780
12. Z.B. Jemihin, S.F. Tan, G.C. Chung, Attribute-based encryption in securing big data from post-quantum perspective: a survey. Cryptography **6**(3) (2022). https://doi.org/10.3390/cryptography6030040

13. K. Li, R. Shi, M. Wu, Y. Li, X. Zhang, A novel privacy-preserving multi-level aggregate signcryption and query scheme for Smart Grid via mobile fog computing. J. Inform. Secur. Appl., 67 (2022). https://doi.org/10.1016/j.jisa.2022.103214
14. R.A. Berry, Z. Han, K. Narayanan, H.V. Poor, C. Verikoukis, O. Yagan, Special issue on communications and networking approaches for combating COVID-19. J. Commun. Networks, 23(5) (2021). https://doi.org/10.23919/jcn.2021.100030
15. S.K. Nukavarapu, T. Nadeem, Securing edge-based IoT networks with semi-supervised GANs. 2021 IEEE International Conference on Pervasive Computing and Communications Workshops and Other Affiliated Events, PerCom Workshops 2021 (2021). https://doi.org/10.1109/PerComWorkshops51409.2021.9431112
16. Y. Ma, E. Kashefi, M. Arapinis, K. Chakraborty, M. Kaplan, QEnclave—a practical solution for secure quantum cloud computing. Npj Quantum Inform. **8**(1) (2022). https://doi.org/10.1038/s41534-022-00612-5
17. S. Sarkar, Quantum machine learning: a review. Int. J. Res. Appl. Sci. Eng. Technol. **11**(3) (2023). https://doi.org/10.22214/ijraset.2023.49421
18. Y. Deldjoo, T. Noia Di, F.A. Merra, A survey on adversarial recommender systems: from attack/defense strategies to generative adversarial networks. ACM Comput. Surv. **54**(2) (2021). https://doi.org/10.1145/3439729
19. K. Wang, N. Deng, X. Li, An efficient content popularity prediction of privacy preserving based on federated learning and Wasserstein GAN. IEEE Internet Things J. **10**(5) (2023). https://doi.org/10.1109/JIOT.2022.3176360
20. L. Malina, P. Dzurenda, S. Ricci, J. Hajny, G. Srivastava, R. Matulevicius, A.A.O. Affia, M. Laurent, N.H. Sultan, Q. Tang, Post-quantum era privacy protection for intelligent infrastructures. IEEE Access, 9 (2021).https://doi.org/10.1109/ACCESS.2021.3062201
21. S.K. Singh, A. El Azzaoui, M.M. Salim, J.H. Park, quantum communication technology for future ICT—review. J. Inform. Proc. Syst. **16**(6) (2020). https://doi.org/10.3745/JIPS.03.0154
22. B. Liu, X. Zhang, R. Shi, M. Zhang, G. Zhang, SEPSI: a secure and efficient privacy-preserving set intersection with identity authentication in IoT. Mathematics **10**(12) (2022). https://doi.org/10.3390/math10122120
23. W. Shafik, Data privacy and security safeguarding customer information in ChatGPT systems, in *Revolutionizing the Service Industry with OpenAI Models*, pp. 52–86. IGI Global (2024). https://doi.org/10.4018/979-8-3693-1239-1.ch003
24. S. Jerald Nirmal Kumar, S. Ravimaran, A. Sathish, Robust security with strong authentication in mobile cloud computing based on trefoil congruity framework. J. Organ. Comput. **33**(6) (2021). https://doi.org/10.4018/JOEUC.20211101.oa11
25. M. Patel, A. Gupta, S. Tanwar, M.S. Obaidat. Trans-DF: A transfer learning-based end-to-end Deepfake detector. *2020 IEEE 5th International Conference on Computing Communication and Automation, ICCCA 2020* (2020). https://doi.org/10.1109/ICCCA49541.2020.9250803
26. C. Wang, W.Y. Kon, H.J. Ng, C.C.W. Lim, Experimental symmetric private information retrieval with measurement-device-independent quantum network. Light: Sci. Appl. **11**(1) (2022). https://doi.org/10.1038/s41377-022-00959-6
27. J.E. Raya, A.S. Yahya, E.K. Ahmad, Protection from a quantum computer cyber-attack. Technium: Romanian J. Appl. Sci. Technol., 5 (2023). https://doi.org/10.47577/technium.v5i.8293
28. A. Mashatan, D. Heintzman, The complex path to quantum resistance. Queue **19**(2) (2021). https://doi.org/10.1145/3466132.3466779
29. M.N. Fekri, A.M. Ghosh, K. Grolinger, Generating energy data for machine learning with recurrent generative adversarial networks. Energies **13**(1) (2019). https://doi.org/10.3390/en13010130
30. S. Lu, X. Li, Quantum-resistant lightweight authentication and key agreement protocol for fog-based Microgrids. IEEE Access, 9 (2021). https://doi.org/10.1109/ACCESS.2021.3058180
31. J. Liu, Y. Yu, H. Wang, H. Zhang, Lattice-based self-enhancement authorized accessible privacy authentication for cyber-physical systems. Secur. Commun. Networks (2022). https://doi.org/10.1155/2022/8995704

32. K. Management, D. Mining, medical knowledge management and data mining in, in *Medical Informatics* (Vol. 8) (2005b). https://doi.org/10.1007/b135955
33. C. He, S. Huang, R. Cheng, K.C. Tan, Y. Jin, Evolutionary multiobjective optimization driven by generative adversarial networks (GANs). IEEE Trans. Cybern. **51**(6) (2021). https://doi.org/10.1109/TCYB.2020.2985081
34. W. Shafik, Deep learning impacts in the field of artificial intelligence, in *Deep Learning Concepts in Operations Research*, pp. 9–26. Auerbach Publications (2024). https://doi.org/10.1201/9781003433309-2
35. H. Sun, T. Zhu, Z. Zhang, D. Jin, P. Xiong, W. Zhou, Adversarial Attacks against deep generative models on data: a survey. IEEE Trans. Knowl. Data Eng. **35**(4) (2023). https://doi.org/10.1109/TKDE.2021.3130903
36. W. Shafik, The role of generative artificial intelligence in e-commerce fraud detection and prevention, in *Strategies for E-Commerce Data Security: Cloud, Blockchain, AI, and Machine Learning*, pp. 430–469. IGI Global (2024). https://doi.org/10.4018/979-8-3693-6557-1.ch018
37. B. Sharma, L. Sharma, C. Lal, S. Roy, Anomaly based network intrusion detection for IoT attacks using deep learning technique. Comput. Electr. Eng., 107 (2023). https://doi.org/10.1016/j.compeleceng.2023.108626
38. S.S. Priscila, A. Sharma, S. Vanithamani, F. Ahmad, R. Mahaveerakannan, A.J. Alrubaie, V. Jagota, B.K. Singh, Risk-based access control mechanism for internet of vehicles using artificial intelligence. Secur. Commun. Networks (2022). https://doi.org/10.1155/2022/3379843
39. W. Shafik, Dissecting the role of women in cybersecurity and information technology: a medical perspective. in Kaushik, K., Sharma, I. (eds.) *Next-Generation Cybersecurity*. Blockchain Technologies. Springer, Singapore (2024). https://doi.org/10.1007/978-981-97-1249-6_15
40. A. Agoub, Y. Filippovska, V. Schmidt, M. Kada, Automatic generation of photorealistic image fillers for privacy enabled urban Basemaps using generative adversarial networks. Adv. Cartogr. GIScience ICA, 1 (2019).https://doi.org/10.5194/ica-adv-1-1-2019
41. W. Shafik, Cyber attacker profiling and cyberbullying overview, in *Cyber Space and Outer Space Security*, pp. 125–149 (2024). River Publishers. https://doi.org/10.1201/9781003558118-5
42. X. Cao, G. Sun, H. Yu, M. Guizani, PerFED-GAN: personalized federated learning via generative adversarial networks. IEEE Internet Things J. **10**(5) (2023). https://doi.org/10.1109/JIOT.2022.3172114
43. J. Mendes, T. Pereira, F. Silva, J. Frade, J. Morgado, C. Freitas, E. Negrão, B.F. de Lima, M.C. da Silva, A.J. Madureira, I. Ramos, J.L. Costa, V. Hespanhol, A. , Cunha, H.P. Oliveira, Lung CT image synthesis using GANs. Expert Syst. Appl., 215 (2023). https://doi.org/10.1016/j.eswa.2022.119350
44. K. Marshall, C.S. Jacobsen, C. Schäfermeier, T. Gehring, C. Weedbrook, U.L. Andersen, Continuous-variable quantum computing on encrypted data. Nature Commun., 7 (2016). https://doi.org/10.1038/ncomms13795
45. S.K. Nukavarapu, M. Ayyat, T. Nadeem, MirageNet—towards a GAN-based framework for synthetic network traffic generation. 2022 IEEE Global Communications Conference, GLOBECOM 2022—Proceedings (2022). https://doi.org/10.1109/GLOBECOM48099.2022.10001494
46. M. Masood, M. Nawaz, A. Javed, T. Nazir, A. Mehmood, R. Mahum, Classification of Deepfake videos using pre-trained convolutional neural networks. 2021 International Conference on Digital Futures and Transformative Technologies, ICoDT2 2021 (2021). https://doi.org/10.1109/ICoDT252288.2021.9441519
47. Y. Ding, N. Thakur, B. Li, Does a GAN leave distinct model-specific fingerprints, in *Proceedings of the BMVC* (2021)
48. J. Yang, S. Xiao, A. Li, G. Lan, H. Wang, Detecting fake images by identifying potential texture difference. Future Gener. Comput. Syst., 125 (2021).https://doi.org/10.1016/j.future.2021.06.043
49. H. Abulkasim, A. Mashatan, S. Ghose, Quantum-based privacy-preserving sealed-bid auction on the blockchain. Optik, 242 (2021).https://doi.org/10.1016/j.ijleo.2021.167039

50. J. Liu, J. Wen, B. Zhang, S. Dong, B. Tang, Y. Yu, A post quantum secure multi-party collaborative signature with deterability in the Industrial Internet of Things. Future Gener. Comput. Syst., 141 (2023). https://doi.org/10.1016/j.future.2022.11.034

Quantum Computing and Cybersecurity: Systematic Review of Algorithms, Challenges, and Emerging Solutions

Chitra Ravi

1 Introduction

Quantum computing harnesses quantum mechanics principles of superposition, entanglement, and interference to perform complex computations much quickly than classical computers for some problems. While bits 0 and 1 are utilized in classical computers, quantum bits (qubits) are employed in quantum computers. Quantum gates use operations, as in classical logic gates, to manipulate qubits, capable of performing more complex transformations. Quantum algorithms specifically tailored for quantum computers like Shor's to factor large numbers and Grover's for search in unsorted databases were developed.

Cybersecurity prevents damage to communication systems and services. It protects data and information to maintain availability, integrity, confidentiality, non-repudiation, and authentication [1, 2]. Key aspects of cybersecurity include cryptography, network security, Identity and Access Management, incident response, and endpoint security.

Quantum computing poses both threats and opportunities to cybersecurity. Quantum computers have the ability to compromise often-used conventional RSA and ECC cryptographic algorithms. Shor's algorithm can achieve exponential speed up to factorize large integers compared to known classical algorithms. This means that once sufficiently large quantum computers are available, they could break widely used encryption schemes, potentially exposing sensitive data worldwide. This is the quantum threat to cryptography, as it is a vulnerability of public key cryptography. Quantum algorithms like Grover's algorithm could search through large datasets quickly than traditional algorithms, reducing the effective security of hash functions by offering quadratic speedup. Quantum Random Number Generator (QRNG) produces random numbers that are vital in cryptographic protocols, thus improving

C. Ravi (✉)
Cybersena (R&D) India Pvt. Ltd., Bangalore, Karnataka, India
e-mail: drchitraravi@cybersenarnd.net

S. B. Goyal et al. (eds.), *Quantum Computing, Cyber Security and Cryptography*,
https://doi.org/10.1007/978-981-96-4948-8_16

the encryption key. These are already being used in cybersecurity systems to generate truly random numbers for cryptographic keys, making encryption more secure against traditional attacks. Information in a quantum system cannot be copied or read by an unauthorized party, due to a concept called the No Cloning Theorem [3].

New cryptographic algorithms considered as resistant to quantum attacks are being developed in order to mitigate quantum threats. Post-quantum cryptographic algorithms that are quantum-safe and difficult for both classical and quantum computers to break are being developed by researchers. These algorithms rely on mathematical challenges thought to withstand quantum-based attacks. This provides quantum-safe cyber solutions and is known as post-quantum cryptography. Quantum computing provides enhanced security protocols, namely, Quantum Key Distribution, that secures communication channels that are theoretically immune to eavesdropping on account of the no-cloning theorem and detection of any attempted interception. In Quantum Key Distribution (QKD), any attempt to intercept or listen in on the key exchange is immediately detectable due to the distinct attributes of qubits. This feature makes QKD theoretically impervious to computational attacks.

The National Institute of Standards and Technology (NIST) has led initiatives to standardize post-quantum cryptographic algorithms, concentrating on areas like lattice-based, code-based, hash-based, multivariate polynomial-based, digital signature-based and isogeny-based cryptography to safeguard data from quantum threats. Adopting quantum-safe cryptographic methods presents significant challenges, requiring substantial updates to current infrastructure, including hardware, software, and protocols. Some quantum-resistant algorithms are computationally demanding, raising concerns about the efficiency and performance of future security systems. Global coordination is essential to ensure a seamless transition to quantum-safe cryptography and to establish uniform standards across countries and industries.

The following sections will cover topics such as overview of quantum computing, overview of cybersecurity, impact of quantum computing on cybersecurity, vulnerabilities in quantum computing, post-quantum cryptography, real-world case studies in fields like healthcare, financial services, and space, as well as issues of standardization, international cooperation, challenges, and future directions.

1.1 Overview of Quantum Computing

Quantum computing represents a new approach to computation, utilizing the principles of quantum mechanics to perform tasks in ways that differ significantly from classical computing. It holds the potential for extraordinary computational power, offering the ability to solve intricate problems that are beyond the reach of traditional computers. Quantum mechanical principles like superposition, entanglement are leveraged to execute complex computations. While quantum computing introduces new challenges and vulnerabilities, it also offers opportunities to strengthen cybersecurity through the deployment of quantum-secure communication technologies and the exploration of quantum-inspired cybersecurity approaches.

1.1.1 Key Quantum Concepts

Quantum bit: It serves as the basic unit of information, having a value of "0," "1," or simultaneously both "0" and "1." It is also known as a qubit. A qubit is represented as $a|0\rangle + b|1\rangle$ where a and b are complex numbers that represent probability amplitudes of "0" and "1" states.

This representation permits quantum parallelism, leading to increased computational power.

Superposition: A particle could be in several states simultaneously, such as spinning both clockwise and counterclockwise. However, when measured or when it interacts with its environment, it collapses into one state, with a probability of either spinning clockwise or counterclockwise. This superposition of qubits opens up a vast computational space, allowing the efficient solving of complex problems that require significant computational power [4].

Entanglement: Multiple particles could get intrinsically linked, making it impossible to describe them as independent entities. Observing one particle will affect the result of a corresponding measurement on the other particle. This phenomenon occurs when two or more particles, like photons or electrons, become so intricately connected that the state of one particle instantaneously influences the state of the other, no matter how distant they are. It serves as a crucial resource for both quantum computing and communication.

It is essential for quantum computation, enabling qubits to interact within quantum algorithms [5]. Entanglement can also be used for dense coding and simulating correlated systems in quantum contexts. This property aids quantum computers in executing highly interconnected operations and achieving unprecedented computational power [6] (Fig. 1).

Quantum interference: Quantum interference is a key quantum mechanics principle that enables quantum computers to enhance the probability of obtaining correct solutions while suppressing the probability of obtaining incorrect solutions. The wave-like behavior of quantum particles leads to interference between various probability states, which can produce either constructive or destructive interference, similar to the interference patterns observed in waves. This interference phenomenon is essential for quantum algorithms to amplify desired outcomes and mitigate errors.

Quantum Gates: These gates perform operations on qubits to process quantum information. There are various gates such as the Pauli gate, Hadamard gate, CNOT gate, Phase gate, SWAP gate, Quantum Fourier transform gates, and many more.

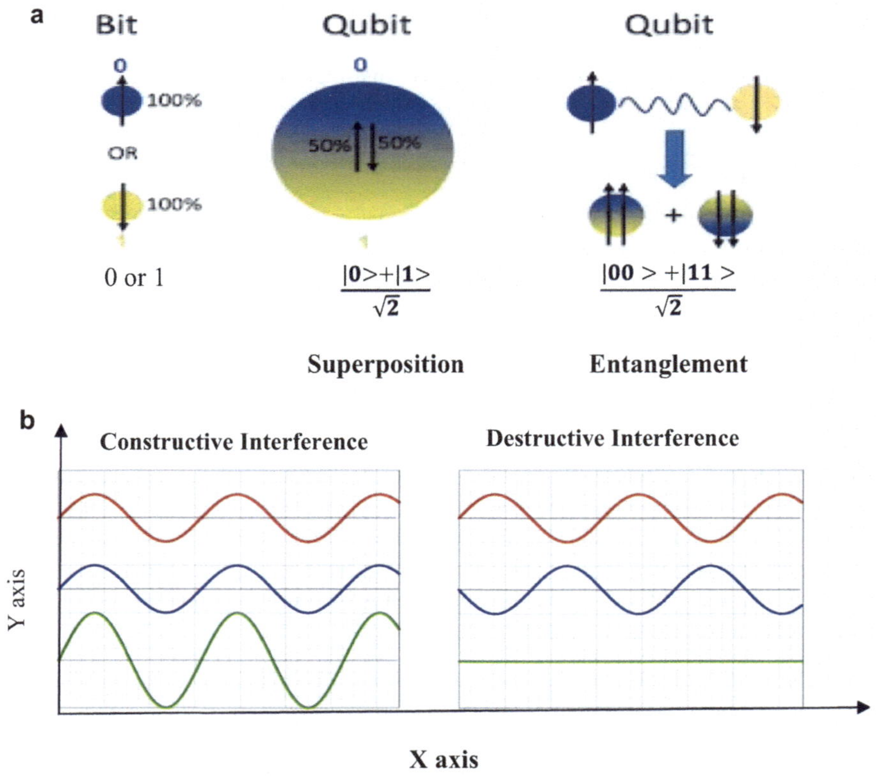

Fig. 1 **a** Quantum principles—qubit, superposition, entanglement **b** Quantum constructive interference versus destructive interference

Quantum Circuits: Quantum circuits consist of sequences of quantum gates that alter the quantum state of qubits for performing computational tasks and executing particular quantum algorithms. The creation and operation of these circuits significantly enhance the computational capabilities of quantum computers [7].

Quantum communication: This includes technologies for transmitting quantum bits over long distances, facilitating communication between quantum computers, and parallelizing quantum computations [7].

Quantum parallelism: It facilitates quantum computers to assess numerous potential outcomes of a computation at once. This capability contributes to the speed advantages of quantum algorithms in comparison with traditional algorithms for certain problems [8].

1.1.2 Quantum Algorithms

Quantum algorithms are crafted to operate on quantum computers and leverage quantum properties to solve particular problems efficiently. They achieve this by managing interactions between qubits through quantum gates. Examples include Shor's algorithm used for factoring large integers, Grover's algorithm for searching through unsorted databases, and quantum simulation algorithms for modeling complex quantum systems. These algorithms leverage quantum parallelism, entanglement, and interference to achieve computational speedups over classical algorithms for certain tasks, revolutionizing areas such as Cryptography, Chemistry, Optimization, Artificial Intelligence, and Machine Learning.

Shor's algorithm [8]: Peter Shor proposed this in 1994. It can effectively factor large composite integers and calculate discrete logarithms. It can solve classically challenging problems at an exponential speedup over classical algorithms. Shor's algorithm leverages quantum parallelism and the periodic characteristics of modular exponentiation to achieve polynomial time complexity, offering an exponential speedup over the most efficient classical algorithms. Shor's algorithm underscores the groundbreaking potential of quantum computing in cryptography and computational number theory. It has a major impact on cryptography because numerous public key encryption methods, including RSA, ECC, and DH Key exchange, depend on the difficulty of factoring large integers computationally.

Rivest-Shamir-Adleman (RSA) is a commonly used public key cryptosystem crucial for protecting data transmission and guaranteeing the confidentiality and authenticity of digital communications. Its security relies on the mathematical challenge of factoring large composite numbers, making it suitable for cryptographic protocols like secure communications in banking, e-commerce, digital signatures, and key exchange. Elliptic Curve Cryptography (ECC) is a type of public key cryptography based on the algebraic properties of elliptic curves over finite fields. It provides strong security with smaller key sizes than traditional systems like RSA, making it both secure and efficient for contemporary cryptographic applications. The Diffie-Hellman (DH) key exchange enables two parties to securely establish a shared cryptographic key over a public channel, which can be used for encrypting and decrypting communications. It is built on modular exponentiation and depends on the complexity of solving the discrete logarithm problem. Shor's algorithm presents a major threat to public key cryptosystems like RSA and ECC, as it can solve the mathematical problems that form the basis for security. It can also be used to break cryptographic schemes built on the discrete logarithm problem, such as the Diffie-Hellman key exchange.

Grover's algorithm [8]: It was proposed by Lov Grover in 1996. It can efficiently search an unsorted database or perform a search, quadratically faster compared to the best classical algorithm. Grover's algorithm employs the principle of quantum parallelism, allowing it to simultaneously evaluate multiple database entries using superposition. Each database item is depicted as a quantum state, allowing for parallel evaluation of all items in the database through quantum operations. Grover's algorithm has applications in various fields, including cryptography, database search, optimization, and pattern recognition. It can solve certain combinatorial optimization problems and improve the efficiency of cryptographic protocols, such as key search algorithms.

Grover's algorithm can be applied to perform brute-force attacks on symmetric cryptography algorithms such as DES and AES. While symmetric encryption schemes like AES are much more resistant to quantum attacks than public key schemes, they can lower the security of an n-bit key from 2^n operations to $2^{n/2}$ operations. For instance, a 128-bit AES key would offer the security level of a 64-bit key when facing a quantum attack. A brute-force attack on AES-128 using classical computers would require 2^{128} operations, but with Grover's algorithm, a quantum computer could perform it in roughly 2^{64} operations, which is significantly faster. To counter this, increasing the key length to 256 bits (AES-256) enhances encryption strength, providing greater resistance to quantum attacks.

1.1.3 Current Quantum Computing Scenario

Significant progress has been achieved in building quantum hardware, with various approaches such as topological qubits, superconducting qubits, photonic qubits, and trapped ions. Companies like Google, IBM, IonQ, Rigetti, and others have demonstrated quantum processors with increasing numbers of qubits and improved coherence times and gate fidelities. Quantum processors are now able to handle large amounts of qubits; however, achieving and maintaining coherence and entanglement is a challenge. Notable Quantum computing hardware that incorporates a range of technologies developed to control and utilize qubits are superconducting qubits, trapped ions, and topological qubits. Superconducting qubits utilize superconducting circuits to maintain quantum coherence. Trapped ions involve single ions confined in electromagnetic traps. Topological qubits investigate unique material characteristics, such as Majorana fermions. To improve quantum computation reliability, advancements in quantum error correction are critical. Superconducting qubits offer fast gate speeds but face difficulties with error rates, while trapped ions have prolonged coherence periods but are constrained by extended gate operation times. Topological qubits examine the special properties of certain materials, potentially providing benefits in error suppression and scalability. Quantum Volume is a performance evaluation metric of a quantum computer that takes into account factors like qubit count, error rates, and gate fidelities. Coherence time, which indicates the duration a qubit can preserve its quantum state, is a crucial parameter. Lengthening coherence times enhances the stability of quantum computations, allowing for the execution of complex algorithms and applications. Open-source quantum software frameworks, like IBM Qiskit, offer tools for designing and executing quantum circuits, including simulation capabilities and access to IBM Quantum Experience hardware. Similarly, Cirq, a Google open-source framework, enables the writing, modification, and execution of quantum circuits, with a particular emphasis on near-term quantum computers and integration with Google's quantum processors. These frameworks promote experimentation and collaboration in the field [7].

1.1.4 *Challenges of Quantum Computing* [4]

Quantum computing has the potential to solve specific problems much quickly than classical computers. However, it faces significant challenges shown below:

Quantum Decoherence: This is a phenomenon arising due to quantum systems being highly sensitive to their environment. Qubits can only perform calculations while preserving their quantum properties. Decoherence occurs when qubits lose their quantum characteristics and revert to classical particles, thereby losing their ability to execute quantum computations. Decoherence causes quantum states to lose their coherence and become classical-like, which can lead to errors in computations. The technical challenge lies in prolonging the lifespan of qubits, as an extended decoherence time translates to increased computational time [4].

Error Correction: Quantum systems are inherently noisy and error prone. Quantum error correction is essential to mitigate errors caused by decoherence, imperfect gates, and other sources. However, implementing error correction in quantum systems without significantly increasing resource requirements is challenging.

Scalability: It is a challenge to build large-scale quantum computers with hundreds or thousands of qubits due to technical limitations in qubit coherence, gate fidelity, and connectivity between qubits. Scaling up quantum systems while maintaining low error rates and high coherence times is a formidable engineering task.

Qubit Connectivity: Qubits need to communicate with each other to perform quantum operations and share information. However, achieving long-range qubit connectivity while maintaining low error rates is challenging. Many quantum architectures face limitations in achieving scalable and programmable qubit connectivity.

Noise and Fault Tolerance: Quantum computations are vulnerable to noise and errors from various sources, including environmental interactions, imperfect gates, and hardware imperfections. Development of error correction techniques to mitigate errors and fault-tolerant quantum algorithms is important for reliable quantum computing.

Resource Requirements: Quantum algorithms often need a huge number of qubits and quantum gates to achieve a substantial advantage over classical algorithms. Meeting the resource requirements for practical quantum algorithms poses a significant challenge given current technological limitations.

Engineering environment: Maintaining near absolute zero temperatures for quantum computing, to avoid decoherence, is very challenging. Additionally, the cost of the various components required for quantum infrastructure is another significant challenge [4].

Algorithm Development: Designing efficient quantum algorithms that leverage unique quantum mechanics properties to solve real-world problems remains a challenging task. This requires a deep understanding of quantum mechanics and computational complexity theory.

Quantum computing poses many prospective threats and vulnerabilities to the security of endpoint devices. In spite of full disk encryption of the contents of endpoint devices, an adversary employing quantum algorithms may have the ability to decrypt. They may exploit vulnerabilities in secure remote access protocols to create unauthorized connections between devices, potentially introducing additional points of compromise within an enterprise network. The deployment of quantum algorithms may compromise the encrypted information in network infrastructure [9].

Addressing these challenges requires advances in quantum hardware, algorithm development, error correction methods, and system integration. Despite these challenges, rapid progress is being made in the quantum computing field, with researchers working toward building scalable and practical quantum computers.

1.2 Overview of Cybersecurity

Cybersecurity adopts a wide range of technologies and effective practices to safeguard systems, networks, and data from attacks or unauthorized access. This includes deploying intrusion detection systems (IDS), firewalls, antivirus software, encryption, and secure coding practices. It is crucial in safeguarding the confidentiality, availability, and integrity of digital information and ensuring the resilience of critical infrastructure, businesses, governments, and individuals against cyber threats [1].

1.2.1 Cyber Threats and Vulnerabilities

Cyber threats include a broad spectrum of malicious activities, such as malware infections, phishing schemes, ransomware, data breaches, insider threats, social engineering scams, viruses, Trojans, worms, rootkits, hackers, spyware, adware, scareware, joke programs, ransomware, Denial of Service (DoS) attacks, bots, passwords and man-in-the-middle attacks (eavesdropping, session hijack attack, SSL stripping), Distributed Denial of Service (DDoS), advanced persistent threats, etc. Cyber threats can arise from a wide variety of sources, including individuals, criminal groups, nation-states, and even members within an organization [10]. The complex network infrastructure, growing network capacity, and breaches of security protocols all contribute to challenges in cybersecurity [1]. Cyber attacks resulting from existing system flaws can be categorized into three types: those triggered by hardware weaknesses, those due to software bugs, and those stemming from vulnerabilities in computer networks. Cyber attacks come in various forms, each crafted to target distinct vulnerabilities in systems, networks, and users. Due to new-generation attacks, like fileless malware attacks, AI-powered attacks, IoT-based attacks, Cloud-based attacks, deepfake attacks, quantum-resistant threats, and blockchain attacks. Conventional protection measures like Intrusion Detection Systems (IDS), firewalls, and antivirus software are no longer sufficient to detect these advanced attacks. As a result, there is an urgent need to develop innovative, more practical, and effective cybersecurity solutions to prevent such threats [11].

The pyramid shown in Fig. 2 represents different layers of security vulnerabilities in quantum cybersecurity. This pyramid emphasizes the multifaceted nature of security, where vulnerabilities exist at various levels, from human users to cryptographic algorithms, with quantum computing posing a significant threat at the foundational level.

User—Poor passwords or phishing attacks: This layer focuses on vulnerabilities related to human error, such as weak passwords or falling victim to phishing attacks.

Admin—Misconfigured or unpatched systems: This layer highlights vulnerabilities due to administrative mismanagement, including systems that are not correctly configured or those lacking the latest security updates.

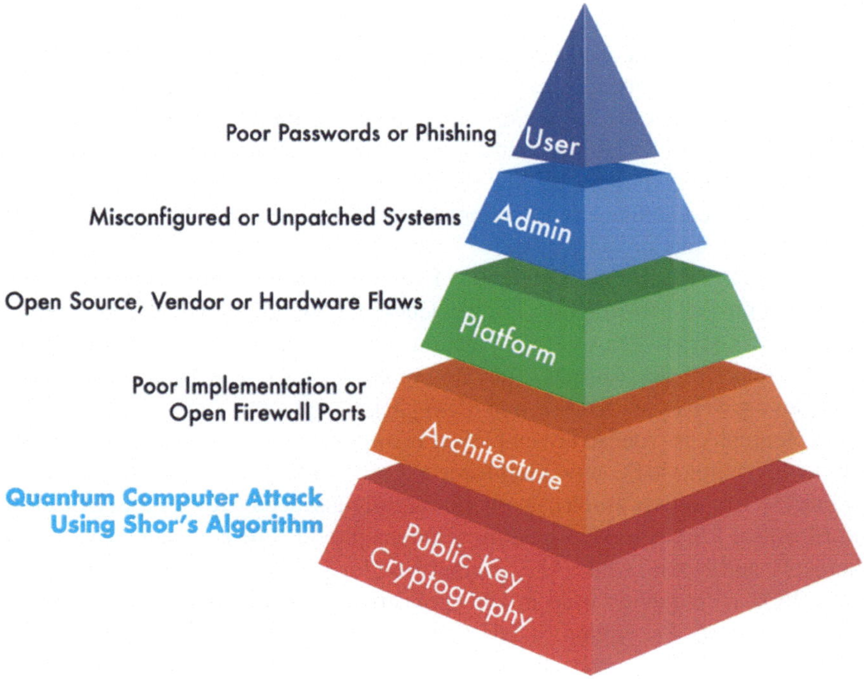

Fig. 2 Potential causes of security vulnerabilities in quantum cybersecurity

Platform—Open-source, vendor, or hardware flaws: This refers to potential vulnerabilities in the platform or underlying hardware/software that could arise from flaws in open-source code, vendor software, or hardware design.

Architecture—Poor implementation or open firewall ports: Vulnerabilities at this level are related to weak architectural design or open ports that can be exploited.

Quantum computer attack by Shor's algorithm: The foundational layer focuses on vulnerabilities in public key cryptography, particularly the Quantum computer attack by Shor's algorithm, which threatens the security of various classical encryption methods.

1.2.2 Cryptographic Algorithms

Most internet-based applications use cryptography to ensure data integrity, confidentiality, and authenticity. Cryptography relies on the premise that some mathematical problems are exceedingly hard to solve and cannot be resolved by a computer within a feasible time frame.

However, it has been found that some of these problems could be solved efficiently by a powerful quantum computer.

Cryptographic techniques are fundamental to securing information and communications. They can generally be categorized into three types, on the basis of how keys are used: symmetric (secret key), asymmetric (public key), and keyless (hash functions) systems [12]. In symmetric cryptography, the same key is employed for encryption and decryption. This key should be kept confidential within the parties involved in the communication. Examples are DES, AES, and Triple AES. Asymmetric cryptography involves a pair of keys: a public key that is openly distributed and a private key that is kept confidential. Examples of asymmetric cryptography are Rivest-Shamir-Adleman (RSA), Elliptic Curve Cryptography (ECC), and Digital Signature Algorithm (DSA). Hash functions convert input data of any length into a fixed-size hash value (digest). They do not use keys for their primary operation. Examples are SHA-256, SHA-3, and MD5 [12].

Cryptography is extensively employed in applications, like finance, healthcare, Internet of Things (IoT), network infrastructure, cloud services, and critical infrastructure [5]. Commonly utilized encryption algorithms include RSA, AES, ECC, DH, and Blowfish. RSA is used in security solutions provided by companies like Cisco, Microsoft, and Nokia, while AES is employed for client/server encryption of web traffic. ECC is frequently used in IoT applications, and Blowfish is an alternative to RSA and Data Encryption Standard (DES) with a key of varying size. Diffie-Hellman is used for securely sharing secret keys among users. The complexity of keys, based on intricate mathematical operations, increases with key length, making it more challenging to crack and thereby enhancing network security [1].

1.2.3 Cybersecurity Controls and Measures

Cybersecurity Controls and Measures can be adopted to mitigate cyber threats and vulnerabilities as well as protect against cyber attacks.

- **Access Controls**: Limiting access to sensitive data and systems according to user roles, privileges, and authentication mechanisms for preventing unauthorized access.
- **Intrusion Detection Systems (IDS) and Firewalls**: Deploying network security devices like firewalls and IDS to filter incoming and outgoing traffic for signs of malicious activity.
- **Endpoint Security**: Installing antivirus software, intrusion prevention, Endpoint Detection and Response system to protect endpoints and devices from security threats.
- **Encryption**: Encrypting data, at rest and in transit, using cryptographic algorithms to safeguard sensitive information from tampering or unauthorized disclosure.
- **Patch Management**: Consistently applying software patches, updates, and security patches to handle known vulnerabilities and reduce the risk of exploitation by attackers.
- **Incident Response and Disaster Recovery**: Developing plans for handling incident responses, incident handling procedures, and disaster recovery strategies to effectively detect, address, and recover from cyber events and data breaches.

1.3 Impact of Quantum Computing on Cybersecurity

Quantum computing can both enhance and challenge cybersecurity in significant ways. The ever-transforming nature of cybersecurity threats presents a significant challenge to traditional cybersecurity practices. Quantum computing has a positive and negative impact on cybersecurity. Security controls that are highly susceptible to quantum attacks are cryptosystems based on the mathematical challenges of Integer Factorization and Discrete Logarithms, such as RSA, DSA, ECDSA, DH, and ECDH. This also applies to any security protocols that use these public key ciphers and products or systems using the protocols [13]. However, it also offers the capability to develop new quantum-resistant cryptographic protocols, like Quantum Key Distribution (QKD) and lattice-based cryptography, that can offer enhanced security against quantum attacks.

Quantum cryptography aims to develop secure encryption methods by leveraging quantum mechanical properties. Quantum devices with cutting-edge technology can improve security through secret key expansion. They can efficiently solve critical integer factoring and discrete logarithms problems, fundamental to the security of many widely used protocols. Cryptosystems can be broken by a large, universal, and fault-tolerant quantum computer [14]. Both RSA and the Diffie-Hellman key exchange are utilized for creating secure (https) connections on the Internet. Public key cryptography depends on the computational difficulty of factoring large prime number products and solving the discrete logarithm problem, which quantum computers can address efficiently. If these key exchange algorithms are compromised, data confidentiality is threatened, and signature forgery becomes feasible [15].

Quantum Key Distribution (QKD) facilitates the secure distribution of encryption keys between parties with quantum mechanics principles. QKD allows for the detection of any eavesdropping attempts, providing a higher level of security for communication channels. Advancements in QKD technology have led to increased distances over which keys can be securely distributed, higher key generation rates, and improved robustness against various attacks. Quantum computing can potentially enhance cybersecurity by speeding up threat detection and analysis processes. Quantum algorithms can efficiently search large datasets for patterns or anomalies, enabling faster threat identification and response [1].

Quantum algorithms can break the widely deployed classical cryptographic algorithms. This could lead to compromise of sensitive data. Shor's algorithm, a quantum computing method, efficiently factors large integers and calculates discrete logarithms, which underpin many cryptographic systems like ECC and RSA. If used on large-scale quantum computers, Shor's algorithm could undermine these widely used cryptographic methods, threatening the integrity and confidentiality of the sensitive information they protect. Meanwhile, Grover's algorithm could reduce the key length of symmetric encryption algorithms (such as AES) by a square root. Although this does not directly break symmetric encryption, it weakens its security, necessitating longer key lengths to preserve the same level of protection [16].

Quantum-inspired algorithms and techniques, such as quantum-inspired optimization and machine learning, could be employed to enhance threat detection, anomaly and intrusion detection, and prevention in cybersecurity operations. Quantum computers possess capabilities to develop novel cryptanalytic techniques exploiting quantum principles to break classical cryptographic schemes more efficiently than classical methods. These techniques could target vulnerabilities in cryptographic algorithms or protocols that were previously unknown or impractical to exploit using classical computation. Quantum computing facilitates collaboration of several parties to compute a function over inputs, allowing for Secure Multi-Party Computation (SMPC) [17].

Quantum computing algorithms are breaking existing cryptographic schemes, weakening security guarantees, and posing challenges for migrating to quantum-safe cryptographic solutions, all have a great impact on Cybersecurity.

1.4 Quantum Cryptographic Vulnerabilities

In RSA and ECC, the private key can be derived by factoring the product of two prime numbers. The key length is crucial to algorithm security. For instance, RSA employs 2,048-bit keys, which current computing systems cannot break, but quantum computers could potentially crack 4,096-bit key pairs within hours using Shor's algorithm. Both Shor's and Grover's algorithms pose significant threats to ECC. Shor's algorithm simplifies the factoring process, making it nearly certain that an intruder could reveal the private key. Meanwhile, Grover's algorithm enhances brute-force attacks by creating a balanced superposition of all possible inputs, eliminating invalid states, and efficiently identifying inputs that meet the specified criteria [1]. AES uses a key size of 128 bits. Though it is difficult for conventional computers to crack, it can be compromised by Grover's algorithm. Symmetric cryptographic block cipher Blowfish has the ability to create long keys, and each key can, in turn, generate subkeys that vary from one another. Thus, a much longer key, say a 256-bit key, can be used instead of 128-bit in classical thus increasing the complexity and preventing any attack. In the Diffie-Hellman key exchange, two users agree on a key pair and then share the public key to communicate over an insecure channel. Neither the sender nor the receiver needs to know the other party's identity on the network. However, the lack of user authentication makes the system vulnerable to eavesdropping, including man-in-the-middle attacks and other insider or outsider threats. Shor's algorithm can undermine this by quickly breaking the factorization [1, 5]. Additionally, quantum computers also affect cryptographic hash functions and symmetric key cryptography, which are commonly used for data encryption [15].

1.5 Quantum Resistant Cryptography (Post-Quantum Cryptography)

The advent of quantum computing poses a security threat to traditional cryptographic schemes based on mathematical problems such as integer factorization and discrete logarithms, susceptible to quantum attacks. To ascertain data security in the quantum age, cryptographic algorithms resistant to quantum computing must be developed and implemented. Quantum-resistant cryptography/Post-Quantum Cryptography (PQC) involves creating algorithms that are secure against the capabilities of quantum computers [18]. They are designed to overcome attacks from classical and quantum adversaries and ensure long-term security and integrity of digital information in the quantum age. To address this vulnerability, developers are tasked with creating quantum-safe cybersecurity solutions through the design of algorithms and protocols, resilient to quantum computer attacks. Quantum-safe cryptography finds applications in securing financial and banking transactions, protecting government and military communications, maintaining the confidentiality of Electronic Healthcare Records (EHR), protecting personal cloud data, providing restricted access to confidential data, and many more. Post-quantum cryptography includes a range of cryptographic techniques, such as digital signatures, hash functions, and key exchange protocols. Quantum computers might employ Grover's algorithm to discover hash function collisions, potentially enabling the forging of digital signatures [19]. Cryptographic algorithms and protocols designed to resist quantum attacks are developed to protect digital information against all threats, ensuring long-term security and integrity in the quantum era. These quantum-resistant cryptographic methods use a diverse range of mathematical and computational techniques that do not rely on factorization or discrete logarithm problems, which are vulnerable to Shor's algorithm [8]. Post-Quantum Cryptography (PQC) seems to be the most suitable approach for creating quantum-resistant encryption. However, many PQC protocols face challenges related to confidence, efficiency, and usability [4].

Some of the approaches are:

Lattice-based cryptography: It utilizes the geometric properties of lattices in multi-dimensional spaces to create cryptographic components. Lattices are geometric constructs consisting of infinite sets of points arranged in a regular, repeating pattern across multiple dimensions. Lattice-based cryptography depends on the difficulty of specific mathematical problems defined within these lattice structures. Two prominent lattice problems used in this field are the Shortest Vector Problem (SVP) and the Learning With Errors (LWE) problem. Lattice-based cryptography supports public key encryption, key exchange protocols, digital signatures, and identity-based encryption. The security of lattice-based signatures depends on the difficulty of solving lattice-related problems in high-dimensional spaces and is considered resilient to quantum attacks [20]. The robustness of lattice-based cryptography is attributed to the complexity of these lattice problems, which are thought to be resistant to both classical and quantum computing threats. The advantage is its efficiency, but the drawback is large key lengths as compared to current cryptography. Lattice-based cryptographic schemes are recognized for efficiency and performance regarding key sizes, computational complexity, and resistance to quantum attacks. They offer cryptographic solutions with compact parameters and fast cryptographic operations [8].

Code-based cryptography: Security is built upon the mathematical properties of error-correcting codes. Error-correcting codes are mathematical constructs utilized to identify and rectify errors that arise during data transmission or storage. The code-based scheme security hinges on the assumed difficulty of decoding random linear codes. This task is challenging for both classical and quantum computers, making these schemes resistant to quantum attacks. The McEliece Cryptosystem, one of the earliest and prominent code-based cryptographic methods, leverages the complexity of decoding random linear codes, a challenge that remains formidable even for quantum computers. Its advantages include relatively small key sizes and efficient encryption and decryption processes [8].

Hash-based cryptography: Also referred to as hash-based signatures or hash-based authentication, this cryptographic scheme relies on properties of hash functions to ensure security. It utilizes features such as collision resistance and pre-image resistance inherent in hash functions. Hash functions are deterministic, meaning it is computationally impractical to deduce the original input from a hash value (pre-image resistance) or to identify two different inputs that yield the same hash value. Hash-based cryptography is valued for its simplicity and efficiency, offering practical cryptographic solutions with small key sizes. It provides a strong basis for ensuring data integrity and authenticity, rendering it highly suited for environments with limited resources and applications [8].

Multi-variate-based cryptography: It depends on solving systems of multivariate polynomial equations, specifically Multi-variate Quadratic Equations (MQ), over finite fields to ensure security. This approach provides key exchange protocols, encryption, and digital signatures by relying on the complexity of certain mathematical problems. Multi-variate Polynomial Cryptography (MPC) involves polynomial equations with multiple variables and coefficients. It offers efficient cryptographic operations and small key sizes. Challenges are related to parameter selection and resistance to specific algebraic attacks [8].

Digital signature-based cryptography: This involves creating digital signature algorithms to withstand quantum attacks. These signatures are critical for the verification of authenticity, integrity, and non-repudiation of digital messages and documents. A digital signature is a cryptographic scheme that allows an entity to be widely distributed and utilized for verifying digital signatures, whereas the private key is maintained confidential and employed for signing messages. Digital signature schemes frequently utilize a cryptographic hash entity (the signer) to sign a message, document, or transaction using the private key. The signature verification can be done by using the signer's public key to confirm the authenticity and integrity of the signed data. This typically depends on a public key infrastructure, where users have public-private key pairs. The public key functions to generate a fixed-length hash value of the message before signing it. This ensures that the message size does not affect the signature and provides additional security against collision attacks.[8]

Isogeny-based cryptography [19]: It relies on the hardness of computing isogenies among elliptic curves possessing special properties to offer cryptographic primitives like key exchange, encryption schemes, and digital signatures, to prevent classical and quantum computer attacks. Isogenies are mathematical mappings among elliptic curves that preserve specific properties, such as the group structure. It sends points from one elliptical curve to the other while preserving the group structure. The Supersingular Isogeny Diffie-Hellman (SIDH) protocol is among the well-known isogeny-based cryptographic schemes that provide a key exchange protocol enabling two parties to agree upon a common secret key over an insecure communication channel, relying on the challenge of computing isogenies among supersingular elliptic curves. The Supersingular Isogeny Key Encapsulation (SIKE) protocol facilitates encryption of messages using a public key and decryption using the corresponding private key, on the basis of the isogeny computing difficulty factor.

Table 1 gives a summary of the quantum-resistant algorithms in terms of type, examples, strengths, weaknesses (limitations), and current development status.

Table 1 Summary of quantum-resistant algorithms/post-quantum cryptographic algorithms [21, 22]

Algorithm type	Algorithm example	Strengths	Weaknesses	Current development status
Lattice based	–PKE algorithms Crystals-Kyber –Digital signature algorithms Crystals-Dilithium	–Strong resistance to both classical and quantum attacks –Efficient key generation and encryption/decryption –Can be employed efficiently in hardware	–Larger key sizes and ciphertexts compared to RSA or ECC –Implementation errors could lead to vulnerabilities	–NIST finalists for public key encryption and digital signatures –Actively tested in real-world scenarios
Code based	Classic McEliece, BIKE, HQC	–Long-established with strong theoretical security (since the 1970s) -Quantum attacked resistant due to the difficulty of decoding random linear codes	–Very large public key sizes (in megabytes) –Slower encryption/ decryption for certain applications	–NIST finalist McEliece –Alternatives are BIKE and HQC
Hash based	**–SPHINCS + , LMS, XMSS**	Strong security based on well-understood cryptographic hash functions –Suitable for digital signatures with quantum resistance	Large signature sizes compared to classical algorithms –Limited number of signature uses (stateful versions like XMSS)	–SPHINCS + NIST finalist for stateless hash-based signatures –Well understood but can be slow
Multi-variate based	**Rainbow, GeMSS**	–Fast signature generation –hard for quantum and classical computers due to the difficulty of solving systems of multivariate polynomials	–Very large signature and public key sizes –Vulnerable to certain algebraic attacks in some configurations	–Rainbow was a NIST finalist, and vulnerabilities have been found –GeMSS is still under evaluation

(continued)

Table 1 (continued)

Algorithm type	Algorithm example	Strengths	Weaknesses	Current development status
Digital signature based	Lamport Signatures (One-time signatures) Lamport OTS Hash-based Signature-eXtended Merkle Signature Scheme)	*Quantum-resistant –Simple and secure –Provides one-time use, reducing risks *Quantum-safe –Efficient for post-quantum systems –Scalable for high security	–Key can be used only once –Inefficient for multiple signatures –Large signature sizes –Requires careful implementation to prevent reuse of key	–Mainly in research phase for quantum-safe cryptography –In development for post-quantum cryptography, for long-term security
Isogeny based	–SIDH –SIKE	–Smallest key sizes among quantum-resistant algorithms –Suitable for bandwidth-constrained environments	–Slow encryption and decryption processes –Significant vulnerability was found, casting doubt on security	–**SIKE** was a promising NIST candidate but suffered from a critical attack in 2022. –Development continues with patches

1.6 Quantum Cryptography

Quantum computing also presents opportunities for enhancing cybersecurity through the use of quantum communication technologies. Quantum Cyber Security encompasses the study of facets that impact the security and privacy of computations and communications, by the advancement of quantum technologies [14]. Quantum cryptography exploits the quantum mechanics principles, namely, superposition and entanglement, to achieve secure communication between users over the network. In quantum cryptography, any attempt to observe the quantum carrier alters its state, leaving detectable marks. Consequently, the entire process is scrutinized, interventions are assessed, and any intrusions into the communication channel are identified and addressed [12]. Quantum communication security is ensured by Heisenberg's Uncertainty principle, asserting that a quantum system is disturbed while measuring it, thereby allowing the detection of eavesdropping. The uncertainty principle indicates that in the microscopic realm, a particle's position cannot be precisely determined, and it exists in various locations with different probabilities [23]. Additionally, as per the No Cloning theorem, it is not possible to create a copy of an arbitrarily unknown quantum state, further safeguarding the quantum keys. No cloning theory means cannot produce another completely identical quantum state in another system [24]. Information cannot be copied or read without authorization [25]. The primary objective of researching quantum cryptography is to develop cryptographic algorithms and protocols designed to resist attacks from quantum computing [26]. The challenges are that specialized quantum communication infrastructure, including photon sources, detectors, and often quantum repeaters, is required as quantum signals degrade over long distances. The best known form of quantum cryptography is Quantum Key Distribution QKD, which leverages quantum mechanical properties to securely exchange cryptographic keys. Additionally, quantum cryptography includes other methods such as Quantum Random Number Generators (QRNGs) [5].

Quantum key distribution (QKD): It is a secure communication technique that leverages quantum mechanics to create a shared secret key between parties. During the exchange of quantum keys, clients can exchange keys while restricting unauthorized users from accessing the key [27]. QKD facilitates the creation of secure communication channels that are theoretically impervious to eavesdropping attempts, utilizing the quantum mechanics principles of superposition and entanglement to generate and distribute encryption keys, thus ensuring key distribution security [1]. It exploits the fundamental quantum phenomenon that observing a quantum system disrupts it, enabling detection of eavesdropping attempts [28]. In QKD, the transmitter and receiver exchange quantum signals, typically encoded in the polarization states of a single photon. These signals contain random quantum bits (qubits), which are made use of to establish a secret key that is shared between transmitter and receiver. QKD protocols typically use two or more quantum states to encode information, such as photon polarization states. Common quantum states in QKD include Vertical ($|0\rangle$) and horizontal ($|1\rangle$) polarization states, Diagonal ($|+\rangle$) and anti-diagonal ($|-\rangle$) polarization states. Some common QKD protocols include the BB84 Protocol that uses two mutually unbiased bases (rectilinear and diagonal) to encode qubits and the E91 Protocol that uses entangled photon pairs to set up a mutually shared secret key between distant parties.

BB84 protocol [29] uses photons to transmit data over a fiber-optic wire. Polarization refers to the orientation of the oscillations of a wave, especially with respect to light. Photons can be polarized in specific directions, and qubits can be implemented using the polarization of photons.

A polarization filter or polarizer selectively allows transmission of light waves with specific polarization orientation/states while blocking light waves with different orientations. The BB84 protocol uses four polarization states: linear states (vertical, horizontal), diagonal states (+45°, −45°). Figure 3a shows the polarization of the photon used in the BB84 protocol.

Working of the BB84 protocol: Fig. 3b shows the schematic diagram for the working of the BB84 protocol. The fundamental principle involves the sender (Alice) generating a random sequence of ones and zeroes using polarized photons. As this sequence is transmitted, each bit is represented by a photon in one of the polarization states: horizontal, vertical, or diagonal. The specific polarization states are determined by the sender [1]. The receiver (Bob) must randomly choose between two measurement bases—rectilinear or diagonal—to measure the incoming qubits

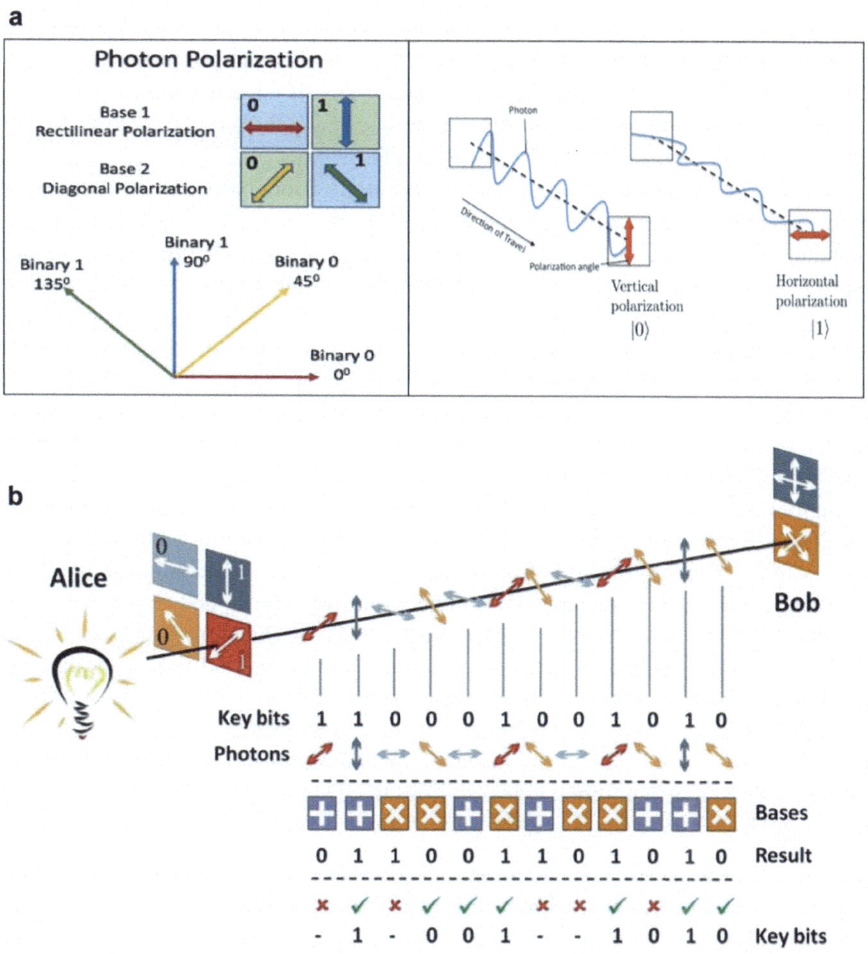

Fig. 3 **a** Polarization of a photon in the BB84 protocol. **b** Schematic diagram of the BB84 protocol

(photons) sent by Alice. At the receiving end, two beam splitters generate quantum superposition by splitting the photon beam. This process effectively determines whether Bob will measure the photon in the rectilinear or diagonal basis. The beam splitters enable the separation of photons with different polarization states, allowing Bob to check if the photon aligns with the selected basis. If an eavesdropper (Eve) attempts to intercept and measure the polarization, the inherent disturbances caused by quantum mechanics will result in random measurement outcomes at Bob's end. This inconsistency can be detected through the use of beam splitters to split and measure various states, thereby exposing any potential eavesdropping [29].

When a match occurs, the receiver obtains the sender's exact message. If there is no match, the receiver will not be able to deduce the sequence or attempt to determine the polarization again, because an attempt would destroy the information. The users then communicate over a public channel to discuss which settings the receiver used, but they do not reveal the specific polarization states of the photons. The values that align at both ends become the shared secret key. If eavesdropping occurs, the polarization of the photons transmitted is altered, causing the secret keys to no longer match, which alerts the users to the intrusion [1].

QKD can be utilized to secure quantum communication, particularly in scenarios where high levels of security are required, such as government communications, financial transactions, and critical infrastructure protection. It provides a secure method for establishing cryptographic keys over potentially insecure communication channels. The BB84, E91, BBM92, and SARG04 are some of the secure quantum key distribution protocols [30]. An eavesdropper trying to capture the quantum key will create noticeable disturbances, which will alert the genuine users to a possible breach of security. Additionally, the effectiveness of Quantum Key Distribution (QKD) is affected by the distance separating the parties involved in communication. Quantum signals may experience attenuation and loss over long distances, requiring the use of quantum repeaters or other methods to extend the range of secure communication. In the Internet of Things IoT frameworks that use enabling technologies like Wi-Fi, Zigbee, Bluetooth, and IoT communication through the quantum channel are a challenge due to short-distance communication [30]. The speed of quantum key exchange (Key exchange rate) limits the practical use of Quantum Key Distribution (QKD). Swift key exchange rates are essential for secure real-time communication, but current technological constraints may restrict the speed of quantum key generation and exchange [4]. The technical limitations of QKD are that the transmission source is not verified, the need for special hardware, escalated infrastructure expenses, and insider threat vulnerabilities like Denial-of-service compromising security [31].

Quantum Random Number Generators (QRNG): Random numbers are utilized for generating keys in cryptography using quantum mechanics, by exploiting the inherent randomness at the quantum scale. This arises from Heisenberg's Uncertainty Principle, which asserts that it is impossible to measure certain pairs of physical properties, such as position and momentum, simultaneously with precision. Detectors are used to measure the polarization of individual photons, producing a stream of random numbers. It is theoretically impossible to predict the randomness, even with complete knowledge of the system [5]. The Mersenne Twister generates random numbers using 624 numbers with a repetition period called a Mersenne prime, making it difficult to find the pattern. The use of Quantum random walks enhances security as there are no repetitive cycles, causing decryption to be difficult [32].

1.7 Classical Cryptography Versus Quantum Cryptography

Classical Cryptography is an established, widely used system that integrates seamlessly into existing systems but has vulnerabilities that quantum computers could exploit, especially in asymmetric methods. **Quantum Cryptography** provides theoretically unbreakable security based on quantum principles but is currently limited by hardware requirements and scalability challenges. It represents a future-proof solution against quantum attacks, especially for key distribution. The comparison between Classical cryptography and quantum cryptography [33] provides a broad view of the strengths and drawbacks of both classical and quantum cryptography techniques, showcasing performance metrics and practical considerations as shown in Table 2.

Table 2 Comparison of classical cryptography and quantum cryptography [33]

Sl. No	Criteria	Classical cryptography	Quantum cryptography
1	Security basis	Mathematical computation hardness (Factorization, Discrete Logarithms)	Quantum mechanics (No closing theorem, Entanglement)
2	Key types	Symmetric (AES) Asymmetric (RSA, ECC)	Quantum key distribution
3	Resistance to quantum attacks	–Symmetric key encryption is relatively resilient –Asymmetric key encryption is vulnerable	Designed to be secure against quantum attacks (QKD)
4	Key management	Requires secure key exchange methods and regular key updates	Key distribution is secure through quantum entanglement and requires less frequent key updates
5	Scalability	Scales well with current hardware and infrastructure	Limited scalability due to limited quantum hardware and quantum communication infrastructure
6	Implementation complexity	Mature with well-established standards and widespread support	High complexity due to the need for specialized quantum hardware and infrastructure
7	Performance metrics:	–Encryption/Decryption speed fast (AES) –Key Size is smaller for asymmetric methods but increases with security level –Computational Resources: Variable and efficient	–Encryption/Decryption speed is slow –Key Size is typically smaller for QKD, as quantum keys are not the same as classical keys –Computational Resources: Specialized quantum hardware and infrastructure
8	Current use cases	All digital security areas like banking, communications, and data protection	Mostly experimental, secure communication research, financial institutions, and government agencies
9	Technology	Established technology—ongoing improvements against quantum attacks	Emerging technology—quantum hardware and protocols mature, potential for widespread use

Table 3 Comparison of computational complexity of classical algorithms versus quantum algorithms for various cryptographic tasks [34, 35]

Sl. No	Cryptographic task	Computational complexity-classical algorithm	Computational complexity-quantum algorithm
1	Integer factorization	–Classical factorization algorithm GNFS used in RSA –Sub-exponential: $O(e^{1.923(\log N)1/3}(\log \log N)^{2/3})$	–Shor's algorithm breaks RSA: –Polynomial: $O((\log N)^3)$
2	Discrete logarithm	–Index Calculus algorithm –Sub-exponential: $O^{(sqrt(elogploglogp))}$	–Shor's algorithm breaks Diffie-Hellman and ECC –Polynomial: $O((\log p)^3)$
3	Symmetric key search	–Brute force search –Exponential: $O(2^n)$	–Grover's algorithm –Quadratic speedup: $O(2^{n/2})$
4	Hash function collision (Birthday attack)	SHA-256, SHA-3, Hash-based message authentication code (HMAC) O(n)	Grover's algorithm Quadratic speedup: $O(2^{n/2})$
5	Elliptic curve factorization	Pollard's rho algorithm Sub-exponential: O(sqrt(p))	Shor's algorithm Polynomial: $O((\log p)^3)$
6	Random number generation	–True random number generators –Pseudo random number generators complexity: polynomial	Quantum random number generator Complexity: truly random O(1)

Cryptographic tasks are critical for securing communications and data. Classical algorithms, which run on traditional computers, and quantum algorithms, which operate on quantum computers, differ fundamentally in their computational models. Table 3 lists the comparison of computational complexity of classical and quantum algorithms for various important cryptographic tasks.

1.8 Real World Case Studies of Quantum Computing in Cybersecurity

Healthcare: Healthcare providers face various vulnerabilities due to quantum computers. These include breaches of Electronic Health Records of patients, through inadequately encrypted endpoint devices or poorly secured data connections among healthcare centers in a regional network. Additionally, unauthorized access to personal patient information could occur through non-privacy-preserving data extraction techniques in the post-quantum world. There is also a risk of acquiring patient files fraudulently, through channels lacking proper authentication, as well as, potential release of vulnerable scientific information such as specific genetic patterns, which could compromise biometric security markers. Protecting medical and healthcare-related information using quantum-safe solutions is crucial, given the necessity for long-term confidentiality, especially concerning the patient's life expectancy and potentially extending beyond, particularly the genetic data [22]. A quantum blockchain security framework incorporates Consultative Transaction Key Generation and Management (CTKGM) for producing unique key pairs using random values, multiplicative operations, and timestamps. This method securely stores patient data within distinct blocks of hash values on a blockchain [36]. Additionally, the framework includes the Quantum Trust Reconciliation Agreement Model (QTRAM), which calculates trust scores from feedback data to ensure secure and reliable data transmission [36] (Tables 4 and 5).

Financial Services: Quantum cybersecurity is critical to ensuring the future security of financial services, particularly as quantum computing poses new risks. By adopting quantum-safe solutions, financial institutions can safeguard transactions, communication, and data, protecting the global financial system from emerging cyber threats [9]. Internal Communications in the organization network or across data centers, essential for information exchange, disaster recovery, and backup, must be safeguarded. The systems rely on AES encryption, which can be susceptible to attacks. To ensure secure financial communication between banks for transferring payment orders globally, quantum-safe encryption methods and digital signatures are essential. While credit card information is currently protected using symmetric key cryptography, the exchange of keys via public key cryptosystems necessitates a quantum-resistant alternative. Online banking security, reliant on Transport Layer Security protocol to apply encryption on web traffic, faces vulnerabilities in server authentication using RSA public keys, X.509 certificates, and session key establishment. Hence, adopting quantum-safe methods for encryption, key generation, and decryption is imperative [38].

Space Industry: The space industry anticipates significant advantages from a range of quantum technologies that are employed in quantum metrology and sensing. Although quantum key distribution (QKD) offers secure quantum communications, it doesn't guarantee absolute security for space communications due to various intrusion vectors that can introduce vulnerabilities [39]. Most quantum computing processing is expected to be conducted via the cloud, prompting cloud providers to explore new security strategies to foster trust among potential enterprise clients. Theorized approaches like quantum homomorphic encryption aim to ensure that cloud providers have no knowledge of either the program or the data, thereby enhancing security effectively [31].

Table 4 Real-world applications of quantum cybersecurity in healthcare [22, 37]

S.N	Application	Challenge	Solution
1	Secure medical data transmission	Data integrity and unauthorized access of sensitive data of patients	QKD for secure sharing of patient medical records, diagnostic images, and treatment information across large networks, communication cannot be intercepted without being detected
2	Quantum-enhanced electronic health records	Storage of sensitive patient data in EHR systems that are tamper-proof and secure to maintain trust and patient safety	Post-Quantum Cryptographic algorithms to protect EHRs from future quantum attacks
3	Quantum-secure cloud storage	Cloud-based storage systems are widely used for patient data storage and sharing, medical images, and research data Encryption methods may be vulnerable once quantum computers can break classical encryption schemes	Cloud service providers offering Quantum Encryption-as-a-Service QaaS can store patient medical records, genomic data, or clinical trial information securely, protecting them from unauthorized access
4	Privacy-preserving genomic data sharing	Genomic data containing highly sensitive personal information should be shared securely, and privacy should be maintained	Quantum cryptographic techniques—quantum homomorphic encryption and quantum SMPC allow secure analysis of encrypted genomic data
5	Quantum-resistant telemedicine	Transmitting sensitive data over the internet, including video consultations, diagnostic results, and personal health information, must be protected from interception or tampering	QKD can ensure end-to-end encryption for telemedicine platforms, which would protect video calls, file sharing, and messaging from quantum computer threats

(continued)

Table 4 (continued)

S.N	Application	Challenge	Solution
6	Secure medical device communication and IoT in healthcare	Medical devices (critical life saving) and wearables that generate real-time health data, often transmitted over wireless networks to be protected from cyber attacks	Quantum cryptography for IoT (Internet of Things) can enhance security in medical devices by using quantum-resistant algorithms to secure communication channels between devices like Pacemakers, insulin pumps, and smart health monitors and healthcare systems
7	Quantum-resistant blockchain for drug supply chain	Secure tracking of medications and their ingredients in the pharmaceutical industry that faces challenges with counterfeit drugs and ensuring the authenticity of drugs along the supply chain	Use quantum-safe blockchain technology to securely track medications from production to the patient, ensuring authenticity and preventing counterfeit drugs from entering the supply chain

2 Standardization

Globally, efforts are underway to standardize post-quantum cryptographic systems. In the field of Post-Quantum Cryptography, the US NIST has a key role in the process of soliciting, assessing, and standardizing post-quantum cryptographic primitives [28]. The NIST Post-Quantum Cryptography Standardization project tries to establish universal standards for quantum-resistant cryptographic methods, ensuring global security and interoperability [26]. ETSI Quantum-Safe Cryptography Working Group has the goal of making assessments and then recommendations for quantum-safe cryptographic primitives protocols. They also work on implementation, considering the present state of research on cryptography and quantum algorithms and industrial requirements for real-world deployment [9]. The ISO and German Federal Office for Information Security are focused on fostering a secure digital landscape in Germany, with an emphasis on cybersecurity and digital transformation. At present, no single post-quantum cryptographic system matches the overall performance (key size, message size, speed) of most advanced pre-quantum solutions [40].

3 International Collaborations

Collaborations in the cybersecurity community are crucial for developing standardized quantum-resistant cryptographic techniques that could be widely implemented. Cybersecurity standards must be jointly created to integrate quantum-safe cryptographic techniques, quantum key distribution, and guidelines to securely adopt

Table 5 Real-world applications of quantum cybersecurity in financial services [9]

S.N	Application	Challenge	Solution
1	Secure financial transactions	Financial transactions, including payments and money transfers, are prone to fraud	QKD to secure communications for international money transfers, interbank settlement systems, or high-frequency trading platforms, ensuring that transactions remain safe from quantum computer-based attacks
2	Banking systems	Cryptographic algorithms used in banking and financial services, like RSA and ECC, are prone to future quantum computing attacks	Banks, payment processors, and financial service providers can adopt PQC to secure online banking portals, credit card transactions, and digital wallets, ensuring their systems are protected in the post-quantum era
3	Quantum-safe blockchain for financial transactions	Current cryptographic methods securing blockchain used in financial services for cryptocurrency, smart contracts, and decentralized finance (DeFi) applications are vulnerable to quantum attacks	Quantum-resistant blockchain technology can be adopted by Financial institutions involved in cryptocurrency, cross-border payments, or supply chain finance
4	Mobile payments and digital wallets	Cyberattacks targeting sensitive financial information in mobile banking, contactless payments, and digital wallets	Quantum-safe cryptographic algorithms to secure payment data and transactions can be used by payment platforms and digital wallets to secure their assets, prevent fraud, and ensure transparency
5	Risk modeling and fraud detection	Fraud detection and risk modeling are prone to cyber attacks	Financial institutions can deploy quantum-enhanced algorithms to detect fraudulent transactions, money laundering, or insider trading, enhancing their ability to respond quickly to potential risks

(continued)

Table 5 (continued)

S.N	Application	Challenge	Solution
6	Identity verification and authentication	Current Financial services rely on robust identity verification and authentication processes, such as two-factor and multi-factor authentication methods, to protect user accounts and prevent unauthorized access. These may be vulnerable to future quantum attacks	Quantum digital signatures and quantum-resistant public key infrastructure (PKI) provide enhanced protection for verifying identities and securing access to financial systems
7	Financial communication networks	Financial losses, market manipulation, or fraud due to a breach in financial communication networks utilized for sharing confidential information, such as trading data, risk assessments, and regulatory reports	Stock exchanges, banks, and regulatory bodies could deploy QKD-based networks for secure communication of sensitive trading data, financial reports, or regulatory information

quantum technologies, ensuring global cybersecurity resilience. Governments and international organizations should work together to establish policies and standards addressing legal, ethical, and security challenges posed by quantum systems. Regulatory frameworks must cover data protection, privacy, and the prudent use of quantum-driven cybersecurity tools. International collaboration is key to establishing effective quantum security standards [41]. Collaboration is critical in order to address systemic risks posed by quantum computing. Both private and public sectors can collaborate to integrate countermeasures into the design and deployment of quantum-related technologies for cybersecurity.

4 Challenges

The paradigm shift to quantum-safe cryptography demands meticulous planning to guarantee the security and integrity of digital communication in the post-quantum era. Implementing quantum-resistant cryptographic algorithms presents technical challenges due to their structural differences from conventional methods. Incorporating these algorithms in current cryptography systems needs extensive testing and validation. There should be collaboration among industry, academia, and standardization bodies. Organizations must develop strategies, like as evaluating the effects on existing systems, revising protocols, and maintaining backward compatibility, to shift seamlessly to quantum-safe cryptography.

5 Quantum Cybersafe Solutions

Cybersecurity is crucial for safeguarding all sectors of critical infrastructure, encompassing both physical and digital systems and assets. The Information Security Management System (ISMS) provides a global, risk-oriented framework that supports the creation of cybersecurity strategies through the implementation of security policies, controls, and measures to protect against information threats and vulnerabilities. Quantum computing can enhance incident response management by identifying incidents, their associated threats and controls, and the optimal solution for resolving incidents as efficiently as possible. However, even with such security management frameworks in place, incidents and breaches can still occur. To address these challenges, various cybersecurity solutions are being developed using quantum technologies. Additionally, the MARISMA framework, which utilizes an automated cloud-based eMARISMA tool, leverages quantum algorithms to handle incidents and restore system security [42]. KeyShield offers a scalable and quantum-resistant key management solution, providing robust security by leveraging the difficulty of solving underdetermined linear systems of equations uniquely. Unlike systems that depend on pairwise secure channels, KeyShield utilizes a single broadcast message over an open channel for rekeying [43]. Organizations should adopt proactive measures to evaluate quantum vulnerabilities and devise effective strategies for transitioning to encryption methods resistant to quantum attacks. Quantum computers decrypting and compromising security can be mitigated by migrating applications to encryption techniques resistant to quantum attacks, with future advancements. System security must be ensured to span across supply chains, complete operational processes, and shared resources to foster robustness against evolving threats posed by quantum computing. Quantum-safe protocols, including NTRUEncrypt and Hash-Based Message Authentication Code (HMAC), provide novel solutions to the vulnerabilities created by quantum computers [17].

6 Conclusion

In this chapter, an overview of quantum computing and cybersecurity is provided, cyber attacks and vulnerabilities are explored, cryptographic algorithms and quantum algorithms are elaborated, and cybersecurity controls are elaborated. The convergence of these fields represents both an opportunity and a challenge. While quantum computing holds the promise of revolutionizing various scientific and technological domains, it also necessitates a fundamental rethinking of cryptographic frameworks for cybersecurity. The cybersecurity landscape must evolve to address these emerging threats, ensuring that our data remains secure in an increasingly complex digital world. The post-quantum cryptography is explored as a critical defense mechanism designed to resist the computational power of quantum computing systems. Additionally, real-world case studies in the healthcare, financial services, and space industries are discussed, highlighting practical implications and adaptations required in these sectors. Standardization efforts and international collaboration are vital in developing and deploying quantum-resistant cryptographic solutions. The challenges in this endeavor are significant but not insurmountable. By fostering a cooperative global effort, we can develop robust standards and strategies to secure our digital future.

The convergence of quantum computing and cybersecurity underscores the need for continuous research and innovation. By understanding and anticipating the challenges presented by quantum advancements, we can harness their potential while protecting the integrity and confidentiality of our digital infrastructure. The journey toward quantum-resilient cybersecurity is essential, ensuring that as technology evolves, our security measures advance in tandem.

7 Future Scope

Quantum computing presents thrilling prospects for the evolution of healthcare, artificial intelligence, machine learning, and communications. As this technology advances, networks will increasingly integrate classical and quantum devices, resulting in more complex systems. This integration brings opportunities and threats to the computing and communication landscape, making security a top priority [5].

Cybersecurity experts, policymakers, and researchers need to remain alert to advances in quantum computing and implement strategies to address new threats. It is important to foster education and awareness about the impact of quantum computing on cybersecurity. Building a skilled workforce with expertise in quantum technologies will be vital for maintaining the future robustness of systems [41].

For businesses, the advent of quantum computing necessitates a proactive stance on cybersecurity. Preparing for a future where quantum systems are prevalent involves implementing several strategic measures to guard against quantum-specific vulnerabilities and threats. The initial step is to perform a thorough quantum risk assessment, analyzing the sensitivity and weaknesses of existing cryptographic systems. Companies must identify the areas most vulnerable to quantum attacks and prioritize upgrading these systems to quantum-resistant solutions. Cybersecurity in manufacturing, industrial environments, and robotics focuses on protecting industrial control systems (ICS), robotic systems, and operational technology (OT) from cyber threats. As these environments become more interconnected through the Industrial Internet of Things (IIoT), cyber threats increase, making it essential to adopt robust cybersecurity measures [44].

The shift to quantum-resistant cryptographic solutions should begin promptly. Implementing post-quantum cryptography (PQC) algorithms into current security systems can be done in phases, starting with the most crucial systems. It is important to stay updated on recent advancements and adjust cybersecurity strategies as needed, which includes ongoing education and training for IT and cybersecurity professionals. Quantum as a Service QaaS can be used to explore new cryptographic protocols and test the resilience of existing ones. Zero vulnerability computing refers to a state where computer systems and networks are completely secure from all cyber threats and vulnerabilities. It aims for absolute protection against potential security breaches and unauthorized access. It includes advanced threat detection, robust encryption, and multi-factor authentication [45]. Post-Quantum Cryptography solutions are to be applied to Critical Infrastructure networks, which are interconnected systems and assets, vital for public health and safety, economic stability, and national security [19]. Navigating the transition to quantum technology will necessitate collaboration across industries and adherence to new standards and regulations. Participating in industry consortiums and following guidelines from regulatory bodies such as NIST will be essential. For sectors dealing with highly sensitive data, adopting advanced solutions like Quantum Key Distribution (QKD) may be advantageous, offering a quantum-safe approach to secure key distribution. Preparing for the era of quantum technology presents businesses with a chance to spearhead technological innovation. Embracing a proactive cybersecurity strategy will help safeguard against emerging threats and provide a competitive advantage. Given that quantum computing systems are expected to be hybrid—integrating CPU-hosted and cloud-based or managed APIs—the demand for robust, secure services and architectures is crucial. Implementing effective security measures for critical applications and data managed by these systems is essential [46].

There are some emerging trends, potential breakthroughs, and upcoming research initiatives that could impact cybersecurity. China's Micius satellite demonstrated long-distance QKD, enabling global-scale quantum-secure communication. Progress has been made to integrate QKD over existing fiber-optic networks, extending the range of secure communication over land-based infrastructure. Researchers are exploring ways to increase the efficiency and range of QKD systems, with a focus on using entangled photons for faster and more secure key distribution.

NIST is at the forefront of standardizing Post-Quantum Cryptographic algorithms. In July 2022, NIST selected the first group of PQC algorithms for standardization, with a focus on lattice-based cryptography (e.g., CRYSTALS-Kyber and CRYSTALS-Dilithium). Ongoing research focuses on improving the security and efficiency of these algorithms, as well as developing hardware accelerators to speed up PQC operations. Authentication methods based on quantum mechanics, such as quantum digital signatures and quantum fingerprinting, offer potential breakthroughs in secure identity verification. These techniques exploit the no-cloning theorem of quantum mechanics to make it difficult for an adversary to duplicate authentication tokens. There is an increasing focus on developing user-friendly quantum-secure authentication systems, especially for IoT and critical infrastructure security, where traditional methods may fail to scale. Companies like ID Quantique and Quintessence Labs have developed commercial QRNG devices that can be integrated into cybersecurity systems. Recent breakthroughs in quantum memory and quantum repeaters could significantly extend the range of QKD, allowing for secure quantum communication over thousands of kilometers. Advances in quantum repeater technology could help solve the distance limitations in quantum communication, facilitating the development of extensive quantum networks (quantum Internet). Hybrid methods that integrate both classical and quantum cryptographic techniques seek to provide immediate security improvements while quantum infrastructure is still evolving. Developing protocols that utilize the advantages of both approaches enables the creation of more robust systems capable of protecting from classical and quantum vulnerabilities.

Some Global Research Initiatives and Collaborations are:

EU Quantum Flagship Program that focuses on building a quantum-secure communication infrastructure across Europe.

U.S. National Quantum Initiative that encourages public–private partnerships to advance quantum computing, cryptography, and communication technologies.

China's Quantum Development Program where there is heavy investment in quantum communication technologies, particularly in QKD and quantum satellites.

In the next 5–10 years, quantum cryptography is set to witness significant advancements across several research areas and technologies, with collaborative efforts and funding opportunities driving these innovations. The key areas of focus are:

Quantum internet and Quantum Networks [47]: Quantum internet that enables ultra-secure communication over quantum networks adopts Quantum Repeaters that are vital for extending the distance of quantum communication over large networks, Quantum Routers that can manage the routing of quantum information and Entanglement Distribution that expands the ability to distribute entangled states over long distances, essential for robust quantum communication. The potential impact is that the quantum internet will revolutionize sectors like banking, national security, and healthcare by enabling unbreakable encryption for critical communications. The collaborative efforts are the Quantum Internet Alliance (part of the EU Quantum Flagship) working on creating a prototype of the quantum Internet, and U.S. Department of Energy- Quantum Internet Blueprint involving universities, national labs, and private companies.

Quantum-Enhanced Cybersecurity for IoT and Edge Devices [48]: Securing IoT and Edge devices using quantum cryptographic solutions includes key technologies such as Lightweight Quantum-Safe Encryption for small, resource-constrained devices in IoT networks, Quantum Random Number Generators (QRNG) ensuring randomness in cryptographic keys, even on IoT devices, Quantum Fingerprinting for quickly verifying whether two quantum states are the same, which can be useful in IoT security. Quantum solutions for IoT will prevent large-scale cyber attacks targeting smart homes, connected cars, and industrial systems. The collaborative efforts are the U.S. National Quantum Initiative (NQI) funding research into quantum-safe solutions for IoT, and the EU Quantum Flagship Initiative and the UK National Quantum Technologies Programme are also focusing on quantum-enhanced security for IoT and critical infrastructure.

Quantum-Secure Cloud and Data Storage: This involves protecting cloud infrastructure and data storage systems using quantum cryptography. In Quantum

Encryption-as-a-Service (QaaS), the cloud providers offer quantum-secure encryption for data stored in the cloud. Quantum-Secure Multi-Party Computation (SMPC) [49] can be employed to equip parties to collaboratively process a function based on inputs, while preserving their privacy. It will be key to securing sensitive corporate and government data, offering superior privacy for cloud-based services. The collaborative efforts are companies like IBM, Google, and Microsoft are exploring quantum-secure cloud solutions, with partnerships across academia and the public sector. The European Horizon 2020 Programme is funding quantum-secure cloud computing initiatives.

These emerging technologies and collaborative efforts will help quantum cryptography mature, offering transformative security solutions over the next decade.

Acknowledgements I wish to thank Dr. Dinesha H A, Founder Director, Cybersena (R&D) India Pvt. Ltd., Bangalore, for his constant guidance and encouragement in carrying out this work.

References

1. A. Vaishnavi, S. Pillai, Cybersecurity in the quantum era-a study of perceived risks in conventional cryptography and discussion on post quantum methods. J. Phys.: Conf. Ser. 1964 (2021), IOP Publishing https://doi.org/10.1088/1742-6596/1964/4/042002
2. A.A. Abushgra, How quantum computing impacts cyber security, in *2023 Intelligent Methods, Systems, and Applications (IMSA)*, Giza, Egypt, (**2023**), pp. 74–79. https://doi.org/10.1109/IMSA58542.2023.10217756
3. Q. Hussein, E. Taha, New trends for securing cybersecurity data. J. Madenat Alelem Coll.E **14**, 36–43 (2022)
4. Understanding the Strategic and Technical Significance of Technology for Security Implications of Quantum Computing within the Cybersecurity Domain, Hague Security Delta
5. L. Pupillo, A. Ferreira, V. Lipiainien, C. Polito, Quantum Technologies And Cybersecurity: Technology, Governance and Policy Challenges, Task Force Report, Centre for European Policy Studies, Brussels (2023)
6. S.S. Gill, A. Kumar, H. Singh, et al., Quantum computing: a taxonomy, systematic review and future directions. Softw: Pract Exper. **52**(1), 66–114 (2022). https://doi.org/10.1002/spe.3039
7. O.A. Ajala, C.A. Arinze, O.C. Ofodile, C.C. Okoye, A.I. Daraojimba, Exploring and reviewing the potential of quantum computing in enhancing cybersecurity encryption methods. Magna Sci. Adv. Res. Rev. **10**(01), 321–329 (2024). https://doi.org/10.30574/msarr.2024.10.1.0038
8. A. Mishra, Towards quantum-proof cybersecurity: challenges and progress. Cybersecur. Insights MagazineCSIM **5** (2022)
9. Y. Pei, The potential impact of quantum computing on the strategy of the financial service industry. Econ. Manag. Stud. (2) (2024). Future Science Press
10. D. Dave, G. Sawhney, P. Aggarwal, N. Silswal, D. Khut, The new frontier of cybersecurity: emerging threats and innovations, in *2023 29th International Conference on Telecommunications (ICT)*, Toba, Indonesia, (2023), pp. 1–6. https://doi.org/10.1109/ICT60153.2023.10374044
11. Ö. Aslan, S.S. Aktuğ, M. Ozkan-Okay, A.A. Yilmaz, E.A. Akin, Comprehensive review of cyber security vulnerabilities, threats, attacks, and solutions. Electronics **12**, 1333 (2023). https://doi.org/10.3390/electronics12061333
12. A. Bilgehan, U. Mehtap, A. Sedat, S. Seref, A study on the use of quantum computers, risk assessment and security problems, in *IEEE 2018 6th International Symposium on Digital*

Forensic and Security (ISDFS) - Antalya (2018.3.22–2018.3.25)], 1–6 (2018). https://doi.org/10.1109/ISDFS.2018.8355318
13. Quantum Safe Cryptography and Security An introduction, benefits, enablers and challenges ETSI Whitepaper 08, 2015, ISBN No. 979–10–92620–03–0 https://www.etsi.org/technologies/quantum-safe-cryptography
14. P. Wallden, E. Kashefi, Cyber security in the quantum era. Commun. ACM **62**(4) (2019)
15. U. Ghosh, D. Das, P. Chatterjee, *A Comprehensive Tutorial on Cybersecurity in Quantum Computing Paradigm* (Howard University, 2023)
16. R. Azhari, A.N. Salsabila, Analyzing the impact of quantum computing on current encryption techniques. IAIC Trans. Sustain. Digit. Innov. (ITSDI) **5**(2), 148–157 (2024). https://doi.org/10.34306/itsdi.v5i2.662
17. B. Dash, S. Ullah, Quantum-safe: cybersecurity in the age of quantum-powered AI. World J. Adv. Res. Rev. **21**(01), 1555–1563 (2024). https://doi.org/10.30574/wjarr.2024.21.1.2640
18. Quantum Computing: Revolutionizing Cyber Security In 2024 (omexsecurity.com)
19. J. Oliva del Moral, A. deMarti iOlius, G. Vidal, P.M. Crespo, J.E. Martinez, *Cybersecurity in Critical Infrastructures: A Post-Quantum Cryptography Perspective* (2024). arXiv:2401.03780v1 [cs.CR]
20. D.J. Bernstein, T. Lange, Post-quantum cryptography. Nature **549**(7671), 188–194 (2017)
21. K. Kan, M. Une, Recent trends on research and development of quantum computers and standardization of post-quantum cryptography. Monet. Econ. Stud. (2021)
22. M. SaberiKamarposhti, K.W. Ng, F.F. Chua, J. Abdullah, M. Yadollahi, M. Moradi, S. Ahmadpour, Post-quantum healthcare: a roadmap for cybersecurity resilience in medical data. Heliyon **10**(10) (2024)
23. C.S. Xin, B.K. Payne , Special issue on interdisciplinary cybersecurity research: a critical high-impact practice in cybersecurity education. OUR J.: ODU Undergrad. Res. J. **7**, Article 1 (2020). https://doi.org/10.25778/cs5t-y658
24. H.F. Md Jobair, T. Sharaban, T. Masrura, S. Hossain, S. Nazmus, *A Review of Quantum Cybersecurity: Threats, Risks and Opportunities* (2022). https://doi.org/10.1109/ICAIC53980.2022.9896970
25. Q. Hussein, E. Taha, New trends for securing cybersecurity data. J. Madenat Alelam Coll.E **14**(3), 36–43 (2022)
26. Z. Tianqi, S. Jian, L. Xiong, W. Chen, S. Jun, Quantum cryptography for the future internet and the security analysis. Hindawi, Feb 18
27. J. Kaur, K.R. Ramkumar, The recent trends in cyber security: a review. J. King Saud University—Computer Inf. Sci. **34**, 5766–5781 (2022). https://www.sciencedirect.com/journal/journal-of-king-saud-university-computer-and-information-sciences
28. J. Tibbetts, Quantum computing and cryptography: analysis, risks, and recommendations for decisionmakers. Cent. Glob. Secur. Res., Lawrence Livermore Natl. Lab. (2019)
29. M. Sujay Kumar Reddy, S. Mandal, B. Chandra Mohan, Comprehensive study of BB84, a quantum key distribution protocol. Int. Res. J. Eng. Technol. (IRJET) **10**(03) (2023). e-ISSN: 2395–0056, p-ISSN: 2395–0072
30. D. Chawla, P.S. Mehra, A survey on quantum computing for internet of things security. Procedia Comput. Sci. **218**(2191), 2200 (2023). https://doi.org/10.1016/j.procs.2023.01.195
31. M. Lee, Quantum Computing and Cyber Security, in *Harvard Kennedy School, Belfer Centre for Science and International Affairs*, Report July 2021
32. K.-K. Ko, E.-S. Jung, Development of cybersecurity technology and algorithm based on quantum computing. Appl. Sci. 11, 9085 (2021). https://doi.org/10.3390/app11199085
33. P.A. Patil, R. Boda, Analysis of cryptography: classical verses quantum cryptography. Int. Res. J. Eng. Technol. (IRJET), **3**(5) (2016). e-ISSN: 2395-0056, p-ISSN: 2395–0072
34. A. Montanaro, Quantum algorithms: an overview. Npj Quantum Inf. **2**(1), 1–8 (2016). https://doi.org/10.1038/npjqi.2015.23
35. J.-F. Biasse et al., Quantum algorithms for attacking hardness assumptions in classical and post-quantum cryptography. IET Inf. Secur. **17**(2), 171–209 (2023). https://doi.org/10.1049/ise2.12081

36. S. Selvarajan, H. Mouratidis, A quantum trust and consultative transaction-based blockchain cybersecurity model for healthcare systems. Sci. Rep. **13**, 7107 (2023). https://doi.org/10.1038/s41598-023-34354-x
37. R. Ur Rasool, H.F. Ahmad, W. Rafique, A. Qayyum, J. Qadir, Z. Anwar, Quantum computing for healthcare: a review. Futur. Internet **15**, 94 (2023). https://doi.org/10.3390/fi15030094
38. J. Hoffstein, j. Pipher, J.H. Silverman, NTRUEncrypt: Digital Signatures and Public-Key Cryptosystems. Springer (2019)
39. R. Meraz, L. Vahala, Application of quantum cryptography to cybersecurity and critical infrastructures in space communications. OUR J.: ODU Undergrad. Res. J. **7**, Article 5 (2020). https://doi.org/10.25778/kmha-zs76
40. S. Galbraith, D. Liu, S. Nepal, S. Ruj, J. Pieprzyk, J. Liu, R. Steinfeld, A. Sakzad, M. Esgin, V. Kuchta, W. Susilo, T. Plantard, D. Dung, *The Quantum Threat to Cybersecurity: Looking Through the Prism of PostQuantum Cryptography* (CSIRO, Canberra, Australia, 2021)
41. O.A. Ajala, C.A. Arinze, O.C. Ofodile, C.C. Okoye, A.I. Daraojimba, Exploring and reviewing the potential of quantum computing in enhancing cybersecurity encryption methods. Magna Sci. Adv. Res. Rev. **10**(01), 321–329 (2024)
42. M.A. Serrano, L.E. Sánchez, A. Santos-Olmo, D. García-Rosado, C. Blanco, V.S. Barletta, D. Caivano, E. Fernández-Medina, Minimizing incident response time in real-world scenarios using quantum computing. Softw. Qual. J. **32**, 163–192 (2024). https://doi.org/10.1007/s11219-023-09632-6
43. M.Y. Al-darwbi, A.A. Ghorbani, A.H. Lashkari, Keyshield: A scalable and quantum-safe key management scheme. IEEE Open J. Commun. Soc. **2**, 87–101 (2021)
44. D. Ghelani, Cyber security, cyber threats, implications and future perspectives: a review. Am. J. Sci., Eng. Technol. **3**(6), 12–19 (2022)
45. F. Raheman, The future of cybersecurity in the age of quantum computers. Futur. Internet **14**, 335 (2022). https://doi.org/10.3390/fi14110335
46. N. Kilber, D. Kaestle, S. Wagner, *Cybersecurity for Quantum Computing* (2021). arXiv:2110.14701
47. A. Abelem, D. Towsley, G. *Vardoyan, Quantum Internet: The Future of Internetworking* (2023). https://doi.org/10.5753/sbc.5033.7.2
48. A. Hosam, Enhancing IoT security: quantum-level resilience against threats. Comput., Mater. & Contin., Tech Sci. Press. ISSN:1546–2218(print)ISSN:1546–2226(online)
49. Y. Luo, Y. Chen, T. Li et al., Cloud-SMPC: two-round multilinear maps secure multiparty computation based on LWE assumption. J Cloud Comp **13**, 22 (2024). https://doi.org/10.1186/s13677-023-00586-5

Securing the Future: Quantum Computing in Cybersecurity

R. Thangamani, M. Vimaladevi, G. K. Kamalam, and K. M. Subramanian

1 Introduction

In recent years, quantum computing has emerged as a revolutionary technology with the potential to transform various fields, including cybersecurity. Traditional cryptographic protocols, which form the backbone of digital security, face a significant threat from quantum algorithms such as Shor's algorithm. These quantum algorithms can break widely used encryption methods, posing a formidable challenge to data security [1] in the digital age. The intersection of quantum computing and cybersecurity presents both challenges and opportunities. On one hand, the vulnerabilities exposed by quantum algorithms raise concerns about the integrity and confidentiality of sensitive data. On the other hand, quantum computing also offers innovative solutions to enhance cybersecurity defenses and mitigate emerging threats. In this, we delve into the dynamic role of quantum computing in cybersecurity and explore the innovations that are shaping the future of digital security. We examine the fundamental principles of quantum computing, including quantum bits (qubits) and quantum superposition, to understand how these concepts underpin the computational power of quantum algorithms [2].

The proposed work in quantum computing and cybersecurity encompasses several key areas. One area focuses on developing and implementing quantum-safe cryptographic algorithms and protocols to protect sensitive data and communication channels from quantum attacks. Another area involves the deployment of quantum key distribution (QKD) technologies for secure key exchange in communication

R. Thangamani (✉) · M. Vimaladevi · G. K. Kamalam
Kongu Engineering College, Perundurai, Tamil Nadu, India
e-mail: thangamani.ai@kongu.edu

M. Vimaladevi
e-mail: vimaladevi.ai@kongu.edu

K. M. Subramanian
Shadan College of Engineering Technology, Hyderabad, India

S. B. Goyal et al. (eds.), *Quantum Computing, Cyber Security and Cryptography*,
https://doi.org/10.1007/978-981-96-4948-8_17

networks. The research contribution in this domain lies in advancing the understanding of quantum-resistant cryptography, quantum key distribution, and other quantum-safe security measures. This includes developing innovative encryption schemes, digital signatures, authentication mechanisms, and secure communication protocols resilient to quantum attacks.

Use cases of quantum computing in cybersecurity extend to secure communication, data encryption, threat detection, and secure authentication. Quantum technologies can enhance the security of critical infrastructure, financial systems, healthcare networks, and government communications by leveraging quantum-safe security solutions. The proposed methods aim to solve the fundamental problem of securing sensitive data and communication channels in the quantum computing era. By developing quantum-safe cryptographic solutions, deploying QKD technologies, and advancing quantum-resistant security measures, the goal is to ensure the integrity, confidentiality, and authenticity of digital transactions and communications in a quantum-enabled world.

Furthermore, we analyze the impact of quantum algorithms on traditional cryptographic protocols, such as RSA and ECC, and discuss the urgent need for quantum-safe cryptographic solutions. These solutions, based on mathematical structures resistant to quantum attacks, are essential for securing sensitive data and ensuring the integrity of digital transactions in a quantum-enabled world. Moreover, we explore the application of quantum computing in key areas of cybersecurity, such as secure communication, data encryption, and threat detection. We highlight the role of quantum-resistant algorithms, quantum key distribution (QKD), and quantum random number generation (QRNG) in fortifying cybersecurity defenses and safeguarding critical digital assets. As we navigate the complexities of quantum computing and its implications for cybersecurity, it becomes evident that collaboration between researchers, industry experts, and policymakers is crucial.

1.1 Importance of Quantum Computing in Cybersecurity

Quantum computing holds significant importance in the field of cybersecurity due to several key factors [3]:

- Breaking Traditional Encryption: Quantum computers have the potential to break widely used cryptographic algorithms, such as RSA and ECC, which are the foundation of digital security. This capability poses a significant threat to sensitive data and communications, highlighting the need for quantum-resistant cryptographic solutions.
- Quantum-Safe Cryptography: Quantum computing also offers solutions in the form of quantum-safe cryptographic algorithms. These algorithms are designed to withstand attacks from quantum computers, ensuring data confidentiality, integrity, and authenticity in a quantum computing era.

- Secure Communication: Quantum computing enables the development of secure communication protocols based on quantum key distribution (QKD). QKD uses the principles of quantum mechanics to secure communication channels, offering unprecedented levels of security against eavesdropping and data interception.
- Advanced Threat Detection: Quantum computing can enhance threat detection capabilities by leveraging quantum algorithms and machine learning techniques. These advancements enable real-time anomaly detection, predictive analytics, and adaptive cybersecurity measures to combat evolving cyber threats effectively.
- Future-Proofing Cybersecurity: As quantum computing continues to advance, integrating quantum-safe solutions into cybersecurity strategies becomes imperative. Organizations must prepare for the quantum computing era by investing in research, development, and implementation of quantum-resistant technologies.
- Quantum Randomness and Entropy: Quantum computing introduces new sources of randomness and entropy, which are critical for generating secure cryptographic keys and ensuring unpredictability in cryptographic operations. This randomness enhances the overall security posture of cryptographic systems.

Securing the future of quantum computing in cybersecurity presents several challenges that need to be addressed to ensure robust and resilient digital security:

- Quantum-Safe Cryptography Adoption: One of the primary challenges is the widespread adoption of quantum-safe cryptographic algorithms and protocols. Organizations must transition from traditional cryptographic methods to quantum-resistant ones to protect sensitive data from future quantum attacks effectively.
- Post-quantum Cryptography Standardization: There is ongoing work in standardizing post-quantum cryptographic algorithms to ensure interoperability and compatibility across different systems and platforms. However, achieving consensus on standardized quantum-resistant algorithms and implementing them universally remains a complex challenge [4].
- Quantum Key Distribution Implementation: While QKD offers secure key distribution based on quantum principles, its practical implementation and scalability pose challenges. Overcoming technical hurdles and ensuring the reliability and efficiency of QKD systems in real-world environments are ongoing challenges.
- Cybersecurity Risks in Quantum Computing: The development of quantum computers introduces new cybersecurity risks, such as the potential for quantum-based attacks targeting quantum algorithms and protocols. Understanding and mitigating these risks require comprehensive security strategies and continuous threat monitoring.
- Education and Awareness: Building awareness and educating stakeholders about the implications of quantum computing on cybersecurity are essential. This includes training cybersecurity professionals, policymakers, and the general public about quantum-safe practices, risk mitigation strategies, and the need for proactive cybersecurity measures in the quantum computing era.
- Addressing these challenges requires collaborative efforts from researchers, industry leaders, government agencies, and regulatory bodies to navigate the complex intersection of quantum computing and cybersecurity effectively.

Here are some applications of quantum computing in cybersecurity are shown in Table 1.

Motivations

The motivation behind quantum computing in cybersecurity stems from the need to address the challenges posed by traditional cryptographic methods in the face of advancing quantum technologies. Here are some key motivations:

- Quantum Computing Threats: The development of quantum computers introduces new threats to cybersecurity by potentially breaking widely used cryptographic algorithms, such as RSA and ECC, which form the basis of digital security. Motivated by the urgency to protect sensitive data and communication channels from quantum attacks, researchers and industry experts are exploring quantum-safe cryptographic solutions.
- Post-quantum Security: Quantum computing motivates the exploration and development of post-quantum cryptographic algorithms and protocols that are resistant to quantum attacks. The goal is to ensure that cryptographic systems remain secure in the presence of powerful quantum computers, thereby safeguarding critical digital assets and information.
- Quantum Key Distribution Advantages: Quantum computing also motivates the deployment of quantum key distribution (QKD) technologies, which offer unique advantages in secure key exchange. QKD leverages quantum principles to enable

Table 1 Applications of quantum computing in cybersecurity [5]

Application	Description
Quantum-safe cryptography	Developing and implementing cryptographic algorithms and protocols resistant to quantum attacks
Post-quantum cryptography	Standardizing and deploying cryptographic algorithms that can withstand attacks from quantum computers
Quantum key distribution	Using quantum principles to securely distribute cryptographic keys, protecting communication channels
Quantum-safe software development	Integrating quantum-resistant encryption, digital signatures, and authentication mechanisms into software
Quantum computing simulations	Simulating quantum algorithms and evaluating their impact on cybersecurity to identify vulnerabilities
Education and awareness	Raising awareness, training cybersecurity professionals, and educating stakeholders about quantum security
Collaborative research and development	Collaborating across academia, industry, and government to advance research and innovation in quantum cybersecurity

secure communication channels, protecting against eavesdropping and interception. The motivation lies in leveraging quantum capabilities to enhance the security and confidentiality of digital communications [6].

- Security Innovation: Quantum computing motivates innovation in cybersecurity by driving research and development in quantum-resistant encryption schemes, digital signatures, authentication mechanisms, and secure communication protocols. This innovation is essential for staying ahead of emerging cyber threats and ensuring the resilience of digital systems in a quantum-enabled world.
- Future-Proofing Cybersecurity: Motivated by the rapid advancements in quantum technologies, organizations, and researchers are focused on future-proofing cybersecurity strategies. This includes preparing for the quantum computing era by adopting quantum-safe practices, investing in quantum-resistant technologies, and developing robust security measures to mitigate quantum-related risks.

To connect more directly with **current cybersecurity challenges**, this section will explore recent, high-profile cybersecurity incidents and discuss how **quantum computing** and **quantum-safe technologies** could offer solutions to these issues.

1. **Ransomware Attacks**

Cybersecurity Challenge

In recent years, ransomware attacks have surged, targeting critical infrastructure and high-value industries like healthcare, finance, and energy. Notably:

- The **Colonial Pipeline** attack in 2021 disrupted fuel supplies across the U.S., leading to millions in ransom payments.
- **JBS**, one of the world's largest meat suppliers, also suffered a ransomware attack that threatened global food supplies.

Ransomware attacks typically encrypt an organization's data, demanding a ransom in exchange for the decryption key. These attacks exploit vulnerabilities in **data encryption systems** or through **social engineering techniques**, and as encryption remains a key aspect of ransomware, attackers have become more sophisticated in leveraging strong cryptographic methods to hold data hostage.

Quantum Computing Solution

Quantum computing, when combined with **quantum-resistant encryption algorithms**, could help mitigate ransomware threats by making it far more difficult for attackers to infiltrate and encrypt sensitive data [7].

- **Quantum-Safe Encryption**: Quantum-resistant encryption algorithms (e.g., **lattice-based cryptography** and **multivariate cryptography**) can help secure sensitive data against future quantum decryption attempts. These algorithms would prevent quantum computers from easily breaking the encryption used to protect data, even when compromised by ransomware.
- **Post-quantum Signatures**: Ransomware often penetrates systems via phishing attacks, delivering malware through fake digital certificates. Quantum-safe digital

signatures, such as **hash-based signatures** (e.g., SPHINCS +), could ensure that certificates and updates come from legitimate sources, making phishing attempts easier to detect and avoid.

- **Threat Detection and Response**: Quantum machine learning algorithms could enhance threat detection systems by rapidly analyzing patterns and anomalies in network behavior, potentially identifying ransomware before it encrypts files. The use of **quantum-enhanced AI** would enable faster and more accurate identification of malicious software, enhancing response time to prevent damage.

2. **Fundamentals of Quantum Computing**

The fundamentals of quantum computing lie at the intersection of quantum mechanics, information theory, and computer science. Here are key aspects that form the foundation of quantum computing [8–10]:

- Quantum Bits (Qubits):

Classical Bits: In classical computing, information is processed using bits that can exist in one of two states: 0 or 1.

Quantum Bits (Qubits): Qubits are the quantum counterpart to classical bits. Unlike classical bits, qubits can exist in multiple states simultaneously due to quantum superposition. This property allows qubits to represent and process vast amounts of information in parallel, leading to exponential computational power.

- Quantum Superposition:

Principle: Quantum superposition is a fundamental principle in quantum mechanics that allows particles like qubits to exist in multiple states simultaneously.

Example: A qubit in superposition can represent both 0 and 1 simultaneously, with varying probabilities assigned to each state. This enables quantum computers to explore multiple computational paths simultaneously, drastically increasing computational efficiency.

- Entanglement:

Principle: Entanglement is another key concept in quantum mechanics where the state of one particle (such as a qubit) becomes correlated with the state of another particle, even if they are physically separated.

Application: Entanglement enables quantum computers to perform highly interconnected computations and communication protocols, leading to enhanced processing capabilities and secure quantum communication.

- Quantum Gates:

Similarity to Classical Gates: Quantum gates are analogous to classical logic gates used in traditional computing.

Operations: Quantum gates manipulate qubits to perform operations such as superposition, entanglement, measurement, and quantum interference.

Example Gates: Some common quantum gates include Hadamard gate (creates superposition), CNOT gate (creates entanglement), and Pauli-X gate (flips the qubit state).

- Quantum Algorithms:

Shor's Algorithm: A quantum algorithm developed by Peter Shor that efficiently factors large integers, threatening classical RSA encryption.

Grover's Algorithm: A quantum search algorithm developed by Lov Grover that provides quadratic speedup in searching unsorted databases or solving unstructured search problems.

- Quantum Computing Models:

Gate-Based Quantum Computing: In gate-based quantum computing, qubits are manipulated using quantum gates to perform computations.

Adiabatic Quantum Computing: Adiabatic quantum computing involves evolving a quantum system from an initial state to a final state that represents the solution to a computational problem.

- Quantum Error Correction:

Challenge: Quantum systems are susceptible to errors due to decoherence, noise, and environmental interactions.

Solution: Quantum error correction techniques, such as quantum error-correcting codes and fault-tolerant quantum computing, are developed to mitigate errors and preserve quantum information.

Understanding these fundamentals is crucial for grasping the capabilities, principles, and potential applications of quantum computing, paving the way for advancements in computational power, cryptography, optimization, and scientific simulations.

- Applications:

Quantum Computing: Qubits and quantum superposition are foundational to quantum computing, enabling exponential speedups in solving certain computational problems, such as factoring large numbers, optimization, cryptography, and simulating quantum systems.

Quantum Communication: Quantum superposition plays a crucial role in quantum communication protocols, such as quantum key distribution (QKD), where qubits in superposition are used to securely exchange cryptographic keys over quantum channels.

Quantum Algorithms: Shor's Algorithm and Grover's Algorithm

Quantum algorithms are computational procedures designed to be executed on quantum computers, leveraging the unique properties of quantum mechanics to solve specific problems more efficiently than classical algorithms. Two notable quantum algorithms are Shor's algorithm and Grover's algorithm, each with distinct applications and computational advantages.

Shor's Algorithm:

Shor's algorithm is a quantum algorithm designed to factorize large integers efficiently, posing a direct threat to cryptographic systems like RSA, which rely on the difficulty of factoring large numbers.

Mathematical Principle:

The RSA encryption method's security depends on the difficulty of factoring large composite numbers into their prime factors. Classical algorithms for factoring take exponentially longer as the number grows, but Shor's algorithm can achieve this in polynomial time using quantum Fourier transform (QFT) and modular arithmetic.

Steps Involved in Shor's Algorithm:

1. Quantum Parallelism: Using qubits, Shor's algorithm performs parallel computations on multiple values simultaneously. The problem of factoring is reduced to finding the period of a function that can be calculated with modular exponentiation.
2. Modular Exponentiation: Shor's algorithm uses the function f(x) = a^x mod N where N is the number to be factorized, and a is a randomly chosen number. The goal is to find the period r, which satisfies f(x + r) = f(x).
3. Quantum Fourier Transform (QFT): The heart of Shor's algorithm is the QFT, which efficiently identifies the period of a function. The QFT converts the periodicity of the function into peaks in the frequency domain, allowing the algorithm to extract the period r.
4. Continued Fractions: Once the period is found using QFT, the continued fraction algorithm is applied to recover the factors of N.

Proof of Correctness:

The success of Shor's algorithm hinges on the efficiency of the QFT in identifying the period r. Given a function f(x) = a^x \mod N, where a is co-prime to N, the modular arithmetic guarantees that this function will be periodic with a period r. The QFT then efficiently computes this period, from which the factors of N can be extracted by solving:

$$\gcd(\mathrm{a}^{\{r/2\}}, \pm 1, \mathrm{N})$$

This approach works in polynomial time $O((\log N)^2 (\log \log N) (\log \log \log N))$, whereas classical algorithms, like trial division or elliptic curve factorization, take sub-exponential or exponential time.

Implications for Cryptography:

RSA: Shor's algorithm can efficiently factor the product of two large primes, thus breaking RSA encryption. For example, RSA-2048, which is currently considered secure with classical computers, would be vulnerable to a quantum computer running Shor's algorithm.

Grover's Algorithm:

Grover's algorithm is designed for unstructured search problems. It provides a quadratic speedup over classical algorithms, making it applicable for breaking symmetric encryption systems by searching through keyspaces.

Mathematical Principle:

Classical search algorithms require O(N) operations to search through N unsorted entries. Grover's algorithm reduces this to O(sqrt{N}) using quantum superposition and amplitude amplification.

Steps in Grover's Algorithm:

1. Initialization: The algorithm starts by placing the quantum state in an equal superposition of all possible entries, represented as frac{1}{\sqrt{N}} \sum_{x = 0}^{N-1} |x. This is achieved using a Hadamard transformation on the qubits.
2. Oracle Function: An oracle is used to mark the correct answer. The oracle function O(x) flips the sign of the state corresponding to the correct solution (i.e., the target element is marked with a phase shift of -1).
3. Grover Diffusion Operator: Grover's diffusion operator amplifies the probability amplitude of the correct answer by reflecting all states around the average amplitude. This operation increases the likelihood of observing the correct answer when the system is measured.
4. Iteration: Grover's algorithm repeats the oracle and diffusion operations O(\sqrt{N}) times to ensure that the probability of measuring the correct answer becomes significant.

Proof of Correctness:

Grover's algorithm leverages the principle of amplitude amplification. Initially, all states have an equal amplitude of frac{1}{\sqrt{N}}. The oracle marks the correct state, and the diffusion operator reflects around the mean, amplifying the amplitude of the marked state.

The key idea is that after approximately O(\sqrt{N}) iterations, the amplitude of the correct state will dominate, making it highly probable that measurement will yield the desired result. The quadratic speedup means that searching through a keyspace of size N takes O(\sqrt{N}) operations, which is a significant improvement over classical brute-force methods.

Implications for Cryptography:

- Symmetric Key Cryptography: Grover's algorithm halves the security of symmetric cryptographic algorithms. For instance, a 128-bit key (used in AES) would have a classical security level of 2^{128}, but Grover's algorithm reduces the effective security to 2^{64}, which is much more feasible for a quantum computer to break.

Conclusion and Cryptographic Relevance:

- Shor's Algorithm: A critical threat to asymmetric encryption systems like RSA and ECC (Elliptic Curve Cryptography) due to its ability to factorize large integers and compute discrete logarithms in polynomial time.
- Grover's Algorithm: Though not as devastating as Shor's algorithm, it weakens symmetric encryption by reducing the effective key length, requiring larger keys to maintain current security levels.

Both algorithms showcase the immense potential of quantum computing in cryptanalysis, emphasizing the need for the development and implementation of quantum-resistant cryptographic systems.

2 Impact of Quantum Computing on Cybersecurity

The impact of quantum computing on cybersecurity is multifaceted, presenting both challenges and opportunities that reshape the landscape of digital security. Here are key aspects of how quantum computing impacts cybersecurity [3]:

- Cryptographic Vulnerabilities:

Challenge: Quantum computers have the potential to break widely used cryptographic algorithms, such as RSA and ECC, by efficiently factoring large numbers or solving the discrete logarithm problem using algorithms like Shor's algorithm.

Opportunity: This challenge motivates the development of post-quantum cryptographic algorithms that are resistant to quantum attacks, ensuring long-term security for sensitive data and communication channels.

- Quantum Key Distribution:

Challenge: Quantum computers can also threaten classical key exchange protocols, prompting the need for quantum-safe key distribution mechanisms.

Opportunity: Quantum key distribution protocols leverage quantum principles to securely exchange cryptographic keys, offering provably secure communication channels immune to quantum attacks.

- Improved Encryption and Hashing:

Opportunity: Quantum computing enables the development of stronger encryption schemes and cryptographic hash functions that leverage quantum-resistant algorithms, enhancing data confidentiality and integrity in digital transactions.

- Faster Threat Analysis and Encryption Breaking:

Challenge: Quantum computers have the potential to accelerate threat analysis, including breaking symmetric encryption keys and brute-forcing cryptographic hashes.

Opportunity: Quantum-resistant algorithms and encryption methods mitigate the risk of rapid encryption breaking, maintaining the confidentiality of sensitive information.

- Quantum-Safe Protocols and Standards:

Opportunity: The emergence of quantum-safe protocols and standards ensures interoperability and compatibility across quantum and classical systems, facilitating a seamless transition to quantum-resistant security measures.

- Securing Internet of Things (IoT) Devices:

Opportunity: Quantum computing enables robust security solutions for IoT devices, such as quantum-resistant authentication protocols and secure communication channels, addressing vulnerabilities in interconnected systems.

- Advanced Threats and Defenses:

Challenge: Quantum computing introduces new threats, such as quantum-based attacks targeting quantum-resistant algorithms and protocols.

Opportunity: Advanced cybersecurity defenses leverage quantum technologies, such as quantum machine learning and quantum-resistant firewalls, to detect and mitigate emerging quantum threats effectively.

- Global Cybersecurity Collaboration:

Opportunity: Quantum computing fosters international collaboration in cybersecurity research and standards development, promoting information sharing, best practices, and coordinated responses to cyber threats in a quantum-enabled world [11, 12].

2.1 *Vulnerabilities Exposed by Quantum Algorithms*

Quantum algorithms, while offering immense computational power and capabilities, also expose vulnerabilities in traditional cryptographic schemes and protocols. Here are vulnerabilities exposed by quantum algorithms [11, 12]:

Factorization Vulnerability:

Exposed by: Shor's algorithm.

Impact: Shor's algorithm can efficiently factorize large composite numbers into their prime factors, threatening the security of cryptographic schemes based on integer factorization problems, such as RSA and ECC.

Consequence: The ability to factorize large numbers quickly undermines the confidentiality and integrity of encrypted data and digital signatures, potentially leading to data breaches or unauthorized access.

Discrete Logarithm Vulnerability:

Exposed by: Quantum algorithms like Shor's algorithm.

Impact: Quantum algorithms can solve the discrete logarithm problem efficiently, compromising cryptographic protocols that rely on the difficulty of computing discrete logarithms, such as Diffie–Hellman key exchange and DSA (Digital Signature Algorithm).

Consequence: The vulnerability in discrete logarithm-based schemes can result in compromised key exchange, forged digital signatures, and unauthorized access to sensitive information.

Grover's Algorithm Search Speedup:

Exposed by: Grover's algorithm.

Impact: Grover's algorithm provides a quadratic speedup in searching unsorted databases or solving unstructured search problems, including brute-force attacks on symmetric encryption keys or hash functions.

Consequence: The increased search speed enabled by Grover's algorithm reduces the computational effort required for brute-force attacks, potentially weakening the security of symmetric encryption and hashing algorithms.

Quantum Key Distribution Vulnerabilities:

Exposed by: Quantum technologies and quantum channel vulnerabilities.

Impact: Quantum key distribution protocols, while offering provably secure key exchange, can still face vulnerabilities related to hardware exploits, implementation flaws, or side-channel attacks.

Consequence: Vulnerabilities in QKD systems can compromise the confidentiality and integrity of cryptographic keys exchanged over quantum channels, leading to key compromise or interception.

Entanglement-Based Attacks:

Exposed by: Quantum entanglement properties.

Impact: Quantum entanglement, while enabling secure communication and quantum teleportation, can also be exploited for quantum-based attacks, such as entanglement eavesdropping or quantum channel manipulation.

Consequence: Entanglement-based attacks can breach the security of quantum communication protocols, compromise quantum key distribution, or intercept quantum states without detection.

Post-quantum Cryptography Adoption:

Exposed by: Transition challenges and vulnerabilities in adopting post-quantum cryptographic algorithms [4]

Impact: The transition to post-quantum cryptographic algorithms faces challenges related to algorithm standardization, interoperability, implementation security, and resistance to new types of attacks.

Consequence: Vulnerabilities in the adoption process, such as insecure implementations or misconfigurations, can undermine the effectiveness of post-quantum cryptographic solutions and leave systems susceptible to quantum attacks.

3 Transition to Quantum-Safe Cryptographic Solutions

The transition to quantum-safe cryptographic solutions is a critical step in mitigating the vulnerabilities exposed by quantum algorithms and ensuring long-term security in the era of quantum computing. Here are key considerations and strategies for transitioning to quantum-safe cryptographic solutions [13]:

Algorithm Evaluation and Standardization:

Evaluate Post-quantum Cryptographic Algorithms: Assess and select post-quantum cryptographic algorithms that are resistant to quantum attacks, such as lattice-based, code-based, multivariate, and hash-based algorithms.

Standardization Efforts: Engage in standardization efforts with organizations like NIST (National Institute of Standards and Technology) to establish benchmarks, guidelines, and certification processes for quantum-safe algorithms.

Interoperability and Compatibility:

Ensure Interoperability: Ensure that quantum-safe cryptographic solutions are interoperable with existing protocols, systems, and applications to facilitate a smooth transition.

Compatibility Testing: Conduct thorough compatibility testing to verify the functionality and performance of quantum-safe algorithms in diverse environments.

Key Management and Lifecycle:

Secure Key Generation: Implement secure key generation mechanisms for quantum-safe cryptographic keys, ensuring randomness, entropy, and resistance to quantum attacks.

Key Lifecycle Management: Establish robust key lifecycle management practices, including key distribution, storage, rotation, and revocation, to protect against key compromise and unauthorized access.

Implementation Security and Best Practices:

Secure Implementation: Implement quantum-safe cryptographic algorithms securely, following best practices in software development, cryptographic module design, and secure coding standards 19,22].

Security Audits: Conduct regular security audits, code reviews, and penetration testing to identify and remediate vulnerabilities in cryptographic implementations [3].

Migration Planning and Roadmap:

Develop Migration Plan: Develop a comprehensive migration plan and roadmap for transitioning to quantum-safe cryptographic solutions, including timelines, milestones, resource allocation, and risk management strategies.

Phased Approach: Adopt a phased approach to migration, starting with pilot implementations, testing, and validation before full-scale deployment across the organization.

Education and Training:

Awareness and Training: Provide education, training, and awareness programs to stakeholders, including IT personnel, developers, and end-users, on quantum-safe cryptography, its importance, and best practices for implementation and usage.

Knowledge Sharing: Foster knowledge sharing and collaboration within the cybersecurity community, academia, and industry to stay updated on emerging threats, vulnerabilities, and solutions in quantum-safe cryptography.

Regulatory Compliance and Certifications:

Compliance Requirements: Ensure compliance with regulatory requirements and industry standards related to cryptographic security, data protection, privacy, and risk management.

Certifications and Assurance: Seek certifications, assurance validations, and third-party audits for quantum-safe cryptographic solutions to demonstrate their security, reliability, and adherence to standards.

Continuous Monitoring and Adaptation:

Monitoring and Evaluation: Implement continuous monitoring, threat intelligence, and risk assessment processes to detect and respond to evolving quantum threats, vulnerabilities, and attack vectors.

Adaptive Security Measures: Adopt adaptive security measures, resilience strategies, and contingency plans to address emergent quantum challenges and maintain cybersecurity resilience over time.

By proactively addressing these considerations and implementing a structured approach to transitioning to quantum-safe cryptographic solutions, organizations can enhance their cybersecurity posture, safeguard sensitive information, and mitigate the risks associated with quantum computing advancements.

3.1 Overview of Quantum-Safe Cryptographic Algorithms

Quantum-safe cryptographic algorithms, also known as post-quantum cryptographic algorithms, are designed to resist attacks from quantum computers, offering security in the era of quantum computing [13]. Here's an overview of some key quantum-safe cryptographic algorithms are shown in Fig. 1:

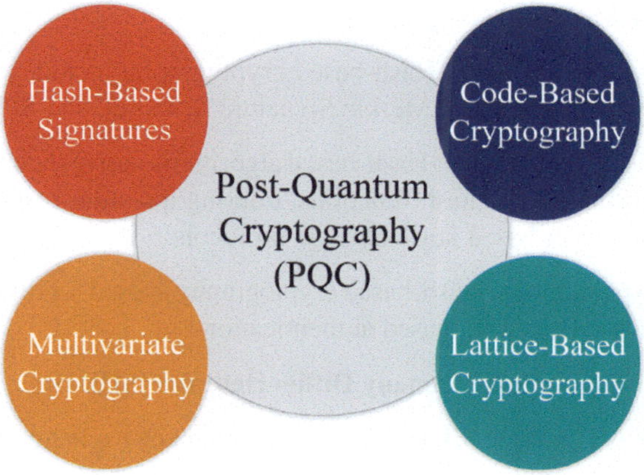

Fig. 1 Post-quantum cryptography algorithms

Lattice-Based Cryptography:

Key Algorithms: Lattice-based cryptography includes algorithms such as NTRUEncrypt, NTRU, and Ring-LWE (Learning with Errors) schemes.

Security Features: Lattice-based algorithms rely on the hardness of certain lattice problems, making them resistant to quantum attacks, particularly against Shor's algorithm.

Applications: Lattice-based cryptography is used for digital signatures, key exchange protocols (e.g., NTRUEncrypt), and encryption schemes (e.g., NTRU).

Code-Based Cryptography:

Key Algorithms: Code-based cryptography includes algorithms like McEliece cryptosystem and BIKE (Bit-flipping Key Encapsulation).

Security Features: Code-based algorithms are based on error-correcting codes, offering resilience against quantum attacks by leveraging the difficulty of decoding linear codes.

Applications: Code-based cryptography is used for digital signatures, key encapsulation mechanisms (e.g., BIKE), and secure communication protocols.

Multivariate Polynomial Cryptography (MPC):

Key Algorithms: MPC includes schemes such as Rainbow, Unbalanced Oil and Vinegar (UOV), and Hidden Field Equations (HFE) cryptosystems.

Security Features: MPC relies on the complexity of solving systems of multivariate polynomial equations, providing resistance against quantum attacks.

Applications: MPC algorithms are used for digital signatures, encryption, and key exchange in quantum-resistant cryptographic protocols.

Hash-Based Cryptography:

Key Algorithms: Hash-based cryptography includes algorithms like Merkle trees, XMSS (Extended Merkle Signature Scheme), and SPHINCS (SPHINCS +).

Security Features: Hash-based algorithms derive their security from the properties of cryptographic hash functions, offering quantum-resistant digital signature schemes and hash-based key exchange protocols.

Applications: Hash-based cryptography is used for digital signatures, time-stamping, and secure hash-based authentication.

Supersingular Isogeny Diffie–Hellman (SIDH):

Key Features: SIDH is a post-quantum key exchange protocol based on isogenies of supersingular elliptic curves.

Security Features: SIDH leverages the computational hardness of isogeny problems, providing quantum-resistant key exchange mechanisms.

Applications: SIDH is used for key exchange in quantum-resistant cryptographic protocols and secure communication channels.

Quantum Key Distribution:

Key Algorithms: QKD protocols include BB84, E91, and B92 protocols.

Security Features: QKD leverages quantum principles, such as quantum entanglement and uncertainty, to establish provably secure key exchange mechanisms immune to quantum attacks.

Applications: QKD is used for secure key distribution in quantum-resistant cryptographic systems and quantum-safe communication networks.

These quantum-safe cryptographic algorithms play a crucial role in ensuring the security, confidentiality, and integrity of digital communications and data in the face of advancing quantum computing technologies and potential quantum-based threats [13].

3.2 Comparison with Traditional Cryptographic Methods

Here's a tabulated comparison between quantum-safe cryptographic methods and traditional cryptographic methods shown in Table 2:

Table 2, summarizes the key differences between traditional cryptography and quantum-safe cryptography across various aspects related to security, performance, key management, interoperability, and future-proofing against quantum attacks.

Quantum-Safe Implementation Using Lattice-Based Crypto Scheme

Implementing quantum-safe cryptography [13] using a lattice-based cryptographic scheme involves several key steps and considerations are shown in Fig. 2:

Table 2 Comparison with traditional cryptography and quantum safe cryptography

Aspect	Traditional cryptography	Quantum-safe cryptography
Resistance to quantum attacks	Vulnerable to quantum attacks, e.g., RSA, ECC	Designed to resist quantum attacks, e.g., lattice-based, code-based schemes
Key length and security margins	Longer key lengths for security, increased complexity	Shorter key lengths while maintaining security margins
Algorithmic diversity	Relies on a few established algorithms	Offers a diverse range of algorithms for resilience
Performance and computational complexity	Performance degradation with longer key lengths	Maintains efficient performance with shorter key lengths
Key management and distribution	Requires secure key management practices	Incorporates efficient key management mechanisms
Interoperability and standards	Adheres to established standards and protocols	Undergoes standardization for interoperability
Future-proofing security	May require periodic updates and replacements	Provides future-proofing against quantum attacks

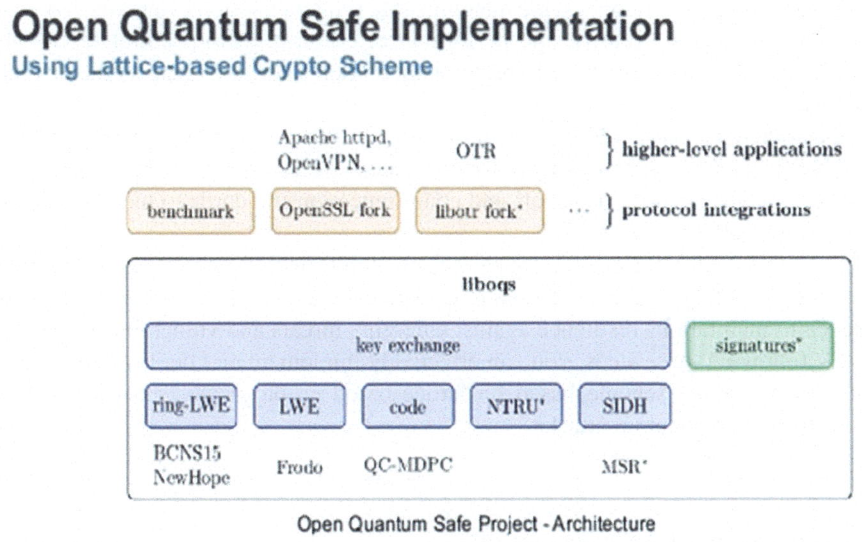

Fig. 2 Implementation of quantum-safe lattice-based crypto scheme

Choose a Lattice-Based Cryptographic Scheme: There are various lattice-based cryptographic schemes available, such as the Learning With Errors (LWE) problem, Ring Learning With Errors (Ring-LWE), and others. Select a scheme that suits your specific security requirements and computational constraints.

Generate Keys: Like traditional cryptographic systems, lattice-based schemes require key generation. This involves creating public and private keys using algorithms specific to the chosen scheme. The security of the scheme should be analyzed based on parameters like key length and error distribution.

Encryption: The encryption process in a lattice-based scheme typically involves encoding messages into lattice points and adding random noise to obscure the original message. The ciphertext is then generated using the public key and sent to the recipient.

Decryption: Upon receiving the ciphertext, the recipient uses their private key to decrypt it. Decryption involves finding the closest lattice point to the received ciphertext and recovering the original message.

Security Analysis: Conduct a thorough security analysis of the chosen lattice-based scheme. This includes evaluating its resistance against known quantum algorithms like Shor's Algorithm and Quantum Fourier Sampling, which can break traditional cryptographic schemes like RSA and ECC.

Implementation Considerations: Implement the scheme using appropriate programming languages and libraries that support lattice-based cryptography. Ensure that the implementation follows best practices for security, such as secure key storage, proper random number generation, and protection against side-channel attacks.

Testing and Validation: Test the quantum-safe implementation extensively to validate its correctness, security, and performance. Use test vectors, simulate various attack scenarios, and benchmark the implementation to ensure it meets the desired standards.

Deployment: Once validated, deploy the quantum-safe lattice-based cryptographic scheme in your applications or systems where quantum resistance is a priority. Update security protocols and communication standards to incorporate the new scheme effectively.

Continuous Monitoring and Updates: Keep monitoring advancements in quantum computing and cryptographic research. Update the quantum-safe implementation as needed to maintain its resilience against emerging threats and vulnerabilities.

By following these steps, you can effectively implement and deploy a quantum-safe cryptographic scheme based on lattice-based cryptography, ensuring robust security against quantum attacks.

4 Lattice-Based Cryptography

Lattice-based cryptography is a branch of cryptography that leverages the complexity of problems defined on lattices, which are geometric structures with repeating patterns extending indefinitely in all directions. Lattice problems form the basis of cryptographic algorithms that are resistant to attacks from quantum computers, making them crucial in the development of post-quantum cryptographic solutions. Here's an overview of lattice-based cryptography:

Lattice Problems:

Basis: Lattice-based cryptography relies on the hardness of solving mathematical problems on lattices, such as the Shortest Vector Problem (SVP), Closest Vector Problem (CVP), and Learning With Errors (LWE) problem.

Computational Complexity: These problems are believed to be computationally difficult, even for quantum computers, due to their inherent structure and complexity.

Key Components:

Polynomial Rings: Lattice-based schemes often operate in polynomial rings, where computations involve polynomials and modular arithmetic.

Noise Addition: Many lattice-based schemes introduce noise or errors into computations, adding a layer of security against attacks.

Security Features:

Quantum Resistance: Lattice-based cryptography is designed to be resistant to attacks from quantum computers, particularly against algorithms like Shor's algorithm that threaten traditional cryptographic schemes.

Hardness Assumptions: The security of lattice-based schemes is based on the assumed hardness of solving lattice problems, which are believed to require exponential time and resources to solve.

Lattice-Based Cryptographic Schemes:

NTRUEncrypt: A lattice-based encryption and key exchange scheme that relies on the difficulty of the CVP in certain lattices.

Key Exchange and Encryption: NTRUEncrypt is a lattice-based cryptographic algorithm used for both key exchange and encryption. It operates in a lattice defined by polynomial rings.

Security Features: NTRUEncrypt relies on the hardness of the "Closest Vector Problem" (CVP) in lattices, which involves finding the lattice point closest to a given vector.

Applications: NTRUEncrypt is used in various protocols for secure key exchange and encryption in a quantum-resistant manner.

Ring-LWE (Learning With Errors): A variant of the LWE problem defined on polynomial rings, used in key exchange protocols and digital signatures.

Key Exchange and Digital Signatures: Ring-LWE is a variant of the Learning With Errors problem defined on polynomial rings.

Security Features: Ring-LWE security is based on the difficulty of solving the Ring-LWE problem, which involves finding a secret polynomial given noisy linear equations.

Applications: Ring-LWE is used in cryptographic schemes for key exchange, digital signatures, and other cryptographic primitives.

Lattice-Based Digital Signatures: Schemes like BLISS and Dilithium use lattice techniques for generating and verifying digital signatures, offering quantum-resistant authentication mechanisms.

Structure: Lattice-based digital signatures, such as BLISS and Dilithium, use lattice-based constructions for generating and verifying digital signatures.

Security Features: These schemes rely on the hardness of lattice problems like "Shortest Vector Problem" (SVP) or "Learning With Errors" (LWE), ensuring resistance against quantum attacks.

Applications: Lattice-based digital signatures are employed in secure authentication, document signing, and digital identity verification.

Fully Homomorphic Encryption (FHE): Lattice-based FHE schemes enable computations on encrypted data without decryption, providing privacy-preserving data processing capabilities.

Capabilities: Fully Homomorphic Encryption allows computations on encrypted data without decrypting it, enabling privacy-preserving data processing.

Security Features: Lattice-based FHE schemes, such as TFHE (Homomorphic Encryption over Torus), use lattice techniques to achieve computational security.

Applications: FHE has applications in secure cloud computing, privacy-preserving data analysis, and secure outsourcing of computations.

Standardization and Adoption:

NIST Post-quantum Cryptography Standardization: Lattice-based cryptographic schemes are part of ongoing standardization efforts by organizations like NIST to identify and evaluate post-quantum cryptographic algorithms.

Adoption in Practice: Lattice-based cryptography is being adopted in various applications requiring quantum-resistant security, such as secure communications, data protection, and cloud computing.

Lattice-based cryptographic schemes offer strong security guarantees against quantum attacks and are actively researched and standardized as part of efforts to develop quantum-resistant cryptographic solutions for various applications in cybersecurity, data privacy, and secure communications [14, 15].

5 Hash-Based Cryptography

Hash-based cryptography is a branch of cryptography that relies on hash functions for securing data and verifying integrity. Unlike traditional encryption schemes that use keys for encryption and decryption, hash-based cryptography operates by generating fixed-size hash values or digests from input data [8, 16, 17]. These hash values are unique representations of the input data and are used for various cryptographic purposes. Here's an overview of hash-based cryptography:

Hash Functions:

Definition: Hash functions are mathematical algorithms that take input data of arbitrary size and produce fixed-size output values, known as hash values or digests.

Properties: Hash functions should be deterministic, producing the same hash value for identical inputs, and should exhibit the avalanche effect, where a small change in input results in a significantly different hash value.

Security: Cryptographic hash functions have additional properties such as preimage resistance, second preimage resistance, and collision resistance, making them suitable for security applications.

Hash-Based Cryptographic Schemes:

Message Digest Algorithm (MD): MD algorithms like MD5 and SHA-1 generate hash values of fixed sizes (128 and 160 bits, respectively) and were widely used but are now considered insecure due to vulnerabilities.

Secure Hash Algorithm (SHA): SHA-2 (including SHA-256, SHA-384, and SHA-512) and the newer SHA-3 are cryptographic hash functions that produce hash values of varying lengths and are designed to be resistant to attacks.

Merkle Trees: Merkle trees use hash functions to construct a hierarchical data structure, enabling efficient verification and authentication of large datasets or files.

Security Applications:

Data Integrity: Hash functions are used to ensure data integrity by generating hash values for data and comparing them to verify if the data has been altered or tampered with.

Digital Signatures: Hash functions are integral to digital signature schemes, where a hash value of the message is signed to provide authenticity, integrity, and non-repudiation.

Password Storage: Hash functions are employed in password hashing, where passwords are hashed before storage, preventing direct exposure in case of data breaches.

Blockchain Technology:

Block Hashing: Blockchain technology relies heavily on hash functions for creating block hashes, linking blocks together, and maintaining the integrity and immutability of the blockchain ledger.

Proof of Work (PoW): Hash functions play a crucial role in PoW consensus mechanisms used in cryptocurrencies like Bitcoin, where miners compete to find a hash value below a target threshold to validate transactions and create new blocks.

Challenges and Considerations:

Collision Resistance: Ensuring collision resistance (two different inputs producing the same hash value) is critical in hash-based cryptography to prevent attacks like hash collisions.

Algorithm Selection: Careful selection of cryptographic hash algorithms is essential, considering factors such as security strength, performance, and compatibility with security standards.

Hash-based cryptography provides essential tools and techniques for securing data, verifying authenticity, and supporting various cryptographic protocols and applications. However, it's crucial to use modern, secure hash functions and follow best practices to mitigate potential vulnerabilities and ensure robust cryptographic security.

5.1 Overview of Hash-Based Cryptographic Algorithms

Hash-based cryptographic algorithms play a fundamental role in securing data integrity, authentication, and various cryptographic protocols. Here's an overview of some key hash-based cryptographic algorithms:

Message Digest Algorithm (MD):

MD5 (Message Digest Algorithm 5): Produces a 128-bit hash value and was widely used but is now considered insecure due to vulnerabilities like collision attacks.

SHA-1 (Secure Hash Algorithm 1): Generates a 160-bit hash value and has also been deprecated due to vulnerabilities. It is no longer recommended for cryptographic security.

Secure Hash Algorithm (SHA):

SHA-2 (SHA-224, SHA-256, SHA-384, SHA-512): SHA-2 family includes hash functions that produce hash values of different lengths (224, 256, 384, 512 bits). These algorithms are widely used and considered secure against known attacks.

SHA-3: Introduced as the winner of the NIST hash function competition, SHA-3 is based on the Keccak algorithm and offers a different design from SHA-2, providing additional diversity in hash functions.

Whirlpool:

Features: Whirlpool is a cryptographic hash function that produces a 512-bit hash value. It is designed to be resistant against various cryptanalytic attacks and offers a high level of security.

BLAKE2:

Variants: BLAKE2 comes in two variants, BLAKE2b (for 64-bit platforms) and BLAKE2s (for smaller embedded systems). It provides high-speed hashing with security features suitable for various applications.

RIPEMD (RACE Integrity Primitives Evaluation Message Digest):

RIPEMD-160: Produces a 160-bit hash value and is designed as an alternative to SHA-1. It is used in digital signatures and message authentication codes (MACs).

Skein:

Flexible Hash Function: Skein is a cryptographic hash function that allows for flexible configurations, including hash sizes and security parameters. It is designed for efficiency and security.

GOST Hash Function:

Russian Standard: The GOST hash function is based on the Russian national standard and is used in various cryptographic applications, particularly in Russia and surrounding regions.

Hash-Based Message Authentication Code (HMAC):

Construction: HMAC is a construction that combines a cryptographic hash function (such as SHA-256 or SHA-3) with a secret key to generate a MAC for verifying data integrity and authenticity.

These hash-based cryptographic algorithms are essential components in secure communication protocols, digital signatures, password hashing, blockchain technology, and various cryptographic applications. It's crucial to use modern and secure hash functions while considering factors such as security strength, performance, and compatibility with cryptographic standards and protocols [18, 19].

5.2 *Use Cases and Advantages in Quantum-Resistant Cryptography*

Quantum-resistant cryptography, also known as post-quantum cryptography or quantum-safe cryptography, is designed to withstand attacks from quantum computers, which have the potential to break traditional cryptographic algorithms. Here are some use cases and advantages of quantum-resistant cryptography [8–10]:

Secure Communication Protocols:

- Use Case: Quantum-resistant cryptographic algorithms are crucial for securing communication channels, such as SSL/TLS protocols for web encryption, VPNs, and secure email communication.
- Advantages: They provide confidentiality, integrity, and authenticity of transmitted data, ensuring that sensitive information remains protected against quantum attacks.

Digital Signatures and Authentication:

- Use Case: Quantum-resistant digital signatures are used in electronic transactions, document signing, and identity verification to guarantee the authenticity and non-repudiation of digital assets.
- Advantages: These algorithms prevent malicious actors from forging digital signatures or impersonating legitimate users, maintaining trust and accountability in digital transactions.

Data Encryption and Storage:

- Use Case: Quantum-resistant encryption schemes are employed to encrypt sensitive data stored in databases, cloud storage, and IoT devices, safeguarding data confidentiality even in the presence of quantum threats.
- Advantages: They ensure that encrypted data remains secure, even if an adversary gains access to quantum computing capabilities, mitigating the risk of data breaches and unauthorized access.

Blockchain and Cryptocurrencies:

- Use Case: Quantum-resistant cryptographic algorithms are essential for securing blockchain networks and cryptocurrencies, such as Bitcoin and Ethereum, protecting transactions, digital assets, and smart contracts [4].
- Advantages: These algorithms prevent quantum attacks that could compromise the integrity of blockchain data, disrupt consensus mechanisms, or enable double-spending attacks.

Key Management and Exchange:

- Use Case: Quantum-resistant key management protocols and key exchange mechanisms, such as Quantum Key Distribution (QKD) and quantum-resistant key agreement protocols, ensure secure generation, distribution, and storage of cryptographic keys.
- Advantages: They protect cryptographic keys from quantum attacks, maintaining the confidentiality and integrity of encrypted communications and data.

Government and Military Applications:

- Use Case: Quantum-resistant cryptography is vital for securing classified information, national security communications, and military command and control systems.
- Advantages: These algorithms offer robust protection against quantum-enabled adversaries, ensuring the secrecy and reliability of sensitive government and military operations.

The advantages of quantum-resistant cryptography lie in its ability to provide long-term security assurances in a post-quantum computing era, where traditional cryptographic schemes may become vulnerable. By adopting quantum-resistant algorithms and protocols, organizations can proactively mitigate the risks posed by quantum computing advancements and safeguard their digital assets and communications against future threats.

6 Multivariate Cryptography

Multivariate cryptography is a branch of cryptography that involves the use of mathematical problems related to multivariate polynomial equations. Unlike traditional cryptographic algorithms that rely on number theory or elliptic curves, multivariate cryptography operates in finite fields and involves solving systems of multivariate polynomial equations. Here's an overview of multivariate cryptography [2, 3]:

Mathematical Basis:

Multivariate cryptography is based on the hardness of solving systems of multivariate polynomial equations over finite fields.

These equations involve variables, coefficients, and operations like addition, multiplication, and inversion in finite fields.

Components of Multivariate Cryptography:

Multivariate Quadratic Equations (MQ): Involves quadratic equations over finite fields, where the security relies on the difficulty of solving systems of quadratic equations.

Multivariate Polynomial Equations (MP): Extends to higher-degree polynomial equations, such as cubic or quartic equations, for increased security.

Key Components:

Public-Key Cryptography: Multivariate cryptography is often used in public-key cryptography schemes, where the public key is derived from the solution to a system of polynomial equations.

Private Key Generation: The private key is generated based on secret information used to create the multivariate polynomial equations.

Security Features:

Hardness Assumptions: The security of multivariate cryptography schemes relies on the presumed difficulty of solving systems of polynomial equations, which is considered computationally hard.

Resistance to Quantum Attacks: Multivariate cryptography is designed to be resistant against attacks from quantum computers, providing post-quantum security assurances.

Applications of Multivariate Cryptography:

Digital Signatures: Multivariate cryptography is used to create digital signatures, where the private key corresponds to a solution of multivariate equations, ensuring authentication and non-repudiation.

Key Exchange: It can be applied in key exchange protocols, enabling secure communication channels and data encryption using keys derived from multivariate equations.

Secure Hash Functions: Multivariate hash functions can be used for data integrity verification and hash-based message authentication codes (HMACs).

Advantages of Multivariate Cryptography:

Quantum Resistance: One of the primary advantages is its resistance to attacks from quantum computers, offering long-term security against quantum adversaries.

Efficiency: Multivariate cryptography can offer computational efficiency compared to some other post-quantum cryptographic schemes, making it suitable for certain applications.

Challenges and Considerations:

Key Sizes: To maintain security, multivariate cryptographic systems may require larger key sizes compared to traditional cryptographic algorithms.

Cryptanalysis: Efficient cryptanalysis techniques and advancements in algebraic attacks may pose challenges to the security of some multivariate schemes.

Multivariate cryptography represents a valuable area of research and development in the field of post-quantum cryptography, offering promising solutions for achieving quantum-resistant security in various cryptographic applications.

Analysis of Its Effectiveness in Protecting Against Quantum Attacks

The effectiveness of multivariate cryptographic schemes in protecting against quantum attacks is a crucial aspect to consider, especially in the context of post-quantum cryptography. Here's an analysis of its effectiveness:

Resistance to Shor's Algorithm:

Shor's algorithm, a quantum algorithm, can efficiently factor large numbers and solve the discrete logarithm problem, which is the basis of breaking many traditional cryptographic schemes.

Multivariate cryptographic schemes, however, do not rely on these mathematical problems. Instead, they are based on the complexity of solving systems of multivariate polynomial equations, which Shor's algorithm does not directly address.

This makes multivariate cryptography potentially resistant to attacks from quantum computers employing Shor's algorithm.

Search Space Complexity:

Multivariate cryptographic schemes typically have a large search space, requiring attackers to solve systems of polynomial equations with many variables and coefficients.

Quantum computers would need to efficiently handle this large search space and solve complex algebraic equations, which remains a challenging task even for quantum algorithms.

Quantum Cryptanalysis Challenges:

While quantum computers have the potential to perform certain operations exponentially faster than classical computers, they also face challenges in implementing and scaling quantum algorithms effectively.

Quantum cryptanalysis techniques for breaking multivariate cryptographic schemes are still under development and may require significant advancements in quantum computing capabilities.

Parameter Selection and Security Levels:

The security of multivariate cryptographic schemes depends on the choice of parameters, such as the degree of polynomial equations, key sizes, and algebraic structures.

Proper parameter selection is essential to ensure adequate security levels against both classical and quantum attacks.

Research and Development:

Ongoing research and development in multivariate cryptography aim to enhance its resistance to quantum attacks while optimizing performance and efficiency.

New algorithmic approaches, parameter optimizations, and cryptanalysis techniques contribute to improving the overall effectiveness of multivariate cryptographic schemes.

7 Real-Life Applications of Quantum Computing in Cybersecurity

Quantum computing holds significant promise for transforming cybersecurity by introducing powerful computational capabilities that can both enhance security measures and pose new challenges. Here are some real-life applications of quantum computing in cybersecurity:

Cryptographic Algorithms:

Quantum-Safe Cryptography: Quantum computing threatens traditional cryptographic algorithms like RSA and ECC by leveraging Shor's algorithm for prime factorization and discrete logarithm problems. Quantum-safe cryptographic algorithms, also known as post-quantum cryptography, are designed to withstand attacks from quantum computers. Examples include lattice-based cryptography, hash-based signatures, multivariate cryptography, and code-based cryptography.

Quantum Key Distribution: QKD protocols, such as BB84 and E91, leverage quantum principles to enable secure key exchange between parties. Quantum properties like entanglement and superposition ensure that any eavesdropping attempts disrupt the quantum state, alerting the parties to potential security breaches.

Quantum Key Distribution (QKD) is a method that uses the principles of quantum mechanics to securely exchange cryptographic keys between two parties, ensuring that any attempt at eavesdropping is detected. The most well-known QKD protocols are **BB84** and **E91**, both of which rely on quantum properties such as **superposition** and **entanglement** to protect the key exchange process. However, while QKD holds great promise for secure communication, its practical deployment faces several challenges [20].

1. **The Operation of QKD:**

QKD works by transmitting quantum states—typically photons—between two parties, conventionally called **Alice** (the sender) and **Bob** (the receiver). Any third party attempting to intercept or measure the quantum states will inevitably disturb them due to the fundamental principles of quantum mechanics, such as the **Heisenberg Uncertainty Principle** and **quantum no-cloning theorem**. This ensures that Alice and Bob can detect eavesdropping by monitoring the error rate in the transmitted quantum states.

BB84 Protocol (1984):

The **BB84** protocol, introduced by Charles Bennett and Gilles Brassard, is the most widely used QKD scheme. It operates as follows:

- **Encoding**: Alice generates a random string of bits and encodes each bit using one of two bases: the **rectilinear basis** (0°) or the **diagonal basis** (45°). These bases are chosen randomly for each bit.
- **Transmission**: Alice sends a photon to Bob, encoding the bit in either the rectilinear or diagonal basis.
- **Measurement**: Bob randomly chooses a basis (either rectilinear or diagonal) to measure each photon.
- **Basis Reconciliation**: After transmission, Alice and Bob publicly compare the bases they used (but not the actual bit values). For the photons where they used the same basis, they keep the corresponding bits; for the others, they discard the data.
- **Error Rate Checking**: Alice and Bob check a subset of their remaining bits to ensure there has been no eavesdropping. If the error rate is below a certain threshold, they conclude the key exchange was secure.

E91 Protocol (1991):

The **E91** protocol, proposed by Artur Ekert, leverages quantum entanglement. Entangled particles are generated and shared between Alice and Bob. The security of the protocol is guaranteed by the **Bell inequality** violation:

- **Entanglement Distribution**: A source generates entangled photon pairs and sends one photon to Alice and the other to Bob.
- **Measurement**: Alice and Bob choose random measurement angles and record the results.
- **Bell's Theorem**: The correlation between their measurements is checked to ensure it violates Bell's inequality, confirming the presence of entanglement.
- **Key Extraction**: From the correlated measurement results, a secure key is generated.

2. **Technical Challenges in Practical Deployments:**

Despite the theoretical security of QKD, its real-world deployment faces significant practical challenges that hinder widespread adoption. These include limitations related to **distance**, **cost**, **speed**, and **scalability**.

2.1. **Distance Limitations:**

- **Signal Loss**: In fiber-optic cables, quantum states (photons) degrade over long distances due to **attenuation** (signal loss). This limits the distance over which QKD can be performed effectively. For example, in standard optical fibers, QKD is usually limited to distances of about 100–150 km before the signal becomes too weak to be reliably detected.
  - **Solution**: Researchers are exploring the use of **quantum repeaters**, which could extend the range by periodically restoring the quantum state without violating the principles of quantum mechanics. However, practical quantum repeaters are still in the early stages of development.
- **Free-Space QKD**: In free-space QKD, quantum signals are transmitted through the air or space instead of optical fibers. **Satellite-based QKD** has been tested successfully over thousands of kilometers. For instance, **China's Micius satellite** demonstrated long-distance quantum communication across more than 1,200 km in 2017, using a satellite to distribute entangled particles between two ground stations. This opens up possibilities for global QKD networks, though atmospheric disturbances and weather conditions remain major challenges.

2.2. **Cost and Infrastructure:**

- **Expensive Hardware**: The detectors used in QKD, such as **single-photon detectors (SPDs)**, are highly sensitive and expensive. Maintaining such equipment requires sophisticated technology to detect individual quantum states without introducing significant noise.
  - **Solution**: Advances in hardware design, such as more efficient single-photon detectors and cheaper quantum devices, are needed to make QKD commercially viable. Ongoing research aims to reduce the cost of quantum hardware.
- **Dedicated Quantum Networks**: QKD requires dedicated quantum communication channels separate from classical networks. Implementing these

networks on a large scale would necessitate significant investments in new infrastructure, especially when building fiber-optic links for quantum communication.

2.3. **Speed and Key Generation Rate:**

- **Low Key Generation Rates**: QKD protocols, especially those based on single-photon detectors, often suffer from low **key generation rates**. This is a challenge in high-demand environments where secure communication needs to be fast and continuous.
  - **Solution**: Researchers are working on improving **high-speed quantum random number generators** and exploring **multi-photon schemes** to increase the rate of secure key generation.

2.4. **Vulnerabilities in Realistic Implementations:**

- **Photon Detectors and Side-Channel Attacks**: While QKD is theoretically secure, real-world implementations are vulnerable to practical attacks such as **side-channel attacks**. For example, a hacker could exploit imperfections in photon detectors, known as **detector blinding attacks**, to gain information about the key exchange without being detected.
  - **Solution**: Security loopholes like this are being addressed by developing **device-independent QKD** (DI-QKD), which aims to remove trust in the devices used for QKD and ensure security purely from the principles of quantum mechanics.
- **Decoy-State Protocols**: These protocols have been developed to counter some practical attacks, such as the **photon-number splitting (PNS) attack**, where an eavesdropper intercepts multi-photon signals. Decoy-state QKD reduces the risk by sending "decoy" signals to detect these attacks.

2.5. **Integration with Classical Systems:**

- **Hybrid Systems**: QKD systems must be integrated with existing classical networks, raising issues related to **interoperability**. Classical cryptography, like RSA, can easily operate at high speeds and across vast distances, but quantum systems require dedicated infrastructure.
  - **Solution**: Researchers are exploring hybrid systems that combine classical cryptography with QKD. For example, **post-quantum cryptographic algorithms** can be used in conjunction with QKD to secure long-distance communication.

3. **Recent Developments in QKD:**

Several advancements have been made in recent years to address the challenges faced by QKD:

- **Measurement-Device-Independent QKD (MDI-QKD)**: This approach eliminates the vulnerabilities of detectors, making the system more robust against attacks such as detector blinding. MDI-QKD has been tested over hundreds of kilometers, improving the security of practical QKD implementations.

- **Trusted Nodes**: Some QKD networks, such as the **European Quantum Communication Infrastructure (EuroQCI)**, are using **trusted nodes** to extend the reach of QKD systems. These nodes act as intermediaries that re-encrypt the quantum key, allowing it to travel longer distances while preserving security.

Quantum Key Distribution (QKD) offers a powerful method for secure key exchange using the principles of quantum mechanics, but its practical deployment is hindered by several challenges such as distance limitations, cost, speed, and device vulnerabilities. Recent advances in satellite-based QKD, measurement-device-independent QKD, and decoy-state protocols are helping to overcome these issues, bringing QKD closer to practical, large-scale use. As quantum technologies evolve, the development of hybrid systems that combine classical and quantum cryptography will likely play a crucial role in the future of secure communications.

Secure Communication Protocols:

Quantum-Secure Communication Networks: Quantum networks ensure secure communication by utilizing quantum key distribution (QKD) and quantum-resistant encryption algorithms. These networks protect data confidentiality and integrity against quantum attacks, enhancing the security of sensitive communications in industries like finance, healthcare, and government [4, 21].

Authentication and Access Control:

Quantum Authentication Methods: Quantum technologies, such as quantum random number generators (QRNGs) and quantum entropy sources, enhance authentication mechanisms by providing truly random and unpredictable data for cryptographic operations. Quantum-resistant authentication protocols ensure secure access control in digital systems and networks.

Malware Detection and Intrusion Prevention:

Quantum Machine Learning for Security: Quantum computing's computational power accelerates machine learning algorithms, enabling more efficient and accurate detection of malware, anomalies, and cybersecurity threats. Quantum-enhanced AI systems improve intrusion prevention and threat response capabilities.

Digital Signatures and Blockchain Security:

Quantum-Secure Digital Signatures: Quantum-resistant signature schemes, such as XMSS and SPHINCS +, protect digital signatures from quantum attacks, ensuring the authenticity and integrity of digital transactions. Quantum-safe blockchain protocols, like qTESLA and Dilithium, safeguard distributed ledgers against quantum threats, preserving the immutability and trustworthiness of blockchain records [22, 23].

Cyber Risk Assessment and Vulnerability Analysis:

Quantum Computing in Cyber Risk Management: Quantum algorithms and simulations enable faster and more comprehensive cyber risk assessments, vulnerability analyses, and penetration testing. Quantum-enhanced risk management tools improve cybersecurity decision-making and threat mitigation strategies.

Biometric Security and Quantum Cryptography:

Quantum Biometrics: Quantum computing enhances biometric security measures by processing large-scale biometric data and improving biometric authentication accuracy. Quantum cryptography protocols, like quantum fingerprinting, enhance the security of biometric data storage and verification processes.

These real-life applications demonstrate the diverse ways in which quantum computing is revolutionizing cybersecurity, from securing communication networks and cloud environments to enhancing authentication methods, threat detection, and blockchain security. As quantum technologies continue to advance, their impact on cybersecurity will further evolve, requiring ongoing research, innovation, and adaptation in security practices and technologies [21, 24].

Real-World Case Studies Demonstrating Quantum Computing's Practical Impact on Cybersecurity

1. **Google's Quantum Supremacy and Its Implications for Cybersecurity**

In 2019, Google achieved a milestone in quantum computing by demonstrating **quantum supremacy**. Their 53-qubit processor, **Sycamore**, completed a task in 200 s that would have taken the world's most powerful supercomputers approximately 10,000 years. Although the task had no direct cryptographic implications, this achievement demonstrated the potential of quantum computers to solve problems exponentially faster than classical computers.

Cybersecurity Implications:

- **Breaking RSA Encryption**: While current quantum computers are not yet powerful enough to break RSA encryption, Google's experiment highlighted the inevitability of reaching that stage in the near future. Shor's algorithm, once implemented on a large-scale quantum computer, would efficiently factor large numbers, directly threatening RSA-based encryption, which underpins much of the internet's security infrastructure.
- **Call to Action for Post-quantum Cryptography**: Google's success in demonstrating quantum supremacy accelerated the urgency for organizations to invest in **post-quantum cryptography (PQC)**, which involves developing encryption algorithms resistant to quantum attacks. Google's experiment marks the beginning of a race to secure digital infrastructure against future quantum threats [13].

2. **IBM's Quantum Computing Research and Quantum-Safe Cryptography**

IBM has been a pioneer in quantum computing and has made significant contributions to cybersecurity through its **IBM Quantum Network**. IBM provides access to its quantum computers via the cloud, allowing researchers and organizations to experiment with quantum algorithms, including those related to cryptography [25].

IBM's Contributions to Quantum-Safe Cryptography:

- **Quantum-Safe Roadmap**: IBM is actively involved in developing **quantum-safe cryptographic protocols**. They are working on algorithms like **lattice-based cryptography** and **hash-based cryptography**, which are designed to resist attacks from quantum computers. This is essential for securing the long-term confidentiality and integrity of data.

- **Collaborative Efforts**: IBM has partnered with global financial institutions, such as **JPMorgan Chase**, to test and implement quantum-safe algorithms in financial systems, which are particularly vulnerable to quantum threats. These institutions rely on strong encryption to protect sensitive financial data, making quantum-safe cryptography an urgent priority.

Cybersecurity Implications:

- **Proactive Adoption of Post-quantum Cryptography**: IBM's active research and cloud access to quantum technology are helping organizations prepare for the transition to **quantum-safe encryption systems** before large-scale quantum attacks become feasible.
- **Partnerships in Cybersecurity**: By collaborating with other industries, IBM helps mitigate future quantum risks in sectors like banking and healthcare, which rely heavily on secure data transmissions.

3. **ID Quantique's Quantum Key Distribution (QKD) Systems**

ID Quantique, based in Switzerland, is a global leader in **Quantum Key Distribution (QKD)** systems, which use the principles of quantum mechanics to enable secure communication. QKD offers **unconditional security** by ensuring that any attempt to eavesdrop on the communication would disturb the quantum state, alerting the communicating parties to the breach.

Practical Applications:

- **Banking and Financial Sector**: ID Quantique has implemented QKD systems in financial institutions, ensuring that sensitive transactions and communications are protected. For example, they partnered with **Swiss banks** to secure their data networks, demonstrating how quantum technology can protect high-value assets from cyber threats.
- **Government Communications**: ID Quantique's QKD systems have been deployed in **government networks**, ensuring secure communication channels that are resistant to classical and quantum attacks. For instance, their technology was used to secure the **Swiss elections**, making sure the voting process was tamper-proof [4].

Cybersecurity Implications:

- **QKD Adoption**: With QKD systems already in place, ID Quantique has shown that practical quantum-safe cryptography is not only a theoretical construct but also an operational reality. QKD ensures that cryptographic keys are exchanged securely, preventing interception by current and future quantum adversaries.
- **Quantum-Safe Communications**: By securing financial institutions and government communications, ID Quantique has set a precedent for how quantum-safe technologies can protect critical infrastructures globally, ensuring resilience against emerging quantum threats.

4. **Volkswagen's Use of Quantum Computing for Cybersecurity and Optimization**

Volkswagen has ventured into quantum computing, using it to enhance cybersecurity and optimize processes in automotive manufacturing. While not directly cryptographic, their work is paving the way for **quantum-enhanced cybersecurity measures**.

Quantum Applications in Cybersecurity:

- **Traffic Optimization**: Volkswagen used quantum computers to develop algorithms for optimizing traffic flow in real-time. Though this is not directly linked to cryptography, it demonstrates how quantum computing can manage large datasets and solve complex problems quickly—skills that are also essential in **cyber defense**, where quantum algorithms could help detect and respond to cyber threats more efficiently.
- **Quantum-Enhanced AI for Threat Detection**: By applying quantum machine learning algorithms, Volkswagen is also investigating quantum-enhanced **threat detection** systems. These systems can potentially predict and identify cyberattacks by analyzing patterns in vast datasets at speeds that classical systems cannot match [21].

Cybersecurity Implications:

- **Quantum-Enhanced Threat Detection**: Volkswagen's work illustrates how quantum computers can enhance cybersecurity in areas such as **threat detection and anomaly identification**. This capability is crucial for future cybersecurity frameworks, where detecting advanced persistent threats (APTs) will require significant computational power.
- **Securing Autonomous Vehicles**: As vehicles become increasingly autonomous, quantum computing could play a vital role in securing the software systems that control these vehicles, ensuring that they remain safe from both classical and quantum-enabled cyberattacks.

These real-world case studies demonstrate how quantum computing is already beginning to influence cybersecurity, from **quantum supremacy** to **QKD** systems and **quantum-safe encryption**. Organizations such as Google, IBM, ID Quantique, and Volkswagen are at the forefront of integrating quantum technology into cybersecurity, highlighting the need for industries and governments to begin transitioning to **quantum-resistant cryptographic systems**. As these technologies continue to evolve, they will play a pivotal role in defending against the next generation of cyber threats.

Benefits and Outcomes of Implementing Quantum-Safe Solutions

Implementing quantum-safe solutions offers several benefits and outcomes across various domains, especially in the context of cybersecurity and data protection. Here are some key benefits and outcomes of implementing quantum-safe solutions [13]:

Resistance to Quantum Attacks:

Benefit: Quantum-safe solutions are designed to withstand attacks from quantum computers, which have the potential to break traditional cryptographic algorithms.

Outcome: Enhanced security and long-term protection of sensitive data against emerging quantum threats, ensuring confidentiality, integrity, and authenticity in digital communications and transactions.

Secure Key Exchange and Encryption:

Benefit: Quantum-safe key exchange protocols, such as lattice-based cryptography and code-based cryptography, ensure secure generation, distribution, and management of cryptographic keys resilient to quantum attacks.

Outcome: Strengthened encryption mechanisms, data confidentiality, and secure communication channels, reducing the vulnerability of sensitive information to quantum-enabled attacks like key extraction and decryption.

Long-Term Data Security:

Benefit: Quantum-safe solutions offer long-term data security by addressing the cryptographic vulnerabilities exposed by quantum algorithms, such as Shor's algorithm for factorization and Grover's algorithm for search problems [1].

Outcome: Preservation of data integrity, protection against data breaches, and sustained confidentiality of stored and transmitted data assets throughout their lifecycle, even in the presence of powerful quantum adversaries.

Innovation and Technological Leadership:

Benefit: Embracing quantum-safe solutions reflects technological leadership, innovation readiness, and strategic foresight in addressing future cybersecurity challenges posed by quantum computing advancements.

Outcome: Competitive advantage, market differentiation, and resilience in dynamic cyber threat landscapes, positioning organizations as trusted partners and pioneers in quantum-safe cybersecurity solutions.

Overall, the benefits and outcomes of implementing quantum-safe solutions encompass heightened security, risk mitigation, regulatory compliance, innovation readiness, and future-proofing against quantum threats, fostering a secure and trusted digital ecosystem for businesses, governments, and individuals alike.

Detailed Vision of Future Research Paths and the Evolution of Quantum Cybersecurity

Quantum cybersecurity is poised to evolve significantly as both quantum computing technologies and cryptographic defenses advance. While the current discussion surrounding quantum technologies is optimistic, it is essential to provide a more structured vision for future research and development to address the emerging threats and opportunities. The following sections outline key research paths and the potential future evolution of **quantum cybersecurity**.

1. **Development and Standardization of Post-quantum Cryptography (PQC)**

    1.1 **Algorithm Selection and Optimization**

One of the most critical areas for future research is the continued **selection, optimization, and standardization** of post-quantum cryptographic algorithms. The **NIST Post-quantum Cryptography Standardization Process** is in its final stages, and the selection of standardized algorithms will shape the next decade of cryptography. The algorithms under consideration, such as **lattice-based**, **hash-based**, **code-based**, and **multivariate polynomial** schemes, need further refinement for practical implementation.

Key Research Directions:

- **Security Analysis**: There is a need for ongoing research into the cryptographic security of PQC algorithms against **quantum attacks** and **classical side-channel attacks**. This includes developing robust techniques to handle side-channel threats like power analysis and timing attacks, which can be used to break otherwise secure quantum-resistant cryptography.
- **Efficiency Improvements**: Further work is required to optimize PQC algorithms for **performance and efficiency**, ensuring that quantum-resistant encryption can be deployed in real-time applications without significant overheads. Research into **hardware accelerators** and **parallelized implementations** could enhance the speed of these algorithms, making them suitable for use in environments like the **Internet of Things (IoT)** and **mobile devices**.

1.2 Hybrid Cryptographic Systems

The transition from classical to quantum-safe cryptography will not be immediate. Hybrid systems that combine **classical cryptography** with **PQC** will play a crucial role during this transition phase. These systems will need to be carefully designed to ensure backward compatibility and security during the period when quantum computers are not yet powerful enough to break classical encryption but the threat is looming [16, 17].

Key Research Directions:

- **Interoperability**: Developing **interoperable hybrid systems** that work seamlessly across different cryptographic standards is critical. This ensures organizations can adopt quantum-safe measures while continuing to use existing systems until the full migration to PQC is feasible.
- **Incremental Transition Frameworks**: Researchers should focus on creating structured frameworks for **incremental migration** to post-quantum cryptography, helping industries and governments gradually transition their cryptographic infrastructures without compromising security or incurring high costs.

2. Regulatory and Policy Frameworks for Quantum Security

2.1 Global Cooperation and Standardization

The international landscape for quantum computing is highly competitive, with nations investing heavily in **quantum research** for both civilian and military applications. Future research must focus on developing **global standards and policies** to regulate the deployment of quantum technologies, ensuring **cross-border cooperation** and preventing the misuse of quantum computing for malicious purposes.

Key Research Directions:

- **Quantum Export Controls**: Just as current encryption technologies are subject to export controls, quantum technologies—particularly **quantum cryptography tools** and **quantum computers**—will require strict regulatory frameworks to prevent them from falling into the hands of adversaries. Research into **regulatory frameworks** that balance innovation with national security is essential.
- **Quantum Cybersecurity Governance**: Governments and international organizations should focus on developing a **universal quantum cybersecurity governance framework** that includes best practices for deploying quantum-safe cryptography and quantum networks. This framework must ensure the security of critical infrastructure, financial systems, and military operations against future quantum attacks.

2.2 Legal and Ethical Considerations

The development of quantum computing also brings up **legal and ethical challenges** regarding the privacy and security of encrypted data. Governments and organizations must carefully navigate the **retroactive decryption** of data that was previously encrypted with classical methods once quantum computers become powerful enough to break existing encryption.

Key Research Directions:

- **Data Sovereignty and Privacy**: Researchers need to explore how **quantum decryption** will affect personal privacy and data sovereignty. Legal frameworks must be established to ensure that previously secure data, such as medical records and financial information, remains protected even after quantum computers are capable of breaking classical encryption.
- **Intellectual Property Protection**: As quantum-safe algorithms and systems are developed, ensuring that **intellectual property rights** for quantum technologies are protected will become a key issue. Collaboration between governments and private companies is required to protect innovations in quantum cryptography from theft and misuse.

The future of quantum cybersecurity lies in a multifaceted approach that integrates **technical advancements**, **policy frameworks**, and **international cooperation**. Research into **post-quantum cryptography**, **quantum networks**, and **quantum-enhanced AI** will drive the evolution of the field, ensuring that security systems are prepared for the quantum future. Additionally, quantum technologies must be developed responsibly, with robust legal and regulatory measures in place to prevent their misuse. By focusing on these research areas, we can ensure a secure and resilient digital ecosystem, capable of withstanding the challenges posed by quantum computing [26, 27].

Innovations and Ongoing Research Initiatives in Quantum Computing and Cybersecurity

Innovations and ongoing research initiatives in quantum computing and cybersecurity are driving advancements in digital security, threat mitigation, and privacy protection. Here are some key innovations and research areas shaping the field [13]:

Quantum-Safe Cryptography:

Ongoing research focuses on developing and optimizing quantum-resistant cryptographic algorithms, such as lattice-based cryptography, code-based cryptography, and hash-based signatures.

Innovations include the exploration of quantum-resistant encryption schemes for securing data at rest and in transit, quantum-safe digital signatures, and quantum-resistant key exchange protocols.

Post-quantum Cryptography Standards:

Collaborative efforts by academia, industry, and standardization bodies aim to establish post-quantum cryptography standards to replace vulnerable cryptographic algorithms.

Innovations include the evaluation, selection, and validation of quantum-safe cryptographic primitives and protocols through international competitions and peer-reviewed research.

Quantum-Secure Blockchain and Distributed Ledger Technologies (DLTs):

Research explores the integration of quantum-resistant cryptographic primitives into blockchain and DLT platforms to protect against quantum attacks [18, 19].

Innovations include quantum-safe consensus algorithms, quantum-resistant smart contract languages, and quantum-enhanced auditability and transparency in decentralized applications (dApps).

Quantum Machine Learning (QML):

Advances in quantum machine learning algorithms and quantum computing hardware enable the development of QML models for cybersecurity applications [5, 28, 29].

Innovations include quantum-enhanced anomaly detection, intrusion detection systems, and adversarial machine learning defenses against sophisticated cyber threats.

Quantum-Safe Cloud Computing:

Research focuses on implementing quantum-resistant encryption and authentication mechanisms in cloud computing environments.

Innovations include quantum-safe cloud storage solutions, quantum-aware access control policies, and quantum-enhanced data privacy and compliance frameworks for cloud services.

Quantum-Secure Authentication and Access Control:

Innovations in quantum-resistant authentication methods, such as quantum-resistant digital signatures, biometric authentication, and zero-trust security models.

Research explores the integration of quantum-safe authentication protocols into identity management systems, secure access gateways, and multi-factor authentication solutions.

Quantum-Safe AI and Autonomous Systems Security:

Research initiatives focus on developing quantum-aware AI models, quantum-resistant machine learning algorithms, and secure AI-driven decision-making frameworks.

Innovations include quantum-enhanced cybersecurity analytics, threat intelligence platforms, and quantum-aware security orchestration for autonomous systems and IoT devices.

These innovations and ongoing research initiatives underscore the collaborative efforts across academia, industry, and government sectors to address the cybersecurity challenges posed by quantum computing advancements. By advancing quantum-safe solutions, exploring new security paradigms, and fostering interdisciplinary collaborations, researchers and practitioners aim to ensure the resilience, integrity, and trustworthiness of digital infrastructures in an era of quantum-enabled cyber threats.

Perspectives from Cybersecurity Professionals on Operational Challenges of Adopting Quantum Technologies

While the technical aspects of quantum computing and cryptography are central to understanding the future of cybersecurity, insights from **cybersecurity professionals** provide a practical view of the operational challenges that organizations face when adopting quantum technologies. These professionals are tasked with integrating new technologies into complex infrastructures, managing security risks, and ensuring that existing systems remain secure during and after the transition to quantum-safe solutions [4, 13].

1. **Operational Readiness and Expertise Gap**

One of the primary challenges that cybersecurity professionals highlight is the **lack of operational readiness** and expertise in quantum technologies. The complexity of quantum systems, coupled with the novelty of the field, means that many organizations lack the necessary skills and understanding to deploy quantum-safe cryptography or quantum-based security measures effectively.

Key Insights:

- **Skill Shortage**: Many organizations are struggling to find professionals with the required knowledge in both quantum computing and cybersecurity. Cybersecurity teams are already tasked with protecting systems from classical cyber threats, and the addition of quantum security measures adds another layer of complexity.
  - Professional Insight: "Quantum technologies introduce a steep learning curve. It's not only about understanding cryptography but also the underlying quantum mechanics. We need to build a workforce that can handle both the quantum and classical aspects of security."—Chief Information Security Officer (CISO), Financial Services Sector

- **Training and Development**: Cybersecurity professionals recommend that organizations begin **training** their teams on quantum-safe practices and technologies. This includes understanding post-quantum cryptographic algorithms, implementing quantum key distribution (QKD), and recognizing potential vulnerabilities unique to quantum systems.

Proposed Solutions:

- **Educational Initiatives**: Industry leaders advocate for increased investment in **quantum cybersecurity education**. This includes university programs, certifications, and on-the-job training programs that focus on quantum cryptography, quantum network security, and quantum-safe algorithm implementation.
- **Cross-disciplinary Teams**: Many professionals recommend forming **cross-disciplinary teams** that bring together experts from quantum computing, traditional cybersecurity, and cryptography. Such teams would ensure that the integration of quantum technologies is done securely and efficiently.

2. **Integration with Existing Security Frameworks**

Cybersecurity professionals emphasize the difficulty of integrating quantum technologies into existing security frameworks. Many organizations rely on established systems for securing communications, data, and infrastructure, which are built around classical cryptographic methods. Transitioning to quantum-safe systems presents operational challenges, particularly in terms of compatibility, interoperability, and scalability [25].

Key Insights:

- **Compatibility Issues**: Existing infrastructure, such as network security devices, databases, and encryption systems, is often not designed to handle quantum technologies. This means that organizations face significant **compatibility issues** when attempting to introduce quantum-safe encryption or QKD into their systems.
  - Professional Insight: "Legacy systems aren't built for quantum technologies. Even just implementing post-quantum cryptography can cause disruptions, especially in large-scale environments like cloud networks or IoT systems." – Head of Security Operations, Telecom Industry
- **Scalability Concerns**: Many professionals raise concerns about the **scalability** of quantum security solutions. For example, QKD systems have distance limitations, and the infrastructure required to deploy QKD on a large scale—such as **quantum repeaters**—is still under development. Similarly, quantum-resistant algorithms often come with performance overheads that may slow down network operations, especially in high-traffic environments.
  - Professional Insight: "Quantum technologies are great on paper, but when it comes to deploying them across distributed networks, it's a different story. We need solutions that scale without causing latency or requiring massive infrastructure overhauls." – Network Security Architect, Global Enterprise

Proposed Solutions:

- **Hybrid Solutions**: Cybersecurity experts suggest adopting **hybrid cryptographic systems** that combine classical and quantum-safe encryption methods. This phased approach allows organizations to maintain compatibility with existing systems while preparing for future quantum threats.
- **Pilot Projects**: Professionals recommend running **pilot projects** to test quantum-safe solutions in controlled environments before fully integrating them into critical infrastructure. This approach helps identify compatibility issues and operational challenges before scaling.

3. **Cost and Resource Allocation**

Another major challenge that cybersecurity professionals face is the **cost** associated with adopting quantum technologies. Quantum hardware, such as **quantum key distribution systems** or **quantum random number generators (QRNGs)**, is expensive and requires specialized infrastructure. Additionally, the cost of **migrating existing systems** to post-quantum cryptography is significant, especially for large enterprises and government agencies.

Key Insights:

- **High Initial Costs**: The cost of deploying quantum security solutions, particularly QKD, is currently prohibitive for many organizations. The equipment needed for quantum-safe communication, such as single-photon detectors, quantum repeaters, and secure fiber-optic networks, is expensive and often requires specialized expertise to maintain.
  - Professional Insight: "The costs involved in rolling out quantum key distribution are substantial, especially when you're talking about securing communications across global operations. Not every organization has the budget for that right now." – CISO, International Banking Group
- **Resource Allocation**: Organizations must allocate significant resources to research, pilot testing, and deployment of quantum technologies, which can strain existing cybersecurity budgets. This is particularly challenging for industries that already face tight margins, such as **healthcare** and **public services**.

Proposed Solutions:

- **Government Subsidies and Incentives**: Many professionals advocate for **government subsidies** and **incentives** to help organizations adopt quantum-safe technologies. Governments can provide grants or tax incentives to encourage the development and deployment of quantum-safe infrastructure, particularly in critical industries like finance, healthcare, and defense.
- **Collaboration and Shared Resources**: Professionals suggest that organizations collaborate on **shared quantum infrastructure** to reduce costs. For example, multiple organizations within a sector could share access to **quantum communication networks** or **quantum cloud computing** services, distributing the cost across multiple entities.

4. **Security Risks in the Transition Phase**

Cybersecurity professionals warn that the transition to quantum-safe systems poses its own set of security risks. During this phase, both classical and quantum systems may be in use simultaneously, creating vulnerabilities that attackers could exploit. The coexistence of different encryption standards, especially as **post-quantum cryptographic algorithms** are rolled out, introduces complexities in key management and data protection.

Key Insights:

- **Interoperability Issues**: The biggest risk in the transition phase is **interoperability** between quantum-safe and classical cryptography systems. The use of hybrid systems can create **attack surfaces** that did not exist when only classical encryption methods were used. For example, attackers might exploit weaknesses in how classical and quantum keys are managed together.
  - Professional Insight: "The transition period could be a goldmine for attackers. The more complex our encryption ecosystem becomes—mixing classical and quantum approaches—the more difficult it is to ensure everything is airtight." – Lead Cybersecurity Engineer, Energy Sector
- **Data Breach Risks**: As organizations begin to store data using quantum-safe algorithms, they must ensure that older, classically encrypted data is not vulnerable to quantum decryption. However, the process of migrating historical data to post-quantum encryption systems can expose it to attackers during the transition.
  - Professional Insight: "Migrating sensitive data is always risky. During the transition, we're especially vulnerable to breaches because attackers know we're trying to secure older, weaker encryption systems." – Cybersecurity Analyst, Healthcare Industry

Proposed Solutions:

- **Comprehensive Transition Plans**: Cybersecurity professionals recommend developing **comprehensive transition plans** that address the management of both classical and quantum systems. These plans should include detailed guidelines on **key management**, **secure data migration**, and **hybrid encryption** techniques.
- **Real-Time Monitoring**: During the transition, it is crucial to implement **real-time monitoring** of both quantum and classical systems to detect any vulnerabilities that could arise. This includes monitoring **key exchanges**, **data transfers**, and **encryption–decryption processes** for signs of tampering or compromise.

While the technical promise of quantum technologies is significant, cybersecurity professionals face substantial operational challenges in deploying these solutions. From **skill shortages** to **integration issues**, **cost constraints**, and **transition risks**, organizations need to carefully plan how they adopt quantum-safe measures. By considering these operational challenges and implementing strategies such as hybrid cryptographic systems, government incentives, and cross-disciplinary teams, organizations can better manage the transition to quantum cybersecurity. These insights from professionals on the front lines of cybersecurity help bridge the gap between technical theory and real-world practice, ensuring a more secure future in the face of quantum advancements.

Risk Assessment of Quantum Technologies in Cybersecurity

The rise of quantum technologies introduces significant risks to cybersecurity, necessitating a thorough and systematic risk analysis. This analysis evaluates the **likelihood** and **impact** of potential threats, helping organizations prioritize their responses and allocate resources effectively. The risks associated with the **adoption of quantum technologies**, such as **quantum computing** and **quantum cryptography**, span across technical, operational, and strategic domains.

Risk Assessment Framework

This risk assessment framework follows a structured approach, assessing risks across the following key dimensions:

- **Likelihood**: The probability of a risk occurring in the near term.
- **Impact**: The potential severity of the damage if the risk materializes.
- **Urgency**: The priority for mitigating the risk, based on both likelihood and impact.

For each risk, we provide a **risk rating** (high, medium, low) and recommend strategies to address or mitigate the issue.

1. Risk: Breakage of Classical Cryptography by Quantum Computers

Quantum computers, once fully realized, will be capable of running **Shor's algorithm**, which can efficiently factor large numbers. This threatens the security of widely used classical cryptographic algorithms like **RSA**, **Elliptic Curve Cryptography (ECC)**, and **Diffie–Hellman key exchange**, potentially compromising the confidentiality of encrypted communications, data at rest, and data in transit [20, 30].

Likelihood: Medium

- While current quantum computers do not yet have the capacity to break classical cryptography, advances in quantum hardware are steadily moving toward this capability within the next decade.

Impact: High

- The ability to break RSA and ECC would lead to widespread **data breaches**, the decryption of sensitive information, and the potential compromise of **government**, **military**, **financial**, and **personal data**.

Urgency: High

- Transitioning to **quantum-safe cryptographic algorithms** is urgent to prevent future quantum-based attacks. Organizations must begin migrating now to avoid retrospective decryption attacks (where adversaries intercept and store encrypted data, waiting for quantum technology to decrypt it).

Mitigation Strategies

- **Adopt Post-quantum Cryptography (PQC)**: Begin migrating to **quantum-resistant cryptographic algorithms** such as **lattice-based**, **hash-based**, or **code-based cryptography**. Follow the **NIST standards** for post-quantum cryptography once they are finalized.
- **Hybrid Cryptographic Systems**: Use hybrid approaches that combine classical and post-quantum cryptography during the transition period.

2. **Risk: Quantum Adversarial Attacks on Machine Learning and AI Systems**

Quantum computing could be used to launch **adversarial attacks** against machine learning (ML) systems, compromising **AI-driven cybersecurity defenses**. Quantum adversaries could manipulate input data to trick AI models into making incorrect classifications, thereby bypassing traditional and AI-based defenses.

Likelihood: Low to Medium

- The use of quantum algorithms to generate adversarial examples is still theoretical but could become feasible in the near future as quantum computing progresses.

Impact: Medium to High

- Adversarial attacks on AI-based systems could disrupt **automated cybersecurity defenses**, **fraud detection**, and **threat intelligence platforms**, leading to financial loss and system downtime.

Urgency: Medium

- While quantum adversarial attacks are not yet an immediate threat, research should begin on developing **quantum-safe AI systems** to prepare for future risks.

Mitigation Strategies

- **Develop Quantum-Safe AI Models**: Research into AI algorithms that can resist quantum adversarial attacks is essential. AI models must be robust against quantum-generated perturbations.
- **AI Adversarial Defense**: Strengthen existing AI systems with adversarial defense mechanisms, ensuring that they can detect and counter quantum-based manipulation attempts.

This risk analysis highlights the significant challenges and vulnerabilities introduced by quantum technologies in cybersecurity. The most urgent threats revolve around the **breakage of classical cryptography** and the **interoperability issues** faced during the transition to quantum-safe systems. Quantum **adversarial attacks on AI systems** and the security risks of QKD implementations also require ongoing research and development to ensure they can be deployed safely. Addressing these risks through **post-quantum cryptography**, **hybrid cryptography systems**, and **improvements**

in quantum hardware security will be critical to protecting digital infrastructure in the quantum era.

8 Conclusion

In conclusion, the integration of quantum computing into cybersecurity represents a transformative paradigm shift with profound implications for securing the digital future. Quantum computing's unparalleled computational power and potential offer both opportunities and challenges in safeguarding critical data, communications, and systems against evolving cyber threats. The adoption of quantum-safe cryptographic solutions is imperative to mitigate the risks posed by quantum-enabled attacks, such as Shor's algorithm for breaking traditional encryption. Research and innovation in post-quantum cryptography, quantum key distribution, and quantum-resistant protocols are essential for establishing robust defense mechanisms in the face of quantum adversaries.

Furthermore, the synergy between quantum computing and cybersecurity extends beyond encryption to encompass quantum-enhanced threat intelligence, secure multiparty computation, and quantum-aware AI-driven defenses. Collaborative efforts among researchers, industry leaders, and policymakers are essential to drive forward quantum-safe cybersecurity standards, best practices, and regulatory frameworks. As organizations navigate the complexities of quantum computing's impact on cybersecurity, strategic investments in quantum-ready infrastructure, talent development, and risk management strategies are crucial. Embracing a holistic approach that combines technical expertise, regulatory compliance, and continuous innovation will be pivotal in securing the digital ecosystem and ensuring trust, resilience, and integrity in the quantum era. Ultimately, securing the future of cybersecurity in the age of quantum computing requires a proactive and adaptive mindset, a robust cybersecurity posture, and a commitment to staying ahead of emerging threats through collaborative research, education, and strategic partnerships.

References

1. V. Jha, S.K. Jha, Quantum cryptography for data security: principles, techniques, and solutions. J. Quantum Inf. **12**(2), 150–165 (2021)
2. H.A. Alsalamah, Quantum computing and cyber security: a new era for computing. J. Cybersecur. **15**(2), 45–58 (2021)
3. D. Goyal, R. Mishra, Quantum cryptography for cybersecurity: implement quantum cryptography in practice. J. Quantum Inf. Secur. **8**(1), 20–33 (2021)
4. C. Di Maio, Quantum computing for networking and telecommunications: QKD, quantum cryptography, post-quantum cryptography, and the quantum internet. Int. J. Quantum Commun. **10**(3), 112–125 (2021), Springer
5. V. Chitra, G. Seetharaman, Applications of quantum cryptography. Quantum Inf. Process. **25**(4), 789–802 (2021)

6. S. Mukhopadhyay, Quantum cryptography and quantum key distribution. J. Quantum Cryptogr. **14**(3), 180–195 (2021)
7. S. Bai, S. Galbraith, F. Hao, Efficient lattice-based cryptography: challenges and solutions. J. Cryptogr. Eng. **13**(1), 45–62 (2023)
8. S. Liao, Quantum cryptography for cybersecurity: secure information exchange in a quantum world. J. Secur. Quantum Commun. **18**(4), 301–315 (2021)
9. K. Ozawa, Quantum computing and information theory: recent advances and future directions. Inf. Sci. **35**(5), 400–415 (2021)
10. J. Peng, X. Wang, W. Xu, Quantum cryptography and quantum information theory. Quantum Inf. Process. **30**(3), 250–265 (2021)
11. Y. Allouche, A. Bauspiess, Quantum computing and quantum information processing: theory and applications. Quantum Inf. Process. **28**(4), 320–335 (2020)
12. S. Choudhury, R. Chakraborty, Quantum cryptography and quantum information security. J. Quantum Inf. Secur. **7**(3), 190–205 (2020)
13. M. Nandi, Quantum cryptography: an introduction to quantum safe communication. J. Quantum Commun. **7**(4), 250–265 (2021)
14. N. Nain, P. Kundu, Quantum computing and quantum cryptography: theory and applications. Int. J. Quantum Commun. **8**(3), 200–215 (2020)
15. A. Pal, M. Ghosh, Quantum cryptography and quantum computing: fundamentals and applications. Quantum Inf. J. **6**(4), 280 (2020)
16. X. Liu, Y. Huang, S. Wang, Quantum cryptography for secure data transmission: principles, methods, and applications. Int. J. Quantum Secur. **6**(2), 80–95 (2021)
17. S.K. Sharma, Quantum cryptography for secure communication: theories and applications. J. Secur. Commun. **12**(3), 180–195 (2021)
18. S.K. Negi, S. Choudhary, A. Aggarwal, Blockchain for cybersecurity: a comprehensive review. J. Blockchain Technol. **3**(2), 120–135 (2021)
19. Y. Zhou, M. Gu, S. Wang, Blockchain technology for cybersecurity and privacy. Int. J. Blockchain **7**(2), 90–105 (2021)
20. J. Preskill, P. Shor, The future of quantum algorithms for cryptography. Quantum Inf. Process. **23**(4), 389–410 (2024)
21. A. Jayabalan, A. Kumar, Quantum cryptography and quantum communication. J. Quantum Commun. **6**(2), 120–135 (2020)
22. S.K. Das, A. Choudhary, Quantum computing and blockchain technology. Int. J. Blockchain 5(1), 30–45 (2020), Springer
23. X. Gu, Z. Zhang, Blockchain technology for cybersecurity and privacy protection. J. Blockchain Technol. **4**(3), 180–195 (2020)
24. Y. Huang, Y. Huang, Blockchain technology and its applications in cybersecurity. J. Blockchain Technol. **3**(1), 45–60 (2020)
25. M.G. Raymer, C. Monroe, The quantum internet: secure communication with quantum cryptography. Science **379**(6632), 119–127 (2023)
26. M. Kjaergaard, M.E. Schwartz et al., Superconducting qubits for quantum computing. Annu. Rev. Condens. Matter Phys. **15**, 187–216 (2023)
27. S. Pirandola et al., Advances in quantum cryptography and quantum key distribution. Adv. Opt. Photonics **15**(1), 25–70 (2023)
28. K. Matsuura, T. Matsumoto, Quantum computing: principles, technologies, and applications. Quantum Inf. J. **9**(1), 45–58 (2021), Springer
29. R.K. Raju, M. Swain, Quantum cryptography: principles and applications. J. Cryptogr. 20(1), 55–70 (2021), springer.
30. A. Gheorghiu, M.J. Hoban, Quantum algorithms for cryptographic applications. Nat. Rev. Phys. **5**, 123–135 (2023)
31. M.R. Khademi, M. Khademi, Quantum cryptography: the complete guide. Quantum Comput. J. 5(3), 210–225 (2021), Springer
32. V. Patidar, V. Jain, Quantum computing and quantum cryptography. J. Quantum Comput. **11**(4), 300–315 (2021)

33. S. Banerjee, S. Roy, Quantum cryptography: from theory to implementation. J. Quantum Secur. **6**(3), 150–165 (2020)
34. S. Bhowmick, P. Bose, Quantum computing and quantum security. Int. J. Quantum Secur. **9**(2), 100–115 (2020)
35. D. Gangopadhyay, S. Kumar, Quantum cryptography and quantum Computing. J. Quantum Comput. **10**(2), 130–145 (2020)
36. M. Hassan, S. Ahmed, Quantum cryptography and quantum computing for cybersecurity. J. Cybersecur. **14**(4), 260–275 (2020)
37. N. Bindel et al., Quantum-safe cryptography: protocols and standards. ACM Comput. Surv. **55**(3), 57–90 (2023)

Quantum Computing and Cybersecurity: Navigating Threats and Opportunities

Asma Malik and Sardar M. N. Islam

1 Introduction

Profound developments in computer technology have significantly influenced both our professional and personal spheres. Continuously, the information technology sector is actively engaged in the development of artificial intelligence, machine learning, and other revolutionary technologies. These technologies have been widely adopted in many industries globally, resulting in significant improvements in productivity and efficiency. The exponential expansion of the IT industry has resulted in empowerment, enabling humanity to more effectively address longstanding challenges accompanied by a wide range of data. Notwithstanding all efforts, even the most advanced supercomputers encounter difficulties in maintaining pace with the increasing diversity of newly developing workloads. Researchers in the fields of science and computer engineering are currently working on creating a new computer system that is capable of effectively handling and solving highly complex problems, as well as managing extraordinarily large files [1, 2].

Safeguarding sensitive data has become a crucial aspect in the era of modern technology as the invention of new technologies has increased rapidly, thus, creating demand for strong security measures for data protection. As the loss of data, illegal access by hackers, and malware outbreaks are some significant threats in case of having unprotected data. Having and being able to maintain confidence in digital systems along with data security is crucial since it is dependent on the principles

A. Malik (✉)
FAST NUCES, Islamabad, Pakistan
e-mail: i191293@isb.nu.edu.pk

S. M. N. Islam
ISILC, Victoria University, Melbourne, Australia
e-mail: sardar.islam@vu.edu.au

S. B. Goyal et al. (eds.), *Quantum Computing, Cyber Security and Cryptography*,
https://doi.org/10.1007/978-981-96-4948-8_18

of fairness, impartiality, and individual liberty. Universities, firms, and other organizations prioritize the protection of digital assets from hackers, fraudsters, and new cybersecurity risks [1–3].

Computer scientists and physicists in the late twentieth century pioneered quantum computing by merging information theory and quantum physics. Superposition distinguishes quantum computers from conventional ones. This special characteristic boosts their cognition, making them powerful tools. However, advanced quantum computers pose cybersecurity risks. Their processing power allows them to decipher public-key cryptography, the foundation of modern security, compromising data privacy. As various organizations strive to safeguard themselves against this increasing threat, the transformation can restructure global connections. Quantum computers, on the other hand, have the potential to improve the security of computer connections. Quantum cryptography enables the use of secure encryption and the creation of random numbers, hence simplifying complex encryption methods. This can enhance worldwide cybersecurity.

Quantum computing may benefit and challenge cybersecurity experts. This chapter discusses how quantum computing may compromise most encryption approaches and cybersecurity. Our goal is to start interdisciplinary discussions about quantum computing's impact on cybersecurity stakeholders by assessing different methods [4, 5].

1.1 History of Quantum Computing

In the early 1980s, Richard Feynman proposed a quantum computer. Traditional computers may struggle to replicate quantum systems, Feynman argued. In 1981, he demonstrated that quantum computers could surpass conventional computers [6]. Deutsch expanded Feynman's theories in 1985 to propose the first quantum computer. He generated interest in quantum algorithms and paved the way for their development [7].

Quantum computing experiments began in the 1990s. Peter Shor created a quantum technique for integer factorization in 1994, which conventional computers cannot handle. Shor's method showed that quantum computers tackle complicated problems faster [8]. Peter Shor and Andrew Steane introduced quantum codes that corrected errors at the same time. These codes are designed to protect quantum information from errors, which is an essential phase in building functional quantum computers [9]. In 1998, Isaac Chuang and his team at IBM effectively implemented Shor's algorithm on a 2-qubit NMR quantum computer. This achievement was remarkable in the domain. Since then, quantum computing error-correcting algorithms, hardware, and methodologies have advanced rapidly. Google, IBM, and D-Wave are developing functioning quantum computers, while university research investigates theoretical concepts and applications [10].

2 Fundamentals in Quantum Computing

Quantum computing is built on the principles of quantum mechanics. Quantum objects are two-dimensional particles that act like waves when not observed. This dual nature results in fascinating natural events. The state of a quantum object is determined by a coherent wave function, which can experience interference. Observing quantum objects, known as quantum measurement, gathers information but can also disturb the quantum state, leading to information loss. These principles are crucial for qubits in quantum computing [11].

Schrodinger's Equation and Quantum Systems

Quantum systems are governed by Schrödinger's equation, which explains how the system's wave function changes based on its energy environment, represented by the Hamiltonian. Keeping a quantum system isolated from external influences and limiting energy exchange is crucial to reducing decoherence. These systems operate under conceptual rules and mathematical principles that define the behavior of quantum objects.

Quantum Mechanics in Computing

Quantum computers leverage three main features of quantum mechanics to store, represent, and analyze data, making them significantly faster than classical computers. These features are.

Superposition

Superposition allows a quantum system to be in multiple states at once. Unlike classical bits, which are either 0 or 1, qubits can represent both 0 and 1 simultaneously. This is expressed by the following equation:

$$|\psi\rangle = \alpha|0\rangle + \beta|1\rangle$$

Here, α and β are the complex numbers and the probabilities of measuring the qubit as 0 or 1 are given by $|\alpha|2|\alpha|^2|\alpha|2$ and $|\beta|2|\beta|^2|\beta|2$. A qubit in superposition can process information in parallel, enabling faster solutions for certain problems compared to classical computers. This is the foundation of quantum parallelism, a significant advantage of quantum computing [12].

Entanglement

Entanglement links the states of multiple qubits so that the state of one cannot be described independently of the state of the other. The combined state of entangled qubits is represented by

$$|\psi\rangle = \alpha|00\rangle + \beta|01\rangle + \gamma|10\rangle + \delta|11\rangle$$

Here, α, β, γ, and δ are the complex numbers representing probability amplitudes, and $|00\rangle$, $|01\rangle$, $|10\rangle$, and $|11\rangle$ are the basic states of the two qubits. Entanglement enhances the performance of quantum algorithms, such as Grover's search algorithm and Shor's algorithm for factoring large numbers, beyond the capabilities of classical computers [13].

Interference

Quantum objects exhibit wave-like behavior, leading to interference, a key aspect of quantum computing. Quantum states interfere based on probability amplitudes. This interaction might be positive or negative, altering state measurement probabilities. For example, combining two qubits in superposition results in a new state,

$$|\Psi\rangle = \alpha\gamma|00\rangle + \alpha\delta|01\rangle + \beta\gamma|10\rangle + \beta\delta|11\rangle$$

Here, the parameters α, β, γ, and δ influence the probability of finding the qubits in particular states. Constructive interference occurs when phases align, increasing measurement probability, while destructive interference decreases it. Grover's algorithm exploits interference to amplify the probability of the solution state, improving efficiency compared to classical algorithms [14].

3 Quantum Computing in Cybersecurity: A Double-Edged Sword

3.1 Pros of Quantum Computing in Cybersecurity

Quantum computing is a game-changer in the tech world. Companies are racing to achieve quantum supremacy because it promises to slash computing times from years to mere minutes. This power is especially beneficial for data-heavy industries like pharmaceuticals, AI, industrial design, logistics, and national security [15].

Revolutionizing Cybersecurity: Quantum computing is transforming cybersecurity by making risk management more reliable. Traditional models use classical probability methods, but quantum principles, grounded in Hilbert space mathematics, offer a significant upgrade. In cybersecurity, risk involves the likelihood and impact of threats on digital assets. Quantum cognition, replacing classical probability with Hilbert space axioms, provides new and effective mental models. This shift enhances adversary understanding and tracking, improving decision-making. Quantum cognition allows for the simultaneous analysis of multiple states, enabling a more comprehensive risk assessment than classical methods [16].

The Internet of Things (IoT) is another rapidly expanding sector. Billions of connected devices transmit vast amounts of personal data, making security crucial. IoT is used in residential systems and hospital patient care. Quantum computing enhances device security by adding a "Quantum Security Layer," creating a Hybrid IoT Network Infrastructure. This layer bolsters communication through quantum cryptography. When combined with cloud technology, it enhances cloud server security. This innovation enables secure mass communication networks and cloud-based quantum virtual computers, fostering quantum technology adoption [16].

Protecting Cyber-Physical Systems: Critical infrastructure, such as power and oil delivery systems, are Cyber-Physical Systems (CPS) that require continuous efficiency and availability, making security paramount. Quantum algorithms, nearly impossible to breach, secure CPS, and ensure data encryption. Quantum encryption prevents hacking and eavesdropping by using principles like Heisenberg's uncertainty principle. Unlike classical cryptography that relies on software-generated random numbers, quantum cryptography uses the laws of physics to generate random numbers.

Advances in Quantum Cryptography: Quantum cryptography and Quantum Key Distribution (QKD) are significant advancements in quantum computing. Quantum cryptography generates random numbers through physics, unlike classical cryptography which relies on software. Heisenberg's uncertainty principle is applied in quantum encryption, replacing bits with photons. These photons are sent in various polarized directions, and the receiver detects and interprets them as keys. Any eavesdropping attempt alters the state of the photons, alerting both the transmitter and receiver. This makes quantum cryptography a groundbreaking method of encryption. As quantum computing and cybersecurity converge, we can expect to see increasingly secure systems and infrastructures. Quantum technologies will integrate more deeply into IoT and CPS, providing unmatched security and efficiency. The era of quantum-enhanced cybersecurity heralds a safer, more robust digital world [16, 17].

3.2 *Cons of Quantum Computing in Cybersecurity*

Quantum computing not only holds the promise of revolutionizing cybersecurity but also introduces significant risks. One of the most concerning possibilities is "Q-Day"—a hypothetical future point when quantum computers could potentially break current internet encryption. This would spell disaster for the cryptographic systems that secure our online activities. The idea that all existing cryptographic techniques could fall prey to quantum attacks has many experts worried about a possible global crisis.

Google's announcement in 2019 of achieving quantum supremacy—where they solved a complex problem in just three minutes that would take a classical computer a thousand years—confirmed the potential of quantum computing. However, despite this significant advancement, constructing a quantum computer with millions of qubits that can effectively compromise contemporary cryptography remains an extremely difficult task. Should quantum computers be successfully constructed, they have the potential to compromise the majority of existing public-key cryptography techniques. Therefore, it is of utmost importance to create and apply quantum-safe cryptography algorithms to protect our conventional information and communication technology infrastructure [18]. Quantum computing poses hazards that go beyond mere data breaches, encompassing national security implications. A quantum computer with adequate computing power can unravel RSA, AES, DSA, ECDSA, and password encryption in hours. This could lower global trust, increase geopolitical tensions, and trigger war [17].

The combination of quantum computing and AI could pose unprecedented risks. AI can offer solutions and analyze options, but it struggles with massive datasets. Quantum computers excel in this field, which could pose risks when AI is combined with quantum technology. AI and quantum computers may outpace human control, threatening world security. Inappropriate use of this powerful technology could harm humans and the environment. Quantum computing can solve difficult mathematical equations that underlie many encryption schemes, making it a cybersecurity hazard. Quantum computing has many cybersecurity risks.

Quantum Computing and Cybersecurity Risks

Quantum computing, especially Shor's and Grover's algorithms, threatens cybersecurity systems. Shor's method undermines RSA encryption, a key component of asymmetric cryptography, by calculating the factors of huge numbers and compromising encrypted communications. To mitigate these problems, post-quantum cryptography must be developed [19, 20]. Grover's rapid unstructured search method threatens symmetric encryption techniques like AES by shortening the key length needed for security. Even brute-force-resistant AES-256 would require only a 128-bit key to provide equal security in a quantum context [20, 21].

Grover's quantum method, which allows quantum computers to handle unstructured search problems quicker than classical methods, is challenging. Grover's approach searches an unsorted database of N items in $O(\sqrt{N})$ operations, while conventional computing requires O(N) operations in worst-case scenarios. This development could drastically reduce the key length for symmetric encryption algorithms like AES [22]. Its large key space, AES-256, contains a 256-bit key that normal computers cannot brute force. For similar security, Grover's technique decreases the key length to 128 bits, lowering encryption strength. Symmetric encryption with long key lengths that resist quantum computing is needed to defend against hazards. Security against quantum attacks is improved by using key lengths that quantum computers cannot brute force [22–24].

An increasing number of methods, such as those based on lattices, hashes, or codes, are being developed to mitigate these risks; collectively, they constitute quantum-safe cryptography. Lattice-based cryptography, exemplified by CRYSTALS-KYBER and CRYSTALS-Dilithium, employs intricate lattice issues to develop strong encryption and digital signature techniques that are impervious to quantum attacks [25]. Hash-based methods such as XMSS and SPHINCS utilize hash functions to ensure the security of signatures. NIST is now working on standardizing these algorithms, which are resistant to quantum attacks [26]. Code-based cryptography, which was developed by McEliece and Niederreiter, is still considered a highly viable option for achieving post-quantum security. This is because it has proven to be resistant to attacks by quantum computers [27]. Furthermore, the development of quantum-safe networks is of utmost importance, necessitating the implementation of advanced cybersecurity architecture to safeguard against potential breaches from quantum computing [28]. The objective of these combined endeavors is to create cryptographic systems that can withstand the advancing quantum dangers, thereby guaranteeing the reliability and secrecy of digital communications in a future where quantum computing becomes more widespread.

Quantum computing poses substantial cybersecurity vulnerabilities, especially affecting well-established cryptographic systems. RSA, an asymmetric encryption technique that depends on the challenge of factoring big prime numbers, may be vulnerable to Shor's algorithm, a quantum computing algorithm that easily factors enormous numbers. The vulnerability of RSA encryption has led to the emergence of post-quantum cryptography solutions [19, 20]. Similarly, Grover's technique poses a challenge to systems like as Diffie–Hellman and the Digital Signature technique (DSA) by calculating discrete logarithms at a faster rate than traditional approaches. As a result, this reduces the level of security provided by symmetric encryption methods like AES. For example, the level of security provided by AES-256 would be comparable to that of a 128-bit key when facing quantum attacks. This means that greater key lengths are required to offer strong resistance against such attacks, as stated in references [22–24]. Quantum Key Distribution (QKD) utilizes principles of quantum mechanics to detect any attempts of eavesdropping on encryption keys, hence providing heightened security. Nevertheless, QKD is susceptible to possible advancements in quantum computing that could exploit its fundamental principles. This emphasizes the necessity for continuous development in quantum-resistant cryptography [26, 29]. Hence, quantum computing not only has the potential to bring about significant changes but also highlights the pressing requirement for cryptographic solutions that can withstand and adapt to these technological breakthroughs.

4 Quantum-Safe Cryptography

Quantum computing introduces significant cybersecurity risks by challenging the security of current cryptographic systems. Traditional methods like RSA and ECC, which rely on the complexity of certain mathematical problems, may become vulnerable as quantum computers can solve these problems exponentially faster [24, 30]. In response, the field of quantum-safe cryptography has emerged, aiming to develop algorithms resilient to both quantum and classical computational threats. Among these, code-based cryptography, pioneered by researchers such as McEliece and Niederreiter in the late 1970s and the early 1980s, uses error-correcting codes to create cryptographic systems that are particularly robust against quantum attacks [27]. The focus of ongoing research is to design these systems to be both secure and efficient. Additionally, as quantum computing technology progresses, there is an urgent need for quantum-safe network designs. These designs will enhance infrastructure to withstand potential breaches from quantum threats, ensuring that organizations can maintain their cybersecurity defenses against this evolving risk [28].

5 Real-World Applications of Quantum Computing in Cybersecurity

Quantum computing, with its extraordinary computational capabilities, is poised to revolutionize the field of cybersecurity. Below are a few prominent real-world applications that demonstrate how quantum technologies are being utilized to improve security measures and tackle potential threats.

Quantum computing can revolutionize cybersecurity by leveraging its unmatched computational capabilities to provide novel solutions for present-day security obstacles. QKD Secures cryptographic key communication using superposition and entanglement and is considered one of the most groundbreaking uses. Unlike the conventional systems, interrupting messages can bring about changes in the quantum state of the encryption key, thus, alerting parties regarding breach of potential security. QUESS also known as Chinese Quantum Science Satellite has exemplified a secure transmission between the cryptographic key and space and ground stations with the help of QKD. This achievement unlocked a global pathway for quantum communication networks [31]. Quantum-enhanced security systems demonstrate a highly successful advancement as these systems take advantage of faster data processing capabilities for attaining anomaly identification as well as advanced threat detection. Through the utilization of techniques regarding quantum machine learning which can help in the detection of sensitive information, these systems can point out potential threats that conventional methods may seem to overlook. For the advancement and development of quantum-enhanced Security Information and Event Management (SIEM) systems, efforts are on full focus. As these efforts objectify the enhancing ability for prompt identification of system along with showing response to attacks through monitoring logs and alerts from multiple sources [32]

Security protocols must use post-quantum cryptography to protect against classical and quantum vulnerabilities in the future. Quantum computers may break RSA and ECC encryption as they advance. To counter this, the National Institute of Standards and Technology (NIST) is spearheading efforts to standardize post-quantum cryptographic methods, with lattice-based, hash-based, and code-based cryptographies emerging as leading candidates [24].

In the realm of secure voting systems, quantum cryptography promises to safeguard electoral processes through tamper-proof encryption and privacy protection. Various nations and academic institutions are exploring these technologies to enhance the integrity and security of electronic voting [35]. Similarly, quantum-resistant encryption is becoming crucial for cloud storage providers. Companies like Google and IBM are developing encryption methods that can withstand quantum attacks, ensuring data protection even as quantum computing evolves [35, 36] (Table 1).

This table illustrates the various applications of quantum computing in cybersecurity, demonstrating how quantum technologies are being used to improve security systems, identify advanced dangers, and prevent potential weaknesses. It offers a comprehensive summary of the advantages each domain of cybersecurity derives

Table 1 Key applications of quantum computing in cybersecurity

Domain	Quantum computing application	Details	Examples
Quantum Key Distribution (QKD)	Securing cryptographic key communication using quantum properties like superposition and entanglement	- Quantum key is altered when intercepted, alerting parties to security breaches - Enables secure communication	**QUESS (Chinese Quantum Science Satellite)**: Demonstrated secure key transmission between space and ground stations
Quantum-enhanced security	Leveraging faster quantum data processing for enhanced threat detection and anomaly identification	- Quantum machine learning can identify sensitive information - Detects threats overlooked by conventional methods	Efforts are underway to develop **Quantum-Enhanced Security Information and Event Management (SIEM)** systems for rapid identification and response
Post-quantum Cryptography	Protecting against future quantum vulnerabilities that may break classical encryption methods	- RSA and ECC encryption may be broken by future quantum computers - NIST is standardizing post-quantum cryptography methods	Leading candidates for post-quantum cryptography include **lattice-based, hash-based, and code-based cryptographies**
Voting systems	Securing electoral processes through quantum cryptography	- Ensures tamper-proof encryption and protects voter privacy - Enhances the security and integrity of electronic voting	Various nations and academic institutions are researching quantum cryptography for secure voting systems
Quantum-resistant encryption	Developing encryption that can withstand quantum attacks	- Critical for cloud storage providers to protect sensitive data - Ensures long-term data security	**Google and IBM** are leading efforts in creating quantum-resistant encryption methods to safeguard cloud storage and future data protection

from quantum breakthroughs, along by examples of practical applications and current progress.

6 Current Publications in Quantum Cybersecurity

Quantum computing, which was first proposed in theory in 1980 [41], has had a substantial influence on the field of cybersecurity. Presented here is an overview of the most recent advancements and discoveries in the field of quantum cybersecurity.

Recent research [48] has emphasized the increasing necessity of safeguarding quantum computing infrastructures. The study analyzes possible risks to quantum systems and suggests methods for ensuring the security of both software and hardware elements. The statement highlights the correlation between the progression of quantum computing technology and the increasing vulnerability to digital attacks, underscoring the importance of implementing strong security measures. A detailed study [42] examines the influence of post-quantum cryptography on blockchain technology. Post-quantum cryptosystems including code-based and lattice-based methods are evaluated to mitigate Grover's and Shor's vulnerabilities. This research highlights the effectiveness of cryptosystems in protecting blockchain public-key encryption and digital signatures.

Edwards and Rawat [46] suggest unique techniques to quantum adversarial machine learning problems in machine learning. Quantum probabilistic data issues are addressed by their quantum machine learning research. It helps overcome limits and adversaries in quantum-assisted systems. Suryotrisongko and Musashi [47] developed a hybrid quantum–classical deep learning model to detect botnets. Their method achieves 94.7% accuracy using quantum circuits and classical models. The study uses multiple datasets and noise conditions to demonstrate the model's botnet detection accuracy.

KeyShield, a quantum-resistant key management system, is fully described in the current research [45]. This method solves key management problems by providing forward/backward secrecy and quantum attack resistance. KeyShield boasts exceptional computational and storage efficiency. Banded matrices improve quantum resistance while reducing expenses. AL-Darwbi [44] studies how error-correcting codes can strengthen digital signature systems against classical and quantum cryptanalysis. Niederreiter-like transformations and McEliece-based approaches are tested for attack protection and complexity. Althobaiti and Dohler [43] use post-quantum methods and 5G networks to improve IoT security. Malware and brute-force attacks are vulnerabilities, and the paper advises employing lattice-driven cryptography to improve quantum resilience. Their analysis stresses the need for IoT security advancements to combat quantum attacks. Ronald P. Uhlig et al. [48] explore the challenge of recruiting graduate students to quantum computing and cybersecurity. Their approach promotes quantum computing in academic courses and incorporates hands-on exercises to spark enthusiasm. This strategy helps students adapt to future technology advances and understand quantum cryptography.

Table 2 Comparison of quantum-resistant algorithms: security, efficiency, and deployment challenges

Algorithm	Type	Security level	Efficiency	Deployment challenges
Lattice-based [42, 43]	Post-quantum cryptography	High (resistant to Grover & Shor)	Moderate	Complex implementation, higher computational cost
Code-based [44]	McEliece, Niederreiter	High	High	Large key sizes, storage limitations
Multivariate [42]	Post-quantum signatures	Moderate	Moderate	Requires optimization for scalability
Hash-based [42, 45]	Digital signatures	High	High	Efficient, but not fully quantum-resistant
Symmetric-key quantum [42]	Grover's resistant	Low (against Shor)	High	Efficient, but not fully quantum-resistant

To enhance the literature review, the following table provides a comparative analysis of quantum-resistant algorithms based on their security level, efficiency, and deployment challenges (Table 2).

This table highlights the relative strengths and weaknesses of various quantum-resistant algorithms, providing insight into their practicality and the challenges they present for real-world deployment.

7 Future Research Directions

While quantum computing presents revolutionary potential in cybersecurity, several challenges and gaps remain, particularly in developing robust quantum-safe cryptographic algorithms. To ensure the field progresses effectively, the following areas are suggested for future research.

Development of Hybrid Cryptosystems

Current cryptosystems face the challenge of transitioning from classical to quantum-resistant algorithms. Future research should focus on hybrid cryptosystems that combine classical and Post-quantum Cryptography (PQC). This approach would allow a gradual transition to fully quantum-safe environments, providing security in both classical and quantum contexts. Research projects could explore best integrating lattice-based, code-based, and hash-based algorithms with existing systems, ensuring compatibility while maintaining security.

Optimization of Post-quantum Algorithms for Efficiency

One major issue with quantum-safe algorithms is their computational cost, especially when deployed at scale. Future research could focus on optimizing the efficiency of post-quantum algorithms like lattice-based cryptography, code-based cryptography (such as McEliece), and multivariate schemes to make them more suitable for real-time applications. Exploring algorithmic improvements that reduce key sizes and computational overhead without sacrificing security could be a key research focus.

Quantum Key Distribution (QKD) with High-Dimensional Qubits

Although research on QKD has enormously paid attention to two-level qubit systems; the d-level systems or high-dimensional qudits tend to offer escalated data capacity and security. Studies in the future should be aimed to design as well as implement QKD protocols making use of high-dimensional quantum states, thus resulting in improved security measures like against eavesdropping along with reducing vulnerability to quantum attacks. Moreover, for the integration of high-dimensional QKD into presently available quantum communication networks, experimental work remains a critical aspect for practical deployment.

Exploration of Quantum-Resistant IoT Security

The rising popularity of IoT devices makes it necessary for security against quantum threats. Future research should consist of lightweight quantum-resistant cryptographic solutions that remain tied up to IoT devices resource constraints. Furthermore, projects can also look into lattice-based or hash-based cryptography for the protection of IoT networks, keeping their attention towards the minimization of computational overhead along with maintaining strong security guarantees.

Quantum-Enhanced Machine Learning for Threat Detection

Cybersecurity can be completely revolutionized through Quantum Machine Learning (QML) as it carries the ability to improve threat detection and response times. With focus on the application and implementation of QML algorithms, future research will be able to detect threats in advance such as those of quantum-enhanced malware or quantum-resilient ransomware. To make quantum systems even more resilient, a collaborated project among the quantum computing and machine learning experts will eventually lead in the development of more efficient models that will continue to address potential threats in real time.

Error-Correcting Codes in Quantum Cryptography

Error-correcting codes play a vital role in both; the classical and quantum cryptography. However, there is still a need for detailed further research into more effective and scalable quantum error-correcting codes that will help in protecting from both natural noise and cyberattacks in quantum systems. Projects can pay attention towards the testing of numerous strategies regarding error-correcting, such as topological codes, along with their implementation for securing quantum key distribution systems and signatures.

Standardization of Post-quantum Algorithms

Due to the lack of standards in quantum-safe cryptographic algorithms there remains a key gap in the modern body of knowledge. Therefore, research should be focused towards contributing to global efforts to develop cryptographic standards for the quantum era. This involves participation in a collaborated research to figure out the strengths and weaknesses of candidate algorithms under real-world conditions, hence contributing to institutions like NIST, and working on the long-term deployment strategies for these cryptographic standards.

8 Conclusion

Quantum computing is revolutionizing cybersecurity, offering both opportunities and challenges. This chapter discusses quantum computing's newest advances and their implications on cybersecurity, concentrating on how they might improve security. The internet has revolutionized how we communicate, work, and play. The digital world has benefits, but it is unregulated and vulnerable to viruses, hackers, and harmful software. Quantum computing targets discrete logarithm encryption methods, increasing susceptibility. This emphasizes the need for quantum-resistant cryptography. Quantum computing has distinct advantages over traditional systems. The uncertainty principle in quantum physics prevents precise particle positioning. Quantum cloning can recover lost data. Quantum teleportation transfers data without physical displacement, but quantum entanglement requires shared measurements to retrieve it. These properties make quantum cryptography a promising future cybersecurity option for impenetrable security and superior eavesdropping detection.

However, access to advanced quantum technology is limited. Quantum machine learning is showing promise in addressing cybersecurity difficulties, while quantum computing instruments are helping solve complex quantum physics and other problems. Quantum computing can revolutionize cybersecurity. Technology poses new challenges, yet its unique qualities can improve security like never before. Research and development are needed to build strong cryptography systems that can resist quantum assaults and use quantum technology to address emerging threats. These advancements are intended to transform cybersecurity and protect digital data.

The advent of quantum computing is poised to have a transformative impact on the field of cybersecurity, bringing both unparalleled prospects and formidable obstacles. The chapter began by presenting the latest advancements in quantum computing and their possible influence on cybersecurity, in particular emphasizing their ability to enhance security standards. Where the digital environment appears to provide many advantages, it also remains very susceptible to the swiftly growing cyber risks. Hence, to preserve conventional encryption techniques, it is imperative to develop quantum-resistant cryptographic solutions, as quantum computing inherently undermines their effectiveness.

The distinctive features of quantum computing, including quantum uncertainty, cloning, and teleportation, provide promising opportunities for the advancement of invincible security techniques, such as cryptography. In the future, these features have the potential to significantly enhance the detection of eavesdropping and improve overall data security. Nevertheless, the accessibility of quantum technology is still restricted, but quantum machine learning is displaying initial potential in tackling cybersecurity challenges. Progressive advancement of quantum computing technologies will ultimately have a pivotal function in addressing intricate cybersecurity challenges, thereby safeguarding digital data against growing quantum-based risks. The future of cybersecurity depends on the emergence of quantum-resistant encryption architectures that can endure quantum attacks and incorporate quantum technologies to enhance defenses. Ongoing research and development will further explore the potential of quantum computing to revolutionize the cybersecurity environment and safeguard the digital world in the future.

References

1. J. Singh, M. Singh, Evolution in quantum computing, in *Proceedings of the 5th International Conference on System Modeling and Advancement in Research Trends, SMART 2016*, pp. 267–270 (2017). https://doi.org/10.1109/SYSMART.2016.7894533
2. W. Polak, E. Rieffel, Quantum computing: a gentle introduction (Scientific and Engineering Computation), in *Good Paperback*, ed. by W.H. Polak, E.G. Rieffel (HPB-Red, 2014). Accessed: Jun. 15, 2024. https://www.abebooks.com/servlet/BookDetailsPL?bi=31262285781&searchurl=ds%3D20%26kn%3D0262526670%26sortby%3D17&cm_sp=snippet-_-srp1-_-title1
3. *Wielding a Double-edged Sword: Preparing Cybersecurity for a Quantum World*. Accessed: Jun. 15, 2024. https://www.virtualpaper.pro/index.php?route=mag2/article&tracking=5cb932ee673d6&a_id=244
4. *Rethinking cybersecurity for a quantum world—Curious*. Accessed: Jun. 15, 2024. https://www.science.org.au/curious/policy-features/rethinking-cybersecurity-quantum-world
5. *The Quantum Threat to Cyber Security—Centre for International Governance Innovation*. Accessed: Jun. 15, 2024. https://www.cigionline.org/articles/quantum-threat-cyber-security/
6. R.P. Feynman, Simulating physics with computers. Int. J. Theor. Phys. **21**(6–7), 467–488 (1982). https://doi.org/10.1007/BF02650179/METRICS
7. D. Deutsch, Quantum theory, the church–turing principle and the universal quantum computer, in, *Proceedings of the Royal Society of London. A. Mathematical and Physical Sciences* **400**(1818), 97–117 (1985). https://doi.org/10.1098/RSPA.1985.0070
8. P.W. Shor, Algorithms for quantum computation: discrete logarithms and factoring, in, *Proceedings—Annual IEEE Symposium on Foundations of Computer Science, FOCS*, pp. 124–134 (1994). https://doi.org/10.1109/SFCS.1994.365700
9. A.M. Steane, Error correcting codes in quantum theory. Phys. Rev. Lett. **77**(5), 793 (1996). https://doi.org/10.1103/PhysRevLett.77.793
10. I.L. Chuang, N. Gershenfeld, M. Kubinec, Experimental implementation of fast quantum searching. Phys. Rev. Lett. **80**(15), 3408 (1998). https://doi.org/10.1103/PhysRevLett.80.3408
11. T. Albash, D.A. Lidar, Decoherence in adiabatic quantum computation. Phys. Rev. A **91**(6), 1–21 (2015). https://doi.org/10.1103/PhysRevA.91.062320
12. F. Zickert, *Hands-On Quantum Machine Learning With Python Volume 1: Get Started*, vol. 1 (2021)
13. C. Bernhardt, *Quantum Computing for Everyone*. MIT Press Cambridge (2019)
14. M.A. Nielsen, I.L. Chuang, Quantum computation and quantum information. Quantum Comput. Quantum Inf. (2010). https://doi.org/10.1017/CBO9780511976667
15. *The Race Toward Quantum Advantage*. Accessed: Jun. 15, 2024. https://semiengineering.com/the-race-toward-quantum-advantage/
16. W. Barker, W. Polk, M. Souppaya, *Getting Ready for Post-Quantum Cryptography: Exploring Challenges Associated with Adopting and Using Post-Quantum Cryptographic Algorithms* (2021). https://doi.org/10.6028/NIST.CSWP.04282021

17. *Quantum Computing + Cybersecurity | CSA*. Accessed: Jun. 15, 2024. https://cloudsecurityalliance.org/research/topics/quantum-safe-security
18. A. Elaziz, F. Raheman, The future of cybersecurity in the age of quantum computers. Futur. Internet **14**(11), 335 (2022). https://doi.org/10.3390/FI14110335
19. R.L. Rivest, A. Shamir, L. Adleman, A method for obtaining digital signatures and public-key cryptosystems. Commun. ACM **21**(2), 120–126 (1978). https://doi.org/10.1145/359340.359342
20. W. Diffie, W. Diffie, M.E. Hellman, New directions in cryptography. IEEE Trans. Inf. Theory **22**(6), 644–654 (1976). https://doi.org/10.1109/TIT.1976.1055638
21. P.W. Shor, Polynomial-time algorithms for prime factorization and discrete logarithms on a quantum computer **26**(5), 1484–1509 (2006). https://doi.org/10.1137/S0097539795293172
22. L.K. Grover, A fast quantum mechanical algorithm for database search, in, *Proceedings of the Annual ACM Symposium on Theory of Computing*, vol. Part F129452, pp. 212–219 (1996). https://doi.org/10.1145/237814.237866
23. D.J. Bernstein, J. Buchmann, E. Dahmen (eds.) *Post-Quantum Cryptography* (2009). https://doi.org/10.1007/978-3-540-88702-7
24. *Post-Quantum Cryptography|CSRC*. Accessed: Jun. 15, 2024. https://csrc.nist.gov/Projects/post-quantum-cryptography
25. C. Peikert, B. Delft, A decade of lattice cryptography. *Found. Trends® Theor. Comput. Sci.* **10**(4), 283–424 (2016). https://doi.org/10.1561/0400000074
26. R. McEliece, *A Public Key Cryptosystem Based on Algebraic Coding Theory* (1978)
27. M. Finiasz, N. Sendrier, Security bounds for the design of code-based cryptosystems, in, *Lecture Notes in Computer Science (including subseries Lecture Notes in Artificial Intelligence and Lecture Notes in Bioinformatics)*, vol. 5912 LNCS, pp. 88–105 (2009), https://doi.org/10.1007/978-3-642-10366-7_6
28. M. Mosca, Cybersecurity in an era with quantum computers: will we be ready? IEEE Secur. Priv. **16**(5), 38–41 (2018). https://doi.org/10.1109/MSP.2018.3761723
29. H.K. Lo, M. Curty, K. Tamaki, Secure quantum key distribution. Nat. Photonics **8**(8), 595–604 (2014). https://doi.org/10.1038/nphoton.2014.149
30. D. Poulin, A. Qarry, R. Somma, F. Verstraete, Quantum simulation of time-dependent hamiltonians and the convenient illusion of hilbert space. Phys. Rev. Lett. **106**(17) (2017). https://doi.org/10.1103/PHYSREVLETT.106.170501
31. S.K. Liao, et al., Satellite-to-ground quantum key distribution. Nature **549**(7670), 43–47 (2017). https://doi.org/10.1038/nature23655
32. Z. Yan, T. Shi, Y. Fan, L. Zhou, Z. Yuan, Compact InGaAs/InP single-photon detector module with ultra-narrowband interference circuits. Adv. Devices & Instrum. 4 (2023). https://doi.org/10.34133/ADI.0029/ASSET/417A39FD-979F-488F-9E90-96F400B8F1F2/ASSETS/GRAPHIC/ADI.0029.FIG.006.JPG
33. C. Abellán, W. Amaya, D. Mitrani, V. Pruneri, M.W. Mitchell, Generation of fresh and pure random numbers for loophole-free bell tests. Phys. Rev. Lett. **115**(25), 250403 (2015). https://doi.org/10.1103/PHYSREVLETT.115.250403/FIGURES/5/MEDIUM
34. *Quantis QRNG (Quantum Random Number Generator)—ID Quantique*. Accessed: Jun. 15, 2024. https://www.idquantique.com/random-number-generation/products/quantis-random-number-generator/
35. A. Pappa, et al., Experimental plug and play quantum coin flipping. Nat. Commun. **5**(1), 1–8 (2014). https://doi.org/10.1038/ncomms4717
36. *Quantum Safe|IBM Quantum Computing*. Accessed: Jun. 15, 2024. https://www.ibm.com/quantum/quantum-safe
37. C.H. Bennett, G. Brassard, Quantum cryptography: public key distribution and coin tossing. Theor. Comput. Sci. **560**(P1), 7–11 (2020). https://doi.org/10.1016/j.tcs.2014.05.025
38. S. Pirandola et al., Advances in quantum cryptography. Adv. Opt. Photonics **12**(4), 1012 (2019). https://doi.org/10.1364/AOP.361502
39. J. Preskill, Quantum computing in the NISQ era and beyond. Quantum 2 (2018). https://doi.org/10.22331/q-2018-08-06-79.

40. L. Chen, et al., *NISTIR 8105 Report on Post-Quantum Cryptography*. https://doi.org/10.6028/NIST.IR.8105
41. J.D. Hidary, *Quantum Computing : An Applied Approach* (2021)
42. T.M. Fernandez-Carames, P. Fraga-Lamas, Towards post-quantum blockchain: a review on blockchain cryptography resistant to quantum computing attacks. IEEE Access **8**, 21091–21116 (2020). https://doi.org/10.1109/ACCESS.2020.2968985
43. O.S. Althobaiti, M. Dohler, Cybersecurity challenges associated with the internet of things in a post-quantum world. IEEE Access **8**, 157356–157381 (2020). https://doi.org/10.1109/ACCESS.2020.3019345
44. A. Kuznetsov, A. Kiian, V. Babenko, I. Perevozova, I. Chepurko, O. Smirnov, New approach to the implementation of post-quantum digital signature scheme, in *Proceedings—2020 IEEE 11th International Conference on Dependable Systems, Services and Technologies, DESSERT 2020*, pp. 166–171 (2020). https://doi.org/10.1109/DESSERT50317.2020.9125053
45. M.Y. Al-Darwbi, A.A. Ghorbani, A.H. Lashkari, KeyShield: a scalable and quantum-safe key management scheme. IEEE Open J. Commun. Soc. **2**, 87–101 (2021). https://doi.org/10.1109/OJCOMS.2020.3046110
46. D.M. Edwards, D.B. Rawat, Quantum adversarial machine learning: status, challenges and perspectives, in *Proceedings—2020 2nd IEEE International Conference on Trust, Privacy and Security in Intelligent Systems and Applications, TPS-ISA 2020*, pp. 128–133 (2020), https://doi.org/10.1109/TPS-ISA50397.2020.00026
47. H. Suryotrisongko, Y. Musashi, Evaluating hybrid quantum-classical deep learning for cybersecurity botnet DGA detection. Procedia Comput. Sci. **197**, 223–229 (2022). https://doi.org/10.1016/J.PROCS.2021.12.135
48. R.P. Uhlig, P.P. Dey, S. Jawad, B.R. Sinha, M. Amin, Generating student interest in quantum computing, in *2019 IEEE Frontiers in Education Conference (FIE)*, vol. 2019-October, pp. 1–9 (2019). https://doi.org/10.1109/FIE43999.2019.9028378

The manufacturer's authorised representative in the EU is Springer Nature Customer Service Centre GmbH, Europaplatz 3, 69115 Heidelberg, Germany. If you have any concerns regarding our products, please contact ProductSafety@springernature.com

Printed and bound by CPI Group (UK) Ltd, Croydon, CR0 4YY
01/10/2026
02225923-0018